Sergey Artemyevich Aivazian

Quality of Life and Living Standards Analysis

Also of interest

Quality of Urban Life
Frick, Hoefert, Legewie, Mackensen, Silbereisen *(Ed), 2013*
ISBN 978-3-11-010577-3, e-ISBN 978-3-11-088496-8

Journal of Econometric Methods
Abrevaya, Honore, Inoue, Porter, Wooldridge (Eds.)
ISSN 2194-6345, e-ISSN 2156-6674

Journal of Time Series Econometrics
Javier Hidalgo (Editor-in-Chief)
ISSN 2194-6507, e-ISSN 1941-1928

Studies in Nonlinear Dynamics & Econometrics
Bruce Mizrach (Editor)
ISSN 1081-1826, e-ISSN 1558-3708

Sergey Artemyevich Aivazian

Quality of Life and Living Standards Analysis

An Econometric Approach

DE GRUYTER

Author
Prof. Dr. Sergey Artemyevich Aivazian
Russian Academy of Sciences
Central Economics & Mathematics Institute
Nakhimovsky Prosp. 47
Moscow
117418
Russia
aivazian@cemi.rssi.ru

ISBN 978-3-11-031624-7
e-ISBN (PDF) 978-3-11-031625-4
e-ISBN (EPUB) 978-3-11-038206-8
Set-ISBN 978-3-11-031626-1

Library of Congress Cataloging-in-Publication Data
A CIP catalog record for this book has been applied for at the Library of Congress.

Bibliographic information published by the Deutsche Nationalbibliothek
The Deutsche Nationalbibliothek lists this publication in the Deutsche Nationalbibliografie; detailed bibliographic data are available on the Internet at http://dnb.dnb.de.

Translation: Marianna Chirkov
Typesetting: Lumina Datamatics
Cover image: DGLimages/iStock/thinkstock
Printing and binding: CPI books GmbH, Leck
♾ Printed on acid-free paper
Printed in Germany

www.degruyter.com

Preface

> Wealth is not the good we are seeking as it merely useful for the sake of something else. ... It is only a necessary condition for people to lead a prosperous life.
>
> *Aristotle*

Subject of this monograph is immense and inexhaustible. The researcher who sets himself the task of measuring and comparing the quality of life and the living standards of two or a larger number of individuals or entire populations of conglomerates (regions, countries, social groups), with *no further mention of the limits and conditional qualities of proposed approaches, certainly looks* unrealistically ambitious. After all, even the popular and seemingly true saying that "it is better to be rich and healthy than poor and sick" may be questioned if when assessing the quality of life of an individual we give the highest value to the importance of self-realization, to the satisfaction from the process of creating something and to the fact of how much the creation benefits the society in whole (there are multiple examples of poor and sick people who were happy in their own way and have left an invaluable legacy, as well as examples of healthy and rich people who were miserable and unhappy with their lives).

Nevertheless, the task of measuring and comparative assessment of the quality of life of individuals or entire conglomerates of individuals that is formulated in the framework of a given strategy of development within specific historical, socio-economic, geographical, environmental, political and other objective conditions is undoubtedly relevant and can be solved. That is the task that the author set for himself, complementing this set goal with the condition that the proposed methods of assessing the quality and way of life will use both the measurement results of multi-aspect statistical indicators of living conditions of the object as well as subjective evaluations of the object itself. Therefore, the approach described in the book is named **the econometric approach**.

I started to get interest in questions of statistical analysis of the way and quality of life in the late 1960s under the influence of colleagues Karapetyan, Armen Khachaturovich, Rimashevskaya and Natalia Mikhailovna. Later came our collaborative work and publications on the analysis and modelling of the distribution of workers by wage, on the distribution of population by the size of per capita income, as well as on the problem of typology of consumption. The main results of my work at that time were reflected in particular in the publications by Aivazian [Aivazian, 1976] and Типология потребления, под редакцией С.А. Айвазяна, Н.М. Римашевской, 1978 [Типология потребления, С.А. Айвазян, Н.М. Римашевская]. Radical changes in the socio-economic situation of the country in the early 90s served as further motivation and a definite push for a deeper and broader coverage of the subject. Those changes were decentralizing the national economy and prioritizing its competitiveness, and as a result the need to develop approaches to inter-regional and cross-national comparative analysis of the key indicators of social and economic development arose, among which, in accordance with article 7 of the Constitution of the Russian Federation, the priority is given to the concepts of "good life" and "quality of life". Thus, in the second half of the 90s, the Central Economics and Mathematics Institute of the Russian Academy of Sciences (CEMI RAS) received orders from the State Statistics Committee, the Economic Development Ministry and the governments of a number of Russian regions to perform scientific research on the subject of the quality of life. Around the same time there has been a burst of activity (both research and regulatory institutional) at the global level: the production of specialized scientific journals (including the interdisciplinary journal *Social Indicators*); the development and distribution of methods of measurements of various kinds of synthetic latent categories (for example, the so-called Human Development Index) in the form of UN recommendations and ratings of the countries and regions by quality of life, level of corruption, level of development of democratic freedoms, etc.

When I was almost finished with my work on the book, I got the opportunity to get acquainted with interesting materials: Report of the Commission on the Measurement of Economic and Social Progress, to a large extent devoted to the issue of quality of life [Report, 2010–2011]. The work of this authoritative international commission was initiated in 2008 by French President Nicolas Sarkozy and led by Nobel Laureates Joseph Stiglitz and Amartya Sen. I'd like to note with some satisfaction that the basic spirit and recommendations of the report are consistent with what the reader will find in this monograph.

I would like to emphasize that the proposed methodology of constructing integral measurements for the synthetic latent categories of quality and way of life in this book can be extended to a much wider range of situations. It can, for example, be applied to a problem of reduction of multi-criteria schemes to the ones with much smaller dimension or multi-criteria schemes to one criterion one (as it was repeatedly demonstrated by the author and other researchers [Айвазян, 1974, 1976, 2000; Айвазян, Бежаева, Староверов, 1974]). For the past 10 years I've been teaching this subject through my lecture course for the students of economic specialization (for example for the post-grad Masters Degree students of Higher School of Economics – HSE of the Economic Department of the National Research University).

In conclusion: my gratitude. I'd like to express my sincere gratitude to my dear colleagues, Makarov, Valery Leonidovich, and Eliseeva, Irina Ilinichna, who took the trouble to review the manuscript of this monograph and expressed a number of valuable comments. I am grateful to the colleagues of CEMI of Russian Academy of Sciences, Moscow School of Economics of Moscow State University after Lomonosov and Higher School of Economics for fruitful discussions of various aspects of the problems presented in the book and useful feedback. I am also grateful to many graduates of HSE for their interest in the subject and sometimes unexpected questions that contributed to the refinement and rethinking of a number of positions. My thanks go to the researchers of the Laboratory of Probability and Statistical Methods and Models for Economics of CEMI, Stepanov V. S, Volkova M. I., Zhidko D. S. and Britikov V. E.: they had to work with arrays of basic statistical data and perform calculations related to the implementation of the above-mentioned contract work. Some results of these calculations are presented in this book. I am also grateful to some other employees of the same laboratory: Grohotova A. P. and Grohotova G. U. for their highly professional and hard work in preparing the original layout of the book.

Village of Pluzhkovo, Moscow Region, 2011

Contents

Introduction

I.1 On the Meaning of the Concept “Quality of Life”

Today, the concept of “quality of life” is widely used in economics, sociology, medicine, politics and many other social sciences. The problem of assessing the quality of life (QOL) arises as we are trying to find a solution to various socio-economic objectives, such as a comparison of individual territories and social groups in terms of socio-economic development as well as establishing a criterion for the effectiveness of social and economic policies in certain area. It is obvious that QOL belongs to both categories: synthetic (that is, combines the various aspects of living conditions and perceptions of these conditions by the individual), and latent (that is, not directly measurable). It is this synthetic latent category that is the subject of this book. Mentioning of econometric approach in the book’s name means that this approach is supposed to use a specific model for the problem of measurement of QOL based on the specific statistical macro- or micro data. The model is to be built on the basis of certain theoretical conceptual principles (we’ll talk about it later, see Chapter 1).

In this context, it seems daunting to me to start my conversation with the readers with the confession that there are numerous theoretical concepts of QOL (of individual and certain conglomerates of the population) highlighting the various aspects of life (happiness, health, the ability to lead a worthy life, etc.), **but there is no single universal definition of synthetic latent category**.

This makes sense because, speaking of QOL, one should take into account the wide range of spheres of human life and the environments in which they take place, as well as the difference (in time and space) of mental attitudes of people in the interpretation of this concept. Therefore, one’s view of the question “what is good and what is bad”, of course, depends on the time and the place, the specificity of the national mentality and accepted system of values and many other factors. After all, the importance of certain aspects of the concept of private “QOL” can vary with time and space depending on the specific global historical conditions of the society. Indeed, the main properties of the QOL are formed and manifested in its ability to adapt to the world, and in its interaction with the “external objects” (production, social institutions, the natural environment, etc.) and with each other. The set of objects through which QOL interacts with the population historically revolves because the needs of the population and the productive forces constantly develop, changing man’s place in nature and society itself.

The anthropogenic factor in the environment and, in particular, the ability of the population to bring nature to the irreversible destruction or on the contrary prevent it serves as an example of the complexity of human interaction with the environment and the emergence of new components in the category of “QOL” today. It is unlikely that this component played any significant role in the spectrum of properties of the concept of "QOL" 60 years ago. However, at any given moment in history, this set of "external objects", in interaction with which the basic properties of the QOL are formed and manifested, infinitely varies. Therefore, the number and composition of the properties that are combined (synthesized) in the integral characteristics of "QOL" also depend on subjective factors, including the level of our knowledge.

I'd like to stress that the problem of establishing and especially *measuring* such a general category without additional explanations **seems overly ambitious**, a sort of a *pretentious claim to the "discovery of the philosopher's stone"*. After all, knowing what "QOL" is, and placing a formalized methodology to measure it, built on the basis of relevant statistical indicators and more specific properties of this category, we are able to determine the strategic objectives in the development of human society. We are able to compare certain indicators of different cells of the society in time and space, and finally, we can design target criteria of social welfare, conditional optimization of which (in different kinds of climatic, political and resource constraints) will allow us to determine the optimal path of socio-economic, ecological and demographic development.

In this context, the position of those scholars who reject in principle the possibility and feasibility of constructing any consolidated, integrated gages of synthetic latent categories of QOL seems to be quite reasonable and convincing. That's exactly the position that a well-known Russian–American researcher J. Birman takes [Бирман И., 2007]. I would call it a comfortable position of irony and scepticism towards the possibility of measuring the latent synthetic categories of QOL of the population. In his work he makes an interesting attempt to analyse a wide range of aspects of QOL by presenting random examples of studies of various aspects of life in Russia and the USA. In my opinion it's convenient because it is difficult to dispute, if the ambitions of such an approach imply a universal character and are not the subject to well-defined finite applied research with objectives (see below in this Introduction, as well as in Paragraph 3.1).

Nevertheless, people cannot (and will not) abandon solving problems related to the interpretation, comparison (spatial and temporal), assessment and measurement of QOL. I think in this context it is appropriate to quote here one of the main points of the report prepared by the Commission on Accessing Economic Performance and Social Progress, led by Nobel Laureates Joseph Stiglitz and Amartya Sen [Report, 2010–2011]: "And though to assess the QOL requires a number of different indicators, **there is an urgent need to establish a single consolidated assessment. Depending on the nature of the issues and the approach we can develop a range of consolidated measurements of quality of life**" (Highlighted by me, S. A). In the Commission's Report one of the official recommendations to the National Statistical Services of different countries was devoted to the same idea: "Recommendation 9: Statistical Offices should provide the information necessary for aggregation on different parameters of QOL, which in turn would allow to develop various indices".

So, it seems to me that the selection of existing basic theoretical concepts of QOL and the main methodological approaches to measuring and modelling QOL as well as tracking emerging trends in the major concepts of the development of human society in post-industrial world, and accordingly which factors affect the QOL of individual members of the society or certain conglomerates (social groups, regions, countries) the most, is a foreseeable task.

Thus, in this book the category of QOL is understood in a synthetic and complex sense, far beyond those of more familiar and more particular (but also synthetic latent) concepts such as "standard of living", "environmental quality", "the level of social security", "human development index", "health index" or "quality of the population". By definition, such a general category, QOL, must integrate smaller more specific aspects of social life, while taking into account the prevailing social attitudes to the system of values and the specificity of the historical moment.

However, each of the synthetic categories that we want to measure in the analysis of QOL can ultimately be characterized by some set of statistical indicators $x^{(1)}, x^{(2)}, \ldots, x^{(p)}$ (we'll call these indicators the partial criteria of analysed synthetic category). Number p of partial criteria can be quite large, but in accordance with the Management Theory's known "threshold of complexity" definition when evaluating an event or phenomenon a person can only take into account no more than 6–8 parameters characterizing this phenomenon. Hence the need arises for a transition from a plurality of individual criteria $x^{(1)}, x^{(2)}, \ldots, x^{(p)}$ to a relatively small number of integrated indicators (and under certain conditions – to a single integral indicator), each of which is constructed in the form of a certain function (convolution) $f(x^{(1)}, x^{(2)}, \ldots, x^{(p)})$ of individual criteria. This of course requires that the transition from the values of the indicators $x^{(1)}, x^{(2)}, \ldots, x^{(p)}$ to their convolution (convolutions) would minimize the loss of the information contained in the original set of particular criteria.

We'd like to emphasize that while setting up a goal of analysing and measuring the various categories of synthetic QOL (different degrees of generality), we should be aware that the **methodology of measurement and interpretation of integrated indicators of quality of life (II QOL) should be specified for particular types of tasks and their applications**. We'll discuss some types of such problems in the Introduction and in Point 1.2.4.

I.2 Why do we need to and why we can Measure the Quality of Life

The fact that one of the most important priorities is the problem of providing a decent QOL to the population is declared in the Constitutions of most countries. For example, Article 7 of the Constitution of the Russian Federation states: "Russia is a social state whose policy is aimed at creating the conditions for a dignified life and free development of the individual". Therefore the integral indicator of QOL of the population can be used, on one hand, as the criteria for the effectiveness of the policy of the government, and, on the other hand, as instruments for making management decisions when choosing the priorities for the government agencies. World practice has identified a number of approaches to assessing the QOL, each of which solves its own tasks (see Paragraph 1.3). These tasks can be divided provisionally into two groups. The first group is determined by the necessity of compiling the inter-territorial ratings. In this case, we use the integrated indicative criteria of QOL of the population that may not have a clear interpretation, but conclusively assess the dynamics of the QOL of the population and compare the values of the indicators of QOL in certain area with other territories. The second group of tasks aimed at management decisions, and therefore all the indicators, should be interpreted unambiguously. In this case, we use the system of statistical indicators of socio-economic development of the territory (the so-called partial criteria), on the basis of which the above-mentioned integral indicative criteria are constructed.

It can be concluded that the II QOL is responsible, on one hand, for monitoring the QOL of the analysed area (country, region, municipality) by comparing it with those of other areas; based on them, on the other hand, we can get recommendations for the adoption of certain management decisions by the administration of the territory.

The main advantages of the model II QOL are as follows:

- uses only the available information;
- the possibility of inter-country, inter-regional and inter-municipal comparison that clearly defines the place of the analysed area among other considered territories based on assessed synthetic categories;
- promptness of diagnostics of the territory and rapid identification of problem areas;
- the possibility of an integrated (combined) assessment of the area (unlike other more specific criteria);
- the ability to use the dynamics of some of the indicators to assess the effectiveness of specific departments and units of regional management bodies;
- minimize the use of subjective evaluations of experts;
- objectivity and validity of the method of convolution of indicators during the transition to integrated indicators;
- compatibility with the procedures of the leading rating agencies;
- the ability to add and exclude new benchmarks, without prejudice to the model;
- the possibility of independent calculations of indicators.

Thus, designed by using a specific technique II QOL is intended in particular for use as performance indicators in the selection of priorities of socio-economic policy of territorial governments.

Definition of the problem areas of social life of the population of certain territory should be based on analysis, on one hand, of the dynamics of key socio-economic indicators and, on the other hand, of the place of the area in comparison to other analysed territories. Natural to assume that the negative dynamics of the indicator in relationship to its past values and the simultaneous deterioration of some of the territory's specific index in comparison to the same index of other areas indicates the presence of a so-called problem areas (bottlenecks) in the socio-economic governance.

In other words, the deterioration of the territory in terms of their relative prior achievements and at the same time in relation to other areas means a high grade of priority of the assessed indicator in terms of the necessity and possibility of changing it. Conversely, the sustainable improvement of the index values of the territory relative to past values of the regions, including its own, in general, points to the need of maintaining the current trend.

Now, as to why we can measure the quality of life of an individual or entire conglomerate population (population, region, municipality, certain social stratum). Once we define the analysed synthetic category ("quality of life" in the most general sense, or "social tensions", "health index", "level of material well-being", etc.) and deliver the ultimate goal of applied research (construction of ratings in the context of inter-territorial comparisons; identifying the problem areas in socio-economic development of the territory that have the most important impact on the QOL; the study of the mentality of the population of the territory on the basis of special questionnaire surveys, etc.), the experts can agree on a set of statistical indicators that adequately characterize the analysed synthetic category. Integral indicator (or integral indicators) is the measurement (measurements) of the analysed synthetic category. It is constructed in the form of a convolution of these indicators (mostly linear). The technique of constructing a similar convolution is described in Paragraph 1.3 (global experience) and Chapter 2 (our proposals and our experience).

I.3 Two Methological Approaches to Evaluation and Measurement of Quality of Life (Micro- and Macro-Analysis)

Getting to the assessment and measurement of the various categories of synthetic QOL, the investigator should determine the choice of the basic provisions of the QOL concept and the corresponding mathematical tools (on the basic theoretical concepts of QOL, see Chapter 1). Methodology for measuring QOL and the construction of various kinds of *integrated indicators of quality of life* (QOL II) depends on within what type of paradigm the reasoning and analysis are carried out. Accordingly, we distinguish two approaches.

Objectivist (or structural-functionalist) approach is based on the structural-functionalist type of paradigm prescribing a leading role in the social life of the community to the structures of different levels of generality that in accordance with this type of paradigm define the place and the "quality" of the individual elements of these structures – individuals and social groups. In this approach, the interest of researchers is focused on the analysis and measurement of statistical data and II QOL that characterize the whole conglomerate of individuals (social groups, the population of a specific region of the country as a whole).

Information support of the research conducted within the objectivist approach consists of the values of statistical indicators and partially expertly evaluated QOL II, **describing in general the QOL of analysed conglomerate of individuals** (social group, the population of the region, the country) for a number of years. A specific list of these indices and the II QOL is determined in accordance with the theoretical provision of the selected concept of QOL.

Subjectivist (or interactionist) approach is based on the type of interactionist paradigm that prescribes a leading role in the society to the so-called actors (which refers, in particular, to the individual) and to the ability to achieve cooperation in their behaviour. In this formulation, the system of practical possibilities and needs of the individual becomes the direct object of the analysis. As a rule, in this case, QOL is a result of subjective assessments by individuals themselves of different aspects of their life and state of mind. Informational support of research conducted within subjectivist approach requires more time and cost than in the case of objectivist one. It is associated with formulation, distribution, filling and handling of a sufficient number of special profiles (the content and structure of those profiles is again set in accordance with the provisions of the selected theoretical concept of QOL).

While paying the tribute to the importance and effectiveness of both of these approaches, recognizing the desirability of their simultaneous use for the purpose of mutual complement, we would like to point out that in this book we will talk mainly about the objectivist approach and examples of its use in empirical research.

I.4 Conclusions

1. *There is no single, universal definition or a single universal method of measuring the latent synthetic category of "quality of life"*. It just doesn't exist. The main properties of the QOL of an individual or a certain part of the population are formed and manifested in its ability to adapt to the world, in its interaction with the "external objects" – the production, social institutions, the natural environment, etc., and each other. The set of objects through which the population interacts with constantly changes in space and time. These changes are subject to the productive forces and the population's needs, man's place in nature and the society itself.

2. When we set a goal of analysis and measurement of the various categories of synthetic QOL (different degrees of generality), we should be aware that the methodology for the measurement and interpretation of integrated indicators of QOL should be specified for certain types of problems and their applications. In other words, the researcher must clearly define the following: the QOL of what type of conglomerates of population he would be analysing (individual people; certain social layer; population of certain areas – countries, regions, municipalities, etc.). What are the common features of the synthetic QOL categories that the researcher is interested in (material well-being, human development, environmental quality, social tension, etc., or is it QOL in the most general sense)? What is the purpose of measuring and analysing the synthetic category of QOL (inter-territorial ratings, evaluation of the effectiveness of social and economic policy, a study of the features of the mentality of certain conglomerate of the population, etc.)?

3. There are two main methodological approaches to assessing and measuring latent synthetic categories of QOL:

- **macro-approach (or objectivist**), based on the analysis and the convolution of the statistical indicators characterizing the analysed conglomerates of the population on certain synthetic category; theoretically, this approach is based on the conceptual principles of structurally functionalist paradigm of sociology, and empirically it is based on macro-economic data;
- **micro-approach (or subjectivist**), based on the analysis and processing of the results of questionnaire surveys of populations, aimed at the study of the analysed synthetic category; theoretically, this approach is based on the conceptual principles of interactionist paradigm of sociology, and empirically, on micro-economic data.

4. After the specification of the problem of analysis and measurement of the synthetic category of QOL, the general methodological scheme of constructing the corresponding integral indicator (the measurement of this synthetic category) brings us to the following function (convolution):

$$y = f(x^{(1)}, x^{(2)}, ..., x^{(p)})$$

of the particular criteria $x^{(1)}, x^{(2)}, \ldots, x^{(p)}$. This function minimizes the loss of information about the analysed synthetic category that could occur during the transition from the information that is contained in indicators $x^{(1)}, x^{(2)}, ..., x^{(p)}$ to the information delivered by the only indicator (integral indicator) y. In this problem, the particular criteria $x^{(1)}, x^{(2)}, \ldots, x^{(p)}$ are actually the statistical indicators that adequately characterize the viewed conglomerates of the population based on analysed synthetic categories (with the macro-approach) or the results of a questionnaire survey of the population on various items of the questionnaire (with the micro-approach).

There may be situations when a satisfactory solution to this problem with a single integrated indicator does not exist. Then we have to come up with the minimum number k of integral indicators (where k is a "lot less" than p, which is usually denoted by "$k \ll p$"), with which it is possible to achieve a satisfactory solution to this problem (in some sense) and construction of k integral indicators

$$y^{(1)} = f_1(x^{(1)}, ..., x^{(p)}),\ y^{(2)} = f_2(x^{(1)}, ..., x^{(p)}), ..., y^{(k)} = f_k(x^{(1)}, ..., x^{(p)}).$$

These functions in total meet the above-mentioned property of minimal loss of information at the transition from $(x^{(1)}, x^{(2)}, ..., x^{(p)})$ to $(y^{(1)}, ..., y^{(k)})$.

1 Main Theoretical Concepts of Quality of Life (QOL) and Examples of Approaches to its Measurement

1.1 The Choice of a Certain Theoretical Concept of Quality of Life Depends on the Answers to Which Questions?

When analysing the development of thought in the philosophical, economic, sociological and psychological sciences, we can find a very wide range of theoretical conceptual approaches to understanding, evaluation and interpretation of the notions of "quality of life of the individual" and "quality of life of a particular conglomerate of the population". This variety of concepts is generated by differences in understanding the mechanism of formation of quality of life assessment. Is feeling of wealth and security of having all necessary commodities and services the main factor in this mechanism? Or is the feeling of achievability of those goods whose utility (i.e. a measure of willingness to pay for them) we recognize as the highest the decisive factor? Or is the expansion of human potential (in individual and social sense) the major thing? Which is more important in determining the quality of life – environmental conditions or the perception of these conditions?

One example of systematizing the content and structure of questions, the answers to which will allow us to choose a certain theoretical concept of quality of life (QOL), is given below.

(i) *What determines the QOL?*
   - environmental conditions;
   - perception of these conditions;
   - both of these factors in a certain ratio.

(ii) *Which reality parameters surround us and which personality traits affect the assessment of QOL?*
   - parameters of material well-being;
   - parameters of human development;
   - parameters of society;
   - environmental parameters.

(iii) *From whose point of view QOL should be assessed?*
   - researchers;
   - an individual or group whose life is being evaluated.

(iv) *If the assessment of QOL should reflect both the opinion of the researcher and the opinion of the individual (or the conglomerate of the population) whose life is being assessed, then exactly how to synthesize these two views?*

The following is a brief description of the three most common theoretical concepts of quality of life: economic prosperity, utilitarian and expansion of human capabilities, and determining the position of the author in this matter.

1.2 The Basic Theoretical Concepts of Quality of Life

If you do not go into the details and differences in shades of interpretation and of certain auxiliary notions used by numerous authors in the construction of their concepts of quality of life, we can distinguish three main theoretical concepts of this synthetic category.

Proponents of the first concept (today recognized as too narrow and limited) identify the quality of life theory with the theory of economic welfare. In the second concept ("utilitarian") quality of life is determined by the concept of utility received by the person during his lifetime. Finally, the third concept (by far the most recognized and demanded at present time) examines the process of human development and raising the quality of life as a process of empowerment of the people in the choice of their fields of activity, methods of self-fulfilment, access to the accumulation of human knowledge, etc.

Let us briefly consider the essence of each of these three concepts.

1.2.1 The Concept of Economic Well-Being

We can say that this approach, considering the quality of life as a wealth, originates from the works of Adam Smith [Smith, 1776]. According to Smith, "the wealth of the nation – is the answer to the question of how the nation is equipped with all the necessary commodities and conveniences", and the latter is precisely what determines the quality of life of the population.

Smith pointed to the two tasks of political economy (which he regarded as a branch of science intended for statesmen and legislators). *The first task* is to ensure that government revenue is sufficient to enable citizens to provide their own income, necessary for their existence. *The second task* is to ensure that the state income is sufficient for the exercise of public functions.

The concept of economic well-being emerged in the heyday of economic thought, which largely explains the desire of scientists of the time to measure quality of life in monetary units. Proponents of this concept are united in their belief that the quality of human life depends solely on its material prosperity. Differences in their thinking were related only to the choice of economic indicators used to assess the level of material well-being, and ultimately, – to assess the quality of life. All kinds of different things were used as these indicators: for example, the gross domestic product (GDP), or the number of years during which the person received a "decent" income or anthropometric measurements of children under 6 years old or the so-called green net national product, or based on consumption, accumulation of inventories and income inequality "index of economic well-being" (for more on this, see point 1.3.1).

Today, however, the global scientific community as well as acting heads of national socio-economic structures more and more come to a conclusion about the need to expand the concept of economic development. It is emphasized that the purely economic indicators such as GDP, income and consumption should be supplemented by indicators of employment, social conditions and the environment. National statistical offices are encouraged to give more attention not to productivity indicators but to the measurements of well-being and quality of life in its broadest sense [Report, 2010–2011]. One of the areas of modern scientific thought that is developing this idea further is presented, for example, in the work of Easterlin [Easterlin, 2002], in which an attempt is made to explain why the so-called happiness index is often higher in the not most economically developed countries ("Easterlin paradox"). There are also some fragments on this topic in the very interesting, in my opinion, fundamental work of V. L. Makarov "Social clasterizm. Russian challenge" [Макаров, 2010].

1.2.2 Utilitarian Concept of Quality of Life

A look at the quality of life from the perspective of the concept of utility is well represented in the works of Pigou [Pigou, 1952]. Proponents of the theory of utilitarianism recognize that the concept of "utility" has several different meanings. Pigou himself draws a distinction between the concepts of "satisfaction" and "degree (intensity) of desire", which he refers to as "desirability". True to the theory of utilitarianism Pigou argues that the term "satisfaction" should be used in determining the usefulness or quality of life, but at the same time he protects the concept of "desirability" (and such a measure as willingness to pay), arguing that "it is obvious to assume that most goods, especially consumer benefits, such as food and clothing, will be in demand as means to meet the needs and, therefore, are desired to be in proportion to the expected satisfaction from them".

It can be stated that nowadays many proponents of utilitarianism look at the utility theory only from the angle of "desire" without linking it with satisfaction. They evaluate the degree of fulfilment of desires, but not the amount of satisfaction that was brought with this fulfilment. Nevertheless, there is a pretty wide recognition for the view, according to which the subjective assessment of quality of life could be best represented as a concept encompassing three major aspects [Diener, 1984]:

– *overall judgement of the people themselves on the quality of their lives at a given time;*

- *presence of a flow of positive emotions (such as feeling joy or sense of vitality and energy) at different times;*
- *lack of (temporary) negative feelings or emotional reactions (such as feelings of resentment, sadness or depression).*

There is already a database covering all these three aspects and allowing to implement a similar approach (survey "World Gallup poll" conducted in 140 countries, as well as a survey of the population "World Values Survey").

I'd like to point out that the utilitarian concept is used in the design of methodologies for assessing the quality of life only within the micro (subjectivist) approach to assessing the quality of life, or an approach that combines both objectivist and subjectivist criteria (on this, see points 1.3.2 and 1.3.3).

1.2.3 Concept of Expansion of Human Capabilities

The basis of this concept is the understanding of the quality of life as a process of freedom of man in terms of expanding his capabilities in the selection of areas of activity, methods of self-realization and access to the accumulation of human knowledge. In this case, freedom is understood in its positive value, that is, man is free to do anything or be anyone, without harming or disturbing anyone.

Even according to Aristotle: "Wealth is obviously not a fact to which we aspire, because it simply is useful and serves some other purpose. ... It is only a necessary condition for people to lead a prosperous life".

Accordingly, *income and economic growth are the means, not the goals, of development.*

This thesis was buried for more than two millennia, until Arthur Lewis, Theodore Schultz and Gary Becker (50s–60s of the last century), and then Amartya Sen and Robert Lane (at the turn of the 21st century) caught up with it and gave it a modern development.

So, Arthur Lewis in the appendix to his book *The Theory of Economic Growth* [Lewis, 1955] wondered: Is economic growth really desirable? He gives a positive answer to his question, emphasizing that it is the result of "a huge increase in the material and cultural level of the population". Thus, Lewis too, as Aristotle, thinks not only (and not so much) about the material well-being as such, but more about a higher cultural level of the population that allows people to lead a meaningful life, to use their talents and to develop their skills. It is obvious that the strategy of this approach is the development of a human, which differs significantly from the strategy for economic growth.

Conceptually economic thought came to the idea of human development through the concept of "human capital". The impetus for the emergence of the latter was the works of American scientists T. Schultz and G. Becker [Schultz, 1960; Becker, 1962] of the 50s–60s of 20th century. In search for the source of productivity growth they highlighted this concept as a separate factor (which previously was not considered as one) and initially called it the “residual”. Subsequently, Schulz identified it as a person’s ability to acquire and develop skills necessary to work. Through the efforts of their followers, the concept of human capital was formulated, which became understood as a single factor of knowledge and skills, and at the same time serving as a function of means of production and a function of consumer durables. Allocation of the factor of “human capital” enables us to define the fundamentally new sources of economic growth, which are knowledge and competence, and therefore to identify correctly the economic role of education, science and health.

One of the fiercest critics of the interpretation of quality of life as material well-being or life satisfaction is Amartya Sen. According to his theory, the individual performs different activities throughout the lifetime that generate a certain style of life and, in turn, largely determine this individual’s future activities and field of potential life possibilities. The possibilities for an individual are determined by the attainment (feasibility) of alternative activities. Some activities are elementary and designed to meet the primary needs (food, sleep, self-preservation). At the same time, other activities can be complex, for example, the achievement of self-esteem and social integration [Sen, 1984].

In this approach the context of social analysis defines a set of specific indicators with particular distribution activities. For example, in the analysis of extreme poverty one would study a small set of basic and most important activities, such as satisfactory nutrition, safety, prevention of disease, etc. In the analysis of a particular social system, some activities may be recognized as valuable, while others, – insignificant. The problem is that to reveal the usefulness of a particular activity may be difficult.

The concept of quality of life assessment from the perspective of expansion of human capabilities, proposed by Sen, is to define the entire set of real possibilities of an individual to engage in a certain set of activities (functioning) and the way of life, which he considers to be worthy. The process of human development is not seen as an increase in material well-being, but as the empowerment of people. In other words, the purpose of development is not to increase production, but to allow people to have more choices, accomplish more projects, live longer, avoid diseases, have access to the accumulation of human knowledge, etc. With this understanding of the quality of life the increased supply of goods and services can enhance human capabilities, but this is an indirect impact on human life and not an end goal in itself. Moreover, Amartya Sen argues that although the increase in production and income actually contributes to the development of human potential, it happens slower and slower [The Quality of Life, 1995; Sen, 1983, 1997]; [Сен, 2004]. By the way, I would qualify one of Sen’s works (which is a translation from English, *Development as Freedom*. – Oxford, 1999), as a kind of manifesto, summarizing the work, thoroughly and in detail describing the transformation of a narrower version of the concept (which could be called the concept of *human development* or *human capital*) into a modern version of the **concept of expansion of human capabilities**. A thorough analysis of the concept of human capital in the projection on the modern Russian reality is contained in the work of N. M. Rimashevskaya [Римашевская, 2004].

R. Lane is also one of the founders of the human development concept. He believes that the concept of “quality of life” can only be analysed by considering two things together: the quality of living conditions and the personality traits [Lane, 1994]. Lane understands the quality of life as the link between quality of the living conditions and the quality of the individual. Quality of life has intrinsic value and can be regarded as the highest good, because it reflects the qualities of the individual and, in turn, has an impact on them. At the same time a high quality environment or circumstances of life have no intrinsic value. In support of her approach Lane refers to cognitive psychology, which always considers two aspects of causal attribution – external circumstances and disposition of the individual.

No matter what characteristics are taken into account in the analysis of the concept of quality of life whether it’s the external circumstances or personal characteristics, it will always pose a question about the source of value. In this regard Lane holds the position that only the relation between the object and the person makes the object a valuable one. Environmental aspects included in the concept of quality of conditions have no intrinsic value. This representation serves as the basis for understanding that we cannot speak about the quality of life without the assessment and perception of this quality by the subject. In this case, Lane did not even consider it necessary to substantiate the possibility of assessing the quality of life by an outside subject, though it is one of the most fundamental and controversial issues in the theory of quality of life.

Lane identifies nine elements that, in his opinion, describe the quality of the person, and improvement and development of those elements will accordingly improve the quality of the person:

1. *the ability to enjoy life;*
2. *cognitive development;*
3. *awareness of autonomy and capacity;*
4. *self-knowledge;*
5. *self-esteem;*
6. *ease of interpersonal communication;*
7. *moral compass;*
8. *integrity of the person/inclusion in society;*
9. *adequate orientation with respect to labour and income.*

List of personality traits was formed on the basis of studies of mental health and human development.

While highlighting these qualities, Lane does not specify the indicators that can be used to measure them, as well as possible directions of changes in quality of life, depending on the direction and changes of the elements. He does indicate though that **the relationship between the qualities of an individual and their ranking depends on the purpose for which the quality of life is being examined.**

Based on a number of empirical studies and the concept of Maslow's hierarchy of needs, Lane distinguishes nine parameters characterizing the high quality of the living conditions:
1. *adequate financial support*;
2. *physical security*;
3. *presence of friends and social support (including safe and accepted/acceptable participation in society, or at least a few of its fields)*;
4. *opportunity to express love and accept love*;
5. *opportunity for internally stimulating work that was freely chosen*;
6. *presence of various free time/leisure, sufficient to develop the skills of creativity and relaxation*;
7. *available set of values (especially moral) that can give a meaning to one's life*;
8. *opportunities for self-development and availability of assistance necessary for this*;
9. *objective justice (executed by fair and competent judges).*

These favourable conditions can be used only by a person with certain personal qualities, and in the definition of quality of life the personality traits are dominant in relation to the quality of the conditions. In general, human behaviour, condition and quality of life are determined only by a combination of the personality traits (PT) and quality of the conditions (QC):

$$\mathrm{QOL} = f(\mathrm{PT}, \mathrm{QC})$$

where $f(y, z)$ – some function of y and z.

Lane also displays the dynamic component of quality of life: the present quality of life is partially the result of past experience(s), and change of circumstances and dispositions. Consequently, the quality of life on the second stage (QOL_2) depends on the quality of life in the first stage (QOL_1), that is,

$$\mathrm{QOL}_2 = \varphi(\mathrm{QOL}_1) = \varphi(f(\mathrm{PT}_1, \mathrm{QC}_1)).$$

Lane pairs up the quality of living conditions and the quality of the individual that should match up to be important for the quality of life (Table 1.1). It should be noted that even though Lane's conceptual principles by their ideology belong to the concept of human development they provide the necessary basis for the method of measuring quality of life, combining both approaches – *objectivist* (*macro*) and *subjective* (*micro*).

Table 1.1. Qualities of an individual and conditions (per R. Lane)

Possibilities (qualities of the conditions)	Qualities of person necessary to use the possibilities
Adequate financial support	Cognitive development, awareness of autonomy and effectiveness, orientation towards work and income
Physical security	Awareness of autonomy and capacity, self-assessment
Having friends and social support	Ease of interpersonal communication, self-assessment
Opportunities for expression	Ease of interpersonal communication, self-esteem, involvement in society
Opportunity for internally stimulating work	Cognitive development, awareness of autonomy and ability
Free time for creativity and relaxation	Self-exploration
Available set of values	Moral guidelines, being included in society, self-esteem
Opportunities for self-development	Inclusion in society, self-esteem, self-knowledge
Objective justice	Moral guidelines

1.2.4 New pragmatic approach to the concept of Quality of Life

Despite the differences in conceptual principles of theoretical concepts of QOL described above, none of them denies that the basis for social development and the transition of humans “from the realm of necessity to the realm of freedom” is *the dialectical unity and contradiction of needs and interests.* It is the process of raising a bar on human needs and expanding the freedom of choosing the way to satisfy them that constitutes the main driving force and the purpose of the development of modern society. *The more freedom a person has to choose the ways of realizing his or her interests, the higher the quality of life.* Interaction of needs and interests creates a hierarchy of needs based on the fact that each higher level of need becomes more significant and important for a person as the lower level need becomes satisfied. This idea is most clearly and fully implemented in the motivational model of ranking needs proposed by Maslow [Маслов, 1999]. The so-called Maslow pyramid extends from the physiological needs, to personal security, to the needs of being involved in community and society and, further, – to the needs of being recognized and maximum self-realization.

That is why (and bearing in mind that there is no theoretical concept of QOL universally suitable “for all times and all peoples”) we decided to choose some basic conceptual positions and accents in the interpretation of synthetic categories of QOL depending on the level at which we are according to “Pyramid of Maslow” as well as the specific application purposes of the study, in which the category of QOL to be used.

Examples of such purposes can be cited:

- cross-country and cross-regional comparisons, ratings;

- analysing and evaluating of the effectiveness of the socio-economic policies and institutional development;
- identification of "bottlenecks", the critical factors that determine the increase or decrease in quality of life of a social group, the population of the region or the country;
- analysis and assessment of the physical health of the society by using the indicators of the "index of health", "quality of life", "Human Development Index" (HDI);
- study of the peculiarities of the national (regional) mentality of the population.

We call this approach pragmatic (and thus the *concept, – pragmatic*). We will adhere to it. As far as strategic plan goes we like the concept of expansion of human capabilities, in which economic growth and industrialization are not the purposes of the development of society, but only a means of expanding the person's choices, of increasing the accumulation of human capital, understood in a broad sense (education, knowledge, health, freedom, possibility of effective self-realization, etc.).

1.3 Methods of Measuring the Quality of Life (World Experience)

For a long time (since the times of Plato and Aristotle) the topic of "dignified life" and ways to achieve it with the help of government policies have been widely discussed. Only relatively recently that we got the means (both in politics and in science) that allow us to measure the level of "dignified life" and use these measurements to develop proposals for its improvement. Over the past two or three decades a number of different indicators of quality of life were suggested by state and political institutions, agencies, research centres and the media. In my opinion, some of the problems encountered in the design of these indicators that still have not found a satisfactory solution are the following.

(A) **The problem of determining the composition of private criteria that would adequately characterize the analysed synthetic category of QOL.** In this case, the private criteria $x^{(1)}, x^{(2)}, \ldots, x^{(p)}$ is made of statistics that adequately characterize viewed conglomerates of population in relation to analysed synthetic category of QOL (macro approach) or the results of a questionnaire survey of the population on the various items (micro). Thus, if we want to measure and compare the level of economic well-being of the subjects of Russian Federation, we have to determine a set of statistical indicators $x^{(1)}, x^{(2)}, \ldots, x^{(p)}$ that will characterize each of the subjects. This way we can use the values of above indicators to execute our comparison.

(B) **The problem of determining the general form of the function *f* (derived from the set of private criteria) with the help of which we can transition from the multi-criteria scheme $(x^{(1)}, x^{(2)}, \ldots, x^{(p)})$ to one-criterion scheme**

$y = f(x^{(1)}, x^{(2)}, ..., x^{(p)})$. Virtually all of the following ways (taken from international experience) on how to measure quality of life are based on the linear function f, that is

$$f(x^{(1)}, x^{(2)}, ..., x^{(p)}) = \sum_{j=1}^{p} w_j x^{(j)} \tag{1.1}$$

with weights w_j $(j = 1, 2, ..., p)$ everywhere (with macro approach) being defined by experts, which means subjectively (!)

(C) **The lack of a satisfactory solution to the problem of transition from multi-criteria scheme to one-criterion scheme within the class of linear convolutions of the form (1.1).** As we will show later in this book (see Chapters 2 and 3), there are situations when there is no satisfactory (in some sense) solution of the problem in the form (1.1). The way to resolve it is to expand the class of admissible solutions beyond the class of parametric set of functions (1.1) or in the transition to multi-criteria scheme $\{y^{(1)} = f_1(x^{(1)}, ...; x^{(p)}),\ y^{(2)} = f_2(x^{(1)}, ..., x^{(p)}), ..., y^{(k)} = f_k(x^{(1)}, ..., x^{(p)})\}$ of significantly smaller dimension (i.e. $k \ll p$ where the symbol $\ll$ means "much less").

(D) **The problem of taking into account the action of "principle of non-compensated preference" in the transition to the "convolutions" of private criteria, i.e. functions** $y = f(x^{(1)}, x^{(2)}, ..., x^{(p)})$. In the analysis of the quality of life of the population in the vicinity of critical threshold values of individual private criteria $x^{(j)}$we should take into account the specific properties of the integral indicators, which in the theory of organizational management are called "non-compensated preference" or "principle of partial non-compensation". This means, in particular, that there are critical threshold values close to $x^{(j)}_{min}$(at least for some of the private criteria $x^{(j)}$) beyond which further reduction of the values of each of these indicators cannot be offset by an increase in the integral indicator (improvement) of the values of other private indicators . For example, in an integrated assessment of QOL a drop below a certain level of quality indicator of the ecological niche or of physical and asset security of a member of a society cannot be compensated with any improvement of values of other parameters, for example, material well-being. Obviously, II QOL as linear functions $x^{(1)}, x^{(2)}, ..., x^{(p)}$ is not adapted to account for this fact. Therefore, in such cases we should look for II QOL in the wider (than linear) class of transformations $x^{(1)}, x^{(2)}, ..., x^{(p)}$ (e.g. with the help of neural networks approach) or use an approach based on the idea of multi-criteria scheme being substantially less than p (the dimension).

Below are the results of the analysis of the world experience in building various kinds of integrated indicators of QOL.[1] This experience is frankly disappointing. We could not find any work that contains at least some suggestions on how to solve problems (B), (C) and (D). One method is different from another

1 [Quality of Life Indexes, 2001].

by only the set of private criteria $x^{(1)}, x^{(2)}, \ldots, x^{(p)}$ (problem (A)) and the weights assigned to them $w_1, w_2, \ldots, w_p$ in the framework of the linear class of functions $f(x^{(1)}, x^{(2)}, \ldots, x^{(p)})$ (part of the problem (B)). Add to this the fact that the direct expert assignment of the weights $w_j (j = 1, 2, \ldots, p)$ is unfortunately a common practice in macro approach; it is difficult to say that we found a satisfactory solution to the problem.

Some approaches to solving problems (A) – (D) are presented in Chapter 2.

1.3.1 Methods of Measurement of the QOL with Objectivist Approach

So, in the objectivist approach the statistical indicators take the role of private criteria $x^{(1)}, x^{(2)}, \ldots, x^{(p)}$. Form of certain functions derived from these criteria is used to build integrated indicators for the analysed synthetic category of QOL. The statistical indicators adequately characterize viewed conglomerates of population (the population of the region and the country as a whole, the social layer) in this synthetic category. The method of building of these integrated indicators is based on the assumption that the desired assessment of II QOL may be based on an external expert's opinion on which particular conditions, opportunities and statistics that measure them should determine the analysed QOL.

1.3.1.a Indicators based on income

GDP received the greatest recognition among economic indicators that are used to measure quality of life. GDP and GDP per capita as an indicator of the quality of life estimate the volume of consumer goods available to public, and thus (indirectly) – the level of consumption. In other words, GDP characterizes the wealth of the nation, ignoring such important components of quality of life, as the level of education, the environment, health of the nation, the state of the social sphere, the level of criminality and many others. Some proponents of using GDP to measure the quality of life conducted a series of studies to establish the existence of relationship between the size of GDP and indicators of socio-economic development of the country or region. Such a relationship would be a significant argument in favour of this indicator. However, studies have shown the possibility of the opposite situation: it can be growth of wealth in the region expressed by GDP, and at the same time deterioration of the socio-economic exponents and the population's appraisal of their own quality of life.

One of the problems of using GDP as a measure of well-being is related to the nature of this indicator and the methods of assessment of its volume. Imperfection of the measurement's methods, the existence of different approaches and corrections create multiple estimates of the GDP, and therefore there is a problem of comparability of estimates. One of the biggest drawbacks of GDP is the inadequacy of this indicator in cases where a significant amount of transactions are carried out by non-market institutions. In addition, GDP does not account for the distribution of income among the population and the structure of national expenditure, for example, the cost of the military-industrial complex that can carry a significant weight. GDP also does not provide for the possibility of citizens living of foreign assets. Finally, we should keep in mind that GDP is an indicator of current economic activity, and it does not reflect the country's developmental prospects.

Researchers in the field of quality of life made multiple attempts to adjust the GDP so as to make it more effective in assessing the welfare of the nation. In particular, G. Becker, T. Philipson and R. Soars [Becker, Philipson, Soares, 2003] argue that quality of life is determined not only by the current wealth, but also by its "duration", that is, not only the income, but also the number of years during which the person receives the income. By developing this idea, these authors propose an algorithm for calculating the monetary equivalent of changes in life expectancy. Informally this is the equivalent of the amount of money that a person agrees to pay for the increase in the duration of his life for a certain number of years or, conversely, monetary compensation to which a person agrees with life expectancy's reduction. Cash equivalents to changes in life expectancy and the GDP per capita will assess the quality of life, which can be conveniently used to assess the temporal dynamics.

1.3.1.b Indicators based on anthropometric measurements of children

J. Micklewright proposes to evaluate welfare on the basis of data on anthropometric indicators of children under 6 years of age living in the countries studied [Micklewright, 2001]. The author identifies three properties of the physical characteristics of a person that would provide an advantage in the evaluation of quality of life in comparison with GDP. First, the anthropometric characteristics of a person testify about the level of his income and consumption, and at the same time information on the physical indicators of the population is more readily available than data on the amount of income and consumption. Second, the individual character of anthropometric characteristics serves as advantage for using them, in comparison with aggregated data that are used in the calculation of income and consumption. For comparison of quality of life in different countries, as well as to monitor the dynamics of changes in the standard of living in a particular country, Micklewright suggests using two measures: correlation of height to age and weight to height. Third, anthropometric characteristics reflect the quality of nutrition. Lack of any nutrient reduces the physical and mental abilities of the person, increases the level of mortality and morbidity and reduces a person's possibilities, for example, prevents a child's school attendance, sports and extracurricular activities that adversely affect the level of development (within the meaning of the Seine). Thus, the indicators of good nutrition are important factors that are advisable to include in the concept of quality of life.

1.3.1.c "Green net national product"

Previous methods of assessing the quality of life were aimed at solving the problem of formation of indicators that were reflecting the current welfare. At the same time, in practice, sometimes the problem arises on how to assess quality of life as a characteristic that would reflect not only its current level but also the ability to maintain it in the future. The development of such an instrument is related to the theory of sustainable development. There are a number of works devoted to the adaptation of GDP to the described goals and building of indicators that would be reasonable estimates of sustainable quality of life.

One of the theoretical constructs that allows us to solve some of the problems inherent to the gross national product (GNP) is the indicator of "green net national product" – GNNP [Asheim, 1999]. This indicator includes information about the depreciation of capital, growth of human capital, knowledge and natural resources. Despite these advantages, the GNNP is not so much an indicator of sustainable quality of life, as characteristic of the wealth of the region. However, a comparison of regions, such as cross-country comparison on the level of sustainable quality of life that is based on this indicator, is more valid than comparisons based on GNP.

On the basis of GNNP it is possible to define the concept of real investments as a change in wealth that is calculated with the help of this parameter. Thus the increase in wealth of the region is a positive real investment, while its reduction is – negative.

1.3.1.d Index of true progress (development)

As an alternative to GDP an index of true development (Genuine Progress Index – GPI) was developed by a group of scientists led by Cobb when assessing QOL in 1995 [Cobb, Halstead, Rowe, 1995]. GPI has been used in the United States. GPI is designed to provide citizens and politicians with more accurate indicator of the economic situation of the region (country) and its dynamics. This index similar to GDP is primarily a measure of economic well-being, but it also extends the inherent structure of GDP by including in its assessment the contribution of social and environmental factors along with the traditional measurements of economic production. GPI takes into account more than 20 aspects of economic life that GDP ignores. It also differentiates the economic transactions that make a positive contribution to the assessment of welfare from those that reduce the quality of life. GPI integrates all of these factors into a single measure so that the benefits of economic activity could be weighed against the costs.

GPI calculation begins with the definition of personal consumption expenditures that are adjusted to reflect the income distribution system expressed by the index of income distribution. The obtained value increases or decreases by expenditure depending on the direction in which they affect the level of welfare. Classification of parameters by their effect on the measurement of GPI is shown in Table 1.2. All categories included in the calculation of the index are expressed in monetary units; therefore, to calculate the GPI we can apply simple mathematical methods of addition and subtraction.

Table 1.2. Classification of expenditure by the way they influence the amount of GPI (direction of influence)

Factors that are increasing the welfare of population (included in the GPI with a "plus" sign)	Factors that are reducing the welfare of population (included in the GPI with a "minus" sign)
1. Personal consumption	1. The crime rate
2. Housework and parenting	2. Divorces
3. Work as a volunteer	3. Time spent on leisure
4. Being able to use good road service	4. Underemployment
5. Net capital investments	5. The cost of durable goods
6. Net foreign borrowing or lending	6. Traffic accidents
	7. Water pollution
	8. Air pollution
	9. Increased noise level
	10. Water logging
	11. Loss of agricultural land
	12. The depletion of non-renewable resources
	13. Long-term damage to the environment
	14. Thinning of the ozone layer
	15. Loss of old-growth forests

The purpose of the development of GPI was to include many spheres of social life in the assessment of the quality of life that are ignored by GDP. However, just as GDP, GPI is based on economic theory and uses its approaches to determine monetary value of non-market variables, and consequently, the obtained values are not always correct. Moreover, serious problems arise with the assumptions and methods used to evaluate many GPI factors that characterize the resources and the environment. For example, the cost of the loss of wetlands over time becomes more significant and less realistic, setting a strong downward slope index. It should also be noted that all the variables that are included in the GPI are based on objective data, while the methodology for estimating these variables is extremely subjective.

Due to the significant shortcomings of GPI in its present form it is not a reliable measure of quality of life or of genuine development. Socially significant purpose of the index is to establish trends in the development as a whole and its components, allowing government officials to highlight problem areas and implement the necessary corrective actions.

1.3.1.e The index of economic well-being

One of the most advanced and sophisticated methods to assess the material well-being of the population is the index of economic well-being (– IEWB), developed by L. Osberg and A. Sharpe in the early 90s of 20th century [Osberg, Sharpe, 1998]. This index is based on the assumption that economic well-being of society is determined by the average level of consumption, total accumulation of inventories, inequality in the distribution of individual incomes and reliability of future income. Accordingly IEWB is an integral characteristic of all these factors. Experts' assessments are used as weights when calculating this characteristic.

The level of per capita consumption includes the following indicators:
- personal consumption flows that are covered by market, adjusted for the share of non-market economy, the value of the increase in life expectancy, changes in the families' size that contribute to economies by the process of domestic consumption;
- changes in labour conditions;
- government services;
- value of unpaid work.

Aggregate Inventories consist of the following components:
- net capital stocks of physical assets, including household supplies;
- research and development;
- national nature reserves;
- human capital;
- level of foreign debt;
- net change in value of the environment under the influence of carbon dioxide emissions.

Inequality component includes:
- degree of inequality in income distribution, determined by the Gini coefficient for after-tax household income;
- intensity of poverty, characterized by the degree of distribution, i.e. percentage of population living below the poverty line; depth, i.e. difference between the average income of people living below the poverty line; and the value of income that determines the poverty line.

Reliability of future income is determined by time induced changes in the economic risks associated with the following points:
- risk of unemployment, which is determined by the level of employment of population, degree of coverage of employed citizens by unemployment insurance, as well as the characteristics of the social protection of the population;
- risk of disease, measured as the proportion of income spent on health maintenance;
- risk of being in the position of a single parent, characterized by the divorce rate and level of poverty for single-parent families;

- risk of falling into poverty by citizens with the onset of old age, which is determined by the level of poverty among the elderly.

Osberg and Sharpe suggested using the following values of weighing coefficient for the average consumption – 0.4; for the total accumulation of inventories – 0.1; for measures of inequality in the distribution of individual incomes and reliability of future income – 0.25. Subcomponents of average consumption and aggregate accumulation of inventories are expressed in monetary units, and hence, there is no need to weigh. For the components of the index of inequality the following value scales are offered: 0.0625 – for the Gini coefficient and 0.1875 for the value of the poverty intensity (level). Subcomponents of the index of reliability of future earnings are weighed according to the relative importance of the categories included.

IEWB has a well-developed theoretical basis because it is based on economics theory but just as all the previous indicators *it does not go beyond economic concepts.* Additionally, a significant disadvantage of this index is a high degree of subjectivity of expert-appointed weight coefficients.

* * *

All five indicators, as described in point 1.3.1, are consistent with the concept of quality of life based on the *economic well-being*: their composition and interpretation do not go beyond the economic concepts. As the *concept of expansion of human capabilities* gets developed further and further a person with his diverse needs and interests is at the centre of attention. The person himself is both a goal and an essential tool to his own development. Samuelson's *social welfare function* as well as function and quality of life by Liu [Quality of life indexes for national policy, 2001] are regarded to be first attempts of building quality of life characteristic that takes into account not only the indicators of economic well-being. The following are examples of the construction of indicators of QOL, to a greater or lesser extent consistent with the concept of expansion of human capabilities.

1.3.1.f Physical quality of life index

Index of the physical quality of life (– PQLI), developed by a group of scientists led by M. Morris [Morris, 1979] is an analogue and precursor of the more famous (and used in the documents of UN Development Program) HDI (see below). PQLI is a weighed average of the indices of literacy, infant mortality and expected (at the age of 1 year) life expectancy. Index of literacy is a literacy rate at age 15 years and older. The index of child mortality takes a minimum value of zero for the region with the worst value of the corresponding indicator and reaches a maximum of 100 for the region with the best value. Similarly, the life expectancy index is 0 (minimum value) for the region with the shortest life expectancy and is equal to 100 (maximum) for the region with the longest one. To build an index of the physical quality of life, we have to produce an average of three abovementioned indices with equal weights. In most cases, it turns out that it is highly correlated with the first principal component, built on these three indices.

Based on social indicators PQLI has several inherent disadvantages that are typical to them, such as high correlation between the statistical indicators and the high values of the index as a result of strong government administration infrastructure. PQLI methodological problem is the use of literacy rate, which is characterized by discrete changes in values. Furthermore, many researchers believe that the proportion of literate population is not the correct measure of the level of education.

The advantages of this indicator are the simplicity of the calculations and the possibility of cross-country comparisons. PQLI is the most frequently used index in assessing the macroeconomic level, the physical health of the population and the degree of satisfaction of basic human needs. In most cases, it is used in conjunction with other indicators, such as GDP of purchasing power parity, the official exchange rate and various social indicators.

1.3.1.g Human Development Index

This index, also known as the Index of Human Potential Development (HDP), was developed in the framework of the UN Development Program, and first published in the Human Development Report of the United Nations for 1990 [UNHDP, 1990]. According to its developers (among whom was Sen) HDI had to perform the function not only of the metre of the economic well-being, but also of the level of human possibilities. In its structure, HDI is a combination of three indicators: *the index of the expected (at birth) life expectancy, education index and the index of income.* Life expectancy index is calculated as the average life expectancy at birth for both sexes. Education index is made up of the percentage of literate population aged 15 years and older, as well as the indicator of how fully the population is covered by primary, secondary and higher education establishments. Income index initially was estimated as the logarithm of the GDP, but it was subsequently modified by taking into account the comments by Atkinson: “The higher the income relative to the subsistence level, the greater impact its reduction has on human development” [Atkinson, 1987]. In the current method of calculating the HDI, the real GDP per capita, modified (at purchasing power parity), is used as a way to estimate the income. In the most general form the HDI is calculated by using the following formula:

$$\mathrm{HDI} = \sum_{j=1}^{4} w_j \frac{x^{(j)} - \min x^{(j)}}{\max x^{(j)} - \min x^{(j)}}$$

where $x^{(j)}$ – the actual value of the j indicator; max $x^{(j)}$ and min $x^{(j)}$ – the maximum and minimum values of the j indicator; and w_j – weight coefficient of the j indicator. The weight coefficient of average life expectancy is 1/3, adjusted GDP per capita – 1/3, level of literacy – 2/9, completeness of educational establishment enrolment – 1/9. Thus, the weight of each of the three abovementioned components of human development is 1/3.

HDI calculation is carried out in two stages. At the first stage the actual values are normalized, i.e. reduced to a single – [0; 1] – scale of measurement. In the second phase the normalized indicators are aggregated into an index of human development.

Although the HDI is among the most well-known and officially used indicators of quality of life, this technique is not without drawbacks. In particular, critics of the concept of HDI dispute the appropriateness of inclusion the life expectancy indicator in the index in the form in which it is presented today. According to I. Nubler [Nubler, 1995], life has value not in itself, but rather as a reflection of something that can benefit us. There are examples of situations where HDI does not reflect any changes in spite of the fact that in reality there has been an increase in the quality of life in certain country or the decline in it. The disadvantage of HDI methodology is also lack of justification of weight distribution between the indicators. Many researchers have noted correctly that the assigned weights for literacy and completeness of schools' enrolment (2/9 and 1/9, respectively) are simply not true. Nubler shows that the infant mortality rate is negatively correlated with the completeness of school enrolment, but not with the level of literacy, and therefore, the first one is more critical and should be assigned a greater weight. Many other researchers also believe that the indicator of comprehensiveness of school enrolment is more informative.

Also of note is the fact that, as previously stated, real GDP per capita that is used in calculating the HDI is not an accurate measure of the level of income, and that is why its weight in the HDI should be recognized as unreasonably high.

Statistical analysis of the dynamics of changes in the values of HDI and other socio-economic indices reveals that there are significant discrepancies. For example, the values of the basic development indicators for 1960–1990 for countries of Persian Gulf show a marked increase, while the HDI does not show significant changes. Also, there are situations when the rate of change of the values of the major indexes and HDI are not the same.

Nevertheless, many scientists recognize that the HDI serves as *an effective tool in solving the problem of determining the priorities of public policy and provides an information base for comparisons of different time periods*. In addition, the structure of the index is easy to understand and use.

1.3.1.h Index of Social Health

The index of social health (– ISH), developed in the mid-90s of the 20th century by Miringoffs [Miringoff, Miringoff, 1999] is another index that belongs to integral indicators used for assessing the quality of life that is based on objective statistical information. This index is calculated on the basis of statistical data on the values of the following 16 factors:

1. infant mortality;
2. child abuse;
3. the number of children living below the poverty line;
4. youth suicide;
5. teen addiction (the percentage of teens who took any kind of narcotic drugs in the past 12 months);
6. percentage of children not attending school;
7. births by minors;
8. unemployment rate among the population;
9. the average weekly earnings;
10. health insurance coverage;
11. proportion of people aged 65 years old and over, who are below the poverty line;
12. life expectancy at age 65;
13. level of serious crimes (violent crime rate);
14. accidents related to alcohol;
15. housing;
16. gap between rich and poor.

The distinguishing characteristic of ISH is the grouping of indicators based not on aspects of quality of life but on different age groups: the first three indicators characterize the child population, the following four (4, 5, 6, 7) characterize – adolescent population; figures 8, 9, 10 pertain to – adults; and figures 11 and 12 to – the elderly, and finally, the last four are common to all age groups. The values of all 16 factors are normalized and stacked with *equal weights*. In this case, the authors use first method of normalizing of baseline values, at which they are measured at a scale of [0; 1] in such a way that the value of 0 corresponds to the worst situation, and a value of 1 to the best (more on the transformations of this type see Chapter 2).

ISH is a convenient tool for setting public policy objectives and the subsequent monitoring of its effectiveness. Its advantage is the use of official and accessible statistical information. However, a significant drawback of ISH is that, first, the selected 16 factors do not cover all relevant aspects of quality of life, and second, they all integrate with equal *weights, appointed by experts*.

1.3.1.i Index of Social Health by Johnston

An unusual method of estimating the quality of life was proposed in 1988 by Johnston who also developed an index of quality of life [Johnston, 1988]. A distinctive feature of this index is the fact that instead of defining a simple significance of each indicator (i.e. weight), it calculates the relative importance based on the analysis of time series. In accordance with the method of calculation of the index first we have to gather data on the values of 21 indicators included in the model. These indicators are divided into nine socially important areas, which are public health, public safety, education, employment, wages and income, poverty, housing, the stability of the family and equality. As a result, each area is represented by two or three indicators; for example, the health sector is estimated from the values of life expectancy at birth, infant mortality and the number of disability days per person per year. To determine the relative importance of the indicator we choose randomly some base year. Then, the importance of the indicator in the current year is estimated as the change of the indicator in percentage in relation to its value in the base year.

The disadvantage of this index is the lack of a sound theoretical framework. In particular, the author does not offer any support for his choice of social spheres (e.g., in terms of content "wages and income" and "poverty" almost describe the same aspect) and the number of indicators that make up each of the areas. In addition, independent (differentiated) weighting for each indicator separately leads to the fact that degrees of relative changes observed in different areas could not be interconnected and unambiguously interpreted.

1.3.1.j International Index of Living Conditions

International index of living conditions is calculated annually since 1984, and published in the periodical magazine *International Living*. During this time, over 150 countries have received their corresponding rating value based on this index. The index itself is a weighted sum of the following measurements of quality of life (weight percentage is shown in parentheses):

- economy (20);
- health (12);
- culture and leisure (12);
- infrastructure (12);
- cost of living (20);
- freedom (12);
- safety and risk (12).

A large number of indicators are used for each of the above categories to calculate the rating of the level of development of each of these measurements. The *economy* is represented, for example, by values of GDP, real GDP growth and level of inflation; *health care* is represented by calorie consumption as a percentage of the recommended number of calories per person/per day, doctors per capita, the percentage of people with access to clean water, infant mortality, the expected (at birth) life expectancy and health expenditure as a share of GDP; *culture and recreation* are presented by the figures of the number of newspapers, literacy rates, the number of national parks and nature reserves; *infrastructure* is described by density of roads, the number of airports, telephones, television sets per 1000 of population; *cost of living* is determined by the US Department of State as the index of residence abroad that corresponds with western lifestyle (*cost of living – US State Department's Index of Overseas Living Costs for a Western style of life*), and so on. However, these indicators could be changed at the request of the editor or readers. Thus, despite the fact that the rating procedure itself contains elements of subjectivity, it is based on data collected by national statistical offices. For more on this index, see the following [Peterson, Malhotra, 1997; Peterson, Malhotra, Wagner, 1999].

1.3.1.k Integral Social Index by Michalos

The report on the social situation by Michalos (Michalos' North American Social Report (1980–1982)) is a comparison of the quality of life in Canada and the United States from 1964 to 1974 where he analysed 126 social indicators, grouped into 12 categories [Michalos, 1997]:

1. structure of the population;
2. mortality, morbidity and health care;
3. crime and justice system;
4. politics and organizational structure;
5. science and technology;
6. education;
7. leisure;
8. environment and natural resources;
9. transport and communications;
10. housing;
11. economy;
12. moral and social customs.

The annual value and characteristics of the dynamics of this value were presented for each indicator, for example, the current level of crime and characteristic of the dynamics of the level of crime. Integral index was obtained as a weighted sum (equal weights) of all indicators. The country receives one point for the value of each indicator that is better for this country than for the other countries. For example, Canada has received one point for a low crime rate compared to the United States in 1964 and one more point for the slower growth in the crime rate in the period from 1964 to

1965. As a result, Canada received 884 points while the United States received – 775 points during the study period.

1.3.1.l Index of Social Progress by Estes

In a series of publications [Estes R.J., 1984, 1988, 1995, 1996, 1997, 1998] R.J. Estes has developed an index of social progress (– ISP) and applied it to the study of several groups of countries in different parts of the world. The goal of ISP is as follows:

- to identify significant changes in the "adequacy of social security", which is observed around the world;
- to assess national and international action towards a more adequate provision of basic social and material goods for the growing population of the world.

"The adequacy of social security" means the ability of governments to provide primary material, social and other benefits to the people living in this country. ISP consists of 46 social indicators that are grouped into "sub-indices", describing:

- education;
- health;
- status of women;
- Possibility of national defence;
- economy;
- demographics;
- geography of the country;
- participation in world politics;
- cultural diversity;
- welfare.

All 46 indicators are known as conventional, reasonable indicators of social development. Indeed, most of the indicators included in ISP are regularly used by other researchers of social and economic development. All of them are "objective" indicators, such as the proportion of adolescent literacy, life expectancy, real per capita GDP and the index of political rights violations.

In the absence of some of the values of indicators that make up ISP, Estes put them to two-step factor analysis (*a two-stage varimax factor analysis*), in which each indicator and sub-index were analysed in terms of their relative contribution to the explanation of the variance that reflects changes in social progress over time. In his work Estes does not mention of how such criteria as "change of social progress in time" is formalized for the factor analysis. Standardized set of indicators are multiplied by factor loading in order to obtain multiple weighted sub-indices. Then the obtained indices are summed up resulting in a *weighted* index of social progress (WISP).

Estes argues that the WISP is different from other social indices by the number, range and relevance of the indicators that are used in its construction. He also stresses that the WISP is considered more comprehensive, valid and reliable tool for assessing temporal changes in social development than such standard codes of national and international development like GDP, GNP and HDI.

1.3.1.m Method of Interregional analysis of Quality of Life by V. V. Kossov

Out of the very large number of approaches to the analysis of quality of life offered by Russian researchers [Бобков, Масловский-Мстиславский, Маликов, 2000; Косов, 2001; Маликов, 2002; Квалиметрия жизни, 2006] we will stop here on the method proposed by V. V. Kossov. This technique is designed specifically for the assessment of quality of life in the subjects of Russian Federation, and accordingly, it is built to take into account Russian specifics.

The author separates processes occurring in the subjects of the federation into *positive* and *negative*. Positive processes lead to the prolongation of life, and the second to its reduction. We select relevant statistics to describe the processes. *Negative* processes are described by indicators such as murder, crime, unemployment, refugees, the proportion of families with incomes below the subsistence level and the infant mortality rate. *Positive* processes are characterized, for example, by increase in the average life expectancy for newborns, growth in GDP (country's and region's as a whole, and per capita), the growth of real money incomes, increased consumption of protein foods, etc.

Separating processes as negative and positive gives clear guidelines for regional development programmes. The developed approach is closely related to the task of assessing the quality of power, based on the understanding that the purpose of government is to promote the positive processes and to organize action to suppress the negative processes.

It's not enough just to have prerequisites for the development of positive principles in the life of society in order to suppress the negative trends described by these or other indicators: the authorities must take actions that would control the situation. The consequence of these actions must be the securing of the downward momentum of indicator values characterizing the negative processes. For normally developing regions, this would mean the existence of a positive growth rate for the positive processes and negative growth rate for the negative processes. Accordingly, an active social policy should be desirably aimed at supporting positive processes and absolutely necessary aimed at the suppression of negative processes.

In the analysis of the situation on several objects (or on one object, but for several years) it is convenient to use Table 1.3. The table allows us to assign any object of observation, in our case, – the subject of the Russian Federation, to one of four groups according to the ratio between the signs of the rates of the processes that determine the direction of social policy in each region, resulting from the savings of people's lives in Russia.

Table 1.3. The speed of the flow of the processes and recommendations for actions when developing social policy

	Negative processes are negative	**Negative processes are positive**
Positive processes are positive	#1 To keep developing	#3 Focus on suppression of negative processes
Positive processes are negative	#4 Focus on development of positive processes	#2 It is necessary to look for an individual solution to remove obstacles on the way to development

In cell number 1 are the so-called honour regions, i.e. areas in which positive processes are being further developed and negative processes are suppressed. For this group of territories we can assume that the negative processes are under control. In terms of approach to designing programmes of development in the region, it is important to focus on the continuing progress of already reached milestones.

Antipode to cell number 1 is cell number 2, which includes areas that are actually moving to a catastrophe because the positive processes are not being developed, and the negative ones are not being suppressed. Obvious limit to such a situation is a disaster. The best form for people to solve the problem is to change the power in the next elections. The decline in the dynamics of positive processes makes us assume that something is blocking their development, and the authorities are unwilling or unable to organize the suppression of negative processes. This situation must be regarded as extraordinary, and each district must seek its own ways of solving the crisis.

Cell number 3 describes a situation in which the development of positive processes is observed in spite of the growth of negative processes. It can be assumed that the absolute level of negative processes still did not reach the level when it's fully blocking the positive processes. For the districts included in this zone it means that the main objective in the designing the programmes for regional development is to focus on the further development of positive processes.

Cell number 4 is symmetric in respect to cell number 3 with the only difference that the suppression of negative processes no longer contributes to the development of positive processes. It seems that such a scenario is possible in the case where the absolute level of negative processes is already so high that it causes suppression of positive. This situation is described by means of bifurcation in the theory of catastrophes. Kossov observed this situation, for example, in the period of revolutionary upheavals and it is recorded by statistics and perceived as a collapse. This means that when designing the regional development programmes for the areas in this situation the main purpose for the authorities should be to focus on the suppression of negative processes.

The main difference of the methodology proposed by V. V. Kossov is the lack of an integrated assessment of the quality of life, typical for most theories in this field. As a result its drawback is the difficulty of the inter-regional (cross-country) and inter-temporal comparisons. However, this technique offers one of the possible tools to establish directions of socio-economic policy for governments and public authorities.

1.3.2 Methods of Measuring the Quality of Life Under Subjective Approach

Conceptually, a subjective approach is based on a utilitarian concept of quality of life. It looks at this synthetic category from the standpoint of the concept of "utility" and is defined as a measure of the quality of life of the individual degree of wish fulfilment or its degree of satisfaction with their lives. Life satisfaction is defined as a property available in a given individual experience of living life as measured subjectively. In the earliest work in this area the appreciation of life was understood as a degree of adaptation to certain conditions of life. Subsequently, researchers began to examine the internal perception of man as the measure of life satisfaction.

In contrast to the objectivist approaches, subjectivist approach the quality of life has always been regarded as a synthetic category that includes a wide spectrum of conditions of life, based on current economic, political, social, psychological and cultural aspects. The main difference of the subjectivist approach is that in this approach the QOL is the result of subjective assessments of various aspects of their own condition.

Some scholars consider "life satisfaction" and "quality of life" as identical concepts. Among these are Campbell, Converse and Rodgers [Campbell, Converse, Rodgers, 1976]. They proposed the use of the degree of life satisfaction index, which determines the quality of life. Others, such as Kleynpel (Kleinpell), regarded as part of the enjoyment of life quality of life. Others, on the contrary, believed that quality of life can be used to get an idea about life satisfaction. Thus, within the subjectivist theory, the concepts of "life satisfaction" and "quality of life" are, if not identical, then strongly correlated.

1.3.2.a Assessment of quality of life according to Ferrans and Powers

Ferrans and Powers understood the quality of life as satisfaction with the events that took place in it and the importance of these events for the individual. Accordingly, these authors developed a technique of assessing the quality of life [Ferrans, Powers, 1985] that included two stages, each of which required for the information to be gathered through a survey of the population. Formed profiles contained the specified list of the main spheres of human life and the interviewees were asked to evaluate these sectors according to the degree of satisfaction with the state of each area and its relevance to the respondent.

Method of assessing the quality of life offered by Ferrans and Powers was later supported by many researchers. For example, American researchers G. Riecen and U. Yavas offered the methodic of policymaking for the government bodies of the regions on the basis of assessments of the quality of life of the population of the region. These assessments were done using the method of Ferrans and Powers. The above researchers proposed that the state policies in the region, aimed at improving the quality of life of the population, would be formed in accordance with the development priorities that are determined through a survey of the population of the region [Riecen, Yavas, 2001].

Riecen and Yavas created a list of 60 socio-economic aspects that can serve as a basis for the assessment of quality of life, and combined them in the following synthetic categories (spheres):

1. crime (4 indicators);
2. economics (8 indicators);
3. education (9 indicators);
4. infrastructure (7 indicators);
5. health (7 indicators);
6. housing (8 indicators);
7. leisure (6 indicators);
8. transportation (6 indicators);
9. art and culture (5 indicators).

Citizens of the region were asked to evaluate this list of socio-economic aspects according to their importance when deciding on the choice of residence. They also were supposed to evaluate each category separately based on the sum of its indicators for the area. Then the synthetic aggregate score for each category was calculated using a 7-point scale, which result in the matrix shown in Figure 1.1.

Some authors on the basis of the methodology of Ferrans – Powers offered to calculate the index of quality of life as a weighted sum obtained in a survey of assessments of selected areas, using the identified significance of those areas to population as weights.

Such an approach to the evaluation of the quality of life has a number of drawbacks. First, the spheres of public life have an impact on people's judgements about their welfare and are selected based on their intrinsic value, and hence the estimation of individual importance (significance) of these spheres is superfluous.

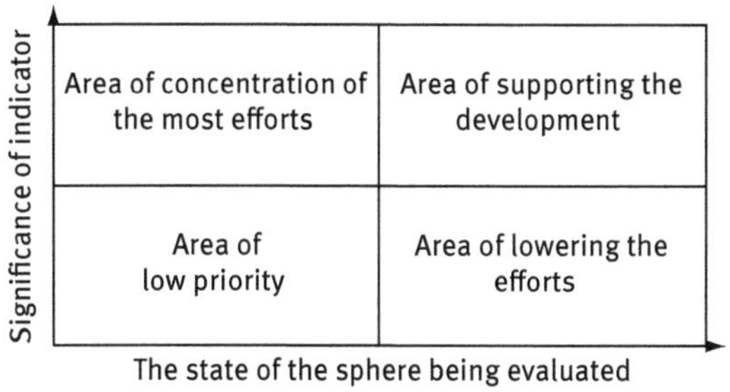

Figure 1.1. Evaluation scores for indicators

Second, the mathematical method used when constructing the index of QOL creates certain difficulties in interpretation: is the high level of satisfaction with a certain sphere that is low on importance list (high satisfaction multiplied by low importance) equivalent to low level of satisfaction with a sphere that is high on importance list (low satisfaction multiplied by high importance)? Third, the values we get can also be influenced by change in measurements' scale: change from 5-point scale [1,5] to a 4-point one for some variable will result in change of correlation between our assesement of QOL and other variables. Finally, correlation analysis and forecast of weighted and unweighted estimates of the quality of life showed no need for such a weighting.

Approach to assessing the quality of life, developed by Ferrans – Powers, is common, but researchers have criticized the appropriateness of direct expert estimation weights of components in partitioning the quality of life for the individual sphere. So, Gerdt and Johanson determine estimates obtained by this method as unreliable, due to the fact that people tend to judge quality as a single category and it is difficult to rank the importance of various aspects of quality of life [Gerdtham, Johannesson, 1997]. As shown by studies conducted by these authors, people are not always aware of the measure of the importance that they attach to a particular sphere of life, and therefore, the method of Ferrans – Powers does not provide reliable information base to obtain correct estimates of welfare (see about the shortcomings of this approach in (b) of the introduction to Chapter 2).

As a solution to this problem Gerdt and Johanson suggest using the same technique, but with reference to the expert assessment of the quality of life as a single category. Such an approach would also include analysis of the quality of life for all the important aspects of an individual without a detailed listing them.

1.3.2.b "Eurobarometer"

As an example, the method of evaluating the quality of life that does not use a measure of the significance of individual factors can lead public opinion survey (Eurobarometer surveys), which the European Union conducted twice a year since 1973 [Euro-barometer, 1998]. The survey involved citizens over 15 years old, who are asked to answer the following questions:

1. Please rate how satisfied you are with your life: very satisfied, relatively satisfied, not very satisfied, not satisfied.
2. Please rate how satisfied you are with the level of democracy in your country: very satisfied, relatively satisfied, not very satisfied, not satisfied.

In addition, each survey contains a number of additional issues related to the time of the study. This structure allows the survey to produce a multilateral assessment of the quality of life that makes this method easy to use by politicians but, on the other hand, presents certain obstacles for detailed work on a theoretical basis.

1.3.2.c Swedish ULF-system

ULF (Undersokningar av Levnads Fordhallanden) – system of population surveys implemented in the framework of the welfare analysis by Statistics Sweden, since 1974, started with a standard population survey; it is further expanded and turned into a system of integrated studies using state statistical registers. Since the late 70s ULF is a yearly data collection on four thematic aspects, each of which lasts 24 months, polling about 15,000 people, aged 16 to 84 years. After eight years, the entire programme is repeated. As a result, each of the four basic research gathers information about 120 social indicators in 12 different social spheres. The Nordic welfare concept has a hierarchical structure and is built around 12 social dimensions, or "areas of well-being" (welfare domains):

1. education;
2. social mobility;
3. employment;
4. working conditions;
5. material wealth (material living standards);
6. housing;
7. transport;
8. leisure;
9. public relations;
10. partnership (participation);
11. harassment (victimization);
12. health.

In general, in the views of the Nordic welfare concept, welfare indicators are usually defined as resources that should be used to improve various spheres of life. The policy in the area of welfare is determined by the the availability of access to these resources and their use. In this case indicators of well-being become an objective basis for the process of planning of social policies [Vogel, Hall, 1997].

1.3.2.d Analysis of the quality of life of the state's population

There are a large number of studies (regional, state, national) to monitor the quality of life in a certain geographical area. In this case we are talking about the study of quality of life in some US states, such as Virginia, North and South Carolina, Wisconsin, Rhode Island, Nebraska, Minnesota, California, New Hampshire, New Jersey, Georgia, Florida and Alabama. Such surveys provide information that can be regarded as information relating to quality of life issues. However, none of these surveys provides information that is conceptually defined. As a typical example of this kind of research, an annual survey of the quality of life of Virginia [University of Virginia, 1996] is considered here.

The Virginia Research Center conducted sample surveys of quality of life at the Technical Institute and State University, Virginia.

Survey within a stratified random sample (about 2,000 persons) occurs on the phone, and less than half of them were successfully interviewed. The study includes an examination of the following components:

- overall quality of life in the state – through the analysis of respondents' answers to a few questions rating (e.g. good/bad place to live, good/bad place to get a college education, good/bad place to spend a holiday, good/bad place for work and retirement, good/bad place for pre-school and school education);
- satisfaction with family life (family relations, the quantity and quality of care services for children, concern for family members who lost their jobs or ill);
- job satisfaction and working conditions (do they take pride in their job? do they have opportunities for fulfilment at their job? did any of their colleagues become their best friends?);
- personal satisfaction from the financial and material condition; the presence of friends;
- satisfaction with health and health care (current health status, quality of care);
- satisfaction with education (the amount of money spent on education, level of education in secondary [public] schools, the level of teaching mathematics and science in public schools);
- satisfaction with law enforcement and public safety (view on the dynamics of crime, the safety of moving through the streets at night);
- satisfaction with the environment (a sense of adequacy of funding for environmental protection, opinion on the quality of air and water, the proportion of respondents recyclable household and other waste);
- opinions on the economy, relations between the races and peoples, religious harmony, attitudes to abortion, death penalty

1.3.2.e Consumer Confidence Index

The consumer confidence index (CCI), originating from the proposals of George Katona (1952) and then improved by several researchers [Zagorski, McDonell, 1995], estimates the economic component of quality of life based on the subjective judgements of citizens with respect to the changes in their financial position and future prospects. Quality of life index is calculated based on respondents' answers to the following questions:

1. Can you tell me whether your financial position (your family) has improved or worsened in the current year compared to the previous?
2. Do you think that next year will be materially worse or better for you or do you think that your financial situation will not change?
3. Do you think that last year's financial situation has worsened, improved or remained unchanged?
4. Do you think that the past five years have been favourable for the country economically or was it a period of high unemployment and depressed?
5. In your opinion, is the current time favourable for purchasing expensive durable goods?

For any question, the share of positive responses is reduced by the proportion of negative responses, which gives the so-called balance favourable opinion. To this 100 is added and the magnitude of the averaging is performed, resulting in CCI. CCI can be used to obtain estimates of the population's expectations regarding changes in financial position. From a social point of view, this index measures the hope estimated on the basis of subjective opinion. In addition, CCI, in a sense, reveals the political preferences of the citizens, and therefore may be useful for policymakers.

1.3.2.f Quality of Life Index, based on an assessment of the health

This index was developed at the Center of Health (USA) and was named HRQOL (Health-Related Quality of Life [Measuring health-related quality of life, 1994; Andersen, 1999]). It was the first American quality of life index, implemented on the basis of a national representative survey of the US population. Over 600,000 people were interviewed since January 1993; questionnaire consisted of four questions (in some states it had more questions). Respondents were asked to answer the following questions:

1. How would you describe your health: excellent, very good, good, not very good, bad?
2. When it comes to your physical health, how many days in the last month your physical health was poor?
3. If you talk about your mental health, how many days in the last month your mental health was poor?
4. About how many days in the last 30 days the poor state of your physical or mental health kept you from carrying out your normal activities, such as taking care of yourself, work, recreation?

HRQOL is calculated as the difference between 30 days and the amount of days on the answers to questions 2 and 3. The 1993–1996 index value was equal to an average of 24.7 days, while the less educated respondents had less "healthy" days per month.

1.3.2.g Quality of Life Index of the World Organization Health

Previous HRQOL-index is not the only indicator of evaluating the health of society. The health status of the population has always been one of the most important components of quality of life, in the first place, because of its relevance to the natural man, and second, because of its strong ties with other private categories of quality of life (material well-being, education level, the degree of criminalization of society, etc.). The characteristics of the health of the population is an indicator to detect the closest relationship with the general level of development of society, which is clearly confirmed by cross-country comparisons and study of the history of mankind. It is therefore not surprising that, in assessing the quality of life, many researchers pay attention to this aspect.

In 1991 and 1992 as a result of a series of conferences in Geneva, the World Health Organization has developed a new tool for assessing the quality of life – WHOQOL – The World Health Organization Quality of Life.

It approved the following definition of quality of life. "Quality of life is defined as the perception of the individual his place in life in the context of cultural values in which he lives, in relation to its goals, expectations, standards and concerns" [The WHOQOL, 1993].

This definition reflects the view that quality of life is subjectively evaluated by categories, and built-in cultural, social and environmental impacts. Therefore, the quality of life cannot be simply identified by one of the terms "health status", "lifestyle", "life satisfaction", "mental condition" or "welfare". Moreover, this concept is multidimensional. It includes an individual perception of the above mentioned aspects of life as well as some other aspects.

The tool describes the six major categories of quality of life: physical condition, psychological state, level of independence, social relationships, environment and spiritual development.

However, the following "bottlenecks" have to be noted:

1. sufficient explanation of the categories of quality of life such as "psychological state" has not been given, while it is intended that all tools will be based on "subjective evaluations"; there is no reasonable explanation for the lack of categories based on the indicators of material well-being or productivity, each of which is usually used in the calculation of similar indices; not indicated as "physical condition" is compatible with "subjective evaluations";
2. the definition of WHOQOL quality of life that should be evaluated solely on subjective individual level, with complete disregard for the important objective indicators.

For more about WHOQOL, – see index [The WHOQOL, 1993, 1995a–g, 1996a,b, 1998a,b].

1.3.2.h Philippine analysis of the social climate

Quality of life survey in the Philippines was carried out by interviewing the population of the Center for Social Climate (Social Weater Station – SWS) [Mangahas, 1976, 1986, 1987, 1995, 1996; Mangahas, Guerrero, 1998].

Quality of life is determined depending on how a person responds to the question: if we compare the quality of your life now and 12 months ago, would you say that it was better for, remained the same, worse? As a result, a person is classified as either "Winner" (gainer), or as a "loser" (looser). Accordingly survey respondents are assigned to one of five "socioeconomic classes": (A) rich, (B) the upper class, (C) middle class, (D) poor, (E) very poor. In the final report classes (A), (B) and (C) are often combined.

The next question is very similar to the previous one: In your opinion, what will be the quality of life over the next 12 months? Will it be better worse, or the same? With this question, respondents were attributed either to the pessimists or to the optimists.

Of course, as expected, quality of life significantly correlated with the general direction of the economy. The greatest number of "losers" was recorded during the depression of 1984–1985. In general, such a subjective approach is approved by specialists in the field of social sciences, because its results are easy to obtain, publish and interpret.

SWS agency also monitors the level of poverty by offering respondents to attribute themselves to any of the following classes: "poor", "non-poor", "border". This technique was first applied in the studies in one of the Philippines provinces. According to their results, since 1983, the percentage of "poor" varied widely from 43% to 74%. In 1989 it was 63%. In the analysis of socio-economic classes, it was shown that 26% of the combined ABC-class consider themselves poor. At the same time, only 84% of the population in Class E (the poorest) consider themselves poor. Of the rural respondents 71% classified themselves as poor, while in Manila and its environs, this figure was 38%.

Regression analysis showed that inflation has the greatest impact on poverty. Unemployment is not so much influenced by the level of poverty and inflation, and GNP per capita is almost not associated with poverty.

Note that these forms of public opinion surveys are still popular and are being used by Russian Centers of Public Opinion.

1.3.3 Methodologies that combine Objectivist and Subjectivist approaches

Experts from different countries agree that the use of only objectivist or subjectivist approach just cannot give fully adequate assessment of the quality of life, and reflects only private (though important in themselves) aspects of such an evaluation.

In other words, only a skilful combination of these two approaches, i.e. simultaneous consideration of the living conditions and perceptions of these conditions individuals, can serve as the basis for a full measure of latent synthetic category of "quality of life".

The following are examples of methods of analysis of QOL using both approaches simultaneously – objectivist and subjectivist.

1.3.3.a Indicator magazine "Money Magazine"

One of the methods that combines subjective approach of Ferrans and Powers with the assessment of the quality of life using objective measures is the indicator of quality of life offered by the magazine "Money Magazine" – the best indicator of the place of residence (index «Best Places») [Kotler et al., 1993; Guterbock, 1997]. If Ferrans and Powers used survey method for assessments of selected areas, the "Best Places" uses data on statistical indicators of socio-economic development for the same purpose at the same time. The purpose of the "Money Magazine" survey is to construct a ranking of American States according to the preference for living. The evaluation procedure is implemented in three stages:

At the first stage the readers of "Money Magazine" are invited to rank the 40 proposed criteria according to their importance when choosing a place of residence. In the second phase, statistical information on the indicators that characterize the criteria used is collected. The full list of indicators of the "Money Magazine" is not disclosed, but following are some examples: the number of doctors per capita, the level of violent crime (violent crime rate), the index of living costs (cost-of-living index), the average price for a house with three bedrooms rooms and paid off his property tax, etc. In the third (final) stage, the indicators are aggregated into nine integrated in-indices.

According to "Money Magazine", the study focuses on the most important categories of readers: the economy, education, culture, recreation and environmental pollution. In this case, "Money Magazine" does not explain the principle of rating within these categories, and no mechanism for their integration in the overall index.

1.3.3.b Trend indicator Myers (Community trend method)

D. Myers [Myers, 1987] is among the supporters of the simultaneous use of both objective and subjective indicators. The author criticizes the objective approach of assessing the quality of life, claiming that the comparison of different countries (regions) only by the values of statistical indicators of socio-economic development is inappropriate for several reasons: first, because of the lack of information on the judgements of the population relative to the level of one's life; second, because of the absence of selection indicators studied and weighed; third, because of the lack of attention to regional specificities. The greatest emphasis Myers makes to the need to define quality of life as past experience of the local population in living their lives. First of all, he drew attention to the fact that by choosing their place of residence, different individuals pay attention to a variety of factors: for someone it can be essential climatic conditions in the region, while for another the greatest significance is the presence of chemical-technological production. Consequently, the population of a region different from that of any other share certain preferences with respect to the place of residence. This means that the factors affecting the perception of quality of life are also different.

In addition, Myers argues that the judgement on the quality of life of the population depends on its past experience, rather than a comparative analysis of statistics from different regions. In this regard, the author proposes to conduct the analysis of objective indicators of well-being not on the basis of their current values but on the basis of their relative changes, which will allow to avoid the problem of different measuring units.

Analysing the quality of life in Austin (Texas), Myers highlights 12 key factors influencing the well-being: transport, crime, health, education, income, inequality, housing, employment, water quality, entertainment, recreation and restaurants/shops. According to the results of statistical studies of these factors Myers divides them into two groups: those with positive dynamics of indicators and indicators with negative dynamics. Method of population survey reveals public opinion regarding the dynamics of the factors, and their importance to citizens. Myers concluded that the subjective and objective assessments of the dynamics of indicators give very similar results, with people giving greater importance to the areas of life that have been deteriorating in the area where they reside.

1.3.3.c Diener's basic and advanced QOL indices

This tool for estimating the quality of life consists of two indices: basic and advanced. The purpose of these indices is to provide information on the quality of life in the countries studied in terms of the objective criteria of quality of life. However, an improved quality of life index contains also a subjective component. Factors applied to the study of quality of life differ in both developing and developed countries [Diener, 1995].

In the process of developing indices, the author took into account the limitations of including different factors with unfounded weights into prior indices and tried to solve this problem. These limitations are:

1. unsystematic codes developed and are non-theoretical in nature, resulting in various different codes allowed assessment of quality of life, although using the same statistical basis;
2. the choice of the weights led to the fact that the quality of life varies greatly between developing and developed countries, rather than across the spectrum of the level of development;

3. indices and weight factors are determined by sequencing (rank-order comparisons), and show that some countries are substantially different from everyone else, rather than when the weights of the index are determined by comparison of parametric (parametric comparison);
4. weights often do not take into account the "universal values".

In order to overcome the above drawbacks Diener turned to Schwartz [Schwartz, 1992], which is published before the list of "ethical values" that are recognized by all cultures. Referring to Schwartz, Diener listed examples of "ethical values": self-discipline, social justice, the ability to enjoy life, success, protection of the environment. "These values represent three universal human needs: the satisfaction of biological needs, orderly social interaction and the desire to live, and safely" [Diener, 1995].

Schwartz made a list of 45 "universal values" and placed them "properly" evenly around the bivariate circular structure. Then he divided the structure into seven segments, each of which, in his opinion, reflects a set of similar values, and gave them names: hierarchy, authority, emotional independence, intellectual independence, commitment to egalitarianism, harmony, conservatism. Next Diener chose from each sector two indicators: one to be included in the main index of quality of life, and the second – to the characteristics of the improved quality of life index. The first group of indicators was selected for the presentation on the concept of Maslow [Maslow, 1999], "the needs of first-order" ("lower order needs"), such as, for example, physical needs or the need for security, while the second group of indicators presented "the needs of higher-order" (higher order needs), for example, respect and a desire to help others.

The two resulting indices thus consist of seven figures. However, some of them are scalar quantities, such as "the number of doctors per capita", while others are composite. For example, a rating of "mastery" (mastery) in the main index is an indicator that is aggregated from five other indicators. The final index values are obtained by adding the standardized values of the seven indicators.

1.3.3.d Cummins's comprehensive indicator of QOL

This indicator was invented by Robert Cummins and covers the following categories of quality of life:
- material well-being;
- health;
- efficiency (productivity);
- knowledge (intimacy);
- security;
- social and emotional well-being.

Each category is described by three factors that add up to the objective value of its condition; the sum of these values in six categories is interpreted as an objective characteristic of the quality of life. Each of these categories of quality of life has two subjective indicators: the first is the level of satisfaction of the appropriate category of quality of life, which is measured in a 7-point scale; second is the – level of significance of the relevant category, which is measured in a 5-point scale. As a result of multiplying the satisfaction of certain categories of life on the corresponding figure of significance subjective characteristics as the test category can be obtained. If the characteristics of subjective quality categories add up to each other, you get a subjective characteristic of well-being [Cummins, 1995, 1996, 1997, 1997, 1999, 2000].

1.3.3.e Ruth Veenhoven's happily lived life indicator

Figure invented by Ruth Veenhoven is not based on any one theory. However, its value is confirmed by both theoretical arguments and empirical research. Very attractive in the works of Veenhoven is that it always presents its readers a fresh (up-to-date) review of theories and hypotheses and achieve them with their empirical studies, theorists usually adding many difficult problems to solve.

Calculation of the index reduces to multiplication "standard" life expectancy in years to measure the "average happiness", which ranges from 0 to 1. For example, suppose that the value of life expectancy in a country is 50 years, and the average value of "happiness" on a scale of 0 to 10 is 5. Thus, the average rate of happiness on a scale from 0 to 1 is 0.5. The product of 0.5 and 50 is 25. Thus, the expected duration of the "happy life" is 25 years. This example is typical for most of the underdeveloped countries of the modern world. If life expectancy is 80 years, and the average value of "happiness" in a scale equal to 8, the expected duration of the "happy life" is 64 years. This example describes the majority of the so-called developed countries of the modern world. For more on this index, see the following – [Veenhoven, 1984, 1994, 1995, 1996, 1998; Veenhoven, Ouweneel, 1995].

1.3.3.f The German system of social indicators

The German system of social indicators has been developed under the direction of Zapf in the mid-70s of the 20th century in the context of the SPES (Sozialpolitisches Entscheidungs und Indikatorensystem) and is defined as a system of socio-political solutions to ratification and monitoring of social indicators (Sociopolitical Decision and Indicator System). Since 1978, this system has been consistently improved by the Department of Social Indicators (Abteilung Soziale Indikatoren) in cooperation with the Center for Questioning, Research methods and Analysis (Zen-trum fur Umfragen, Methoden und Analysen) that is located in Mannheim.

In its modern form, the German system of social indicators covers the following 13 categories of quality of life:
1. population;
2. socio-political status and subjective class identification;
3. labour market and working conditions;
4. income and its distribution;
5. supply and consumption of goods and services;
6. transport;
7. housing;
8. education;
9. health;
10. participation in social and political life;
11. the environment;
12. public safety and crime;
13. leisure and media.

Each category of quality of life is divided into a number of properties, each of which is characterized by a certain number of parameters. Health unit, for example, consists of three properties: state of health, population coverage by health services and living conditions that affect health. Each property is characterized by 4–5 sub-properties; for example, the state of health is characterized by life expectancy, mortality from a number of reasons, resistance to disease and the subjective satisfaction of the overall level of health. Each of the sub-properties will now be described by two or three (and sometimes more) indicators. Currently the system of social indicators consists of about 300 indicators. For each of them there are the time series containing approximately 1,000 values. Thus, the system of social indicators provides the user with empirical information about changing of the living conditions of the population and the transformation of social macrostructures. The observations cover the period from the beginning of the 50s of the last century until the present time. Read more about the system – [Zapf, 1977; Noll, Weigand, 1993; Noll, Zapf, 1994].

1.3.3.g Holland index of living conditions

The Department of Social and Cultural Planning Holland (Netherlands Social and Cultural Planning Office) has developed an index of living conditions (Living Conditions Index – LCI). Each year, since 1974, the values of this index are published in a special report. The main purpose of the index is to – "reflect the conditions of development in areas that fall under the influence of government social policy". LCI is published as monospaced (base index value of 100, was recorded in 1997), but it may decomposed into indices:

- living conditions;
- health;
- purchasing power;
- leisure;
- social mobility;
- social participation;
- the development of sport;
- holidays;
- education;
- employment and working conditions.

Decomposition and weighting components of the LCI focused on maximizing the degree of positive correlation with an average rating of "level of a happy life", produced by special questionnaire surveys of the population.

1.4 Analysing the Properties of Integral Indicators Quality of Life

Sections 1.3.1–1.3.3 represented 28 (!) different methods of estimation of synthetic categories of quality of life and the construction of the corresponding integral indicators, including 13 – in the framework of objectivist (or macro) approach, 8 in the framework of the subjectivist (or micro) approach, and 7 techniques based on the simultaneous use of both approaches. *How to choose the most capable of them? What criteria to evaluate their properties and quality?*

I have already noted (see introduction to Paragraph 1.3) that, unfortunately, the whole experience suffers due to the disadvantages associated with the absence of solving actual problems (B), (C) and (D). Nevertheless, each of the above methods has its advantages, very widely used in practice. The authors of quality of life indices [Quality of life indexes for national policy, 2001] propose a set of criteria for evaluating the usefulness and efficiency of the method of measurement of QOL, which, however, do not contain requirements to address (B), (C) and (D). Below are the criteria and the evaluation of them mostly from the methods described above (Table 1.4; the last row of the table shows the average grade for each II QOL). A summary of each of the proposed criteria was carried out in a 3-point scale, in which 3 corresponds to "excellent", – 2 to "satisfactory" and – 1 to "unsatisfactory". All this activity was carried out by experts of the special commission of the International Society for Quality of Life. They used three round "Delfi-technique" to organize and carry out their examination.

1.4.1 Criteria for Evaluating II QOL

Quality criteria described below for II QOL are presented (with brief comments) as examples of possible system requirements that must be presented to the methods of construction of integral indicators of quality of life. Examples are, in their own way, specific, because they focus on the use of II QOL exclusively in problems and amendments of social and economic policy. Example is, in my opinion, not the best since, first, it ignores the need to address (B), (C) and (D) (see above in 1.3). Second, it detects certain duplication of requirements that were formulated in different criteria. (That's partially the reason, the other one being lack of large percentage of expert assesement of the last four criteria, for the fact that only 10 out of 14 criteria have results in the Table 1.4). Nevertheless, it is useful for the reader to get acquainted with these criteria and demands set out therein. if we have to solve the problem of II QOL.

Criterion 1 ("The management targeting"). Integral indicator of QOL should have a clear practical purpose, namely to measure; it should be useful to managers in the evaluation and adjustments to the social and economic policy [Land, 2000]. Note that this requirement is in line with our "pragmatic concepts" (see point 1.2.4).

Criterion 2 ("Multilevel targeting"). II QOL should be useful to managers – "politikomeykeram" – at all levels of socio-economic programmes: from households to municipal, regional, whole of country and inter-country levels.

Criterion 3 ("Monitoring"). Construction of II QOL should be based on the original data (official statistics and the results of questionnaire surveys) that allow their regular periodic tracking over time (monitoring). It is the raw data in the form of time series that allow to monitor changes (improvement, deterioration) in key areas of social and economic policy and to build the necessary forecasts.

Criterion 4 ("Theoretical justification"). Building of II QOL should be based on specific, proven practices, conceptual provisions of the socio-economic theory. Arising from their system of causal relationships determines how the II QOL conforms to the endogenous and exogenous variables.

Table 1.4. The evaluation results of different methods of analysis of QOL according to ten criteria

Number criterion for evaluating II QOL	Criterion based on evaluation of health (HRQOL)	II QOL of the World Health Organization	II QOL “Money Magazine”	Index of Economic Well-Being	Genuine Progress Index	American Demographic Index	Johnston’s Quality of Life Index	Euro-Barometer Index	Veenhoven’s Index	International Index of Living Conditions	Human Development Index	Index of Social Health	II QOL of the state’s population	Index of social progress by Estes	Index of Diener	II QOL of Commins	Index of Michalos	Philippine Analysis of the Social Climate	Holland Index of Living Conditions	The German System of Social Indicators	Swedish ULF-system	Average score on the II QOL
1	2	3	1	3	2	2	3	3	3	2	2	2	3	3	3	2	3	3	3	3	3	2,6
2	2	1	1	2	2	2	1	3	2	1	1	3	2	3	2	2	2	3	3	3	3	2,1
3	2	1	2	3	3	3	3	3	3	2	2	3	2	3	1	3	2	3	3	3	3	2,5
4	3	1	1	2	1	3	1	3	2	1	1	2	1	3	2	2	2	2	3	3	2	2
5	3	1	1	3	2	2	1	1	2	2	1	1	1	2	3	2	2	2	1	2	2	1,8
6	2	2	1	3	3	3	2	1	2	2	2	3	1	3	3	3	2	1	3	2	2	2,2
7	1	1	1	1	1	3	1	1	2	2	1	1	2	1	1	2	2	1	2	2	2	1,5
8	2	2	2	3	2	1	1	1	2	2	2	1	2	1	2	2	2	2	2	2	2	1,8
9	2	1	2	1	1	2	1	1	3	1	1	1	2	1	1	3	1	1	1	1	1	1,4
10	2	2	2	3	3	3	2	2	3	2	2	2	3	2	2	2	3	2	2	2	2	2,3
Average	**2,1**	**1,5**	**1,4**	**2,4**	**2,0**	**2,3**	**1,6**	**1,9**	**2,5**	**1,7**	**1,5**	**1,8**	**1,9**	**2,0**	**1,9**	**2,5**	**2,0**	**2,0**	**2,1**	**2,1**	**2,1**	

Criterion 5 ("The authenticity and sensitivity"). Components, which are the basis of II QOL, should have properties of authenticity (reliability) and sensitivity. The term "authenticity" (reliability) means complete certainty in the measurement of the respective component and careful adherence to this procedure. The term "sensitivity" refers to the ability of the response in the values of II on QOL "input actions" on social and economic policy.

Criterion 6 ("One-criterion and decomposability"). A method for constructing II QOL should ensure the existence of an indicator of quality of life as a single ("one-criterion") and the possibility of expanding this scalar indicator integrated into a number of indicators measuring QOL synthetic, which are constituent elements of the general concept of "quality of life". Note that there are opponents of this position that are considered improper association (aggregation) of various synthetic QOL categories into a single category of a higher order of generality [Michalos, 1997; Vogel, 1998a,b].

Criterion 7 ("Informativeness"). The composition of the indices (partial criteria), which is constructed on the basis of the II QOL is formed so as to cover all major aspects of the analysed synthetic category. However, when transitioning to Integral Indicators there is a certain loss of information that was contained in the original set of parameters. II QOL in formativeness requirement means that, first, we have to make this change from the minimum losses in the input information, and second, if these losses, however, are large, then the assessment we analysed as a synthetic category should not be limited to one scalar II QOL (see number 4 of point 1.5, as well as our point 2.3.3 below).

Criterion 8 ("Laconic"). This requirement is in contradiction with respect to the previous one. On the one hand, the composition of indicators (partial criteria), which is constructed on the basis of the II QOL, is formed so as to cover all the important aspects of the analysed synthetic category. From this point of view, the more particular criteria we include in the initial set of indicators, the better. On the other hand, many of these indicators are often mutually correlated closely (property multicollinearity). It is this property that should be used to get rid of the redundancy of the initial partial criteria.

Criterion 9 ("Objectivity plus subjectivity"). As part of the initial partial criteria that underpin II QOL indicators must be present, as measured by objective (official statistics) and subjective (the results of questionnaire surveys of the population) scales. This is necessary to allow the comparison of grade synthetic categories analysed, obtained under the "objectivist" and "subjectivist" approaches. Why is this important? Many researchers have demonstrated the quality of life, which is very often the objective indicators in this category are not correlated with their subjective "copies" [Cummins, 1999]. However, it can be argued that the subjective well-being is an important but not sufficient condition for covering all aspects of life. A man may indicate a high level of subjective well-being, despite that the poor quality of the environment has greatly reduced life expectancy, and the objective conditions of quality of life (health, property ownership) may have little to do with the subjective evaluation of quality of life. For example, a person may be rich, but not satisfied with life due to the fact that in his opinion others posses more wealth. In addition, the comparison of results of the assessment of QOL using objectivist and subjectivist approaches is used to clarify the composition of the initial partial criteria and questionnaires distributed among the population.

Criterion 10 ("Relevance to the majority of the population"). Private criteria on which II QOL is based should be applicable to the majority of the analysed conglomerate (the population of the region, country, social class). Features relating only to narrowly specific population groups (e.g. for hospital patients) cannot be used to construct II QOL.

1.4.2 Analysis of the Results of Evaluations of Different Techniques Measurement QOL

The dualism of the results presented in Table 1.4 allows us to analyse both the comparative quality of the considered methods to measure quality of life in terms of their compliance with the criteria put forward (mean scores, which evaluated the methodology given in the last row of the table), and the relative success of the following methods of each of these criteria (estimated average score is given in the last column of the table).

Comparison of methods 1–10 in terms of their eligibility shows that most have proven successful:

- *the Ruth Veenhoven index (average score – 2,5);*
- *QOL index Cummins (average score – 2,4);*
- *the index of economic well-being (average score – 2,4);*
- *American demographic index (average score – 2,3).*

Conditional outsiders can be attributed indicator magazine "Money Magazine"' indicator (1.4 points), Human Development Index (HDI) (1.5 points) and Johnston's index (1.6 points)- all can be attributed to conditional outsiders.

Comparison of criteria (requirements) based on the success of their implementation in above referenced methods shows that Criterion 1 ("The management targeting"; GPA – 2.6), Criterion 3 ("Monitoring"; the average score – 2.5) and Criterion 10 ("Relevance to the Majority of the population"; average score – 2.3) performed the best. Criterion 9 ("Objectivity plus Subjectivity"; average score 1.4) and Criterion 7 ("Informativeness"; average score – 1.5) were the hardest to follow. The other weaknesses of these methods are also not very convincing theoretical basis of II QOL (Criterion 5) and lack of clear procedures in choosing the particular criteria for each synthetic category (Criterion 8). We'd like to add that none of the above methods can solve the problems of major importance such as:

- *the problem of determining the general form of transformation by which we move from the multiobjective scheme to one-criterion (all the above techniques are limited to linear convolution benchmarks – partial criteria); see above item 1.3, problem (B); note that one aspect of this problem is determining the weights given to the particular criteria in their aggregation in the form of linear convolution; but this remains without a solution in all the techniques;*
- *lack of a satisfactory solution of the problem of transition from multiobjective scheme to one-criterion within the bundle of linear partial criteria; see above Paragraph 1.3, problem (B); note that this very fact explains the weakness in adhering to "informativeness" requirement (Criterion 7) by all the methods mentioned above;*
- *the problem of accounting action "principle non-compensated preferences" (see above, Paragraph 1.3, the problem (D)); but this problem cannot be solved by the construction of integrated indicators limited to linear functions of partial criteria.*

The analysis of various synthetic methods of measuring QOL categories shows, on the one hand, a wide variety of such techniques, and, on the other, the presence of a number of important methodological problems in the construction of II QOL, which still await their decision. Some approaches to solving these problems will be described in the following chapter of the book.

1.5 Conclusions

1. Assessment of the quality of life should reflect both the objective conditions of human existence, characterized by the values of the relevant statistics (*objective information*), and the perception of these conditions by the man himself, characterized by the results of the individual responses to the questions of special profiles (*subjective information*).
2. Formation of integrated indicators and measurement of QOL scales should be based on *the concept of expansion of human capabilities*, in order to develop a society which recognizes the needs of the rise and expansion of human freedom to choose how to achieve their own interests, the maximum self-realization.
3. There is no theoretical concept of quality of life, universally suitable "for all times". Therefore, the choice of basic conceptual positions in the formation and interpretation of II QOL should be carried out depending on *what level of motivation in the Maslow's "hierarchy of needs" we are, and he specific application purposes of the use of these integrated indicators* (**pragmatic approach** to the measurement of synthetic categories QOL).
4. The most common and popular targets applied research, using the tools of analysis and measurement of synthetic (latent) categories of quality of life, include:

- *review, evaluate and adjust ongoing socio-economic policy (in the country, region, municipality);*
- *identification of "bottlenecks" and the critical factors that determine the increase or decrease in quality of life of the population defined territory (country, region, municipality);*
- *cross-country and cross-regional comparisons and rankings.*

5. Analysis of international experience in the design and practical use of integrated indicators of quality of life (see point 1.4) shows, on the one hand, the huge variety of approaches to the construction and interpretation of II QOL and, on the other hand, the presence of important methodological problems and deficiencies in the art. These primarily include:
   - *problem specification II QOL, i.e. determine the general form of transformation by which aggregated particular criteria of QOL in the form of a single indicator; almost all known techniques are limited to* **linear functions** *of partial criteria, not justifying the choice of weights that attach to these particular criteria in aggregation;*
   - *the problem of estimating the in formativeness used by the II QOL; the fact is that there may be no satisfactory (in a sense) solution of the problem of transition from many private criteria to a single aggregated II QOL; Then there are questions about the minimum number of II QOL in which our problem can be solved satisfactorily, and the way of constructing these II QOL, as long as these questions remain unanswered;*
   - *the problem of accounting of action of the* **"principle of non-compensatory preferences".** *According to the above principle there are such critically important to QOL criteria with proximity to which the loss in QOL can not be compensated by improving the values of other private criteria.*

2 Macroeconometric Analysis of Quality of Life: Measurement of the Synthetic Latent Categories

Section 1.3 of Chapter 1 provides an analytical review of existing so far world's most well-known methods of modelling and measurement of synthetic (latent) categories of quality of life (QOL) both within the *objectivist* and *subjectivist* approaches including their simultaneous use. There were also the following main drawbacks of almost all of these methods.

(a) *In all the methods it is proposed to construct the integral indicator y, which measures the analysed synthetic latent category QOL* **only in the form of a linear** *function from a set indicators that we have at our disposable (partial criteria)* $x^{(1)}, x^{(2)}, ..., x^{(p)}$, i.e. $y = \sum_{j=1}^{p} w_j x^{(j)}$. *In the* objectivist *approach the statistical indicators that characterize the particular category play the role of these indicators, and in* the subjectivist *approach,* it's the relevant characteristics calculated according to the results of individual survey of the population.

(b) *The weights* $w_j (j = 1, 2, ..., p)$, *involved in the calculation of the integral indicator y are determined by* **direct expert evaluation.** *Meanwhile, numerous experimental studies in economics, sociology and psychology confirm the fact that experts are much more reliable and able to make a concerted integral evaluation of the object rather than to decompose the integral evaluationaccording to the significance of its components.*[1]

(c) *The ability to objectively justify the lack of a satisfactory solution to the problem of transition from multi-objective scheme (with partial criteria* $x^{(1)}, x^{(2)}, ..., x^{(p)}$) to a **one-criterion (y)** *is either ignored altogether, or is not being attempted to deter-*

1 Author was repeatedly faced with similar situations in his personal experience as well. I will give here one of the most impressive examples. While developing the information provision for the World Ice Hockey Championship, Moscow, 1973 [Айвазян, 1974], we, among other tasks, built a so-called *target function of the mastery of the hockey player* $y = \sum_{j=1}^{p} w_j x^{(j)}$, where $x^{(j)}$ is the j-th statistical indicator of the action of the hockey player during the match (the number of goals scored, the number of won power fights or body checks, the number of minutes of penalty time, etc.), and y is the final assessment (in points) of skill level of the player during the game. Then we asked the experts (coaches of the national teams) to complete a two-question survey: 1) *evaluate, please, by using the 10-point system the level of mastery of each forward player (defender) during certain game;* 2) *distribute, please, 100 points among the actions of the hockey player in proportion of this action contributing to the general mastery of this player during the game.*

The answers to the first question were fairly consistent (values of corresponding coefficients of concordance ranged from 0.5 to 0.9) but the *answers to the second question looked absolutely chaotic* (values of concordance coefficients were no higher than 0.2, which statistically is not much different from zero).

mine the number and type of **multiple** criteria $y^{(1)}, ..., y^{(k)}$, *to significantly reduce the dimension p of the original multi-objective scheme (i.e. k ≪ p).*

(d) The integral indicator of quality of life (II QOL) as a **linear** function of $x^{(1)}, x^{(2)}, \ldots, x^{(p)}$ *is not adjusted to take into account the principle of "non-compensated preference" that occurs when the values of at least some of the particular criteria* $x^{(1)}, x^{(2)}, \ldots, x^{(p)}$, *are close to the so-called critical ones (see problem (D) in point 1.3 of the accounting action of "principle of non-compensated preference").*

The following proposed methodology for building II QOL eliminates the disadvantages (a), (b) and (c) and, in some cases lets us gain a significant progress in eliminating the problem of lack of (d) (due to greater than linear class of admissible "bundle" $y = f(x^{(1)}, x^{(2)}, \ldots, x^{(p)})$ and the multicriteria of integral indicators). Recall that this chapter and Chapter 3 are devoted to **macro** econometric analysis and accordingly these chapters reflect on evaluation (measurement) of synthetic categories of quality of life for the *entire population of conglomerates*, including the ones that differ a lot (for the population of the country in the whole and its individual regions – the subjects of the Russian Federation and municipalities). This means that we start from the *objectivist approach*, and our initial information comes from the official international and Russian government statistics.

2.1 Integral Components of "quality of life" category and their analysis

Before describing directly the methodology of constructing the integrated indicators $y = f(x^{(1)}, x^{(2)}, \ldots, x^{(p)})$ that measure one or the other category of latent synthetic QOL it is necessary to define the overall structuring of the entire system of statistical indicators, particular criteria and *basic synthetic categories*, the synthesis of which brings us to the synthetic category of a higher level of generality – "quality of life of the population of the territory" (country, subject of the Russian Federation, municipality).

Described below is a variant of this structuring that was proposed by me at the end of the 90 s of the last century as a result of my work on the subject performed for the State Statistics Committee and the Ministry of Economic Development of Russia [Айвазян, 2000].

In developing this evaluation I proceeded from the assumption that the basic properties of the quality of life of the population are formed and manifested in its interaction with the "external objects" and with each other, in its ability to adapt to the world around it, to survive, to reproduce and to meet its needs. Considering the human being as both biological (operating in accordance with all laws of biological homeostasis) and as a social being (realizes itself only in society, in the processes and mechanisms that operate based on specific "social rules"), I came eventually to the following five basic synthetic categories that form the environment and life support system of the population.

I. *Quality of the population* – integrates such properties as life expectancy, level of education, qualification, births and deaths and marriages [Медков, Качество населения, 1993; Сградов, 1995].
II. *Welfare of the population* – integrates the main indicators of quality of life and reflects the degree of satisfaction of its material needs (real incomes, their differentiation, the level of consumption of goods and services, the level of health care, etc.);
III. *Quality of social sphere* – reflects the level of working conditions, social security, physical and property security of the members of society, the crime and socio-political health of the society, etc.;
IV. *Quality of the environment* (*or the quality of the ecological niche*) – accumulating data on the pollution of air species, soil and water; we have to note that the human population is the only species of the earth that does not provide utilization of its waste, which is environmentally hazardous and not included into natural technology of the nature – "food chain of the biosphere" [Шакин, 1997]
V. *Climatic conditions* – characterized by the composition and volume of natural resources and raw materials, climate, frequency and specificity of force majeure.

It is easy to see that each of the components I–V reflects the conditions under which the processes meet both biological and social needs of members of society. Sequential hierarchical decomposition of each of the basic components would allow us to "go down" to the set of the relevant characteristics of the lowest level, *the vast majority of which can be represented by standard statistical indicators* (as an exception, some of the indicators of the lower level have to be assessed expertly, for example, by using a point scale).

Figure 2.1 represents a general scheme of a hierarchical system of criteria and statistical indicators of the decomposition of the proposed category of "quality of life". We equip this scheme with some comments.

The category of "quality of life" as well as each of its basic integrated components I–V, in one form or another is discussed, and even *evaluated* in various studies (and in the relevant literature) on the analysis of the conditions and the standard of living of the population.

There are also attempts to expertly evaluate the category of "quality of life" [WCY, 1997–2011]. It should be recognized, however, that the degree of subjectivity and hence random spread in the estimates of experts is too high, which is apparently

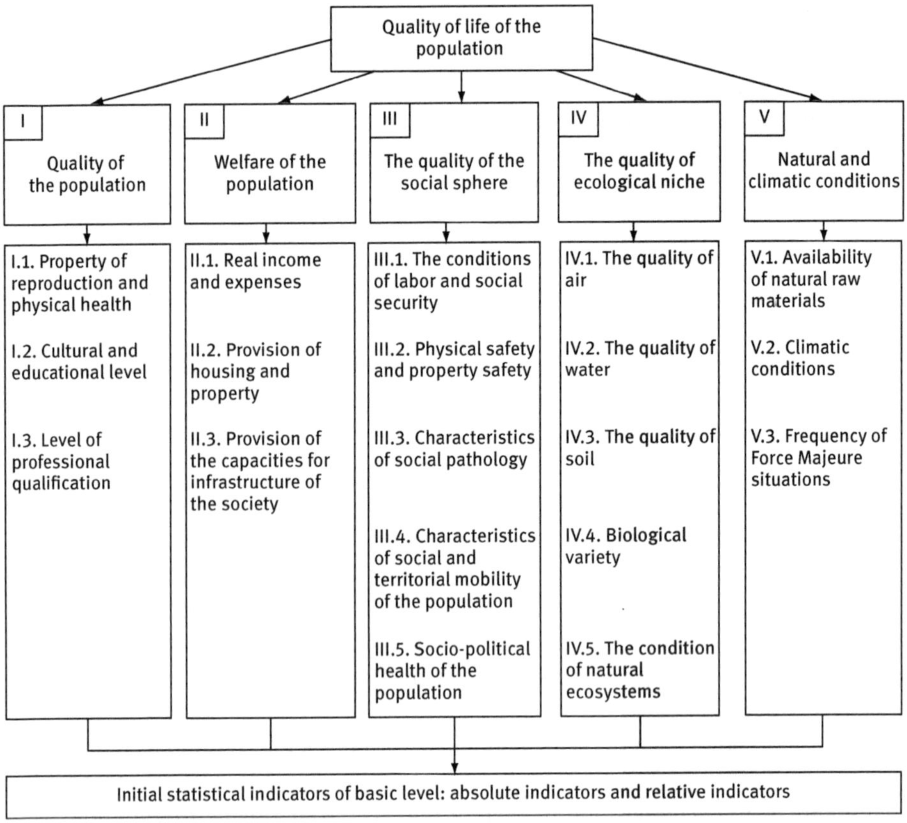

Figure 2.1. Hierarchical system of statistical indicators, particular criteria and integral indicators of the quality of life of the population (country, region)

due to excessive blurring of the conceptual guidelines and criteria that govern experts [WCY, 1997–2011] in the formation of their estimates. A similar observation can be attributed to the indicator “security” (“Security”); we have to add to this our assumption that in the interpretation of this indicator [WCY, 1997–2011], this figure reflects only the *degree of physical and property safety* of members of the society and does not include the working conditions, social security and other necessary characteristics of the category of “social security”.

A certain category of “quality of the population” seems to us as relatively more developed. Our position on the interpretation (*but not on the dimension*) of this integral property is based on conceptual installations close to those given in Sogradov [СограДов, 1995] and in the “Method of calculating the index of human development” proposed by a group of staff and experts of “Development Programme of the UN” [HDR, 2002]. The literature also discusses the characteristics and problems of working with data and experts to assess the quality of the environment [Шакин, 1996; Демография, Экономика, Экология, 1997].

Decoding of the property of "well-being" (or "standard of living") of the population is mainly based on the methodology used by the State Statistics Service of the Russian Federation (Rosstat). When using this technique and related socio-economic indicators we try to possibly avoid or at least reduce the discrepancies and uncertainties arising from the unreliability or lack of representativeness of primary data of Rosstat. So, no doubt that the "bottleneck" of the primary statistics includes: clearly inadequate coverage of income of the population and non-state enterprises, poor representation of the results of surveys of household budgets (the range of surveys misses a significant proportion of the rich and super-rich families, almost all of them) and excessive narrowness of the range of the covered issues (the most important security issues, social relationships, leisure activities and political relations with the authorities are outside of this range); lack of reliable information on the export of capital from Russia; a clear underestimation of production in the country and others.

It is interesting to compare the above criteria and structuring of the private integral indicators of QOL with a modern point of view on the issue of the global scientific community, expressed, in particular, in the report of the *Commission to assess the economic performance and social progress* [Report, 2010–2011]. The commission has identified the following key areas that must be taken into account in the analysis and evaluation of QOL:

(i) the material standard of living (income, consumption, wealth);
(ii) health;
(iii) education;
(iv) personal activities, including work;
(v) political influence and control;
(vi) social ties and relationships;
(vii) the state of the environment (currently and in the future);
(viii) sense of insecurity, both economically and physically.

Projecting these key areas on our scheme (see Figure 2.1), we see:

- **category I "quality of the population"** includes the field (ii) and (iii);
- **category II "well-being"** corresponds to the (i);
- **category III "social quality"** field includes (iv), (v), (vi) and (viii) (taking into account capabilities required due to the availability of a measure of the statistical information);
- **category IV and V "quality ecological niche" and "climatic conditions"** constitute the content area (vii).

We see that there are no fundamental differences with the statements of the report. All the matter is in a *particular filling* of each of the areas (categories) with relevant statistical indicators, and this is more of a technical (and financial!) problem of government statistical services.

Thus, summing up our thinking about the quality of life of the population and its common metric, we can formulate the following points of the starting position adopted in this paper:

(a) it can be varying degrees of integration (and the hierarchy) of the generalized indicator of quality of life – from the reported and recorded statistically indicators (such as life expectancy, the value of per capita expenditures, unemployment or crime) to the integral indicator of top-level that synthesizes all the basic (more specific) aspects of quality of life in it; the choice of the level of generality is dictated by the specific applied research objectives;

(b) for each fixed level of generality, we can postulate the existence of a sufficiently stable in time and space set of basic factors (including latent ones), which together define a quantitative estimate of the analysed categories of integrated quality of life; the composition of these factors depends on the level of generality of the integral indicator and specific applied research objectives; however, the method of constructing the corresponding II QOL (even if we've already selected the level of generality and specific applied research purposes) shall be capable of verifying the underlying factors and the relative densities of their impact on II QOL after a certain period of time and/or the transition from one accepted in the society system of values to another for the purpose of adequate "settings" on the specific time and place;

(c) an integrated indicator of any category of synthetic quality of life is a particular type of convolution of estimates of more specific properties and criteria of this concept (which, in turn, can be presented either in the form of various combinations of reported statistical indicators or in the form of a relevant expert evaluations); it is designed for comparative analysis (in time and space) in this category, as well as to identify "bottlenecks" in the socio-economic development of the country or region.

On the basis of these conceptual provisions that belong to the synthetic categories of quality of life of the population, and in particular to its basic integral properties I–V, let's look at the *main components of each of these properties*.

I. **Quality of the population**. We believe that the quality of the population is not separable from the life and vital functions of the population. It represents a certainty that results from the very fact of its existence, *an integrated set of more specific properties* possessed by the public and manifested in people's interaction with the various phenomena of the surrounding world. The view of the population is formed on the basis of the data on mostly such qualities as:

I.1 ability to change its size and maintain a balance of biological functions of its constituent generations (fertility, mortality, morbidity, life expectancy);
I.2 level of education (proportion of the population with secondary and higher education, the average duration of the period of study);
I.3 skill level or efficiency of the labour that is useful for the society (reflected, for example, in the produced value of the gross domestic product per one employee per unit of time, and calculated while taking into account the purchasing power of the monetary unit).

II. Welfare (standard of living) of the population. This integral property accumulates the following properties of habitats that characterize the degree of satisfaction with material and spiritual needs of the population:
II.1 real income and expenditure;
II.2 provision of housing and other property;
II.3 provision of public health facilities, education, culture (including science), leisure (recreation), infrastructure (energy, means of communication, including modern information technology, etc.).

III. Social security (quality of social services). Considering the social environment of man and his social needs, we distinguish between the properties that characterize the *social conditions of his existence* and the properties that characterize *his behaviour* that is oriented on social homeostasis – supporting the necessary qualities of society. Accordingly, we can note the following main components:
III.1 quality of working conditions (quality of motivation, level of occupational accidents and diseases, the characteristics of employment, racial, religious or political grounds, etc.);
III.2 level of physical safety and property safety of the members of the society (characteristics of social pathology, criminality, organized crime, etc.);
III.3 the quality of the socio-political health of the public and government agencies (the level of corruption in the government and administrative structures, the level of institutional development of the society, the degree of political engagement of media, etc.).

It is obvious that the maintenance on the appropriate level of quality III.1 and III.2 will help improve the efficiency of socially useful work, achieving high rates of GDP and GNP, and create a comfortable socio-psychological climate in society. A high level of component III.3 is the guarantee that the state is providing all the necessary civil rights and freedom, which contributes to the high quality of the socio-economic and socio-political behaviour of the population, and that in turn works on the efficiency of the processes of enhancement of national wealth and its fair distribution among the members of society.

IV. Quality of the environment (ecological niche). Assessment of the state of natural and biological environment in terms of living conditions of the population should characterize the degree of suitability of its elements for life, work and recreation of the population. This could be easily done, if we would have had, at first, *sound system of standards* that characterize the quality of the elements of nature, and, second, the functioning *system of environmental monitoring* in the country, that would have had the capabilities that match the characteristics of the problem at hand.

The first of these problems is solved today through the development of standards for maximum permissible concentrations (MPC) of harmful substances in air and water. The Ministry of Health until recently approved MPC of 114 harmful air substances and 294 harmful waterbody substances that are intended for sanitary and domestic use.

The second problem is the need for the formation and development of the Unified State Environmental Monitoring System [On the organization of work, 1994]. It is aimed at assessing Natural Environment as a factor determining the quality of life of the population. Existing environmental control authorities today carry this task only on a very limited scale. Control over the pollution of water and air pollution is the most developed of their functions today. It is carried out by hydrometeorological, public health and other services. It should be recognized that the national system that would fully ensure the collection and processing of information necessary to monitor the state of the environment as a whole has not yet been created.

So, as of now, means of collecting factual information on the state of the environment as well as an integrated system of assessment of that state is clearly underdeveloped.

Under these conditions, a rough estimate of the quality of the environment is proposed to be made on the basis of the following three private properties:

IV.1 *anthropogenic impact on the air* – it is determined by the mass of harmful substances emitted to the atmosphere per 1 sq. km of land area (t/y) per capita (kg/year) or per $1 million of GDP (t/$ million GDP);

IV.2 *anthropogenic impact on the water basin* – it is determined, in particular, by the volume of polluted water discharged into the surface water bodies per 1 sq. km of land area (cu. m/year), and by the ratio of the volume of wastewater discharged into the surface waters, to the volume of fresh water taken from water bodies;

IV.3 soil conditions – it should be noted that the use of the above indicators for the country as a whole gives us only a very rough estimate. Obviously, for a more accurate assessment of the environmental situation we need the analysis of it at the territorial level.

V. Natural and climatic conditions. Of course, these greatly influence the formation of mentality, values, and living standards of the population. Understanding in this paper the complex climatic conditions that the population of the region lives in and natural resources for the certain region (country) comes from the overall context of analysing and measuring various aspects of *quality of life*. Our understanding of this integral component of the synthetic category of quality of life is based on the thesis, according to which, natural raw materials and climatic conditions are treated as "*body and the forces of nature that at this level of development of the productive forces are essential to the life and activities of human society, but not directly involved in the material production and non-production activities of people*" [Минц, 1972] (italics mine).

Accordingly, the formation of a system of statistical indicators of natural raw materials and climatic conditions are proposed to be done by using the following scheme.

1. Highlight *basic elements* (climate, topography, landscape, natural resources, etc.) that are not the direct object of economic relations in any socio-economic formations, and that are characterized by long-term (at least 25–30 years) stability, and determine the *level of comfort of living conditions of the population* of the region (country) [Resources, the environment, resettlement, 1974, pp. 199–207].
2. Build a rating system of regional climatic conditions, based on the measurements of the previously entered basic elements (for example, using the methods of *assessment mapping*) [Resources, the environment, resettlement, 1974, pp. 189–198].
3. Form integral indicators of the natural resources and natural climatic potentials which are then used as components – correction factors when calculating the integral indicators of quality of life of the population of the region (country). We can do it with the help of special methods based on measurement estimates of basic elements.

2.2. Requirements for the formation of a priori set of statistical parameters for different synthetic categories of quality of life of the population

Each *applied* research that presents the hierarchical system of statistical indicators has to be filled with *specific content*, as we did above, by experts based on:

- *the level* of the analysed population conglomerates (cross-country, inter-regional, inter-municipal);
- *some general requirements* for representativeness (relevance), information accessibility and information reliability of statistics.

With regard to taking into consideration **the level of analysis** (cross-country, inter-regional or inter-municipal), *it is determined by the specifics of the overall strategy of development* of the world, regional (within the country) and municipal (within the region). This means that the composition of initial statistical indicators (partial criteria) that characterize analysed synthetic category, for example, at the country level, will, in general, be different from the composition of the indicators at the regional and municipal levels.

In addition, while forming the initial (a priori) set of statistical indicators (partial criteria), characterizing a certain synthetic category of QOL experts should be guided by the following general requirements:

(A) *representativeness (relevance)*, according to which this list must include (with an exogenous conditions being preset for all other aspects of QOL) all the main indicators that are considered to be a synthetic category; for example, for the category of synthetic QOL of the very top level this list should include the most important quality characteristics of the population (aspect I), material well-being (aspect II), the quality of social services (aspect III) and the quality of the ecological niche (aspect IV);

(B) *the availability of information*, according to which the further analysed indicators and private criteria should be at least available for their statistical registration, and moreover, they must be included in the nomenclature of periodically recorded official statistics (or can be calculated according to the values of the latter);

(C) *the reliability of information* in accordance with which the statistics and the particular criteria used should adequately reflect the condition of the analysed aspects of the quality of life; for example, precisely because of the violation of this requirement, we had to abandon the use of such seemingly essential for the synthetic category of QOL top-level indicator as "the average number of reported violent crimes per year per 100 000 inhabitants": the fact is that the small values of this index and a low crime rate are often not indicators of a great state of affairs in this area, but instead it is due to weak statistical work and poor law enforcement.

2.3. Measurement methodology of synthetic categories of the quality of life and methods of mullticriteria rating of the territories (countries, subjects of russian federation, municipalities)

2.3.1. Statement of the main tasks

So, we have to present and justify the methodology of building the integrated indicators for the various synthetic categories of QOL of a country or a region, based on a special type of convolution of the number of private (*statistically recorded*) criteria of the appropriate synthetic categories of QOL. We also have to demonstrate the efficiency of this methodology in comparative macroeconomic analysis and in the rating of the countries or regions as well as to propose an approach to *multicriteria* classification of countries (regions) according to the analysed integral properties of QOL in situations where the problem of constructing a *single* scalar II QOL has no satisfactory solution. To achieve this goal, we must be able to perform the following tasks.

Task 1 (preanalysis). *Determination of the initial (a priori) list of statistical indicators* $x^{(1)}, x^{(2)}, \dots, x^{(p)}$ that would characterize the analysed synthetic category of QOL adequately (reflecting different aspects). Taking into account the semantic load of the above indicators we will also call them the *partial criteria* of the analysed synthetic category.

Task 2 (preanalysis). Development of specific methods of selection of relatively small number $p'(p'<p)$ of partial criteria that play a major role in shaping the values of the analysed integral indicator of quality of life from a priori list of partial criteria $x^{(1)}, x^{(2)}, \dots, x^{(p)}$, derived from the solution to *problem 1*. In other words, we are talking about a *method of forming a posteriori set of particular criteria* $x^{(j_1)}, x^{(j_2)}, \dots, x^{(j_{p'})}$ selected from a priori list of indicators $x^{(1)}, x^{(2)}, \dots, x^{(p)}$.

Task 3 (preanalysis). *Unification of measuring scales* of all the analysed variables, i.e. transition to [0; *N*]–point scales for the measurements of the analysed partial and integral criteria of QOL so that the value "zero" would correspond to the lowest level of QOL and the value of *N* to the highest (the dimension of the scale *N* is chosen by an investigator).

Task 4 (central). Development of methodology for constructing an integral indicator – a measure of the analysed synthetic category of QOL (II QOL) in each of the two possible formulations of the problems:

- *version "with learning"*: the construction of II QOL using values of partial criteria $x^{(j_1)}, x^{(j_2)}, \dots, x^{(j_{p'})}$ of posteriori set and using some ("teaching") expert information on the comparison of the considered objects (in our case, territories) of the analysed synthetic category of QOL;
- *option "without learning"*: the construction of II QOL *only* by the values of a set of particular criteria $x^{(j_1)}, x^{(j_2)}, \dots, x^{(j_{p'})}$ of posteriori set.

Task 5 (post-analysis in the presence of "learning"). Evaluation of the consistency of the two rankings of the territories in question by analysed synthetic QOL category, one of which is based on the existing expert "training", and the other is based on the values of II QOL constructed by solving the problem 4 (in either of the two options).

Task 6 (post-analysis). The construction and analysis of **auto-** and **inter**-territorial dynamics of II QOL that allows us to compare in time the analysed property of QOL of the territory in relation to itself (*auto-dynamics*) and in relation to other areas (*inter-territorial dynamics*).

Before proceeding to the description of methods for solving the problems stated above, it is necessary to introduce the concepts and forms of "statistical learning" in the general problem of measuring latent synthetic category.

2.3.2. Concept and forms of "statistical learning" in the general problem of measuring latenty synthetic categories

In professional work and in everyday life man is constantly faced with situations where he has to compare some objects with each other and organize or sort them by some of their qualities (not directly measurable). This may include, in particular, a comparison between countries of various aspects of quality of life; companies in the industry, their performance, sophisticated products, generalized characteristics of quality; specialists, effectiveness of their participation in the implementation of certain task; and the participants of team sports, the level of their demonstrated excellence etc. The work of Aivazian [Айвазян, 1974] was devoted to the formalization of the situations arising out of the above-described tasks. It provided recommendation for the construction of a conditional measure of the said property and suggested a so-called expert statistical method for solving this problem.

The method of solving problems of *formation of a-posteriori set of particular criteria and obtaining a certain convolution of these partial criteria as a measure of the analysed integral property essentially depends on the presence or absence of the so-called learning in our initial statistical data and on the form of presentation of this learning*. Let's discuss more aspects of this problem.

So, let's think that the generalized summed up characteristic f of the analysed property of the object is determined by a set of partial criteria defined by identifiable and measurable variables $x^{(1)}, \dots, x^{(p)}$, but this very characteristic is *latent*, i.e. *is not directly quantifiable* (there is no objectively defined scale for it). Natural to assume that its intuitive expert (professional) perception (we'll call it $y_{\text{St.}}$) can be represented as a somewhat distorted value $f(x^{(1)}, \dots, x^{(p)})$, and this distortion δ has an accidental quality to it, and it is due to both the resolution of such a "measuring tool" as an expert (according to our scheme) and also the existence of some particular criteria that have some weak influence on $y_э$, but are not really a part of $X = (x^{(1)}, \dots, x^{(p)})$. Then the model that relates the intuitive idea of a composite index of quality $(y_э)$ with the consolidated exponent $f(X)$ (as a function of X) and a random error $\delta(X)$, can be defined as

$$y_{\text{St.}} = f(X) + \delta(X). \tag{2.1}$$

If we accept the natural, almost not limiting the generality of the scheme assumptions about the first two moments of the residual random component $\delta(X)$:

$$\mathbf{E}\delta(X) = 0, \mathbf{D}\delta(X) = \sigma^2(X) < \infty$$

it is obvious that a *generalized consolidated (integral) characteristic $f(X)$ can be interpreted as regression $y_{\text{St.}}$ no* X.

Note on the objective functions. It is very common in the field of socio-economic research to interpret the generalized consolidated (integral) characteristics of quality $f(X)$ as an *objective function*. Wherein, we consider as objective function of analysed output quality y any transformation f of particular criteria $X = (x^{(1)}, x^{(2)}, \dots, x^{(p)})$ that would preserve a given relation of order between n objects under consideration for y. In other words it has to have the property that from

$$f(X_{i_1}) \geq f(X_{i_2}) \geq \dots \geq f(X_{i_n})$$

always follow the inequalities

$$y_{i_1} \geq y_{i_2} \geq \dots \geq y_{i_n}$$

where $f(X_i)$ and y_i – accordingly is the value of the objective function and a consolidated measure of quality that characterize the i object. Obviously, the definition of objective function given here is ambiguous. Indeed, if $f(X)$ is an objective function and $\varphi(f)$ is any monotonically increasing function, then any function of the form

$$\tilde{f}(X) = \varphi(f(X))$$

will also be objective. This implies that our assumption of a certain scale to measure y in some cases has purely instrumental role and doesn't point us in the direction of identifying it. Indeed, according to this definition, the objective function value does not reflect any real, physically meaningful quantitative regularities. Actual patterns are recorded only as relations of "more" or "less" between the values of this function for different sets of values of the input parameters $X = (x^{(1)}, x^{(2)}, \dots, x^{(p)})$. Thus, these ratios reflect the preference (from the point of view of analysed output quality) of some X values over the others. So, in applications in which it is possible to adjust the value X (in some area that would allow us to do it), the most efficient (optimal) management would be the one that maximizes (with given restrictions for X) values of the objective function.

By the way, the issues of constructing objective functions will be looked at in Chapter 4 in connection with the problem of statistical evaluation of the objective function of consumer preferences based on data from selected surveys of household budget (see point 4.1.2).

Definition. We say that our *initial statistical data contain the "learning"* if when performing the statistical analysis of the model (2.1) we don't only know the values of the partial criteria $x_i^{(1)}, \dots, x_i^{(p)}$ $(i = 1, 2, \dots, n$; n – the total number of statistically surveyed, compared with each other objects), but we also have some information on the comparison of the objects (in our case, territories) by the analysed synthetic category.

This information is usually obtained with the help of specially organized survey of experts and appropriate statistical processing of expert estimations. These data are usually presented by experts in one of the following forms.

Form (i) – the most informative (and therefore the most difficult for experts) option. It involves obtaining the *expert scores* on output quality

$$\begin{array}{c} y_{11_{St.}}, y_{21_{St.}}, \dots, y_{n1_{St.}} \\ y_{12_{St.}}, y_{22_{St.}}, \dots, y_{n2_{St.}} \\ \dots\dots\dots\dots\dots\dots \\ y_{1m_{St.}}, y_{2m_{St.}}, \dots y_{nm_{St.}} \end{array}$$

where $y_{ij_{St.}}$ is the estimation of the output quality of the object O_i, obtained from the j-th expert (here n is the estimated number of objects, and m is the number of participating experts).

Form (ii) – the average in terms of information (and the degree of difficulty for experts) option. It requires the *experts to sort* the inspected objects by the degree of manifestation in them of the property analysed, i.e. rankings of the following type

$$\begin{array}{c} R_{11_{St.}}, R_{21_{St.}}, \dots, R_{n1_{St.}} \\ R_{12_{St.}}, R_{22_{St.}}, \dots, R_{n2_{St.}} \\ \dots\dots\dots\dots\dots\dots \\ R_{1m_{St.}}, R_{2m_{St.}}, \dots R_{nm_{St.}} \end{array}$$

where $R_{ij_{St.}}$ is the rank (or place) assigned to the object O_i by the j-th expert in a line of n inspected objects, sorted by this expert by the degree of manifestation of the analysed property.

Form (iii) – the least informative (and least difficult for experts) option. Information from each (j-th) expert comes in the form $(n \times n)$ – *Boolean matrix of paired comparisons*

$$\Gamma_{St.}(j) = (\gamma_{ik.j_{St.}}), \; i, k = \overline{1, n}, j = 1, 2, \dots, m$$

where $\gamma_{ik.j_{St.}}$ is the result of paired comparison by j-th expert of the objects O_i and O_k, which is defined by the following rule:

$$\gamma_{ik.j_{St.}} = \begin{cases} 1, & \text{if by } j\text{–th expert's opinion } O_i \text{ is better than } O_k, \\ -1, & \text{if by } j\text{–th expert opinion } O_i \text{ is worse than } O_k; \\ 0 & \text{in the case when the } j\text{–th expert can not determine} \\ & \text{which one is better whether it's } O_i \text{ or } O_k. \end{cases}$$

Note. In the inter-territorial analysis of the quality and way of life the most common form of expert learning is perhaps the one when *they split the territory in question into a small number of groups sorted by the synthetic category analysed, that doesn't give any preference though to one territory over the other in the same group* (for example, grouping all the analysed territories into three groups: leaders, "average" and outsiders). But, strictly speaking, such form of expert learning fits both within the frame of the form (ii) (if we assume the existence of so-called connected, i.e. indistinguishable ranks), and also within the form (iii).

Computational difficulties associated with the implementation of algorithm of building II QOL in the form of convolution naturally increase as we move from a more informative options for expert learning to less informative. Below (in point 2.3.5 and 2.3.6) we'll look at these algorithms for each of the above forms of learning.

2.3.3. Unification of scales in the measurements of particular and integral criteria of qol (task 3)

Before we proceed directly to the procedure of convolution of partial criteria, it is necessary to bring these criteria x to the "common denominator", i.e. apply such a transformation to each of them that as a result they all are going to be measured in N-point scale. The zero value of the transformed figure has to correspond with lowest quality of given property, and the maximum (N) value with the the highest. *The particular choice of unifying conversion depends on which of the three types the analysed indicator belongs to.*

1. If the initial value (particular criteria) x is connected with the analysed integral property of the quality of life by *monotonically increasing* dependence (i.e. *the higher the value x, the higher the quality*), then the value of the corresponding unified variable is calculated by the formula

$$\tilde{x} = \frac{x - x_{\min}}{x_{\max} - x_{\min}} \cdot N \tag{2.2}$$

where $x_{\min}$ and $x_{\max}$ are the lowest (worst) and the highest (best) values of the baseline indicator.

2. If the initial value (particular criteria) x is connected with analysed integral property of the quality of life by a *monotonically decreasing* dependence (i.e. the greater the value of x, the lower the quality), then the value of corresponding unified variable is calculated by the formula

$$\tilde{x} = \frac{x_{\max - x}}{x_{\max} - x_{\min}} \cdot N \tag{2.3}$$

3. If the initial value (particular criteria) x is connected with analysed integral property of quality of life by *non-monotonic dependence* (i.e. between $x_{\min}$ and $x_{\max}$ there is an optimal value $x_{\text{opt.}}$, which provides the highest quality), then the value of corresponding unified variable is calculated by the formula

$$\tilde{x} = \left(1 - \frac{|x - x_{\text{opt.}}|}{\max\{(x_{\max} - x_{\text{opt.}}), (x_{\text{opt}} - x_{\min})\}}\right) \cdot N. \tag{2.4}$$

To implement these changes we need to be able to determine the values x_{min}, x_{max} and $x_{opt.}$ for every private criterion x. Since *theoretical-normative approach* in determining these values in most cases can be very complicated (and, in particular, the need to harmonize different points of view), in this paper we propose to use an *empirical* approach for this purpose. In particular, we offer to assign the minimum and maximum values of all the observed values (for different regions or different time steps) of this variable as x_{min} and x_{max}, respectively. Question on the definition of values $x_{opt.}$ is solved in each case individually taking into account the specific situation.

For example, if we have learning in the form of expert scores or rankings of the territories in question by analysed synthetic category, then $x_{opt.}$ can be a result of averaging of this index for three or five territories that *the experts consider to be the leaders in synthetic analysed category.*

Note. It is easy to see that the first two formulas with the help of which we perform the unification of the measurement scales of the analysed parameters are actually special cases of the third formula with $x_{opt.} = x_{max}$ and $x_{opt.} = x_{min}$.

2.3.4. Formation of the reduced (a posteriori) set of particular criteria (task 2)

In forming the initial (a priori) sets of statistic values for each of the analysed latent synthetic categories of QOL (*task 1*) experts are invited to rely on the above-described structuring of particular criteria and integral properties of QOL (see Figure 2.1) and bear in mind the general requirements (A) (B) and (C) (Paragraph 2.2).

The basic idea on which the method of forming the reduced (a posteriori) sets of particular criteria for each of the analysed synthetic categories is built consists of the following.

From the formed (theoretical, substantive considerations at the expert level) a priori set of particular criteria of the specified integral property we want to select (or leave it for further analysis) a relatively small number of indicators so that, first, they really directly characterize this integral property and, second, we should be able to reconstruct fairly accurately the values of all other particular criteria (excluded from the set of a priori) by using the values of this small number of indicators with the help of appropriate regression models.

Method of implementing of this general idea depends on the presence (or absence) of learning in formulating the problem of measurement of latent synthetic category.

a) Selection of the most significant private criteria in the presence of learning. We assume that the model (2.1) is parameterized in the class of *linear* functions of partial criteria $x^{(1)}, \dots, x^{(p)}$, and at the same time we have learning in the form of (i) as a common version of the expert scores $y_{1St.}, y_{2St.}, \dots, y_{nSt.}$ of the analysed synthetic categories for the all considered objects (regions, countries). In this case, for the "point in time" t, we can consider the linear regression model of the following type

$$y_{iэ}(t) = \theta_0(t) + \theta_1(t)x_i^{(1)}(t) + \dots + \theta_p(t)x_i^{(p)}(t) + \delta_i(t). \qquad (2.1')$$

Then, we can do the selection of the most significant partial criteria, for example, through a combination of stepwise regression methods (or "method of all possible regressions") with "addition" and with "exclusion'. Let's recall the general scheme of these procedures.

Step-by-step procedure of successive addition is designed for selection (in the linear multiple regression model (2.1′)) of the most informative set from k of explanatory variables selected from the source (a priori) multitude of explanatory variables $x^{(1)}, \dots, x^{(p)}$ (obviously, $k < p$).

In the first step of this procedure ($k = 1$) we select a variable $x^{(j_1)}$ from the original set of explanatory variables $X = (x^{(1)}, \dots, x^{(p)})^T$ that has the maximum value of the square of the coefficient of pair correlation with the dependent variable y, i.e.

$$r^2(y, x^{(j_1)}) = \max_{1 \le j \le p} r^2(y, x^{(j)}).$$

On the $(\nu + 1)$ *step of this procedure* $(1 \le \nu \le p{-}2, k = \nu + 1)$ we already have the previously selected most informative variables $x^{(j_1)}, \dots, x^{(j_\nu)}$. Next $(\nu + 1)$ informative variable $x^{(j_{\nu+1})}$ is selected so that

$$R^2\left(y; (x^{(j_1)}, \dots, x^{(j_\nu)}, x^{(j_{\nu+1})})\right) = \max_{\substack{1 \le j \le p \\ j \ne j_1, j_2, \dots, j_\nu}} R^2\left(y; (x^{(j_1)}, \dots, x^{(j_\nu)}, x^{(j)})\right),$$

where $R^2(y; (x^{(l_1)}, \dots, x^{(l_N)}))$ – retouched "for unbiased" coefficient of determination between y, on one hand, and the set $(x^{(l_1)}, \dots, x^{(l_N)})$ on the other.

Typically, such a procedure is ended on the step, at which the transition to the next step doesn't give a statistically significant increment to the value R^2.

Step-by-step procedure of successive elimination is designed to remove from the source (a priori) set of explanatory variables $x^{(1)}, \dots, x^{(p)}$ those variables that do not increase the predictive power of the model (2.1′).

In the first step of this procedure from the original set of explanatory variables we remove the variable $x^{(i_1)}$, the exclusion of which from the model (2.1′) leads to a minimum decrease in the value of the coefficient of determination R^2, i.e.

$$R^2(y; X(i_1)) - \max_{1 \le i \le p} R^2(y; X(i)),$$

where $X(l)$ denotes the initial set of explanatory variables, from which the variable $x^{(l)}$ is removed.

Before implementing $(\nu+1)$ *step of the procedure* $(1 \le \nu \le p-2)$ we would have $p-\nu$ of explanatory variables that remained from the initial set. We denote them as $x^{(\nu)} = (x^{(j_1)}, \ldots, x^{(j_{p-\nu})})^T$. Then on the $(\nu+1)$ step we will exclude the variable $x^{(i_{\nu+1})}$ whose removal results in a minimum decrease in the value of $R^2\left(y; (x^{(j_1)}, \ldots, x^{(j_{p-\nu})})\right)$, i.e.

$$R^2(y; X^{(\nu)}(i_{\nu+1})) = \max_{i=j_1,\ldots,j_{p-\nu}} R^2(y; X^{(\nu)}(i)),$$

where $X^{(\nu)}(l)$ denotes the set of explanatory variables $X^{(\nu)}$, from which the variable $x^{(l)}$ is excluded.

Usually, the procedure of successive elimination is carried out till the step, after which the next removal of the variable leads to a statistically significant decrease in the value of the coefficient of determination R^2.

Note that there may be situations where the initial statistics

$$x_i^{(1)}(t), x_i^{(2)}(t), \ldots, x_i^{(p)}(t); \quad y_{i\text{St.}}(t) -$$

$$i = 1, 2, \ldots, n; \quad t = 1, 2, \ldots, T -$$

are *regression-heterogeneous* (in the sense of the criterion of Chow, see [Айвазян, 2010], i.e. the values of regression coefficients $\hat{\theta}_j(t)$ vary statistically significant during the transition from one value t to another one, but the procedure to identify the most important particular criteria in our problems gave, however, as a rule, the same result for different points in time t.

It is possible to summarize the above procedure in relation to cases when learning was presented in the form of (i) or (ii).

b) Selection of the most important particular criteria in the absence of learning. Methodological scheme that the total basic idea relies upon is as follows.

1. *Analysis of multicollinearity of partial criteria of a priori set of indicators.* With above goal in mind, we calculate for the variables of each a priori set the following:
   - *matrix of values of the paired coefficients of correlation,* $r(x^{(l)}, x^{(\nu)})$ $(l, \nu = 1, 2, \ldots, p)$, characterizing the degree of closeness of pair-wise statistical relationships between particular criteria of the analysed property;
   - coefficients of determination R^2 *of each of the particular criteria* $x^{(l)}$ *of the analysed a priori set on the rest of the indicators of this set.*

Analysis of the numerical characteristics allows us to have a preliminary step to identify closely related pairs, triples, etc. groups of variables, and raise the question of selection of one representative from each group. The other evidence that also points to the necessity of reduction of each of the sets of a priori of particular criteria is the fact that the values of the vast majority of the above coefficients of determination R^2 are close to 1 (see examples of the results of calculations based on this scheme in A2.4).

2. *The selection of the most informative particular criteria among the indicators of a priori set for each of the integral property.* Let's say we are given a quantitative content (p') of the reduced set of indicators of the specific integral property $(p' < p)$. A set of partial criteria

$$x^{(l_1^0)}, x^{(l_2^0)}, \dots, x^{(l_{p'}^0)}$$

will be most informative if

$$\sum_{l=1}^{p} R^2\left(x^{(l)}; \left(x^{(l_1^0)}, \dots, x^{(l_{p'}^0)}\right)\right) = \max_{l_1, l_2, \dots, l_{p'}} \sum_{l=1}^{p} R^2\left(x^{(l)}; \left(x^{(l_1)}, \dots, x^{(l_{p'})}\right)\right)$$

where $R^2(y; (x^{(1)}, \dots, x^{(s)}))$ is the coefficient of determination of the dependent variable y on the explanatory variables $x^{(1)}, \dots, x^{(s)}$. Obviously, with such a choice of a posteriori reduced *limited* set of particular criteria we are maximizing the predictive power of the regression models, with the help of which one can restore the **entire** set of an a priori parameters of the analysed integral property based on the values of only those particular criteria that are included in this limited set.

Selecting the quantitative composition (p') of a limited set of particular criteria is made in each case on the base of combination of theoretical (informative) considerations and requirements for the minimum acceptable value $R^2_{\min}$ of the coefficient of determination $R^2(x^{(l)}; (x^{(l_1^0)}, \dots, x^{(l_{p'}^0)}))$.

Note. Selection of the most significant particular criteria both in the presence of learning and without it is based on solving the corresponding optimization problems. In order to leave the possibility of taking into the account the *informative* (and not just formal) considerations, it is proposed along with the optimum composition of the posterior set of variables to also display a so-called near optimum compositions, **i.e. such sets of a predetermined number of variables that the value of the optimized criterion for them differs very little (statistically insignificant) from the optimum.** Then we can make the final choice of the composition of the posterior set of particular criteria (from the options provided by the "optimum" and "near optimum" sets) based mainly on the informative (or meaningful) considerations. I would have recommended the above remark be considered for the general case by including it in the description of the procedures contained in the statistical and econometric packages (SPSS, E-views, STATA, etc.)

2.3.5. Building ii qol with the presence of learning (expert-statistical method)

1. **II QOL at the presence of score based form of learning (see the above form (i))**[2]. In this case, the task of building II QOL comes to statistical estimation of the regression function (convolution) $f(X)$ in model (2.1) based on the initial data of the following type

$$(X_i; y_{i1_{St.}}, y_{i2_{St.}}, \dots, y_{im_{St.}}), \quad i = 1, 2, \dots, n \tag{2.5}$$

i.e. it comes to the standard problem of regression analysis. So if we limit the scope of our search of convolution to parametric family $\{(f(X; \Theta)\}$, then

$$f(X; \hat{\Theta}) = \arg\min_{\Theta} \sum_{i=1}^{n} \sum_{j=1}^{m} c_{ij}^2 (y_{ij_{St.}} - f(X_i; \Theta))^2, \tag{2.6}$$

where the "weighing" coefficients c_{ij}^2 are determined by two factors:
- relative competence of the experts (the dependence on *j*);
- relative importance of the analysed objects or the relative reproducibility (exactness) of expert assessments for each of the objects (the dependence on *i*).

One of the possible methods of determining the coefficients c_{ij}^2 is a factorization of their values using the formula

$$c_{ij}^2 = \frac{1}{s_i^2} \cdot \frac{1}{\sigma_j^2},$$

where

$$s_i^2 = \frac{1}{m} \sum_{j=1}^{m} (y_{ij_{St.}} - \bar{y}_{i.})^2, \quad \bar{y}_{i.} = \frac{1}{m} \sum_{j=1}^{m} y_{ij_{St.}}$$

and

$$\sigma_j^2 = \frac{1}{n} \sum_{i=1}^{n} (y_{ij_{St.}} - \bar{y}_{i.})^2$$

Obviously, the value s_i^2 characterizes the "fuzziness" in the accuracy of the expert evaluation of *i*-th object, and the value σ_j^2 is a measure of deviation in estimates of the *j*-th expert from the opinion of the general group.

2 With the goal of simplifying our text from here on we call the general of particular criteria in **a posteriori** set as a p (and not p').

In the case of a single (group) variant of expert estimation, i.e. when we have data in the form of

$$(X_i; y_{i\text{St.}}), \quad i = 1, 2, \dots, n \tag{2.5'}$$

the II QOL we are seeking will be determined as the solution to the problem

$$f(X; \hat{\Theta}) = \arg\min_{\Theta} \sum_{i=1}^{n} c_i^2 \left(y_i - f(X_i; \Theta)\right)^2, \tag{2.6'}$$

where in the absence of information about the relative importance of the objects it is considered that $c_i^2 \equiv 1$.

Software implementation of the solutions to the problems (2.6) and (2.6') carried out by an ordinary or generalized (weighted) method of the least squares the reader can find in any of the integrated statistical (econometric) software packages (SPSS, Statistica, E-views, etc.).

Note on choosing the general form of II QOL. The choice of a parametric family $\{f(X; \Theta)\}$, as a rule, cannot be backed up by strict theoretical foundation. After all the *scalar* characteristic of the analysed quality may not exist (see above point (c) at the beginning of this chapter), so in this case we are dealing with some approximation that has conventional meaning within a limited period of time and under certain conditions. Therefore, exploiting the idea of expanding any function into a Taylor series, we usually limit ourselves to linear or quadratic approximation of the desired convolution, i.e.

$$f(X; \Theta) = \sum_{k=0}^{p} \theta_k x^{(k)} \, (x^{(0)} \equiv 1)$$

or

$$f(X; \Theta) = \sum_{k=0}^{p} \theta_k x^{(k)} + \sum_{k,l=1}^{p} \theta_{kl} \cdot x^{(k)} x^{(l)}$$

Note that when specifying the problem we can often fill the coefficients θ_k *and* θ_{kl} with real socio-economic sense [Слуцкий, 1963].

2. **Construction of II QOL in the presence of expert rankings of the objects based on the analysed synthetic category (see above, form (ii)).** In this case, we have the data in the form of

$$(X_i; R_{i1\text{St.}}, R_{i2\text{St.}}, \dots, R_{im\text{St.}}), i = 1, 2, \dots, n \tag{2.7}$$

where $R_{ij\text{St.}}$ – rank (place) assigned to the i-th object by j-th expert in a row of n inspected objects, who ordered them based on the analysed synthetic QOL category (in order from best to worst). If we had already have II QOL in the form of $f(X;\Theta)$ (i.e. we had have known the value of a vector parameter Θ), then using the values of $X_i = (i = 1, 2, \ldots, n)$ for $f(X;\Theta)$ and calculating the corresponding values of $f_i = f(X_i;\Theta)$, we could get the ordering of the objects according to the values f_i, i.e. ranks $R_i(\Theta)(i = 1, 2, \ldots, n)$, which, of course, would L depend on the values of the parameters Θ.

But then we can compare the consistency of this ordering with the ordering of each (j-th) of the experts with the help of, for example, the *Spearman rank correlation coefficient*:

$$r_j(\Theta) = 1 - \frac{6}{n^3 - n} \sum_{i=1}^{n} (R_{ij\text{St.}} - R_i(\Theta))^2.$$

In the presence of so-called connected ranks we should use the appropriate modification of the above formula [Айвазян, 2010; point 3.3.4].

It would be natural to "set up" the value of the parameters Θ on expert training, maximizing the consistency of ranking that is based on II QOL $f(X;\Theta)$ with rankings of experts. This can be done, for example, by solving one of two problems:

$$\hat{\Theta} = \arg\max_{\Theta} \sum_{j=1}^{m} q_j r_j(\Theta) \tag{2.8}$$

or

$$\hat{\Theta} = \arg\max_{\Theta} r(\Theta) \tag{2.8'}$$

where $q_1, q_2, \ldots, q_m$ – non-negative weights, proportional to the level of expertise of involved experts (in particular $q_1 = q_2 = \ldots = q_m = 1$), and $r(\Theta)$ – coefficient of Spearman's rank correlation between the ranking of the objects that is based on II QOL $f(X;\Theta)$, and a **single (group)** version of *expert* ordering of the same objects.

Note on weighting coefficients q_j. As one of the possible ways to determine the weighting coefficients $q_j (j = 1, 2, \ldots, m)$ we can recommend a method based on conditional comparative assessment of competence of the j-th expert with the help of values of the coefficients of concordance [Aivazian, 2010, point 3.3.7]

$$W(m) = \frac{12}{m^2(n^3 - n)} \sum_{i=1}^{n} \left(\sum_{j=1}^{m} R_{ij\text{St.}} - \frac{m(n+1)}{2} \right)^2$$

and

$$W_j(m-1) = \frac{12}{(m-1)^2(n^3-n)} \sum_{i=1}^{n} \left(\sum_{j=1}^{m} R_{ij\text{St.}} - \frac{(m-1)(n+1)}{2} \right)^2 .$$

Assuming that the more the coefficient of concordance increases in the rankings of experts W when we exclude the expert with number j from our group of experts (i.e. the greater than the difference $W_j(m-1)-W(m)$), the less competent that expert is, and the values q_j can be calculated, for example, by the following formula:

$$q_j = 1 - \frac{W_k(m-1)-W}{\sum_{l=1}^{m} (W_l(m-1)-W)}, \quad j = 1, 2, \dots, m.$$

Note on the evaluation of the group version of expert ordering. Group ordering $R^{(gr)} = (R_1^{(gr)}, R_2^{(gr)}, \dots, R_n^{(gr)})$ is determined as the solution of the optimization problem

$$R_i^{(gr)} = \arg\min_{\tilde{R}_i} \sum_{j=1}^{m} |R_{ij\text{St.}} - \tilde{R}_i|, \quad i = 1, 2, \dots, n.$$

It is known that the solution of this problem is given by the *median* values of a row $(R_{i1\text{St.}}, R_{i2\text{St.}}, \dots, R_{im\text{St.}})$.

Unfortunately, in terms of computational realization the problems (2.7) and (2.7') are very complex and have no standard software. However, there is a special variant of this problem that gives us a formulation of it with ranking, in which both the expert task and the task of computing implementation are simplified. In this variant, experts are invited to divide multiple objects (in our case, the territories) into *small* number of classes (groups) sorted by analysed synthetic category of QOL. Therein, it does not allow any preference between the objects belonging to any one group. Obviously, such a partition can be formally represented in the form of (ii), in which there is a large number of "connected" (i.e. indistinguishable) ranks. So, in the end, we have data of the following type

$$(x_i^{(1)}, x_i^{(2)}, \dots, x_i^{(p)}; z_i), \quad i = 1, 2, \dots, n \tag{2.7'}$$

where z_i is the number of the class to which the object is assigned by a *single (group) expert partitioning of the objects into* classes[3] (we assume that the total number of classes is equal k, and that the higher the class number, the higher the quality of life of the analysed synthetic category). Then the procedure of multicri-

3 About different ways of building a single (group) expert partitioning see, for example [Миркин, 1976].

teria-ordered classification of objects can be obtained as a solution to the standard problem of multiple choice with an ordinal dependent variable in the form of, for example, the corresponding logit model. In the course of the procedure [Айвазян, 2010, Paragraph 9.2] $x^{(1)}, x^{(2)}, \dots, x^{(p)}$ are latent variables, which in this case play the role of II QOL.

Let us briefly recall the general scheme of the implementation of this procedure. So, let the random variable z take the values $1, 2, \dots, k$ (according to the number of analysed ordered groups) and that the higher the number of the group that owns the region i, the higher the quality of the analysed synthetic category of this region. Since this quality is mainly determined by the values of the partial criteria, it is natural to assume that there is a latent variable in the form of a linear combination

$$y = \sum_{j=1}^{p} w_j x^{(j)} + \varepsilon \tag{2.9}$$

and threshold values $c_0, c_1, \dots, c_k$, such that

$$p_j(X_i) = P\{z = j | X = X_i\} = P\{c_{j-1} < y_i < c_j\} = P\{c_{j-1} - X_i^T W < \varepsilon_i < c_j - X_i^T W\}.$$

The parameters $W = (w_1, w_2, \dots, w_p)^{\mathrm{T}}$ of the linear form (2.9) as well as the thresholds $c_0, c_1, \dots, c_k$, are subject to statistical estimation according to available data (2.7´).

Postulating a linear or logistic form of the distribution $F(u)$ of residues εin the ratio (2.9) and the mutual independence of these residues ε_i (with $\mathbf{E}\varepsilon_i \equiv 0$ and $\mathbf{D}\varepsilon_i \equiv 1$, which doesn't limit the generality of the model), we obtain the following expression for the likelihood function:

$$L(w_1, \dots, w_p; c_0, c_1, \dots, c_k) = \prod_{j=1}^{k} \prod_{i: z_i = j} \left(F(c_j - X_i^T W) - F(c_{j-1} - X_i^T W) \right). \tag{2.10}$$

Maximizing the right side (2.10) on $w_1, \dots, w_p; c_0, c_1, \dots, c_k$, we obtain estimates $\hat{W}$ and $\hat{c}_0, \hat{c}_1, \dots, \hat{c}_k$ for the unknown parameters. Appropriate computational procedures are implemented in standard software packages (SPSS, E-views, etc.).

Note: Normally, without loss of generality, it is believed that: $c_0 = -\infty$ and $c_k = +\infty$.

3. **Building of II QOL in the presence of expert pair-wise comparisons of objects by analysed synthetic category (see above, Form (iii)). In this case we have learning in the form of matrices** $\Gamma_{\mathrm{st}}(j) (j = 1, 2, \dots, m)$. There are different approaches to solving the problem of building an integral indicator based on the data of the following type

$$(X_i; \Gamma_{\mathrm{st.}}(1), \Gamma_{\mathrm{st.}}(2), \dots, \Gamma_{\mathrm{st.}}(m)), \tag{2.11}$$

see, for example, "the method of voting" proposed by Y. Zhuravlev, or a method based on a linear programming [Айвазян, 1974; Киселев, 1980]. However, given the specificity of the tasks of inter-regional multicriteria analysis of QOL, the most appropriate thing to do is to use the opportunity to fit case into the schemes described in the previous paragraph, i.e. the schemes based on the results of the expert ranking of objects (territories), including in the form of dividing them into a small number of ordered classes.

For this purpose it is sufficient for each (i-th) object to calculate its characteristic of preference γ_i from the formula

$$\gamma_i = \sum_{j=1}^{m} \sum_{l=1}^{n} \gamma_{il \cdot j\text{St.}}, \quad i = 1, 2, \dots, n$$

and then rank the objects in descending order of their characteristics. The transition to a variant of the dividing the objects into a small number of ordered classes is done by simple aggregation of objects with relatively similar values γ_i, i.e. by solving a simple one-dimensional problem of cluster analysis.

2.3.6. Construction of ii qol in the absense of learning

In this formulation of the problem we *only* have the unified value of particular criteria of a posteriori set

$$(\tilde{x}_i^{(1)}, \tilde{x}_i^{(2)}, \dots, \tilde{x}_i^{(p)}), \quad i = 1, 2, \dots, n, \tag{2.12}$$

i.e. we do not have any information on the comparison of the objects (territories) by analysed synthetic category. The question arises:

Is it right in such a situation to set a task of measuring the analysed latent synthetic category using convolutions of particular criteria?

Conditionally positive answer to this question is based on the following logic:

(a) The analysed synthetic category is adequately characterized by standardized values of particular criteria $\tilde{x}_i^{(1)}, \tilde{x}_i^{(2)}, \dots, \tilde{x}_i^{(p)}$ of a posteriori set.

(b) If we want to limit ourselves to **only one** (i.e. scalar) aggregate indicator (*integrated indicator*) of the given synthetic category, then it is natural to look for a such an indicator $\hat{y}$ amongst all the other latent indicators that would have the following property: **among all scalar variables that characterize the analysed latent synthetic category, it is this indicator $\hat{y}$ that can be most accurately (in a sense) used to restore the values of all the particular criteria of the a posteriori set,** using for that the appropriate model of pair-wise regression.

But if you limit the view of the desired convolution by class of *linear* functions (i.e. to build II QOL in the form of $\sum_{j=1}^{p} w_j \tilde{x}^{(j)}$), it is precisely this property (b), that the **first principal component** built on particular criteria (2.12) of a posteriori set has, as it is known [Айвазян, Мхитарян, 2001, п. 13.2.5].

But there is another question:

Is it always enough when only one convolution of particular criteria, even if it is best in the sense of property (b), is used for solving the problem of recovering the values of all particular criteria of a posteriori set by using the values of this convolution with satisfactory accuracy, and if not, what is the minimum dimension of the corresponding multicriteria scheme? That is, if using one convolution is not enough then how many of such convolutions should we use and how should we build them?

After all, there are situations where there is no satisfactory solution of the problem of building a scalar indicator of analysed integral category of QOL. These situations happen in particular when considered set of particular criteria has a certain number of mutually weakly correlated variables, although each of them makes a significant contribution to the description and interpretation of the analysed integral category of QOL.

Then the problem of approximation (with reasonable accuracy) of the values of **all** particular criteria while using a **single** scalar value of the indicator may not have a satisfactory solution, even though this scalar indicator is *relatively* the best predictor in this approximation scheme.

And then there is the question of determining the minimum dimension $k < p$ of the multicriteria space within which it is possible to find a satisfactory solution to problems of rating and typology of objects (countries, regions) on the analysed synthetic category. This section of the book offers an approach to finding answers to the above questions.

The overall logical scheme of the proposed approach involves *four main stages* of the research. At the first stage, we solve the question of determining the minimum dimension k of the space of **integral** indicators, sufficient to restore (from the values of these II) with acceptable accuracy the standardized values of all particular criteria of a posteriori set. The criterion of informativeness that is used for that is based on the ideology of factor analysis, in particular, on the known properties of the principal components.

At the second stage (which is necessary only in case of $k > 1$), with the help of special methods of analysing the structure of correlations between particular criteria of a posteriori set, this set is divided into k subsets (blocks) so that each of the resulting subsets of particular criteria serves as an informational base for building of one of k sought integral criteria.

At the third stage, the particular criteria of each of identified at the previous stages subsets fold **linearly** so that when using the obtained liner convolution we can predict reliably (within the linear model of pair-wise regression), the value of each of the particular criteria of the subset.

Finally, the fourth stage (which is the same as the second, is necessary only in case of $k > 1$) is dedicated to building a single (composite) integral indicator in the form of a some **non-linear** convolution of k indicators built at the previous step.

We now describe each of these steps in more detail.

Step 1: **Determining the number of integrated indicators needed to characterize the analysed synthetic category.** First, let's clarify in what sense we understand the accuracy of reconstruction of standardized values of particular criteria $\tilde{x}^{(1)}, \tilde{x}^{(2)}, \dots, \tilde{x}^{(p)}$ of the specified a posteriori set on the values k $(1 \le k < p)$ of auxiliary (integrated) indicators $y^{(1)}, y^{(2)}, \dots \dots, y^{(k)}$.

With this goal in mind let's investigate the linear regression of the type

$$\tilde{x}^{(j)} = b_j(Y) \cdot Y + \varepsilon^{(j)}(Y), \quad j = 1, 2, \dots, p, \tag{2.13}$$

where $Y = (1, y^{(1)}, y^{(2)}, \dots, y^{(k)})^T$, $bj(Y) = (b_{j.0}(Y), b_{j.1}(Y), \dots, b_{j.k}(Y))$ are coefficients of regression $\tilde{x}^{(j)}$ on Y, and $\varepsilon^{(j)}(Y)$ are regression residues in classical linear regression model (2.13).

We are saying that$k$ of integrated indicators $\tilde{Y} = (1; \tilde{y}^{(1)}, \dots, \tilde{y}^{(k)})^{\mathrm{T}}$ provide the best accuracy of reconstruction of the values p of the variables $\tilde{x}^{(1)}, \tilde{x}^{(2)}, \dots, \tilde{x}^{(p)}$ if

$$\sum_{j=1}^{p} \mathrm{E}(\tilde{x}^{(j)} - b_j(\tilde{Y}) \cdot \tilde{Y})^2 = \min_{Y} \sum_{j=1}^{p} \mathrm{E}(\tilde{x}^{(j)} - b_j(Y) \cdot Y)^2.$$

So, it is known [Aivazian, Mkhitaryan, 2001, point 13.2.5] that this exact property is possessed by the first k of principal components built on the variables $\tilde{x}^{(1)}, \tilde{x}^{(2)}, \dots, \tilde{x}^{(p)}$ (i.e. $\tilde{Y}$ is itself the first k of principal components), and the right-hand side (2.14) can be expressed in terms of the ratio $\left(\sum_{j=1}^{k} \lambda_j / \sum_{j=1}^{p} \lambda_j \right)$, where $\lambda_1, \lambda_2, \dots, \lambda_p$ – the eigenvalues of the covariance matrix $\sum_{\tilde{X}}$ of the vector $\tilde{X} = (\tilde{x}^{(1)}, \tilde{x}^{(2)}, \dots, \tilde{x}^{(p)})^{\mathrm{T}}$ positioned in descending order (non-increasing).

Our criterion for determining the minimum number k of integrated indicators that are necessary for the characterization of analysed synthetic categories is based on these very facts. In particular, it is suggested to determine k from the condition:

$$k = \min_{1 \le m \le p-1} \left\{ m : \frac{\lambda_1 + \lambda_2 + \dots + \lambda_m}{\lambda_1 + \lambda_2 + \dots + \lambda_p} \ge q_0 \right\}, \tag{2.15}$$

where q_0 is some number (more than 0.5), the specific value of which depends on the desired accuracy of reconstruction of values $\tilde{X} = (\tilde{x}^{(1)}, \tilde{x}^{(2)}, \ldots, \tilde{x}^{(p)})^T$ based on the values $\tilde{y}^{(1)}, \tilde{y}^{(2)}, \ldots, \tilde{y}^{(k)}$ and on the type of the correlation matrix of vector $\tilde{X}$.

The *empirical* implementation of the rule (2.15) requires:

- *evaluation* $\hat{\Sigma}_{\tilde{X}}$ *of the matrix* $\Sigma_{\tilde{X}}$ *on the observations* (2.12), i.e.

$$\hat{\Sigma}_{\tilde{X}} = \frac{1}{n}\sum_{i=1}^{n}(\tilde{X}_i - \bar{\tilde{X}})(\tilde{X}_i - \bar{\tilde{X}})^T, \text{ where } \bar{\tilde{X}} = \frac{1}{n}\sum_{i=1}^{n}\tilde{X}_i;$$

- *calculation of the eigenvalues* $\lambda_1, \lambda_2, \ldots, \lambda_p$ *of the matrix* $\hat{\Sigma}_{\tilde{X}}$, *i.e. solutions of the equation* $|\hat{\Sigma}_{\tilde{X}} - \lambda I_p| = 0$, where I_p – *the single matrix of dimension p.*

Note that experimentally established acceptable values of the critical constant q_0 ranged from 0.55 to 0.60.

If as a result of the rule (2.15) it was found that $k = 1$, then there is an acceptable scheme of reducing to a **single** integral indicator $\tilde{y}^{(1)}$, and then we can go straight to the final (*in this situation*) Step 3. If $k > 1$, then we should go to step 2.

Step 2: **Breaking the analysed set of particular criteria on** k **disjoint blocks.** As mentioned above, it is the presence in the posterior set of weakly correlated with each other, but significant for the analysed synthetic category of QOL particular criteria that is the reason for not sufficient in formativeness of the *single* integral indicator, built on the ideology of method of principal components.

In contrast, the first principal component will have a high predictive power (in the sense of (2.14)), if *it is built from closely mutually correlated particular criteria.*

This brings us to the following rule of forming blocks into which we should split a posteriori set of particular criteria in the case of a weak performance of a *single* II (i.e. when $k > 1$): *whether a particular criteria belong to the same block shall be determined* by two requirements: they must characterize one aspect of the analysed synthetic category of QOL (for example, real income and expenses in the synthetic category of "welfare", or social pathology in the synthetic category of "quality of social services"), and at the same time have a relatively high level of mutual correlation (usually, but not always, the last property is a consequence of the first requirement).

Splitting a posteriori set of particular criteria into such groups (*blocks*) can be carried out *purely by experts*, i.e. guided by the above-mentioned substantive considerations. However, using special algorithm, namely, *the procedure of extreme grouping of characteristics* [Aivazian, Mkhitaryan, 2001, point 13.4.2] can offer an useful support in solving this problem. This procedure is designed for a partition of the considered set of *p*attributes into a predetermined number $k (k < p)$ of non-overlapping blocks $S_1, S_2, \ldots, S_k$, where features belonging to one block would be relatively strongly correlated, whereas the variables belonging to different blocks would be relatively weakly correlated.

This requirement is formalized in the form of the maximization problem on $S_1, S_2, \ldots, S_k$ and $f^{(1)}, f^{(2)}, \ldots, f^{(k)}$ of criteria

$$F = \sum_{x^{(i)} \in S_1} r^2(x^{(i)}, f^{(1)}) + \sum_{x^{(i)} \in S_2} r^2(x^{(i)}, f^{(2)} + \ldots + \sum_{x^{(i)} \in S_k} r^2(x)^{(i)}, f^{(k)} \tag{2.16}$$

in which $r(\xi, n)$ is understood as an ordinary pair of correlation coefficient between random variables ξиη, and $f^{(l)}$ – the so-called common factor for attributes of the l block, i.e. it is such a variable which maximizes (for a given composition of l block) value $\sum_{x^{(i)} \in S_l} r^2(x^{(i)}, f^{(l)})$.

It can be shown that $f^{(l)}$ is the 1st principal component p_l of characteristics of l block up to a linear transformation. Namely, for a given composition of the block $S_l (l = 1, 2, \ldots, k)$:

$$f^{(l)} = \frac{\sum_{x^{(i)} \in S_l} \alpha_i^{(l)} x^{(i)}}{\sqrt{\sum_{x^{(i)} x^{(j)} \in S_l} \alpha_i^{(l)} \alpha_j^l r(x^{(i)}, x^{(j)}}} \tag{2.17}$$

where $\alpha^{(l)} = (\alpha_1^{(l)}, \alpha_2^{(l)}, \ldots, \alpha_{p_l}^{(l)})^T$ – the eigenvector of the correlation matrix $\mathbf{R}^{(l)}$ of indicators of block S_l corresponding to the maximum eigenvalue of this matrix, i.e. $\alpha^{(l)}$ is a solution of the system of p_l equations

$$\left(\mathbf{R}^{(l)} - \lambda_{\max}^{(l)} \cdot \mathbf{I}\right) \alpha^{(l)} = \mathbf{0}$$

(p_l is the number of indicators included in the block S_l, obviously $f^{(1)}, f^{(2)}, \ldots, f^{(k)}$).

On the other hand, when given the factors $S_1, S_2, \ldots S_k$ it is easy to construct a partition $S_1, S_2, \ldots S_k$ that maximizes the functional (2.16), namely:

$$S_l = \{x^{(i)} : r^2(x^{(i)}, f^{(l)}) \geq r^2(x^{(i)}, f^{(q)}) \text{ for all the } q = 1, 2, \ldots, \bar{m}\}. \tag{2.18}$$

The relations (2.17) and (2.18) are necessary conditions for maximum functional j. Therefore, for the simultaneous determination of the optimal composition of the blocks $S_1, S_2, \ldots S_k$ and for the optimal set of factors $f^{(1)}, f^{(2)}, \ldots, f^{(k)}$ we use the following iterative algorithm: on the υ step of iteration let's build a partitioning of features $x^{(1)}, x^{(2)}, \ldots, x^{(p)}$ into the blocks $S_1^{(\upsilon)}, S_2^{(\upsilon)}, \ldots, S_k^{(\upsilon)}$. Then, for each block of features $S_k^{(\upsilon)}$ we can build factors $f_\upsilon^{(l)}$ according to the formula (2.17), and a new partition $S_1^{(\upsilon+1)}, \ldots, S_k^{(\upsilon+1)}$ in accordance with rule (2.18), i.e. indicator $x^{(i)}$ refers to the block $S_l^{(\upsilon+1)}$, if

$$r^2(x^{(i)}, f_v^{(l)}) \geq r^2(x^{(i)}, f_v^{(q)}) \quad \text{for all the} \quad q = 1, 2, \dots, k.$$

Obviously, at each iteration step the functional does not decrease, so the algorithm will converge to the maximum (although the maximum may be *local*, it is advisable to perform a multiple sweep of the algorithm with different initial conditions).

Step 3: **Building a "block-type" II separately for particular criteria included in each of the blocks defined in the previous step.** So we came to the point where, first, particular criteria included in one block characterize one aspect of the analysed synthetic category of QOL and, second, they are somewhat closely mutually correlated. Now we can form a composition of k-dimensional criterion of the analysed synthetic category (in particular $k = 1$ criterion happen to be one-dimensional, i.e. a scalar one). Let's say

$$\tilde{X}_i(j) = (\tilde{x}_i^{(1)}(j), \tilde{x}_i^{(2)}(j), \dots, \tilde{x}_i^{(p_j)}(j))^T, \tag{2.19}$$

are the values of standardized particular criteria of j-th block ($j = 1, 2, \dots, k$) characterizing the i-th analysed object. So, we'll be building the j-th integral indicator ($j = 1, 2, \dots, k$) in the form of a *unified* (or *modified*, see below) first principal component of variables (2.19). To do this, you must implement the following computational procedure (sequentially for each $j = 1, 2, \dots, k$):

1. Calculate estimates $\hat{\Sigma}_{\tilde{X}}(j)$ of covariance matrices $\hat{\Sigma}_{\tilde{X}}(j)$ of variables (2.19)

$$\hat{\Sigma}_{\tilde{X}}(j) = \frac{1}{n}\sum_{i=1}^{n}\left(\tilde{X}_i(j) - \bar{\tilde{X}}(j)\right)\left(\tilde{X}_i(j) - \bar{\tilde{X}}(j)\right)^T, \quad \text{where} \quad \bar{\tilde{X}}(j) = \frac{1}{n}\sum_{i=1}^{n}\tilde{X}_i(j).$$

2. Determine the eigenvalues $\lambda_1(j) \geq \lambda_2(j) \geq \dots \geq \lambda_{p_j}(j)$ of the matrix $\hat{\Sigma}_{\tilde{X}}(j)$, i.e. solve equations of the form

$$|\hat{\Sigma}_{\tilde{X}}(j) - \lambda \mathbf{I}_{p_j}| = 0.$$

3. Compute the eigenvector vector $C_1(j) = (c_{11}(j), c_{12}(j), \dots, c_{1p_j}(j))^{\mathrm{T}}$ of the matrix $\hat{\Sigma}_{\tilde{X}}(j)$, i.e. solve the following equation

$$\left(\hat{\Sigma}_{\tilde{X}}(j) - \lambda_l \mathbf{I}_{p_j}\right) C_1(j) = \mathbf{0}.$$

4. Build the *first principal component* of the partial criteria $x^{(1)}(j), x^{(2)}(j), \dots, x^{(p_j)}(j)\, j$ of the j-th block:

$$\hat{y}(j) = \sum_{s=1}^{p_j} c_{1s}(j) \cdot (\tilde{x}^{(s)}(j) - \bar{\tilde{x}}^{(s)}(j)), \quad \text{where} \quad \bar{\tilde{x}}^{(s)}(j) = \frac{1}{n}\sum_{i=1}^{n}\tilde{x}_i^{(s)}(j).$$

5. In order to unify the measurement scales (see task 3 in point 2.3.1) construct the II of the variables of the j-th block in the form of:

$$y_i(j) = \sum_{s=1}^{p_j} \tilde{c}_{1s} \cdot \tilde{x}_i^{(s)}(j), \text{ where } \tilde{c}_{1s} = c_{1s} / \sum_{s=1}^{p_j} c_{1s} \tag{2.20a}$$

if all the coefficients $c_{1s}(s = 1, 2, \ldots, p_j)$ *have the same sign* or

$$y_i(j) = \sum_{s=1}^{p_j} c_{1s}^2 \cdot \tilde{x}_i^{(s)}(j), \quad i = 1, 2, \ldots, n \tag{2.20b}$$

if the signs of the coefficients $c_{1s}(s = 1, 2, \ldots, p_j)$ *are different.*

Note. Recall that, by their build, all the particular criteria $x^{(s)}(j)$ are *standardized*, i.e. measured in the same $[0; N]$ scale, where 0 corresponds to the worst quality, and N to best quality. Hence and from the simple analysis of the formulas (2.20a) and (2.20b) we have the following:

(a) II in the form of (2.20a) is also measured in $[0; N]$ scale, and it differs from the first principal component by only a simple linear transformation. So, it does not lose the properties of the highest predictive power with respect to all the particular criteria $\tilde{x}^{(1)}(j), \tilde{x}^{(2)}(j), \ldots, \tilde{x}^{(p_j)}(j)$ (see formulas (2.13) and (2.14)).

(b) *The use of II in the form of* (2.20a) *in the case where the coefficients* c_{1s} *are "with different signs" (although this is very rare in positive mutually correlated particular criteria* (2.19)), **would have contradicted to the natural requirement of monotone non-negative dependence of II on the particular criteria.** *This explains the transition to a form* (2.20b), which, moreover, provides uniformity of the measurement scale $y_i(j)$ (as $\sum_{s=1}^{p_j} c_{1s}^2 = 1$ in accordance with the properties of a component of an eigenvector in Principal Component Analysis).

However, strictly speaking, with that we are losing the property of the highest prognostication force of II in the sense of (2.14). *However, our studies (based on a computational experiment) showed that the predictive accuracy of II in this case is reduced by no more than 10%–12%.*
We'll call the Integral indicator in the form of (2.20a) or (2.20b) *modified* first principal component of the particular criteria.

So, as the "output" of step 3, we have a set of values of block-like integrated indicators (or **multicriteria**)

$$(y_i(1), y_i(2), \ldots, y_i(k)), \quad i = 1, 2, \ldots, n, \tag{2.21}$$

whose components are defined by the formulas (2.20a) or (2.20b). Once again, we note that in the case of $k = 1$ *this multicriteria reduce* to a **scalar one (i.e. one-criterion scheme).**

Step 4: **Building a single (consolidated) II that characterizes the analysed synthetic category QOL.** This step is necessary only in the case of $k > 1$. In essence, we are talking about switching to another, higher level of the hierarchy in the overall scheme of integrated indicators (see Figure 2.1) where the analysis of synthetic category, characterized by a set of II in the form of (2.21) requires us to construct a single (consolidated) measurement tool. This problem may occur during the transition from the 3 rd level of the hierarchical scheme to the 2nd, and from 2nd level to the 1st.

Solution of the problem is based on the following idea. We will use the geometric image of the i-th position of the object being analysed as a point with coordinates (2.21) in the k-dimensional space of integral indicators. Given the fact that all the integral indicators $y(j)(j = 1, 2, \dots, k)$ are measured in N-point scale according to the principle "the more the value of $y(j)$, the higher the quality", it is natural to evaluate the integral level of QOL of the i-th object (O_i) on the analysed synthetic category by the **distance from the point with coordinates (2.21) to the "Standard" (St)**, which is defined by coordinates $(N, N, \dots, N)$. Figure 2.2 shows the scheme for the case $k = 2$.

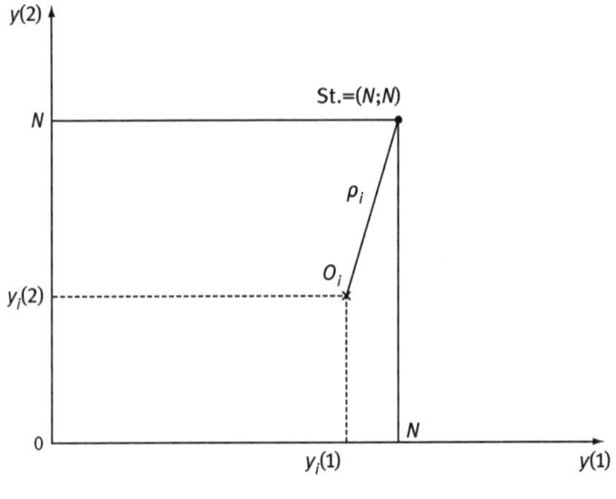

Figure 2.2. Location of the object (territory) O_i with respect to a standard

It remains to answer two questions:
- *how to measure the distance of integral indicators* $y_i(1), y_i(2), \dots, y_i(k)$ *in k-dimensional space*?
- *how to determine the consolidated II – scalar measurement of analysed synthetic category of QOL by the distance* $\rho_i = \rho(O_i; \mathrm{St.})$

To measure distances in space $(y(1), y(2), \dots, y(k))$ it is proposed to use a weighted Euclidean metric in which the weights continue to follow the ideology of principal component analysis (according to which *the higher the variability of the index the higher the informativeness*). We define a proportional spread of data $(\textit{variance}\, s^2(j))$

on each axis $y(j)$ and the number of particular criteria k_j included in the j-th block $(j = 1, 2, \dots, k)$. So:

$$\rho_i^2 = \sum_{j=1}^{k} q(j) \cdot (y_i(j) - N)^2 \tag{2.22}$$

where $q(j) = \frac{p_j \cdot s^2(j)}{\sum_{l=1}^{k} p_l \cdot s^2(l)}$, $s^2(j) = \frac{1}{n} \sum_{i=1}^{n} (y_i(j) - \bar{y}(j))^2$ and $\bar{y}(j) = \frac{1}{n} \sum_{i=1}^{n} y_i(j)$.

Now we can determine the value of y_i – a single (consolidated) integral indicator of the analysed synthetic category for the object (territory) O_i:

$$y_i = N - \rho_i \tag{2.23}$$

Clearly defined this way consolidated II QOL is no longer a **linear** convolution of particular criteria and will be measured in the same $[0; N]$ standardized scale as all other particular criteria and integrated indicators.

Note that the use of same *technique of principal component analysis* that was used in the construction of each block-like II in the task of building of consolidated II is usually not successful because of the "multi-directional" action (and therefore low mutual correlation) of block-like integral indicators.

2.3.7. Analysis of capacity and dynamics of integral indicators of quality of life

Analysis of the capacity of II QOL – it is a matter primarily of following the same quality criteria that apply to the integral indicators while they are being constructed. Ten of these criteria have been discussed in the first chapter of the book (see point 1.4.1). If, in addition we have some form of learning (point 2.3.2), then the capacity of built II QOL can be further evaluated (if it is a quality learning) by comparison of expert evaluations and assessments based on II. At the same time, depending on the form of learning, we can use the coefficients of common pairwise or rank correlation, the percentage of incorrectly (relative to the expert opinion) rated territories, etc.

Analysis of the dynamics of integrated indicators of QOL is essentially used in the evaluation of the effectiveness of social and economic policy and to identify its "problem areas". It requires the introduction and computation of special characteristics that would allow to track improvement or deterioration in the quality of life aspects of the considered territory, both in relation to itself in the previous time cycle (we'll call it **auto-dynamics**) and in relation to its position among the other territorial units (**inter-regional** dynamics).

Speaking about *auto-dynamics*, we note that a simple increment in time even of all particular criteria $\tilde{x}^{(j)}$ of the a posteriori set may not actually mean that the quality of life in the region has improved. That, by the way, can also be confirmed by the corresponding value of II QOL. Due to the fact that these are the *standardized* values $\tilde{x}^{(j)}$ of the particular criteria that participate in II convolution, and because of the simultaneous dynamics of reference regions (i.e. values $x_{\min}$ *and* $x_{\max}$) those standardized values may even decrease with increasing[4] $x^{(j)}$.

Because II as a function of standardized values of particular criteria takes this effect into account, its value can be used to track auto-dynamics.

When measuring the *inter-territorial dynamics* of each individual (i-th) area it is only natural to focus on the dynamics of its position (rank) in a number of other areas under consideration, i.e. by the amount

$$\delta_i(t) = R(y_i(t-1)) - R(y_i(t))$$

where $R(y_i(t))$ _ the rank of the i-th area in the rating of the territories, built in accordance with the values $y_1(t), y_2(t), \dots, y_n(t)$. Obviously, positive values $\delta_i(t)$ are indicative of a positive inter-territorial dynamics of the region i.

2.4. Examples of building integrated indicators of quality of life in cross-country and inter-country analysis

The following examples of building II QOL, in contrast to the results of the empirical analysis of QOL that will be presented in the next chapter, are intended mostly for *purely illustrative purposes*. In particular, we'll use below the statistics by country (point 2.4.1) and by the subjects of the Russian Federation (point 2.4.2) to provide the concrete illustrations of solving the basic problems of methodology for measuring the synthetic categories of quality of life of the population (point 2.3.1.), namely:

- *problem of a selection from a priori set of particular criteria* $x^{(1)}, x^{(2)}, \dots, x^{(p)}$ *of reduced* **(a posteriori)** *set of indicators* $x^{(j1)}, x^{(j2)}, \dots, x^{(jp')}$, *where* $p' < p$ **(objective 2 of preanalysis);**
- *problems of unification (standardizing) of measuring scales of all analysed variables, i.e. such a transition to the N-point scale, where the value of "zero" would have indicated the worst quality as for the analysed variable, and the value of N is the best* **(objective 3 preanalysis);**

4 Indeed, suppose that the value of the particular criterion $x^{(j)}$ has increased over the year by Δ, and the values $x_{\min}$ *and* $x_{\max}$ for the same year increased by 2Δ. Then a unified value $\tilde{x}^{(j)}$ for the same year changed to a value $\tilde{x}^{(j)} = \frac{(x^{(j)}+\Delta)-(x_{\min}+2\Delta)}{(x^{(j)}_{\max}+2\Delta)-(x^{(j)}_{\min}+2\Delta)} \cdot N = \frac{x^{(j)}-\Delta-x^{(j)}_{\min}}{x^{(j)}_{\max}-x^{(j)}_{\min}} \cdot N < \frac{x^{(j)}-x^{(j)}_{\min}}{x^{(j)}_{\max}-x^{(j)}_{\min}} \cdot N.$

- *problems of transition from p'particular criteria $x^{(j1)}, x^{(j2)}, \dots, x^{(jp')}$ to a single one (y) or relatively small number$(y^{(1)}, \dots, y^{(k)})$ of integrated indicators $(k << p')$, which are constructed in the form of some functions (convolutions) of p' particular criteria of the a posteriori set* **(objective 4 – the central objective of the methodology);**
- *finally, the objective of* **post-analysis** *that analyse the performance of the built II QOL and their dynamics* **(objectives 5 and 6).**

With regard to the **substantive** goals of the empirical studies of QOL, in this section, they are limited to the **rating** of the territory in question (countries, subjects of the Russian Federation) on the analysed synthetic categories of QOL.

2.4.1 Inter-country empirical analysis of the synthetic category of "quality of life of population" based on 2009 data

Thus, **the aim of our study** is the ratings of a number of statistically surveyed countries by the synthetic category of "quality of life of population" at the highest level of generality (see *integral characteristic of the highest level* in Figure 2.1). At the same time our rating will be based on values of II QOL, built in the form of a linear convolution of linear particular criteria simultaneously in two versions: with "learning" and without it.

Informational base of the research. Baseline statistics are taken from the yearbook [WCY, 1997-2011]. These annuals include macroeconomic data for a number of countries (the total number of statistically surveyed countries as well as the quantity and composition of indicators undergo some changes over time), structured in blocks and sections, and contain the values of approximately 300 indicators, including *more than 100 indicators that were evaluated by experts.* Appendix 2.1 provides a brief description of the structure and content of the information included in the annual *World Competitiveness Yearbook.*

A priori set of particular criteria of quality of life of the population of the country was formed with the requirements of (A), (B), (C) (see paragraph 2.2) based on the source [WCY, 2009] and contained statistical indicators that characterize all four basic synthetic categories: "***the quality of the population***" (*predicted at birth life expectancy* – a variable $x^{(7)}$ or 4.4.05 encoded [WCY, 2009]; *illiteracy among the population over 15 years old* – $x^{(4)}$ or 4.5.14; *the infant mortality rate* – $x^{(8)}$ or 04.04.07; "***the level of well-being***" (*GDP per capita in USD while taking into account the purchasing power parity* (PPP) – a variable $x^{(1)}$ or encoded 1.1.22 [WCY, 2009]; *private consumption expenditure per capita* in the United States while taking into the account PPP – $x^{(3)}$ or 1.1.23; *consumer price index (CPI)* in percentage – $x^{(6)}$ or 1.5.01; *productivity* in the United States – $x^{(2)}$ or 3.1.04); "***the quality of social services***" (20% *rate funds*, i.e. the ratio of the total income of the richest 20% of the population to the total income of the poorest 20% of the population – a variable $x^{(5)}$, or the ratio of variable 5.2.07 to variable 2.5.06 encoded [WCY, 2009]; the total expenditure on R & D as % of GDP – $x^{(10)}$ or 4.3.02) and ***"quality of the ecological niche"*** (*total CO_2 emissions* in metric tons per 1 million dollars. GDP – variable $x^{(9)}$ or 4.4.17 encoded [WCY, 2009]). List and indexation of particular criteria of a priori set is given in Appendix 2.2. Limiting the a priori set of particular criteria *by only ten indicators* is dictated by considerations of the optimal ratio of n statistically inspected objects (countries) to the number of p variables involved in the model of II QOL, measured either (in the presence of learning) *by linear regression of expert assessment of QOL* $y_{i_{St.}}$ on some subset of particular criteria $x_i^{(j1)}, x_i^{(j2)}, \dots, x_i^{(jp')}$ or in the form of a modified 1st principal component of all ten above-mentioned particular criteria $x^{(1)}, x^{(2)}, \dots, x^{(10)}$ (when building II QOL without learning). It should be noted that, in accordance with the recommendations from point 2.3.3 we did *standardize (unify) the measurement scales* of analysed variables.

It was carried by the formulas (2.2)–(2.4) with $N = 10$, i.e. all variables are measured on a *10-point scale* in such a way that the 0 value of unified indicator shows the worst quality, and a rating of 10 points shows the best quality. Parameters necessary for unification are given in Appendix 2.3b, and standardized values of the analysed variables in Appendix 2.3c.

Building II QOL in the presence of learning. In this case, the II QOL $\hat{y}_i$ of the country i is constructed (see 2.3.5 point 1) based on the initial data of the following type

$$\left(x_i^{(j_1)}, x_i^{(j_2)}, \dots, x_i^{(j_{p'})}; y_{i_{St.}}\right), \quad i = 1,\dots,56$$

in the form of estimation $\hat{y}_i$ of linear regression $y_{i_{St.}}$ on $(x_i^{(j1)}, \dots, x_i^{(jp')})$ in equation

$$y_{i_{St.}} = \theta_0 + \theta_1 x_i^{(j_1)} + \dots + \theta_{p'} \cdot x_i^{(j_{p'})} + \varepsilon_i, \quad i = 1, 2\dots,56$$

where

$$y_i = \hat{\theta}_0 + \hat{\theta}_1 x_i^{(j1)} + \dots \hat{\theta}_{p'} \cdot x_i^{(jp')},$$

$y_{i_{St.}}$ – expert estimation (10-point scale) of QOL of *i-th* country, $(x_i^{(j1)}, \dots, x_i^{(jp')})$ – values of p' of explanatory variables (particular criteria) of the a posteriori set of the same *i-th* country $(p' < p)$, ε_i – random regression remains of the regression model, and $\hat{\theta}_0 + \hat{\theta}_1, \dots, \hat{\theta}_{p'}$ _ estimates of the regression coefficients. The method of estimating the coefficients $\hat{\theta}_p$ depends on the nature of the regression residuals ε_i. In our case, the classical assumptions about the probabilistic nature of residues $\varepsilon_1, \varepsilon_2, \dots, \varepsilon_{56}$ (their mutual statistical independence, homoscedasticity and normal distribution pattern) were not denied as a result of statistical testing of relevant hypotheses, and thus to estimate the parameters $\theta_j (j = 0, 1, \dots, p')$ we used the ordinary method of the least squares.

A posteriori set of particular criteria of QOL and baseline statistics. Formation of the reduced (a posteriori) set of particular criteria for synthetic category of higher level of generality of "quality of life" was aimed at addressing the a *high degree of multicollinearity* (see characteristics of multicollinearity in Tables A2.4a and A2.4c of the Appendix A2.4) that is present in the explanatory variables (particular criteria) $\tilde{x}^{(1)}, \tilde{x}^{(2)}, \ldots, \tilde{x}^{(10)}$. For this purpose, we applied the *method of selecting the most significant particular criteria in the presence of learning* described in point 2.3.4. In particular, based on the extracted from [WCY, 2009] raw data (refer to Appendix A2.2)

$$\left(x_i^{(1)}, x_i^{(2)}, \ldots, x_i^{(16)}; y_{i\text{St}}\right), \quad i = 1, \ldots, 56^5.$$

We have implemented step-by-step procedures for the selection of the most informative explanatory variables in the linear regression model, namely the *procedure of successive addition and successive removal* (they are described above in paragraph 2.3.4 (a)).

Development of the procedure of successive addition is reflected in Table A2.4d of Appendix A2.4. We should add to this that the procedure for sequential removal applied to the same set of predictors $\tilde{x}^{(3)}$–$\tilde{x}^{(10)}$ gave exactly the same result.

So, why did we exclude variables $\tilde{x}^{(1)}$ (GDP per capita) and $\tilde{x}^{(2)}$ (productivity) from the a priori set of particular criteria? This was done for two reasons.

The first reason. These variables are highly correlated with each other (the correlation coefficient $r(\tilde{x}^{(1)}, \tilde{x}^{(2)}) = 0,93$, see Table A2.4a), and each of them is closely correlated with variable $\tilde{x}^{(3)}$ – personal consumption expenditures ($r(\tilde{x}^{(1)}, \tilde{x}^{(3)}) = 0,83$; $r(\tilde{x}^{(2)}, \tilde{x}^{(3)}) = 0,89$). Therefore, simultaneous inclusion of all three variables (or even just the two of them) as predictors into the analysed regression model would result in poor interpretability of results, giving only a minimal gain to the value of the coefficient of determination of the dependent variable $y_{iэ}$ across the *whole* set of particular criteria (the increment R^2 was of the order of 0.045, which *cannot be described as statistically significant* when the number of observations is $n = 56$ and the number of estimated parameters equals to $p + 1 = 11$.

The second reason. Let us ask ourselves: *is gross domestic income of the country as undeniably important as a characteristic of the quality of life of its population?* A number of studies [Сен, 2004; Aivazian, 2008] are convincing examples of situations in which GDP (Gross Domestic product) could not qualify as a determinant of the quality of population. In this regard, I think it is appropriate to cite here some

5 The object of the analysis are 56 countries out of 60 countries and regions included in the yearbook "The World Competitiveness Yearbook", Edition IMD-International, Lausanne, Switzerland, 2009 (see Appendix 2.3). A number of countries and regions were not included in our study either because of omissions in the original data or due to the fact that they did not represent the country as a whole.

excerpts from the report of the authoritative international commission to assess economic performance and social progress, led by Nobel laureate *Joseph Stiglitz* and *Amartya Sen* [Report, 2010–2011]:

1. **(From the introduction to Chapter 1)** "It (GDP – S.A.) is mainly measures total supply, though it has often been treated as if it were a measure of economic well-being. Although the GDP value is linked with many indicators of living standards, the correlation is less comprehensive and much weaker when it comes to certain sectors of the economy. For example, the change in real income of households (income measures that are most closely related to quality of life) in a number of OECD countries is very different from GDP growth. Too much emphasis on GDP as a benchmark measure can lead to misconceptions about the welfare of the people and increase the risk of erroneous political decisions".
2. **(From p. 4 of Chapter 1)** "Most of the public debate about the level of life focused on the performance of the economy as a whole, and more often on GDP. But the ultimate goal of the discussions about the level of life is to look at individuals whose economic situation needs to be evaluated. A look at the evolution of real household income and the change in GDP confirms that, in general, these figures are not interchangeable. Despite the fact that in some countries the real disposable income of households is well correlated with the growth of GDP, there are many countries where the situation is different. Italy, Japan, Korea, Poland, Slovakia and Germany – these are just some of them".

Add to this an interesting (and highly relevant for the modern Russian society) statement of the renowned expert in the field of international economic crime, an economist and a successful entrepreneur John Perkins [Перкинс, 2005, pp. 17–18]: "The current system is governed by indisputable as Gospel belief that economic growth benefits all humanity, and that the higher the growth, the more the people benefit.... Of course, this concept is wrong. We know that in many countries economic growth benefits only a small part of the population, with the majority of people being put in an extremely difficult situation ... When greed is encouraged in people, greed becomes the main motive of decomposition of the society. When we condone reckless devouring of the planet's resources; when we teach our children to emulate people who lead an unbalanced lifestyle; when we think that large groups of the world's population should be taken advantage off by elite minority, – we are asking for troubles ourselves. And we get them."

Thus, in constructing II QOL of the population in a situation where we can use the expert assessment$y_{i_{\text{St.}}}$ of QOL for each of the analysed countries in the presence of learning, we come to the task of estimating the linear regression $y_{i_{\text{St.}}}$ by:

- *personal consumption expenditure per capita*$(\tilde{x}^{(3)})$;
- *predicted (at birth) life expectancy*$(\tilde{x}^{(7)})$;
- *total expenditure on R & D.*

The estimation results of the regression model: the value of regression coefficients $\hat{\theta}_0, \hat{\theta}_1, \hat{\theta}_2$ и $\hat{\theta}_3$, standard errors $s_{\hat{\theta}_j}$ in the estimation of these coefficients, the values of t-statistics ($t_j, -j = 0, 1, 2, 3, 4$), probability P_j of statistically insignificance of j-th predictor, the doctored coefficient of determination (R^2_{adj}) and Durbin-Watson statistic (DW) – all are shown in Table 2.1.

Table 2.1. Results of statistical analysis of regression

$y_{i_{St.}}$ by $(\tilde{x}^{(3)}, \tilde{x}^{(7)}, \tilde{x}^{(10)})$				$R_{adj}{}^2 = 0{,}675$; DW = 2,14	
Number j	**Predictor (particular criterion)**	**Estimate of regression coefficient $\hat{\theta}_j$**	**Standard error of the estimate $s_{\hat{\theta}_j}$**	**t-statistics t_j**	**Values P_j**
0	Free member	2,418	0,584	4,142	0,00013
1	Personal consumption $(\tilde{x}^{(3)})$	0,228	0,084	2,720	0,00886
2	Life expectancy $(\tilde{x}^{(7)})$	0,308	0,105	2,925	0,00510
3	Research and development expenses $(\tilde{x}^{(10)})$	0,216	0,074	2,911	0,00529

Thus, the value of II QOL of the country $i(\hat{y}_i(\text{lear.}))$ constructed and calculated with the use of learning takes the form:

$$\hat{y}_i(\text{lear.}) = 2,418 + 0,228\tilde{x}_i^{(3)} + 0,308\tilde{x}_i^{(7)} + 0,216\tilde{x}_i^{(10)}$$

Appendix 2.5 shows the values $\hat{y}_i(\text{lear.})$ for all the analysed countries (i.e. for $i = 1, 2, \ldots, 56$), corresponding ratings of these countries defined by the values $\hat{y}_i(\text{lear.})$, the measure of consistency (Spearman coefficient of rank correlation $\rho(\hat{y}(\text{lear.}); y_{\text{St.}})$), and countries' rating by $\hat{y}_i(\text{lear.})$ and by $y_{i_{\text{St.}}}$. Note that the value of the last characteristics $\rho(\hat{y}(\text{lear.}); y_{\text{St.}}) = 0,855$ indicates good consistency of ranking the countries according to the values $\hat{y}_i(\text{lear.})$ with expert ranking, and hence, a high informativeness of the three selected particular criteria ($\tilde{x}^{(3)}, \tilde{x}^{(7)}$ and $\tilde{x}^{(10)}$) in the problem of formalization of systems of criteria of experts when ranking countries on QOL!

Building II QOL in the absence of learning. Getting from experts in one form or another, assessments of the analysed synthetic latent categories for each of the objects (countries, regions, conglomerate population) *is expensive and not always practically feasible*. This leads to the relevance of the task of building II QOL only *by obtaining the values of several private QOL criteria*, i.e. **in the absence of learning.** Here's an example of solving this problem by using the same data that is listed in Appendix 2.3, but without the information on values $y_{1_{\text{St.}}}, y_{2_{\text{St.}}}, \ldots, y_{56_{\text{St.}}}$. In this case, we are guided by the methodology and recommendations set forth in point 2.3.6 of this monograph.

Given the fact that the a priori set of particular criteria itself is small (it contains a total of 10 indicators, the reasons were mentioned above), we did not reduce it when converting to the a posteriori set (as it was done in the construction of $\hat{y}(\text{lear.})$ within the appropriate regression model in order to eliminate multicollinearity).

The results of analysis of principal components of a priori set of particular criteria of QOL are listed in Tables 2.2 and 2.3.

Table 2.2. The eigenvalues of the covariance matrix of standardized indicators $\tilde{x}^{(1)}$–$\tilde{x}^{(10)}$ and the percentage of their total variation explained by the first *j* principal components

Number of the main component *j*	Eigen values λ_j	% of variation explained by *j*-th principal component	Accumulated % of explained variation
1	43,38	61,7	61,70
2	10,09	14,4	76,06
3	4,61	6,6	82,62
4	3,58	5,1	87,71
5	3,42	4,9	92,58
6	2,33	3,3	95,90
7	1,20	1,7	97,60
8	0,80	1,1	98,74
9	0,55	0,8	99,53
10	0,33	0,5	100,00

The results presented in Table 2.2, indicate: in the task of cross-country analysis of QOL based on data [WCY, 2009] the building of II QOL *in the absence of learning* **it's ok to use only one integral indicator** (see rule (2.15) in point 2.3.6 at the threshold value of $q_0 = 0,55$).

Since all components of the 1st eigenvector happen to be the same sign, the normalization of the coefficients c_j was made by the formula (2.20a), bearing in mind that in this case we do not divide our set of particular criteria into blocks.

Thus, the value of II QOL of the country i ($\hat{y}_i(w.l.)$), constructed and calculated *in the absence of learning* takes the form of so-called modified *first principal component,* namely:

$$\hat{y}_i(\text{w.l.}) = \sum_{j=1}^{10} w_j \tilde{x}_i^{(j)}$$

where the values of coefficients $w_1, w_2, \dots, w_{10}$ are presented in the last column of the Table 2.3.

Table 2.3. The first eigenvector of the covariance matrix of standardized indicators $\tilde{x}^{(1)}$–$\tilde{x}^{(10)}$ and the weighting coefficients $w_j (j = 1, 2, \dots, 10)$ for II QOL

Number of particular criteria (j)	Particular criteria, $\tilde{x}^{(j)}$	Components of the 1st eigenvector, c_j	Normalized components of the 1st eigenvector w_j
1	GDP per capita, $\tilde{x}^{(1)}$	0,356	0,114
2	Productivity, $\tilde{x}^{(2)}$	0,373	0,120
3	Private consumption, $\tilde{x}^{(3)}$	0,400	0,128
4	How many illiterate, $\tilde{x}^{(4)}$	0,315	0,101
5	Coefficient of the funds, $\tilde{x}^{(5)}$	0,240	0,077
6	The Consumer Price Index, $\tilde{x}^{(6)}$	0,257	0,083
7	Life expectancy, $\tilde{x}^{(7)}$	0,284	0,092
8	Infant mortality, $\tilde{x}^{(8)}$	0,361	0,116
9	Emissions of CO_2, $\tilde{x}^{(9)}$	0,232	0,074
10	Expenditure on R & D, $\tilde{x}^{(10)}$	0,293	0,095

Appendix 2.5 shows the values $\hat{y}_i(\text{w.l.})$ for all the analysed countries (i.e. for $i = 1, 2, \dots, 56$), the respective ratings of these countries defined by the values $\hat{y}_i(\text{w.l.})$, and the measure of consistency (Spearman's coefficient of rank correlation $\rho(\hat{y}_i(\text{w.l.}); y_{\text{St.}}))$ of the rating of countries and on $\hat{y}_i(\text{w.l.})$ and on $y_{i_{\text{St.}}}$. Note that the value of the last characteristics $(\rho(\hat{y}(\text{w.l.}); y_{\text{St.}}) = 0,852)$ shows that the integral indicator $\hat{y}_i(\text{w.l.})$, built *without learning* using ten particular QOL criteria, gives almost as good agreement with the expert ranking of the countries as integral indicator $\hat{y}(\text{w.l.})$, built *in the presence of "learning"* using the most informative three particular criteria of QOL.

Thus the analysis of capacity of the built II QOL (with and without learning) is based on the comparison and evaluation of the consistency of rankings obtained on the basis of expert assessments of quality of life, on one hand, and on the basis of the values of II QOL, on the other.

2.4.2 Inter-regional comparative analysis of latent synthetic categories qol of the population of the subjects of the russian federation according to 2003 data

In this example, **the purpose of empirical research** is a comparative analysis of 78 subjects of the Russian Federation on four basic synthetic categories – "quality of the population (QOP)", "the level of well-being (LWB)", "quality of social services (QOSS)" and "the quality of the ecological niche (QOEN)"– as well as a synthetic category of the highest level of generality "quality of life" on the basis of official statistics from 2003. To achieve the above goal we used *the technique of building II QOL in the absence of learning*, as described in point 2.3.6.

Informational base of the research was formed by sources [I 1], [I 3–I 7] and [I 12 and I 13], the list of which is given in References in the part "Information Materials".[6] The time cycle adopted in it is one year. Econometric analysis described below used the data from 2003.

The a priori set of particular criteria for each of the four basic latent synthetic categories of QOL formed with the requirements (A)–(B)–(C) (see Paragraph 2.2) on the basis of the above sources. A list and indexing of particular criteria of a priori set are given in Appendix 2.6.

A posteriori sets of particular criteria and baseline statistics.

For the formation of the reduced (a posteriori) set of particular criteria for each of the analysed synthetic categories of QOL (QOP, LWB, QOSS and QOEN) we used the technique of selecting the most significant values *in the absence of learning*, as described in point 2.3.4. Selected in accordance with this procedure, the variables are indexed in the penultimate (5th) column of the table of Appendix 2.6.

Thus, out of the 40 particular criteria of a priori set of synthetic category of "Quality of Population" only 14 criteria were selected for the posteriori set, out of the 15 particular criteria of "QWB" 10 were selected, out of the 33 particular criteria of "QOSS" 23 got into the posteriori set, and finally, out of the 27 particular criteria of "QOEN" only 9 got included.

Initial statistics on all the particular criteria (as of 2003) for the 78 subjects of the Russian Federation are given in Appendix 2.7.

Unification of measuring scales of the analysed variables is done by the formulas (2.2)–(2.4) for $N = 10$, i.e. there happens a transition to a 10-point scale measurements, in which a zero value of the unified index indicates the worst quality, and ten – the best. Since in this case we do not have the learning (which is, unfortunately, a typical situation for inter-regional analysis), there are problems with the

6 The database was developed and has been maintained (since 1997 to present) by the staff of the Laboratory of Probabilistic and Statistical Methods and Models in Economics of the Central Economics and Mathematics Institute of Russian Academy of Sciences.

definition of the optimal values ($x_{\text{opt}}^{(j)}$) of the parameters *that are related to the quality of life by* **non-monotonic** *dependence* (and we need these optimal values to implement the unifying transformation (2.4)).

However, taking into account the specifics of the modern Russia, namely, the fact that the range of the *actual* values of these indicators is far from the respective optimum values, the formula (2.4) without any harm to the case can be replaced by the best variant of "monotonous formula" (i.e. by formula (2.2) or (2.3)). In particular, these very considerations were taken into account when performing the unification of the measuring scales of the variables: coefficient of natural increase, migration, unemployment, the total amount of social payments, the area of protected nature reserves.

At the same time, the analysis of international experience allowed us to determine the optimal (for the moment for Russia) coefficient of funds equal to 9.0. Number of the formula of the unifying conversion applied to each of the variables of a posteriori set is specified in the last (6th) column of the table of Appendix 2.6.

The results of principal component analysis of particular criteria of a posteriori set for each of the four analysed synthetic categories (QOP, QWB, QOSS and QOEN). As described in point 2.3.6 (see step 1), the analysis focuses on the answers to two questions:

1. *What is the number* $k(l)$ of *integrated indicators required for the characterization of the analysed (l-th) synthetic category ?*
2. *If it can be limited to* **one** integrated indicator (i.e. if $k(l) = 1$), *what is the procedure of measuring it on the basis of the values of particular criteria of a posteriori set of the 1st synthetic category?*

The answer to the first question is based on the rule (2.15). In particular, if we label

$$I_m(l) = \frac{\lambda_1(l) + \dots + \lambda_m(l)}{\lambda_1(l) + \dots + \lambda_{p(l)}(l)} \cdot 100$$

the accumulated share (in percentage) of the total variation of the particular criteria of a posteriori set of l-th synthetic category, which was explained by the first m of principal components of these particular criteria ($m = 1, 2, \dots, p(l)$); then the proposed *formalized* rule of determining the required number of integrated indicators $k(l)$ would be based on the requirement

$$k(l) = \min_{1 \le m \le p(l)-1} \{m : I_m(l) \ge q_0(l)\}, \quad (l = I, II, III, IV) \tag{2.15'}$$

where p *(l)* – the total number of particular criteria in the a posteriori set of the lth synthetic category, and threshold value $q_0(l)$ is selected from a range of values (0.5; 0.9) while taking into account both the correlation matrix of the particular criteria of the a posteriori set and the required reconstruction accuracy of the values of these particular criteria for the desired values of II QOL.

In this case, it has been selected for all the synthetic categories $q(l) = 55\%(l = I,II,III,IV)$.

Table 2.4 presents the results of calculations of values $I_m(l)$. It leads us in particular to the fact that for the characterization of "*quality of population*" we need to build **three II QOL**; for the characterization of "*well-being*", **two**; for the "*the quality of social services*", **four**; and finally, for the "*quality of the ecological niche*", **four** (the corresponding hierarchical scheme "Consolidated II QOL – II of the major synthetic categories of QOL – divided in blocks II – basic statistical integrators" is shown in Figure 2.3).

Table 2.4. Accumulated share $I_m(l)$ (in percentage) of the total dispersion of particular criteria of a posteriori set of the l-th synthetic category, explained by the first m principal components

Number of synthetic category l	Synthetic category QOL	Number of principal components, m				Required number $k(l)$ of integrated indicators
		1	2	3	4	
I	Quality of the population (QOP)	31,0	53,4	63,8		3
II	Level of well-being (LWB)	38,8	57,5			2
III	The quality of social services (QOSS)	18,5	35,9	48,9	56,6	4
IV	Quality of an ecological niche (QOEN)	32,3	48,0	53,2	55,8	4

The answer to the second question when $k(l) = 1$ is given in formulas (2.20), in which, respectively $p_j(l) = p(l)$ – the total number of particular criteria of the analysed (l-th) synthetic category. However, in our case all $k(l) > 1$, so it is necessary to move to dividing each a posteriori set of particular criteria into blocks and to building block-like II QOL (see steps 2 and 3 of the procedure described in point 2.3.6).

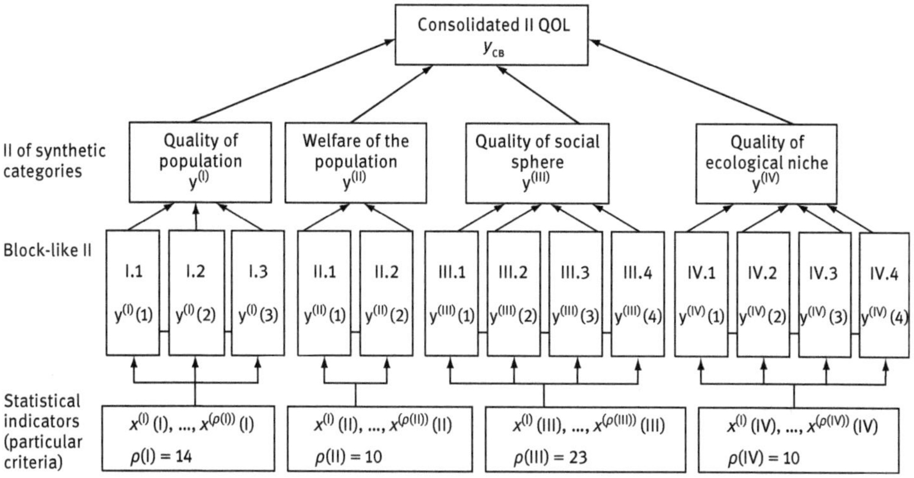

Figure 2.3. Hierarchical scheme of building the integral indicators of QOL in inter-regional analysis

Table 2.5. Block-like particular criteria $(\tilde{x}^{(1)}(j), (\tilde{x}^{(2)}(j), ..., (\tilde{x}^{(p_j)}(j))$ and their weighting coefficients $w_1(j), w_2(j), ..., w_{p_j}(j)$ in block-like II QOL of the synthetic category of "quality of the population"

Number of the p block (j)	Variables of the block (in brackets – corresponding weighting coefficients $w_s(j)$)						
	$\tilde{x}^{(1)}(j)$	$\tilde{x}^{(2)}(j)$	$\tilde{x}^{(3)}(j)$	$\tilde{x}^{(4)}(j)$	$\tilde{x}^{(5)}(j)$	$\tilde{x}^{(6)}(j)$	$\tilde{x}^{(7)}(j)$
1	Life expectancy,	Infant mortality,	Mortality from infectious diseases,	The mortality from respiratory diseases,	The mortality from diseases of digestion,	The mortality from accidents,	Birth defects,
	$w_1(1) = 0,178$	$w_2(1) = 0,130$	$w_3(1) = 0,133$	$w_4(1) = 0,155$	$w_5(1) = 0,150$	$w_6(1) = 0,198$	$w_7(1) = 0,056$
2	Natural increase,	Mortality from cancer,	Mortality from cardiovascular diseases,	Disability,			
	$w_1(2) = 0,262$	$w_2(2) = 0,238$	$w_3(2) = 0,289$	$w_4(2) = 0,211$			
3	Higher education among the employed,	Coverage of youth with education,	Number of the p block *,				
	$w_1(3) = 0,437$	$w_2(3) = 0,162$	$w_3(3) = 0,401$				

* Characterizes the so-called deduced productivity in the region, i.e. calculated for the unified (for all regions) standardized structure of employment.

Table 2.6. Block-like particular criteria $(\tilde{x}^{(1)}(j), (\tilde{x}^{(2)}(j), \ldots, (\tilde{x}^{(p_j)}(j))$ and their weighting coefficients $w_1(j), w_2(j), \ldots, w_{p_j}(j)$ in block-like II QOL of the synthetic category "level of well-being"

Number of the block, *j*	Variables of the block (in brackets – corresponding weighting coefficient) $w_s(j)$)					
	$\tilde{x}^{(1)}(j)$	$\tilde{x}^{(2)}(j)$	$\tilde{x}^{(3)}(j)$	$\tilde{x}^{(4)}(j)$	$\tilde{x}^{(5)}(j)$	$\tilde{x}^{(6)}(j)$
1	GRP per capita, $w_1(1) = 0,093$	The purchasing power of per capita income, $w_2(1) = 0,226$	Share (how many) of the poor, $w_3(1) = 0,360$	Coefficient of fund, $w_4(1) = 0,074$	Automobiles provision, $w_5(1) = 0,112$	Commodity turnover, $w_6(1) = 0,135$
2	Housing provision, $w_1(2) = 0,394$	Housing construction, $w_2(2) = 0,370$	The density of roads, $w_3(2) = 0,128$	Old and dilapidated housing, $w_4(2) = 0,108$		

Table 2.7. Block-like particular criteria $(\tilde{x}^{(1)}(j), (\tilde{x}^{(2)}(j), \dots, (\tilde{x}^{(p_j)}(j))$ and their weighting coefficients $w_1(j), w_2(j), \dots, w_{p_j}(j)$ in block-like II QOL synthetic category of "quality of social sphere"

Number of the block, j	The variables of the block and corresponding weighting coefficients $w_s(j)$									
	$\tilde{x}^{(1)}(j)$	$\tilde{x}^{(2)}(j)$	$\tilde{x}^{(3)}(j)$	$\tilde{x}^{(4)}(j)$	$\tilde{x}^{(5)}(j)$	$\tilde{x}^{(6)}(j)$	$\tilde{x}^{(7)}(j)$	$\tilde{x}^{(8)}(j)$	$\tilde{x}^{(9)}(j)$	$\tilde{10}^{(4)}(j)$
1	Socially significant diseases,	Harmful and dangerous working conditions,	Occupational injuries,	Murders and attempted murders,	Grievous bodily harm,	Rape and attempted rape,	Robbery, theft,	Embezzlement, financial fraud,	Alcoholism,	Suicide,
	$w_1(1) = 0,08$	$w_2(1) = 0,14$	$w_3(1) = 0,11$	$w_4(1) = 0,11$	$w_5(1) = 0,08$	$w_6(1) = 0,07$	$w_7(1) = 1,16$	$w_8(1) = 0,06$	$w_9(1) = 0,04$	$w_{10}(1) = 0,15$
2	Migration,	Depth of poverty,	Unemployment,	Number of unemployed per vacancy,	Long-term unemployment,	Wages not paid,				
	$w_1(2) = 0,06$	$w_2(2) = 0,22$	$w_3(2) = 0,23$	$w_4(2) = 0,15$	$w_5(2) = 0,23$	$w_6(1) = 0,12$				
3	The purchasing power of pensions,	Social benefits,	Social care at home,	Service at welfare centres,						
	$w_1(3) = 0,82$	$w_2(3) = 0,11$	$w_3(3) = 0,01$	$w_4(3) = 0,06$						
4	Funds' coefficient,	Addiction and Substance Abuse,	HIV,							
	$w_1(4) = 0,064$	$w_2(4) = 0,43$	$w_3(4) = 0,50$							

Partitioning of a posteriori set of particular criteria into the blocks and the construction of unit type (block-like) II QOL

Specifying the number of units$k(l)$ for each (l-th) synthetic category and using the results of *the procedure of extreme grouping of features* (see 2.16–2.18 in point 2.3.6) as the main "hint" for the selection of the final variant of partitioning the particular criteria of a posteriori set into blocks, we arrive at the results presented in Tables 2.5–2.8.

Table 2.8. Block-like particular criteria $(\tilde{x}^{(1)}(j), (\tilde{x}^{(2)}(j), \dots, (\tilde{x}^{(p_j)}(j))$ and their weighting coefficients $w_1(j), w_2(j), \dots, w_{p_j}(j)$ in the block-like (unit type) II QOL of the synthetic category of "quality of ecological niche"

The block number, j	The variables of the block and corresponding weighting coefficients $w_s(j)$			
	$\tilde{x}^{(1)}(j)$	$\tilde{x}^{(2)}(j)$	$\tilde{x}^{(3)}(j)$	$\tilde{x}^{(4)}(j)$
1	The volume of wastewater per 1 sq. km area, $w_1(1) = 0,147$	Total emissions from stationary sources of air pollution, $w_2(1) = 0,136$	Toxic waste production (TWP) buried for a year (on 1kv.km territory), $w_3(1) = 0,402$	TWP generated for the year, minus the used one, $w_4(1) = 0,315$
2	Seeded and planted forest, $w_1(2) = 0,570$	Young (new) forest of valuable species category, $w_2(2) = 0,103$	Forest that suffered in forest fires, $w_3(2) = 0,327$	
3	Specially protected nature reserves, $w_1(3) = 1,000$			
4	Remediated and disturbed land, $w_1(4) = 1,000$			

Recall (see (2.20a) and (2.20b)) that the value of block-like II QOL$y_i(j)$ for each (i-th) area (and each synthetic category) is calculated by formula (2.20a) or (2.20b):

$$y_i(j) = \sum_{s=1}^{p_j} w_s(j) \cdot \tilde{x}_i^{(s)}(j),$$

where $\tilde{x}_i^{(s)}(j)$ is unified (standardized) value of s-th particular criterion of j-th block for the i-th territory, and the weighting coefficients $w_s(j)$are determined by the components $c_{1s}(j)$of the first eigenvector $C_1(j) = (c_{11}(j), c_{12}(j), \ldots \ldots, c_{1p_j}(j))$ of the covariance matrix $\Sigma(j)$of a posterior set of unified particular criteria $(\tilde{x}^{(1)}(j), \tilde{x}^{(2)}(j), \ldots, \tilde{x}^{(p_j)}(j))$ according to the formula:

$$w_s(j) = \begin{cases} c_{1s}(j)/\sum_{\nu=1}^{p_j} c_{1\nu}(j), \text{ if all the } c_{1\nu} \text{ are of the same sign } \nu = 1, 2, \ldots, p_j; \\ c_{1s}^2(j) \text{ in the case of the opposite signs.} \end{cases}$$

Calculated according to this formula the values of block-like II QOL for 78 subjects of the Russian Federation for each of the four analysed synthetic categories (according to 2003 data) are given in Appendix 2.8.

The construction of the consolidated integral indicator for the analysed synthetic categories of QOL. Now, following the procedure described in point 2.3.6 (see "Step 4") we must move from $k(l)$ *block-like* II QOL (l-th) of the given (l-th) synthetic category $y^{(l)}(1), y^{(l)}(2), \ldots, y^{(l)}(k(l))$ to the single *consolidated* integral indicator – a scalar gauge (measuring tool) of this synthetic category$\hat{y}^{(l)}$ $(l = I, II, III, IV)$. From (2.22)–(2.23), it follows that

$$\hat{y}_i^{(l)} = N - \left[\sum_{j=1}^{k(l)} q_l(j)(y_i^{(l)}(j) - N)^2\right]^{\frac{1}{2}}, \tag{2.23a}$$

where $$q_l(j) = \frac{p_j^{(l)} \cdot s_l^2(j)}{\sum_{\nu=1}^{k(l)} p_\nu^{(l)} \cdot s_l^2(\nu)},$$

$$s_l^2(j) = \frac{1}{n}\sum_{i=1}^{n} \left(y_i^{(l)}(j) - \bar{y}^{(l)}(j)\right)^2 \text{ and } \bar{y}^{(l)}(j) = \frac{1}{n}\sum_{i=1}^{n} y_i^{(l)}(j).$$

Calculated values of *consolidated* integral indicators for 78 subjects of the Russian Federation $(i = 1, 2, \dots, 78)$ for each of the four analysed synthetic categories of QOL $(l = I, \dots IV)$ are given in Appendix 2.8 (according to the 2003 data).

Building a single consolidated integral indicator for the synthetic category of QOL of the higher level of generality $\hat{y}^{cons}$ using the consolidated II QOL that would characterize the “quality of the population” $(\hat{y}^{(I)})$, “the level of welfare $(\hat{y}^{(II)})$”, “quality of social services$(\hat{y}^{(III)})$” and “the quality of the ecological niche $(\hat{y}^{(IV)})$”, carried out by the same procedure as for the construction of $\hat{y}^{(l)}$block-like II QOL, namely:

$$\hat{y}^{icons} = N - \left[\sum_{l=I}^{IV} \tilde{q}_l \cdot (\hat{y}_i^{(l)} - N)^2 \right]^{\frac{1}{2}}, \qquad (2.23b)$$

where $\tilde{q}_l = \frac{p^{(l)} \cdot \tilde{s}_l^2}{\sum_{l=1}^{IV} p^{(l)} \cdot \tilde{s}_l^2}, p^{(l)} = \sum_{j=1}^{k(l)} p_j^{(l)}$

$$\tilde{s}_l^2 = \frac{1}{n} \sum_{i=1}^{n} (\hat{y}_i^{(l)} - \bar{\hat{y}}^{(l)})^2 \text{ and } \bar{\hat{y}}^{(l)} = \frac{1}{n} \sum_{i=1}^{n} \hat{y}_i^{(l)}.$$

Calculated according to this formula, the values of a single (consolidated) II QOL as well as the values of the corresponding ranks of the regions are given in Appendix 2.9.

Appendices 2.8 and 2.9 contain the data necessary for comparative analysis and ranking of subjects of the Russian Federation on each of the four considered synthetic categories of QOL (QOP, LWB, QOSS and QOEN) as well as on the latent synthetic category of QOL of higher level of generality. This is a task of written description representing a certain independent interest.

However, as it will be shown in Chapter 3, these data when *considered in dynamics* can be used to solve the task of identifying problem areas in socio-economic development of the region, of improving the socio-economic policies and in the performance evaluation of regional administrations.

2.5. Conclusions

1. Integral indicator of the synthetic category of quality of life (II QOL) is built in the form of certain convolution (function) $y = f(x^{(1)}, x^{(2)}, \dots, x^{(p)})$of some *measurable* particular properties of this category, where the role of private properties $x^{(1)}, x^{(2)}, \dots, x^{(p)}$ in the *objectivist* approach play the characterizing this category statistical indicators, and in the *subjectivist* approach, the relevant characteris-

tics calculated from the results of the sampling survey of individual members of the analysed population conglomerate.

2. The main drawback of the vast majority of existing in the world ways of evaluating such convolution as $y = f(x^{(1)}, x^{(2)}, \dots, x^{(p)})$ is the **direct expert estimation of parameters** on which the specific form of the function $f(\cdot)$ depends on. Thus, if the problem is solved in the class of linear convolutions $f(x^{(1)}, x^{(2)}, \dots, x^{(p)}) = \sum_{j=1}^{p} w_j x^{(j)}$ then the unknown weights w_j are *determined directly by experts* as a result of some formalized procedure for obtaining and processing expert opinions.
3. Selection of the parameters of the convolution function $f(x^{(1)}, x^{(2)}, \dots, x^{(p)})$ in the presence of learning (i.e. in the presence of a priori expert information on the comparison of the objects, individuals or the conglomerates, on the analysed synthetic category of QOL) is made in such a way that the result of the comparison of these objects based on the values $f(\cdot)$ (in some sense) would differ the least from the a priori expert results.

   Selection of the parameters of the convolution function $f(x^{(1)}, x^{(2)}, \dots, x^{(p)})$ in the absence of learning is based on the following requirement: among all scalar variables that characterize analysed latent synthetic category, it is the values$y = f(x^{(1)}, x^{(2)}, \dots, x^{(p)})$ that can be used to restore the value of all the particular properties $x^{(1)}, x^{(2)}, \dots, x^{(p)}$ with the most accuracy (in some sense), when applying the appropriate models of pair-wise regression $x^{(j)}$ on $y (j = 1, 2, \dots, p)$.
5. In forming the initial (a priori) set of particular properties that characterize the analysed latent synthetic category QOL, we should follow the requirements of their *relevance, informational availability and the reliability of information.*
6. Before including the particular properties into the content of the arguments of convolution function $f(x^{(1)}, x^{(2)}, \dots, x^{(p)})$ we need, first, to subject them to the *procedure of reduction*, i.e. select while using a specific procedure out of the initial (a priori) set a narrower by composition posteriori set (in order to exclude the effect of multicollinearity), and secondly, by appropriate transformations (see (2.2)–(2.4)), make the unification of their measurement scales so that all of them would be measured in *N*-point scale (choosing the value of *N* is done by the researcher), in which "zero" corresponds to the worst case, and *N*, the best.

3 Macro-Econometric Analysis of Quality of Life Within the Framework of Evaluating the Effectiveness of Socio-Economic Policy

3.1 General logistics of the use of integral quality of life indicators within the framework of effective socio-economic policy

The construction and trending of the dynamics of synthetic categories QOLP (quality of life of population) in addition to descriptive purposes (rating of countries and regions according to these synthetic categories, identifying leaders, outsiders, etc.) can point towards goals related to *evaluating the effectiveness of socio-economic policy and identifying so-called problem areas* in the socio-economic development of the country and separate regions, and therefore to the objectives of improving governance. Indeed, as already mentioned in the Introduction, **the main goal of Russia's state policy is improving the quality of life of population.** This is recorded, in particular, in Art. 7 of the Constitution of the Russian Federation and announced by the federal and regional authorities and administrative structures. That is why these organizations are increasingly turning to the eans of econometric analysis of QOLP. Highly relevant from this point of view is the following general logic applied to the use of these funds.

Obviously, the definition of problem areas in the socio-economic development of a country or region should be based on the analysis of, on the one hand, the dynamics of socio-economic indicators (of the corresponding territory), and on the other, the status of this territory relative to other objects to be analysed (countries or regions of the Russian Federation). It can be assumed that the negative dynamics in the territory's QOLP relative to its past values and the simultaneous deterioration of the area on this indicator in relation to other areas indicate the presence of the problem area ("bottleneck") in the socio-economic governance. Thus, we can use the dynamic of these indicators and their significance in shaping the values of II QOL to identify those areas of social life in need of immediate action by means of public policy.

At the same time, the steady improvement in the value of the indicator relative to the past values of the territories, including its own, generally only indicates the need to maintain the current trend. However, it should be noted that a comparative analysis of II QOL of this area across all the territories in question may not give an adequate assessment due to the significant differences in basic (geographic, resources, scale) characteristics of these territories. Therefore, the ranking should be made primarily in areas within the group, similar to the geographical and other basic features of the analysed group. Thus, identification of problem areas in socio-economic development of the territory, which plays an essential role in the selection of the priorities of management of the respective structures, can be divided into the following steps (Figure 3.1):

- determination of performance indicators of quality of life;
- identification of problem areas and key directions of socio-economic policy of territorial (federal, regional or municipal) government;
- allocation of factors, most significantly affecting the QOLP;
- identifying the preferences of population based on surveys.

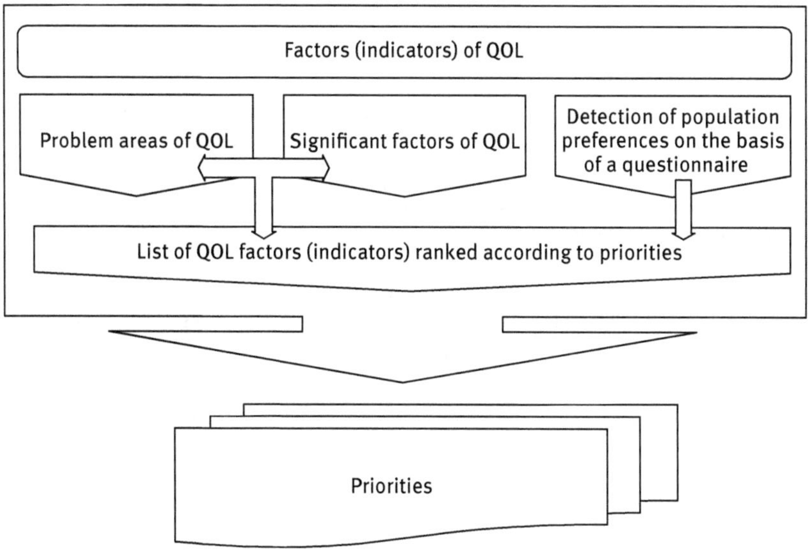

Figure 3.1. The mechanism for the formation of priorities

According to the above logic, indicators, characterizing the level of socio-economic development of the area, should be divided into four groups according to the degree of problematicity (Table 3.1). The first group has the highest level of problematicity, therefore, the values of these parameters should be changed first. Accordingly, the fourth group is the least problematic in terms of changing the values of the indicators. In particular:

- the first group includes the indicators which influence in the most significant way those II QOL whose values deteriorate compared to the ones previously achieved and at the same time lead to a deterioration in the rating of the analysed territory as compared to the others (priority 1);
- the second group consists of indicators which most significantly affect those II QOL, the values of which deteriorate relative to other areas and improve or

remain at the same level with respect to their own values for the previous periods (priority 2);
- the third group consists of indicators that most significantly affecting those II QOL whose values deteriorate relative to their own previous values, but at the same time lead to an improvement in the rating of the analysed territory as compared to the others (priority 3);
- the fourth group consists of indicators which most significantly affect those II QOL whose values improve with respect to their own previous values and at the same time lead to an improvement in the rating of the analysed territory, among other areas (priority 4).

Table 3.1. Priorities of the indicators

		Status as related to other areas	
		Deteriorating	**Improving**
Status of the region relative to its own previous values (dynamics)	Deteriorating	1st (highest) level of problematicity	3rd level of problematicity
	Improving	2nd level of problematicity	4th (lowest) level of problematicity (zone of maintenance of the existing status)

We can propose the following algorithm prioritizing separate indicators characterizing QOL:
- determine medium-term dynamics of change of the values of *integral indicators* of the territory; one of the criteria for the state of deterioration or improvement of the indicator values can be an analysis of the dynamics of the moving average with a "window" corresponding to the indicative planning horizon (usually it corresponds to a window of three or four years);
- in accordance with Table 3.1, determine the levels of problematicity of II QOL under analysis;
- identify indicators (particular criteria) most significantly affecting II QOL of the higher and the following levels of problematicity (for example, by calculating the elasticity of the corresponding II QOL according to the components of the particular criteria);
- according to the results of the analysis, the indicators (particular criteria) are divided into four groups according to their level of problematicity.

Of course, regional governments make their final choice of priorities in carrying out economic and public policies considering the cost of adjustments of one or another particular criteria and the specific political and socio-economic situation in the country.

3.2 Parameters of socio-economic policy and institutional development as determinants of the improved qolp (the results of inter-country econometric analysis)

The study, presented in this section, was conducted in 2005 [Aivazian, 2008] and is dedicated to the identification and assessment of the links between the resulting (criteria bearing) synthetic categories of quality and style of life on the one hand, and the direct or indirect characteristics of current socio-economic policy and the quality of institutions on the other, as well as the analysis of the main trends of the "Russia's path" of development in the period of 1995–2004 in "phase space" of the considered indicators. We will consider *latent synthetic* categories such as levels of development (or effectiveness); the knowledge economy; the legislative, judicial, financial, bureaucratic institutions; the level of bribery and corruption in society; the degree of social responsibility of business leaders; and the degree of safety of the individual and private property as explanatory variables, characterizing the socio-economic policies and institutions, along with *statistically recorded indices* (expenditure on education, health, RD, the ratio of funds, etc.). As the information base for the study (in addition to the official data of Goskomstat of Russia) we used data from *The World Competitiveness Yearbook* (WCY) for 1997–2004. The study is aimed at finding an answer to the question: *Why in spite of the visible signs of economic growth in Russia there are no significant positive trends in improvement of the quality of life of the Russian population, especially if we evaluate the QOL within the framework of cross-country comparative analysis (i.e., within global trends)?*

3.2.1 Motivation, goals and objectives of the study

Analysis of the dynamics (for the period 1998–2004) of the main indicators of Russian economic growth, on the one hand, and indicators of quality of life on the other hand, shows that *in spite of clear signs of economic growth, unfortunately, we do not observe any significant positive trends in key indicators of the quality of life of Russian population.* In the attempt to find an explanation for this situation we need to find answers to more general questions:

- Is it possible to explain this phenomenon exclusively by the specifics of Russian historical conditions prevailing at the turn of the 20th and 21st centuries?

– What are the primary factors in shaping the quality of life: human potential, especially the improvement of educational systems, health, development of science and human freedoms, or economic growth and industrialization?

The fundamental answer to the latter question, of course, depends on the position the researcher is choosing between the two alternative paradigms of socio-economic development that are being discussed in the literature (see, e.g. [Sen, 1997]; [Сен, 2004]).

Paradigm (A). *Development as a process aimed at economic growth and industrialization.* Within this paradigm postulated the need for "hard disciplined approach" to development-centered economic growth (GDP in the first place) and industrialization. This rigidity requires a conscious disregard for various minor, from the point of view of this approach, objectives, such as the development of social services, financial incentives, favouring civil and political rights and benefits of democracy, etc.

Paradigm (B). *Development as a process of expanding human freedoms, which ultimately determines the QOLP.* At the same time, the expansion of freedoms is seen both as *a goal* and as *a means* of socio-economic development. In the former case, we are talking about the **fundamental** role of freedom in the development process, while in the second the freedom plays the *instrumental role.*

We are not going to defend the legitimacy of applying the paradigm (B) to the analysis of Russian reality is the end of 20th and the beginning of 21st century. Moreover, we admit the validity of the possibility of the existence of the whole range of "intermediate" (between (A) and (B)) paradigms of socio-economic development. Choice of a particular version of this spectra depends on the specifics of the country and the historical moment and manifests in, for example, different content (explicit or implicit) of deliberately neglected freedoms within the paradigm (A).

From our point of view, *the first ("hard") approach* to the development, suffers from a certain narrowness of understanding of the objectives of development (which ultimately comes to the understanding of the synthetic category of "quality of life of population"), although it can be very productive in "close to crisis" situations. In the long term, *the second approach to development* is more objectively justified. It is the reality of modern Russian situation that led me to conduct economical analysis aimed to test the hypothesis, suggesting that **the signs of economic growth of a country are not always sufficient to improve the quality of life of its population.**[1]

This thesis, by the way, has been discussed and substantiated by a number of sociologists and economists [The East Asian Miracle, 1993]; [Сен, 2004]. In particular, Noble Prize winner Amartya Sen calls "a living prejudice" the opinion,

1 The acceptance of this provision, in particular, means that the goal of doubling GDP by 2010, declared at the end of the last century by the powerful administrative organs of Russia, really was not *an end, but only one of the means of achieving the goal of improving QOL.*

popular in economic circles, that human development, the development of human potential (especially of the process of improving health, education, and science), is a luxury that only the rich countries can afford. As counterexamples are given: a) the Meiji era in Japan (middle and second half of the 19th century.), when the literacy level among the Japanese was higher than among Europeans, but the standard of living and industrialization at the time was higher in Europe; this, however, served as an impetus for further rapid economic development of Japan; b) the so-called East Asian miracle of the second half of the 20th century, which was largely based on the same causal connections.

Nevertheless, we would repeat that in this book we did not intend to investigate which of the paradigms (A) or (B) is to objectively determine the character and the laws of socio-economic development of modern Russia.

The goal of our study is to identify and assess, using econometric of the analysis, the links between criterial (including *latent, synthetic*) categories of quality and way of life of population on the one hand, and the direct or indirect characteristics of current socio-economic policies and quality of institutions on the other hand, as well as to analyse the main trends of the "Russian path" of development in the period of 1995–2004 in the "phase space" of considered variables (criterial and explanatory).

Due to the fact that when we analyze the explanatory variables (i.e. characteristics of the socio-economic policies and quality of institutions) we will also analyze the factors aimed at the development of human potential, including the improvement of health, education, science and the provision of civil and political rights, then among our results we can ultimately expect to find some empirical confirmation (or refutation?) of thesis of the instrumental role of the latter in a positive socio-economic development of the country, i.e. in realization of the goal of raising QOL (please, see above the definition of of paradigm (B) about the instrumental role of freedoms).

To achieve the above-stated objectives of the research, we will tackle the following tasks:

1. determining a list of resulting (criterial) synthetic categories of quality and way of life of population, i.e. those categories (mostly *latent*) that can be used as a criteria of the level of socio-economic development of the country;
2. for each resulting (criterial) synthetic category, determining *its own set of explanatory variables*, characterizing the level of institutional development of the country, as well as the current socio-economic policies;
3. establishing (evaluating) *regressionional dependency*, existing between the index of each of the resulting (criterial) category and a corresponding set of explanatory variables;
4. according to data from 1995to2004 conducting an *analysis of the trajectories of criterial and key explanatory variables* in order to identify and explain the main trends and *determining major "bottlenecks"*, i.e. those factors from the improved values on which primarily depends further progress in the values of the corresponding resulting category.

Be reminded that the design methodology of constructing measurement of synthetic latent categories of quality and way of life, described in Chapter 2, is founded on the following basic idea. For each synthetic category we formed its own set of particular characteristics (particular criteria) $x^{(1)}, \dots, x^{(p)}$, that lend themselves to a direct statistical observation. After that, depending on the presence or absence of expert information, allowing to compare countries through an analysed synthetic category, one or another *convolution method of particular criteria* was offered, as a result of which the desired integral indicator (y) of this synthetic category was calculated as a function of the values of the particular criteria, $y = f(x^{(1)}, \dots, x^{(p)})$. In other words, particular criteria $x^{(1)}, \dots, x^{(p)}$, which stand for explanatory variables, usually *were not the "cause" of the result "y": they themselves were the result, as only partial or indirect characteristics of the analysed synthetic category*. For instance, analysing the synthetic category of "quality of population", we considered as particular criteria (and as the explanatory variables for "y") such factors as the projected life expectancy at birth, infant mortality, the number of deaths (per year) from a number of major diseases for an average of 1000 people, etc.

The logic of this study is different. We proceed from the fact that we already possess the indexs (i.e. integrated indicators) of the analysed resulting (criterial) synthetic categories. However, we would like to determine how the values of these indicators "y" depend on the explanatory variables $x^{(1)}, \dots, x^{(p)}$, which are in fact the direct or indirect characteristics of the level of the country's institutional development as well as its socio-economic policy. In other words in this formulation of the problem *explanatory variables $x^{(1)}, \dots, x^{(p)}$ are interpreted as the causes of the analysed results* "y", namely as the causes that lend themselves to control and regulation. Guided by this very logic (taking into account the opportunities arising from the availability of information, see point 3.2.2.) we formed compositions of analysed criterial and explanatory variables (see 3.2.3 and 3.2.4).

3.2.2 Information base of research

In the *cross-country* econometric analysis the main sources of information were WCY annuals (Edition IMD-International, Lausanne, Switzerland). The obtained structure and content of macro-economic data are presented in Appendix 3.1. We also used the reports issued by the UN "Human development report", as well as ratings of "Mercer Human Resource Consulting".

In addition to inter-regional econometric analysis of official information publications Rosstat Russia [I1–I13] (see Literature, the section "Information Materials") and specialized sources have been used [I14–I23] (public organization "Support of Russia" Foundation "INDEM" Pension Fund of the Russian Federation, the Ministry of Interior, the Ministry of Finance and others, see Appendix 3.3).

3.2.3 Analysed resulting (criterial) synthetic categories of quality and lifestyle of population

In the selection and interpretation of the resulting synthetic categories of quality and lifestyle of population, i.e. of those categories (mostly latent) which can be used as criteria for the level of socio-economic development, we will proceed from the decomposition of category "QOL", given in Chapter 2, in the form of the general diagram of a hierarchical system of statistical indicators and particular criteria (see Figure 2.1 and the comments thereto). In this case, out of the five integral characteristics of the 2nd level of aggregation *("quality of population", "material well-being of population", "quality of social services", "quality of the ecological niche" and "climatic conditions"*) shown in this diagram we will analyse only the first three. Following the methodology outlined in Chapter 2, we would have to take as a measure of each of these synthetic categories the value of the 1st principal component constructed by a corresponding set of particular criteria. However, as shown in our analysis, the findings in this study are practically unchanged if instead of the values of the 1st principal component we use the value of the particular criterion the most closely correlated with it. Thus, we came to the following resulting (criterial) synthetic categories of quality and lifestyle of population and to their meters of the first three integral characteristics of the 2nd level of aggregation:

Quality of population; estimated with *the index of human development*, ("Human development index"), coded variable 04.04.08 [WCY, *2004*], as a fraction of one (hereinafter referred to as $y^{(1)}$)[2];

- **level of material well-being of population**; measured in US dollars by "private final consumption expenditures per capita" with the purchasing power parity, coded variable 1.1.24 [WCY, *2004*] (hereinafter referred to as $y^{(2)}$);
- **quality of social services**; measured using the method of WCY by the expert evaluation of "level of social cohesion (consent) in society" on a 10-point scale; coded variable2.5.05 [WCY, *2004*] (hereinafter referred to as $y^{(3)}$).
- *Finally, the object of our study also includes a synthetic category of the highest level of generality:*
- **QOL of population**; measured using the method of WCY by the corresponding expert evaluation on a 10-point scale; coded variable 4.4.17 [WCY, *2004*] (hereinafter referred to as $y^{(4)}$).

---

2 The procedure for calculating the value of the index of human development (as of 2009) by the values of GDP per capita, estimated life expectancy, rate of illiterate population in the country among older than 15 years and the rate of the students between the ages of 6 and 24 years is given in 1.3.1.h. Chapter 1.

3.2.4 A Priori Set of Explanatory Variables

The main question that we would like to answer in this study is: *in which way direct or indirect characteristics of quality of institutions and of the country's socio-economic policies affect the values of the indicators of basic criterial categories of quality and lifestyle of population*? Of course, during the formation of the a priori set these direct or indirect characteristics (i.e. a set of explanatory variables $x^{(1)}, x^{(2)}, \ldots, x^{(p)}$), we were limited by the capabilities of informational support of our research, namely, WCY issues. Nevertheless, these yearbooks contain very valuable information for our set task, the analysis of which showed that the indicators – "candidates" for the a priori set of explanatory variables – can be classified into three groups:

- *direct characteristics of current socio-economic policies;*
- *direct characteristics of institutions (or institutional development);*
- *"behavioural" and other "particular resulting" indicators as indirect characteristics of quality of institutions and current socio-economic policies.*

The distribution of the explanatory variables selected from WCY into these groups is presented in Table 3.2. In this table we placed in parentheses the coding according to WCY, [2004] after the notion of the variable $(x^{(j)})$ adopted for this study.

The usefulness of incorporating "particular resulting" indicators (including the ones characterizing the behaviour of various segments of population) in the a priori set of explanatory variables is based on the fact that the causes of the "behaviour" of each of these characteristics are easily explained in terms of various parameters of current socio-economic policies or institutional development. For example, an abnormally high differentiation in the income of population (variable $x^{(12)}$) as well as the low value of the variable $x^{(15)}$ (a degree of social responsibility of business leaders) indicate incorrect social policies and defects in legal and bureaucratic issues of business organization while low values of variables $x^{(3)}$ and $x^{(4)}$ (characterizing the degree of solution of the pollution issues) show the imperfection of the legislature in the environmental protection.

Table 3.2. Classification of explanatory variables of the a priori set

Direct characteristics of socio-economic policies	Direct characteristics of institutional development (quality of institutions)	Particular resulting indicators as indirect characteristics of quality of institutions and socio-economic policies
$x^{(1)}$ (4.4.01) – total health expenditures, % GDP $x^{(5)}$ (4.5.01) – public expenditures on education, % GDP $x^{(6)}$ (4.3.02) – total expenditure on R&D per capita per year, USD $x^{(13)}$ (2.4.02) – quality of policy of protectionism in terms of its impact on business performance	$x^{(8)}$ (2.4.17) – quality of legal regulation of financial institutions $x^{(9)}$ (3.3.18) – the level of transparency of financial institutions $x^{(10)}$ (2.3.16) – efficiency of bureaucratic institutions $x^{(14)}$ (2.5.01) – quality of judicial system	$x^{(2)}$ (4.4.09) – degree of solving the issues of alcohol and drug abuse $x^{(3)}$ (4.4.15) – degree of solving the issues of environmental pollution $x^{(4)}$ (4.4.12) – industrial emissions of carbon dioxide, metric tonnes US$ 1 million GDP $x^{(7)}$ (4.3.10) – level of development of fundamental research $x^{(11)}$ (2.3.17) – effectively addressing bribery and corruption issues $x^{(12)}$ (2.5.08/2.5.07) – characterization of income differentiation $x^{(15)}$ (3.4.09) – degree of social responsibility of business leaders $x^{(16)}$ (2.5.02) – level of safety of individual and private property $x^{(17)}$ (2.5.13) – degree of solving the issues of social unrest and violence

When forming the a priori set of explanatory variables we tried (to the extent possible, due to the available information) to include, in particular, characteristics of each of the five types of "instrumental freedoms" [Sen, 2004][3] namely:

(i) **political freedom**: possibility to choose principles of government and governing bodies; opportunities for public supervision and criticism of authorities; uncensored mass media, etc.;

(ii) **economic freedom**: rules of production and market operation, possibility of economic exchanges, system of distribution of benefits, availability and possibility of access to the finances of various economic agents, etc.;

(iii) **social opportunities**: created by the society systems of education, health and rules of access to the facilities of social infrastructures;

3 In agreement with the above, we understand instrumental freedoms as the fact that freedom can be seen as a means to achieve the desired quality and lifestyle, i.e. they directly or indirectly contribute to the achievement of this goal.

(iv) **transparency guarantees**: freedom of members of the society to cooperate with each other in the conditions of guaranteed clarity and openness of the rules of this cooperation and the "right to expose";
(v) **social protection**: permanently enforced institutional measures such as unemployment benefits, the financial aid to the needy, help to the hungry and creation of jobs during crisis situations, etc.

As a result, we obtained the following "starting" distribution of explanatory variables of the a priori set according to the analysed criterial synthetic categories.

The explanatory variables for the synthetic category of "quality of population" (variable 4.4.08).

$x^{(1)}$ (%) – total health expenditures, % GDP	(4.4.01)
$x^{(2)}$ (points) – degree of solving the issues of alcohol and drug abuse	(4.4.09)
$x^{(3)}$ (points) – degree of solving the issues of environmental pollution	(4.4.15)
$x^{(4)}$ – industrial emissions of carbon dioxide, $\frac{\text{metrictones}}{\text{1mil dollars GDP}}$	(4.4.12)
$x^{(5)}$ (%) – public expenditures on education, % GDP	(4.5.01)

The explanatory variables for the synthetic category "level of material well-being of population" (variable 1.1.24):

$x^{(6)}$ (dollars) – total expenditure on RD per capita per year	(4.3.02)
$x^{(7)}$ (points) – level of development of fundamental research	(4.3.10)
$x^{(8)}$ (points) – legal regulation of financial institutions	(2.4.17)
$x^{(9)}$ (points) – transparency of financial institutions	(3.3.18)
$x^{(10)}$ (points) – efficiency of bureaucratic institutions	(2.3.16)
$x^{(11)}$ (points) – effectively addressing bribery and corruption issues	(2.3.17)
$x^{(12)}$ (times) – 20% funds ratio[4]	(2.5.08/2.5.07)
$x^{(13)}$ (points) – quality of policy of protectionism in terms of its impact on business performance	(2.4.02)
$x^{(14)}$ (points) – quality of the country's judicial system	(2.5.01)

The explanatory variables for the synthetic category "level of social cohesion (consent) in society" (variable 5.2.05).

4 20% funds ratio: we call the ratio of incomes of the richest 20% of population to the income of the poorest 20% of population (as we know, just "funds ratio" is used by the Russian state statistical service for a similar income ratio of upper and lower deciles).

$x^{(12)}$ (times) – 20% funds ratio	(2.5.08/2.5.07)
$x^{(15)}$ (points) – degree of social responsibility of business leaders	3.4.09)
$x^{(16)}$ (points) – level of safety of individual and private property	(2.5.02)
$x^{(17)}$ (points) – degree of solving the issues of social unrest and violence	(2.5.13)
$x^{(11)}$ (points) – effectively addressing bribery and corruption issues	(2.3.17)

The explanatory variables for the synthetic category "quality of life" (*variable* 4.4.17). Variables $x^{(6)}$–$x^{(12)}$, $x^{(14)}$, $x^{(16)}$ and $x^{(17)}$ (see above).

3.2.5 Identification of models of the dependency of indicators of the resulting criterial synthetic categories of quality and lifestyles of population on characteristics of institutional development of society and current socio-economic policies

In this section we will present the results of the solution of the following problem. Let $y_i^{(j)}$ be the value of the indicator of the resulting (criterial) synthetic category "j" of quality or lifestyle of population for the country (region) i ($j = 1, 2, 3, 4,$ – see their description in 3.2.3; $i = 1, 2, \dots, 60,$ – see the list of countries and regions analysed in Appendix 3.1). And let $x_i^{(1)}(j), x_i^{(2)}(j), \dots, x_i^{(p_j)}(j)$ of the value p_j characteristics of the a priori set of explanatory variables of the criterial synthetic category j of the country or region i (a priori sets of explanatory variables for each resulting synthetic categories are listed in the previous section). Using the available data $(x_i^{(1)}(j), x_i^{(2)}(j), \dots, x_i^{(p_j)}(j); y_i^{(j)})$–$i = 1, 2, \dots, 60$, we need to identify a model of dependency of $y^{(j)}$ from $x^{(1)}, x^{(2)}, \dots, x^{(p_j)}$, i.e.:

(a) to determine the parametric family of functions $\{f(x^{(1)}, \dots, x^{(p_j)}; \Theta(j))\}$, within which we search for the dependency in question

$$y_i^{(j)} = f(x_i^{(1)}, \dots, x_i^{(p_j)}; \Theta(j)) + \varepsilon_i^{(j)}, \tag{3.1}$$

where the residual random component $\varepsilon_i^{(j)}$ reflects the impact on the values of $y_i^{(j)}$ factors, not taken into account in the given a priori set of explanatory variables, and, as usual, the model is constructed so that the average value of the residual component is equal to zero (i.e. $E\varepsilon_i^{(j)} = 0$);

(b) within the selected parametric family $\{f(x^{(1)}, \dots, x^{(p_j)}; \Theta(j))\}$ we need to find which of the explanatory variables of the a priori set are in a certain sense decisive in the formation of values $y^{(j)}$, i.e. to select from the a priori set $\{x^{(1)}, \dots, x^{(p_j)}\}$ so-called variables-determinants $x^{(t1)}, x^{(i2)}, \dots, x^{(i_{p'}(j))}$, where $p'(j) < p_j$;

(c) considering the model

$$y_i^{(j)} = f(x_i^{(i_1)}, x_i^{(i_2)}, \ldots, x_i^{(i_{p1(j)})}; \Theta(j)) + \varepsilon_i^{(j)}, \tag{3.1'}$$

we construct estimates $\hat{\Theta}(j) = (\hat{\theta}_1^{(j)}, \hat{\theta}_2^{(j)}, \ldots, \hat{\theta}_{m_j}^{(j)})$ for the unknown values of the parameters $\Theta(j) = (\theta_1^{(j)}, \theta_2^{(j)}, \ldots, \theta_{m_j}^{(j)})$ from the available raw data

$$\{x_i^{(i_1)}, x_i^{(i_2)}, \ldots, x_i^{(i_{p'(j)})}; y_i^{(j)}\}_{i=1,2,\ldots,60.} \tag{3.2}$$

Before presenting the results of the identification of the sought dependencies, we will show only a brief commentary on the course of solving the problems (i)–(iii).

Data support. Solving problems (i)–(iii) may be carried out separately according WCY data per year. However, special analysis showed that the models constructed according to [WCY, t_i] and according to [WCY, t_j] ($t_i, t_j = 1997, 1998,, 2003, 2004; t_i \neq t_j$) differ *statistically insignificantly*. In other words, the data of the form (3.2) taken from the WCY for any two different years are known by various criteria, including the criterion Chow, *regressively homogenous* [Айвазян, 2010; Paragraph 8.3]. Therefore, the following are the results of the constructing the sought models only according to WCY [*2004*] (these data are given in Appendix 3.1).

Unification of measuring scales of analysed variables. As already noted, all expertly evaluated variables are measured on a 10-point scale, where 0 corresponds to the worst case, and 10 to the best. However, among the analysed variables we have seven indicators measured on other scales. These are, by indexation in 3.2.3 and 3.2.4, variables $y^{(1)}, y^{(2)}, x^{(1)}, x^{(4)}, x^{(5)}, x^{(6)}$ and $x^{(12)}$. According to the recommendations in Chapter 2, in order bring variable z, measured on the *arbitrary* scale, to a 10-point scale, we need to transform this variable, i.e., go to the variable $\tilde{z}$ according to the formula

$$\tilde{z} = [1 - \frac{|z - z_{\text{opt}}|}{\max\{(z_{\text{opt}} - z_{\text{min}}), (z_{\text{max}} - z_{\text{opt}})\}}] \cdot 10, \tag{3.3}$$

where $z_{\text{min}}, z_{\text{max}}$ and z_{opt}, respectively, the minimum possible, mazimum possible and optimal (in the sense of this measured variable category) values. Minimum, maximum and optimum values of the analysed variables, necessary to realize this transformations are shown in Table 3.3.

Minimum, maximum and optimal values of the indicator of human development $y^{(1)}$ come from its definition. *The minimum* (*maximum*) value for any of the other variables is defined as is the minimum (maximum) value of this variable among the 60 analysed countries (regions). *The optimal* value of the variable associated with the quality of a *monotonically increasing* (*monotonically decreasing*) dependency is defined as its maximum (respectively minimum) value. Finally, the optimal value of the variable associated with the quality by *non-monotonic* dependency (in this case the variables $x^{(1)}, x^{(5)}, x^{(6)}$ and $x^{(12)}$ and calculated as *the mean value of this variable taken over five countries, the best for the resulting synthetic category for which this variable is used as explanatory*. Moreover, if such a variable is explanatory for *a few* of the resulting synthetic categories, then the average of the average values is taken as its optimal value, calculated by countries, the best for these resulting categories.

Note that a certain modification of the names of the variables in several cases below is due to the applied unifying transformation (3.3). So, the variable $x^{(1)}$ (by [WCY, *2004*] variable 4.4.01) in its *original* form (and in Tables 3.2 and 3.3) means "public spending on health" and is measured in percentage of GDP; after the transformation (3.3), it has to be measured on a 10-point scale in such a way that the evaluation of 10 points shows the *optimal* value of health care costs, and 0 points shows the value, furthest from the optimal level (the same comment applies to the variables $x^{(4)}, x^{(5)}, x^{(6)}$ and $x^{(12)}$). In addition, the variables in the original form related to the concept of quality by a monotonically decreasing dependency (the level of alcohol and drug abuse, the level of bribery and corruption, etc.), required the use of words such as "the degree of solving the problem of ..." or "efficiency of solving problems ..." (see variables $x^{(2)}, x^{(3)}, x^{(11)}, x^{(17)}$).

Table 3.3. Parameters required for transformation of variables, unifying their measuring scales

No	Variable	Explanation (name) of the variable	Minimum value	Maximum value	Optimal value
1	$y^{(1)}$	Human development index, fraction of one	$y^{(1)}_{\min} = 0$	$y^{(1)}_{\max} = 1$	$y^{(1)}_{\text{opt}} = 1$
2	$y^{(2)}$	Private final consumption expenditure per capita, USD	$y^{(2)}_{\min} = 317$	$y^{(2)}_{\max} = 25548$	$y^{(2)}_{\text{opt}} = 25548$
3	$x^{(1)}$	Total health expenditures, % GDP	$x^{(1)}_{\min} = 2,4$	$x^{(1)}_{\max} = 13,9$	$x^{(1)}_{\text{opt}} = 8,6$
4	$x^{(4)}$	Industrial emissions of carbon dioxide, metric tonnes US$ 1 million GDP	$x^{(4)}_{\min} = 120,0$	$x^{(4)}_{\max} = 7677$	$x^{(4)}_{\text{opt}} = 120,0$
5	$x^{(5)}$	Public expenditures on education, % GDP	$x^{(5)}_{\min} = 0,92$	$x^{(5)}_{\max} = 8,80$	$x^{(5)}_{\text{opt}} = 6,4$
6	$x^{(6)}$	State expenditures on R&D per capita per year, USD	$x^{(6)}_{\min} = 0,3$	$x^{(6)}_{\max} = 1060$	$x^{(6)}_{\text{opt}} = 639$
7	$x^{(12)}$	20% funds ration, times	$x^{(12)}_{\min} = 3,37$	$x^{(12)}_{\max} = 33,25$	$x^{(12)}_{\text{opt}} = 5,5$

Choice of a parametric family of feasible solutions. In solving the problem (a) (see above) we considered only two competing options:

$$\tilde{y}_i^{(j)} = \theta_1 + \theta_2\ \tilde{x}_i^{(1)} + \ldots + \theta_{p_j+1} \cdot\ \tilde{x}_i^{(P_j)} + \varepsilon_i^{(j)} - \tag{L}$$

family of *linear* relatively explanatory variable functions with additively superimposed random remainders $\varepsilon^{(j)}$;

$$\tilde{y}^{(j)} = \theta_1 (\tilde{x}_i^1)^{\theta_2} \cdot \ldots \cdot (\tilde{x}^{(p_j)})^{\theta_{p_j+1}} \cdot e^{\varepsilon_i^{(j)}} - \tag{P}$$

family of *power* relatively explanatory variable functions with multiplicative-exponential superimposed random remainders $\varepsilon^{(j)}$. When selecting a version of the model we used as the main criteria of its quality the value of "directed to unbiasedness" *coefficient of determination* R^2_{adj} and some behavioural characteristics of the *estimated remainders*

$$\tilde{\varepsilon}_i^{(j)} = \tilde{y}_i^{(j)} - f\left(\tilde{x}_i^{(i_1)}, \tilde{x}_i^{(i_2)}, \ldots, \tilde{x}_i^{(ip'(j))}; \hat{\Theta}(j) \right)$$

where $p'(j)$, as before, is the number of variables-determinants selected as the result of the solution of problem (b).

Problem-solving methods (a)–(c). Problem solving (a)–(c) is not taking place *consecutively, but simultaneously, in an interactive mode.* In this case, the solution of problem (b) was based on a standard procedure of *single-step regression with the addition* [Aivazian, 2010, 4.4.2], and the solution of (c) was carried out using *the ordinary method of least squares* (ibid., 4.3). A certain modification of the procedure was done as follows. As a result of this procedure we received for each fixed number of explanatory variables their optimum composition. In some cases we tried several versions of this composition, however, only in a small neighbourhood of the optimal values of the optimized criterion R^2_{adj}. If such a change has led to a more natural (from a substantive point of view) interpretation of the analysed dependency, then the resulting composition was chosen as a solution. By this logic we introduced the free members in the equation for $\tilde{y}^{(2)}$ and $y^{(3)}$ (see Table 3.4). According to the formal criteria, these free members didn't have to be included in these equations. In addition, the analysis of random remainders ("residuals") in the equation for $\tilde{y}^{(2)}$ revealed four observations clearly not consistent with the general picture: China and its separate area Zhiyangu, as well as India and its separate district of Maharashtra. Therefore, this equation was calculated not based on 60 but 56 observations.

The results of the identification of sought dependencies. Principal results of calculations, related to the identification of models of dependency of indicators of the resulting (criterial) synthetic categories $y^{(j)}$ of institutional development of society and current socioe-conomic policies, are shown in Table 3.4. Be reminded that a linear form of the function is determined by the ratio (L) and power form by the ratio (P). Explanatory variables-determinants are located in the table for each of the resulting synthetic categories in the descending order according to their impact on this category, determined by the appropriate values of the coefficients $\hat{\theta}_l^{(j)}$ and their relation to its mean-square error (i.e. values of t-statistics). Finally, from (3.1), (L), (P), (3.3) and Table 3.4 we can see that the estimated regression formula according to which we can evaluate (10-point scale) the value of the indicator $\tilde{y}_i^{(j)}$ of the resulting synthetic category j for the i country, which has the following form:

– with the linear form of the dependency of type (L)

$$\tilde{y}_i^{(j)} = \frac{\sum_{l=1}^{p'(j)} \hat{\theta}_{l+1}{}^{(j)} \cdot \tilde{x}_i^{(l)}(j)}{10 \cdot \sum_{l=1}^{p'(j)} \hat{\theta}_{l+1}{}^{(j)}}; \qquad (3.4L)$$

Table 3.4. Main characteristics of the analysed dependencies

Number of resulting synthetic variable, *j*	**Resulting synthetic variable, $\tilde{y}^{(j)}$**	**General type of dependence**	**Main explanatory variables (determinants) and their characteristics**					
			Variable	**Free member**	$\tilde{x}^{(1)}_{(j)}$	$\tilde{x}^{(2)}_{(j)}$	$\tilde{x}^{(3)}_{(j)}$	$R^2_{(adj)}$
			Coefficient of variable	$\hat{\theta}_1^{(j)}$	$\hat{\theta}_2^{(j)}$	$\hat{\theta}_3^{(j)}$	$\hat{\theta}_4^{(j)}$	
			***t*-statistics**	$t_{1(j)}$	$t_{2(j)}$	$t_{3(j)}$	$t_{4(j)}$	
1	$\tilde{y}^{(1)}$ quality of population	Power	Variable's name	Free member	Pollution (4.4.15)	Emissions CO_2(4.4.12)	Health (4.4.01)	0,655
			Numeric value of coefficient	5,7	0,20	0,16	0,03	
			Numeric value of *t*-statistics	19,2	3,5	5,1	2,3	
2	$\tilde{y}^{(2)}$ material well-being	Linear	Variable's name	Free member	Expenditure on R & D (4.3.02)	Funds coefficient (2.5.08/2.5.0.7)		0,908
			Numeric value of coefficient	0,25	0,60	0,23		
			Numeric value of *t*-statistics	1,3	21,7	2,4		

(continued)

Table 3.4. (continued)

Number of resulting synthetic variable, j	Resulting synthetic variable, $\tilde{y}^{(j)}$	General type of dependence	Main explanatory variables (determinants) and their characteristics					
			Variable	Free member	$\tilde{x}^{(1)}_{(j)}$	$\tilde{x}^{(2)}_{(j)}$	$\tilde{x}^{(3)}_{(j)}$	$R^2_{(adj)}$
			Coefficient of variable	$\hat{\theta}^{(j)}_1$	$\hat{\theta}^{(j)}_2$	$\hat{\theta}^{(j)}_3$	$\hat{\theta}^{(j)}_4$	
			t-statistics	$t_{1(j)}$	$t_{2(j)}$	$t_{3(j)}$	$t_{4(j)}$	
3	$\tilde{y}^{(3)}$ social cohesion	Power	Name of the variable	Free member	Business responsibility (3.4.09)	Security(2.5.02)		0,627
			Numeric value of coefficient	0,61	0,71	0,26		
			Numeric value of *t*-statistics	1,0	5,3	3,9		
4	$\tilde{y}^{(4)}$ quality of life	Power	Variable's name	Free member	Security (2.5.02)	Corruption (2.3.17)	Expenditures on R&D (4.3.02)	
			Numeric value of coefficient	0,98	0,28	0,23	0,05	0,881
			Numeric value of *t*-statistics	13,3	5,0	4,3	3,2	

– with the power form of the dependency of type (C)

$$\tilde{y}^{(j)}_i = \left(\frac{\tilde{x}^{(1)}_i(j)}{10}\right)^{\hat{\theta}^{(j)}_2} \cdot \left(\frac{\tilde{x}^{(2)}_i(j)}{10}\right)^{\hat{\theta}^{(j)}_3} \cdots \left(\frac{\tilde{x}^{(p'(j))}_i(j)}{10}\right)^{\hat{\theta}^{(j)}_{p'(j)+1}}. \tag{3.4P}$$

Formulas (3.4L) and (3.4P) were obtained following the rule scale unification using transformations (3.3) considering that in this case:

$$z_1 = \hat{y}^{(j)}_i = \begin{cases} \hat{\theta}^{(j)}_1 + \sum\limits_{l=1}^{p'(j)} \hat{\theta}^{(j)}_{l+1} \cdot \tilde{x}^{(l)}_i(j) \text{ for linear type of dependency;} \\ e^{\hat{\theta}^{(j)}_1} \cdot \left(\tilde{x}^{(1)}_i(j)\right)^{\hat{\theta}^{(j)}_2} \cdot \ldots \cdot \left(\tilde{x}^{(p'(j))}_i(j)\right)^{\hat{\theta}^{(j)}_{p'(j)+1}} \text{for power type of dependency;} \end{cases}$$

$$z_{\text{opt}} = z_{\max} = \hat{y}^{(j)}_{\max} = \begin{cases} \hat{\theta}^{(j)}_l + 10 \cdot \sum\limits_{l=2}^{p'(j)+1} \hat{\theta}^{(j)}_l \text{ for linear type of dependency;} \\ e^{\hat{\theta}^{(j)}_1} \cdot 10^{\sum\limits_{l=2}^{p'(j)+1} \hat{\theta}^{(j)}_l} \text{ for power type of dependency;} \end{cases}$$

$$z_i = \hat{y}^{(j)}_{\min} = \begin{cases} \hat{\theta}^{(j)}_1 \text{ for linear type of dependency;} \\ 0 \text{ for power type of dependency;} \end{cases}$$

and $\tilde{x}^{(1)}(j), \ldots, \tilde{x}^{(p'(j))}(j)$ is a set of variables-determinants (measured in *unified* scale), identified as a result of the solution of (b) for the *resulting* synthetic category *j*. So, for example, if $j = 4$ (which corresponds to synthetic category of QOL) variables-determinants $\tilde{x}^{(1)}_i(4), \tilde{x}^{(2)}_i(4)$ and $\tilde{x}^{3}_i(4)$ are 10-point scale evaluations, respectively, of the level of security (personal and private property), the effectiveness of solving bribery and corruption issues and the degree of optimal costs of R&Din the country *i*.

3.2.6 Results of Empirical Analysis

So, as a result of identification of analysed dependencies (see Table 3.4) we can see that of the 17 explanatory variables of the a priori set (see Table 3.2 above) only 8 can be attributed to variables-determinants. These are: the security level of individual and private property (variable $\tilde{x}^{(16)}$), the degree of optimal spending on scientific research and experimental development ($\tilde{x}^{(6)}$), the efficiency of solving bribery and corruption issues ($\tilde{x}^{(11)}$), the degree of social responsibility of business leaders ($\tilde{x}^{(15)}$), the degree of solutions of environmental pollution issues ($\tilde{x}^{(3)}$), the degree of solving the carbon dioxide issue ($\tilde{x}^{(4)}$), the degree of the optimal level of population income differentiation ($\tilde{x}^{(12)}$) and the degree of the optimal value of the total health expenditures ($\tilde{x}^{(1)}$). What is more, each of the *first* two variables is simultaneously a determinant for **two** resulting (criterial) categories.

Therefore, we will analyse "the Russian path at the junction of the 20th and 21st centuries" in the "phase space" of these eight variables-determinants and, of course, of the four analysed the resulting (criterial) synthetic categories.

Quantitative analysis revealed regression dependencies. We will now use the results of identification of the models (see Table 3.4) for the purpose of their quantitative analysis.

Variation of regression values of **quality of life indicator** $\hat{\tilde{y}}^{(1)}$ for 65.5% is determined by the changes in value of factors-determinant: $x^{(1)}(1) = \tilde{x}^{(3)}$ (degree of solving the issues of environmental pollution), $x^{(2)}(1) = \tilde{x}^{(4)}$ (degree of solving the issues of industrial emissions of carbon dioxide) and $x^{(3)}(1) = \tilde{x}^{(1)}$ (degree of an optimal value of health care costs) using the formula:

$$\hat{\tilde{y}}^{(1)} = 5,71 + 0,196\tilde{x}^{(3)} + 0,158\tilde{x}^{(4)} + 0,031\tilde{x}^{(1)}. \quad (3.5)$$

This means in particular that with an increase of the value $\tilde{x}^{(3)}$ (or $\tilde{x}^{(4)}$, or $\tilde{x}^{(1)}$) the value Δ of the quality of population indicator increases in the average on $0,196 \cdot \Delta$ (respectively on $0,158 \cdot \Delta$ or on $0,031 \cdot \Delta$). In addition, using the known approximate formula for the *elasticities* $e_{\hat{y}/x^{(j)}}$ in the case of the *linear* dependency $\hat{y}$ on the explanatory variables we can determine that when the value $\tilde{x}^{(3)}$ (or и $\tilde{x}^{(4)}$, or $\tilde{x}^{(1)}$) changes by 1%, the value $\hat{\tilde{y}}^{(1)}$ will change by $0,196\,\bar{\tilde{x}}^{(3)}/\bar{\tilde{y}}^{(1)}$ (respectively by $0,158\bar{\tilde{x}}^{(4)}/\bar{\tilde{y}}^{(1)}$ or by $0,031\bar{\tilde{x}}^{(1)}/\bar{\tilde{y}}^{(1)}$), where $\bar{\tilde{y}}^{(1)}$ and $\bar{\tilde{x}}^{(j)}$ the average values of these indicators, obtained by averaging data for all analysed countries.

The variation of the regression values of the indicator of the **level of material well-being** $\hat{\tilde{y}}^{(2)}$ by 90.8% is determined by the changed values of factors-determinants: $x^{(1)}(2) = \tilde{x}^{(6)}$ (the degree of the optimal value of total expenditures on R&D) and $x^{(2)}(2) = \tilde{x}^{(12)}$ (the degree of the optimal level of population income differentiation) using the formula:

$$\hat{\tilde{y}}^{(2)} = e^{0,25}\cdot(\tilde{x}^{(6)})^{0,6}\cdot(\tilde{x}^{(12)})^{0,23}. \tag{3.6}$$

This means in particular that *the coefficients of elasticity* $\hat{\tilde{y}}^{(2)}$ by $\tilde{x}^{(6)}$ and $\tilde{x}^{(12)}$ are, respectively, 0.6 and 0.23.

Variation of the regression values of the indicator $\hat{y}^{(3)}$ *of level of social cohesion (the agreement) in the society by 62.7% is determined by the change in values of factors-determinants:* $x^{(1)}(3) = x^{(15)}$ *(evaluation of the degree of social responsibility of business leaders)* $x^{(2)}(3) = x^{(16)}$ *and (estimation of the level of security of the individual and private property) using the formula*:

$$\hat{y}^{(3)} = 0,612 + 0,710x^{(15)} + 0,264x^{(16)}. \tag{3.7}$$

This means in particular, that by increasing the estimate $x^{(15)}$ (or $x^{(16)}$) by Δ, the value of the indicator of the level of social cohesion will increase in average by $0,710\Delta$ (respectively by $0,264\Delta$). Corresponding *elasticities* $\hat{y}^{(3)}$ by $x^{(15)}$ and $x^{(16)}$ can be calculated using the approximate formulas:

$$e_{\hat{y}^{(3)}/x^{(15)}} = 0,710\cdot\frac{\bar{x}^{(15)}}{\bar{y}^{(3)}},$$
$$e_{\hat{y}^{(3)}/x^{(16)}} = 0,264\cdot\frac{\bar{x}^{(16)}}{\bar{y}^{(3)}},$$

where the bar above the variable denotes averaging of this indicator for all analysed countries.

Variation of regression values of the indicator of quality of life $\hat{y}^{(4)}$ for 88.1% is determined by the changes in the values of factors-determinants: $x^{(1)}(4) = x^{(16)}$ (level of individual and private property security), $x^{(2)}(4) = x^{(11)}$ (assessment of the effectiveness of solution of bribery and corruption in society issues) and $x^{(3)}(4) = \tilde{x}^{(6)}$ (degree of the optimal value of R&D) using the formula:

$$\hat{y}^{(4)} = e^{0.975}\cdot(x^{(16)})^{0,281}\cdot(x^{(11)})^{0,232}\cdot(\tilde{x}^{(6)})^{0,049}. \tag{3.8}$$

This means in particular that the *coefficients of elasticity* are, respectively, 0.281, 0.232 and 0.049.

Note. The question may arise about the legitimacy of the interpretation of the variables that characterize expenditures on health, education and R&D, as the *reasons* in relation to the analysed criterial synthetic categories of quality and lifestyle of population. It is theoretically possible a directly opposite interpretation of these cause-effect relationships: "high standards of quality of life are the reason for fairly generous funding of the mentioned budget expenditures of the country". However, first, the question is not "the more the expenditures the better", but is instead linked to some optimal level of spending. Second, we conducted statistical justification of the postulate, which states that the variables $x^{(1)}, x^{(5)}$ and $x^{(6)}$ should refer specifically to the *reasons*. It is based on the analysis of time series of the considered explanatory and criterial variables uses, in particular, methods of detection "causal relationships by Granger[5]". This analysis, we take out of the scope of this study; however, we can refer, for example, to the work of [Сафонова, 2004], in which it is proven that the variables $x^{(5)}$ and $x^{(6)}$ are informative characteristics of the "economy of knowledge" and that their lagged (with a lag of three to four years) values *significantly affect* socio-economic development of the country and QOLP.

Analysis of the dynamics: trajectories of Russian resultant synthetic categories and their determinants. Let's now equip indicators $y^{(1)}, \ldots, y^{(4)}$ and determinants $x^{(j)}$ $(j = 1, 3, 4, 6, 11, 12, 15$ and $16)$ with the second lower index *t*, so that $y_{it}^{(l)}$ and $x_{it}^{(j)}$ are the values, respectively, of the indicator of the resultant synthetic category *l* and of the determinant *j* registered for country *i* in the year *t* ($t = 1995, 1996, \ldots, 2004$). Speaking about the analysis of the dynamics, we will distinguish *auto-dynamics* (changing the values of the indicator in question $y_{it}^{(l)}$ and $x_{it}^{(j)}$, characterizing a certain country in different years) and *cross-country dynamics* (change in the country's position among other countries). When measuring the dynamics of the cross-country Russia, we will focus on the dynamics of its position (rank $r_t(y^{(l)})$ or $r_t(x^{(j)})$) among the analysed countries using the resultant synthetic category $y^{(l)}$ or the value of the determinant $x^{(j)}$ $y^{(l)}$. Auto- and cross-country dynamics of analysed resultant synthetic categories and their determinants for Russia according to WCY [1997–2004] are shown in Table 3.5. The upper number in each line gives the numerical value of the variable, and the bottom (under the slash) – gives a rank (i.e. ordinal position) of Russia by the given indicators among the 46 analysed countries.[6] The values of all variables are given on a 10-point scale (empty cells are due to a lack of necessary data). However, to adequately assess *auto-dynamics*, it is useful to trace the trajectories of some of the resultant synthetic categories of determinants in their original form measured in *specific physical units*, exactly in terms of these units. Therefore, in Appendix 3.2 we show

5 In general terms, the essence of "causal relationship by Granger" may be formulated as follows: the variable *x* is the *cause* of changes in the variable *y*, if the lagged (i.e. recorded in past times) values of *x* to a large extent determine the current value of *y* and, at the same time, the lagged values of *y* do not have a statistically significant impact on the current value of *x* [Granger, 1969].

6 Despite the presence of 60 compared countries and regions in the issue of WCY [2004], we have to restrict ourselves to analysing the dynamics of those 46 countries which were represented in the issues of WCY, since 1997 (in the Appendix 3.1 marked with an asterisk).

these trajectories for synthetic categories $y_t^{(2)}$ and for variables-determinants $x_t^{(1)}, x_t^{(4)}, x_t^{(6)}$ and $x_t^{(12)}$. In addition, we also give (for each *t*) the values of the parameters required to perform unifying transformation (3.3), namely, minimum, maximum and optimal values for each of the examined variables.

Trajectory of the indicator of human development $\tilde{y}_t^{(1)}$ and its determinant.[7] This synthetic category, as would be expected based on its definition, is the most inertial and less variable. In that, when 46 countries are compared, Russia is steadily ahead of only six countries: India, Indonesia, China, the Philippines, South Africa and Turkey, i.e. of the countries with the lowest *life expectancy* (*at birth*). Of the components used to calculate the value $y^{(1)}$, Russia has a consistently decent position only for one – the *share of literate population*. Only this indicator does not allow the country to slide into the very "tail" of the list of the countries compared by $y^{(1)}$. By another very important component – life expectancy (at birth) – Russia is an obvious outsider. We can see from the resulting dependency (3.5) that *the main reserve to correct the situation is in a radical improvement of the environmental protection policy*, that is, in raising the lowest values (see Table 3.5) values of the determinants $x^{(3)}$ (effectiveness of solving pollution issue, which, however, reveals a certain positive trend over the past three years) and of the $\tilde{x}^{(4)}$ (extent of the solution of the industrial carbon dioxide emissions issue, where Russia ranks last). Provisions of the third determinant – optimal health care costs $\tilde{x}^{(1)}$ – judging by its very conservative estimate (see Table 3.5), are far from being exhausted. Variations $\tilde{y}^{(1)}$ which at 34.5 remained unexplained by the dependency (3.5), caused mainly by the level of GDP per capita, could be analysed by constructing a model similar to ours with the GDP per capita level as the dependent variable. However, given the existence of a very close statistical relationship between the level of GDP per capita and the value of final consumption of households and using derived by us dependency (Table 3.6), it is possible to predict a significant role (in the problem of raising the level of GDP per capita and as a result increasing the value of y(1)) of the factors of the development of the knowledge economy, presented in the Table 3.6 as R&D expenditures.

The trajectory of the indicator of the level of material well-being of Russian population $\tilde{y}_t^{(2)}$ and its determinants. An analysis of the trajectory $\tilde{y}_t^{(2)}$ (see the corresponding row in Table 3.5) shows Russia's strong outsider position on this indicator. Despite (since 2000) observable clear tendency to some increase in the *original* variable $\tilde{y}_t^{(2)}$ (see Appendix 3.2a), Russia remains in 42nd place (out of 46), only ahead of India, China, the Philippines and Indonesia on this indicator and yielding the rest, including South American countries like Colombia, Argentina, Brazil and

7 Here we are using a *unified* value of the standard indicator of human development, which differs from the latter (according to the formula (3.3)) simply by a factor of 10, e.g. $\tilde{y}^{(1)} = 10y^{(1)}$.

Venezuela, all countries of Eastern Europe and the others. From the dependency (3.6)[8] that we derived, we can see that **growth reserves** $\tilde{y}^{(2)}$ *are to be found in the first place in the development of the knowledge economy (presented in (3.6) by the expenditures on R&D) and in measures for lowering differentiation of the population by income* (i.e. increasing value of $\tilde{x}^{(12)}$). On both these variables ($\tilde{x}^{(6)}$ and $\tilde{x}^{(12)}$) Russia is firmly settled on the "near-outsider" positions (on $\tilde{x}^{(6)}$ between 33rd and 39th, and on $\tilde{x}^{(12)}$ between the 37th and 40th place) and it did not show any positive trends in the interval between 1995 and 2002 (see Table 3.5).

Trajectory of the level of social cohesion (agreement) in the Russian society $\tilde{y}_t^{(3)}$ **and its determinants.** Here also the Russia is in an outsider position (placed between the 41st and the last, 46th; see Table 3.5). There is only a timid hint of a positive trend (after the last place in 1999, it moved to the 43rd in 2003 and 2004; only Venezuela, Poland and Argentina or Italy, depending on the year, are on a worse situation). Two determinants explain by the variation of the synthetic category on of 62.7% (see Table 3.4 and equation (3.7)). This is *a degree of social responsibility of business leaders (variable* $x^{(15)}$*) and the degree of safety of the individual and private property* (variable $x^{(16)}$). These determinants are "*behavioural*" (see Table 3.4); however, management specialists are quite capable to identify the urgent (and *realistic!*) necessary improvements of Russian institutions and the ongoing economic and social policies that will dramatically increase the value of variables $x^{(15)}$ and $x^{(16)}$. In the meantime, on both of these variables, Russia is respectively on the 2nd to 3rd and 2nd to 5th places from the rear: in 2004 considering social responsibility of business leaders only Argentina and Poland were behind Russia, and on security (personal and private property), Russia was ahead of Argentina, Venezuela, Poland and Mexico.

Table 3.5. Dynamics of the resultant (criterial) categories of quality and lifestyle of Russian population and of their determinants (values of all variables are given on a 10-point scale)

Name of variable (its code by WCY [2004])	Name of the measured category	Values of variable/Russia's place among 46 countries									
		1995	1996	1997	1998	1999	2000	2001	2002	2003	2004
$\tilde{y}^{(1)}$ (4.4.08)	Index of human development	8,0/ 38		7,9/ 38	7,7/ 41	7,8/ 37	7,8/ 35	7,8/ 35			
$\tilde{y}^{(2)}$ (1.1.24)	Final consumption of households per capita, dollars		0,3/ 42	0,5/ 42	0,6/ 41	0,2/ 43	0,2/ 42	0,3/ 42	0,3/ 42	0,4/ 42	
$y^{(3)}$ (2.5.05)	Level of social cohesion (consent) in society					2,0/ 46	3,2/ 45	4,2/ 43	4,4/ 41	3,6/ 43	4,0/ 43

(continued)

8 Be reminded that this dependency has the highest predictive power: the two of its variables-determinants explain almost 91% of the variation of the dependent variable $\tilde{y}^{(2)}$!

Table 3.5. (continued)

Name of variable (its code by WCY [2004])	**Name of the measured category**	**Values of variable/Russia's place among 46 countries**									
		1995	**1996**	**1997**	**1998**	**1999**	**2000**	**2001**	**2002**	**2003**	**2004**
$y^{(4)}$ (4.4.17)	Quality of life of population			1,0/ 46	2,2/ 46	0,5/ 46	1,3/ 46	1,6/ 45	2,5/ 45	2,2/ 45	3,0/ 44
$\tilde{x}^{(1)}$ (4.4.01)	Optimal total expenditures on health			3,7/ 39	3,8/ 38	5,2/ 36	4,0/ 37	4,5/ 38			
$x^{(3)}$ (4.4.15)	Solution of environmental pollution issue						2,4/ 46	2,5/ 46	3,6/ 44	4,3/ 39	4,8/ 37
$\tilde{x}^{(4)}$ (4.4.12)	Solution of industrial emissions of carbon dioxide	0/ 46	1,4/ 45	1,6/ 45		0/ 46					
$\tilde{x}^{(6)}$ (4.3.02)	Optimal R&D expenditures	1,9/ 33	3,8/ 33	1,1/ 33	1,6/ 36	0,6/ 39	0,3/ 38	0,5/ 33	0,5/ 33		
$x^{(11)}$ (2.3.17)	Solution of bribery and corruption issues					1,5/ 42	1,9/ 39	1,3/ 44	1,6/ 40	1,3/ 43	1,5/ 41
$\tilde{x}^{(12)}$ (2.5.08/2.5.07)	Optimal level population income differentiation	8,8/ 38	6,3/ 39	6,3/ 40	6,4/ 39	6,2/ 39	7,4/ 39	7,9/ 37			
$x^{(15)}$ (3.4.09)	Social responsibility of business leaders			2,9/ 46	3,6/ 46	3,1/ 46	3,4/ 46	3,7/ 46	3,6/ 44	3,0/ 45	3,5/ 44
$x^{(16)}$ (2.5.02)	Individual and private property security			0,9/ 46	2,7/ 42	0,8/ 46	1,6/ 45	2,6/ 43	2,2/ 42	2,0/ 44	3,0/ 42

The trajectory of the synthetic category of "quality of life of population" of Russia and its determinants. Russian trajectory of the synthetic category QOL, to put it mildly, is not encouraging (see Table 3.5): it moved from the last place in 1997–2000 to the penultimate place in 2001–2003 and managed to get ahead of the two countries on this indicator (Venezuela and Argentina) in 2004. However, analysis of the dependence (3.8) allows us to identify ways to increase the level of QOL based on the improvement of socio-economic policies and institutional development. Indeed, from (3.8) and Table 3.4 we can see that the 88.1% of the level of QOL is a result of the ability of the state and the public to ensure the safety of the individual and private property (variable $x^{(16)}$), to reduce the level of bribery and corruption (variable $x^{(11)}$) and to give sufficient impetus to the development of the knowledge economy (variable $\tilde{x}^{(6)}$)! At the same time, the *quantitative* analysis of the dependence (3.8) using the elasticities $\hat{y}^{(4)}$ of $x^{(16)}, x^{(11)}$ and $x^{(6)}$ and standard procedures and expert estimation of values $y^{(4)}, x^{(16)}$ and $x^{(11)}$ allows to obtain the ratios of the "expenditures result" that determine the effectiveness of the institutional and political reforms aimed at increasing the value of determinants $x^{(16)}, x^{(11)}$ and $x^{(6)}$ in order to improve the quality of life of the population.

3.2.7 Conclusion

We must admit that at the stage of *formulation of the problem*, that is, when we formulated the problem of *statistical* research of dependencies between direct and indirect characteristics of institutional quality and of the country's social and economic policies on the one hand, and the main resultant synthetic indicators of the quality of life and lifestyle of population on the other, we had a very little faith in the success, that such dependencies (based on the accessible statistics) *will be statistically significant* and *sufficiently* convincing. The results of the calculations have exceeded our expectations! They explain, in particular, why even against the background of the positive (fragmentary) dynamics of the Russian economy (quite good indicators of overall economic growth, successful exports and foreign exchange reserves) *resultant key indicators of quality and lifestyle of Russian population are very low and do not have any significant positive dynamics.* These results revealed (not on the fictional, but on the quantitative level) those key areas for improvement of Russian institutions and socio-economic policy, on which concentrated efforts should first be applied to improve the quality of life.

In particular, the results of the econometric analysis of the considered dependencies (see Table 3.4) lead to the following conclusions.

1. One of the most important characteristics of the *knowledge economy is that the cost of research and development* (variable $x^{(6)}$ or 4.3.02 on coding [WCY, 2004]) plays the role of determinant (with very high levels of specific values of t statistics!) as both an explanatory variable of the level of material well-being $\tilde{y}^{(2)}$ and an explanatory variable of the QOLP $y^{(4)}$. It is statistically confirmed by the available historical examples (see 3.2.1) which indicate the primacy of the factor of knowledge economy in its causal connection with the QOLP. Conclusion: *we must recognize the priority of the development of knowledge economy, and in particular a significant increase in spending on research and development as an effective means of improving the quality of life of Russian population.*
2. *Security (physical) of a society member and of (institutional) private property* (variable $x^{(6)}$ or 2.5.02 by coding [WCY, 2004]) as well as R&D expenditures (variable $x^{(6)}$), double plays the role of a determinants in the considered dependencies: it is a significant explanatory factor also for the level of social cohesion in society $y^{(3)}$, and for the quality of life of population $y^{(4)}$! Conclusion: **we should immediately recognize the need to focus on socio-economic policies and measures to improve the institutional development of the country, aimed at providing a much higher levels of physical safety of members of society and the strengthening of the institution of private property**.
3. The fact that variables $\tilde{x}^{(1)}$ (total expenditures on health), $\tilde{x}^{(3)}$(degree of solving the issues of environmental pollution), $\tilde{x}^{(4)}$ (industrial emissions of carbon dioxide), $\tilde{x}^{(12)}$ (characteristics of differentiation of the population by income)

$\tilde{x}^{(11)}$ (the level of bribery and corruption) and $\tilde{x}^{(15)}$ (degree of social responsibility of business leaders) are among factors-determinants, significantly affecting the values of criterial indicators of quality and lifestyle, dictates the need for institutional changes and reforms in the field of socio-economic policy, aimed at a *significant improvement of the situation of environmental pollution, reducing the level of corruption and the level of differentiation of the population by income as well as increasing social responsibility of business leaders.*

In conclusion we will discuss the possible directions of development of this study.

Homogenization of the raw data. Raw statistical used in the analysis data covers 46 countries (see the list in Appendix 3.1) which are very heterogeneous in political organization, history and level of socio-economic development. Hypothetically, it can be expected that the results of the evaluation of the sought dependencies, produced only using, for example, the data on developing countries will be significantly different from the same results obtained from the data related to economically developed countries. It would be interesting to statistically test this hypothesis, after dividing the existing data into uniform (in a certain sense) subsamples and estimating the sought regression dependencies separately for each such subsample.

The use of special methods for the analysis of panel data. In the present study the estimate of the sought dependencies was conducted separately using the "cross-section" data, i.e. according to the data characterizing the situation in a given (fixed) year. However, after that, we analysed data to obtain "statistically significant" differences of the regressions (using Chow test), which eventually formed the basis for consideration of the 2004 version as the final result. It may be necessary to *do calculations on the entire set of data for 1997–2004 using the methods of panel data* and identification for so-called fixed and random effects models, etc. (see, e.g. [Baltagi, 1995]). This approach, by the way, will help to "broaden the bottlenecks" which can occur when assessing on a homogeneous subsamples (see previous point) because of the relatively small number of countries comprising each subsample.

Replenishment of a priori set of explanatory variables. The set of variables characterizing (directly or indirectly) current socio-economic policies and the quality of institutions, considered in this study, unfortunately, did not include those of direct performance of investment, tax, tariff and credit policies. Perhaps a special effort to establish the enhanced database (relative to data WCY) will allow to include these characteristics in the general scheme of the analysis. We believe that it could help improve the information value obtained in this analysis results.

3.3 Identifyication of key areas of improvement of social and economic policy in the russian federation regions

Presented in this section are examples of implementation of the general logistics of using II QOL while estimating the effectiveness of current social and economic policy described in Paragraph 3.1 and recommendations for its improvement based on the research carried out by group directed by the author on behalf of the Government of Samara Region, Krasnodar Territory and of Moscow in 2005–2008.

The main goal of the research was to identify "bottlenecks" (major problem areas) in the socio-economic development of Russian Federation subject and (in the case of the Samara region), its municipalities, determining the priorities of the regional and municipal administrations to increase QOL based on the analysis of the dynamics of socio-economic development of the region and its municipalities for four basic synthetic categories QOL and for the synthetic category QOL of higher level of generality. At the same time recommendations to the regional and municipal administrations were based on a special hierarchical scheme of analysis, under which the researcher sequentially moves from *the composite* (*single*) II QOL to II of the *base* synthetic category, and from it to the *block II* of the basic category and, finally, to the *specific statistical indicators*, which stand for particular criteria of the correspondent posteriori set.

3.3.1 Samara Region (Background Data for 2001–2004)[9]

The subject of analysis in this example is a synthetic category of "the quality of population" and the subject – Samara region (SR). We are interested in answering the following questions.

1. What are the dynamics of SR status among all the regions of the Russian Federation and among the 14 regions of the Volga Federal District (VFD) for this synthetic category?
2. What are the main "bottleneck", problem areas that should be addressed first and foremost by the SR government in the process of making and carrying out their social and economic policy?
3. How to identify those variants of threshold (target) values of key performance indicators, the achievement of which would significantly improve the position of SR for a given synthetic category?

9 This part is based on the results of a study commissioned by the administration of the Samara region in 2005 (State Contract with CEMI on 08.02.2005, with the title "Development of the integral indicator reflecting the main trends in the dynamics of quality of life of the Samara region").

Results of calculations. We are using the calculation results in 2.4.2 of the book and, in particular, the decomposition of synthetic categories QOL on the blocks as shown in Figure 2.2., the content of the block of the synthetic category "quality of population", presented in Table 2.5, and a complete description of the analysed statistical indicators (particular criteria) contained in Appendix 2.6.

The general logistics, which we use in such an analysis, is as follows. *During the first step* we analyse the table of movement (in time) of the block and consolidated integral indicators of the considered synthetic category and corresponding ranks of the region among all the subjects of Russia, as well as only among the regions of the Volga Federal District (see Table 3.6).

The table shows a generally positive dynamics on the block, as well as on the consolidated II of population quality. But clearly the indicators of the 1st and 2nd blocks are "pulling back": on II of these blocks Samara region, although progressing, but still remains only on the 20th and 33th places respectively. To understand the causes of this situation, we will *transition to the 2nd step* – *"going down one floor"* in the hierarchical system of the indicators and analysing in detail the statistics included in these two blocks (see Table 2.5).

Table 3.6. Dynamics of the three blocks indicators and a consolidated integral indicator

	$y^{(1.1)}(1) / R^{(1.1)}(1)$	$y^{(1.1)}(2) / R^{(1.1)}(2)$	$y^{(1.1)}(3) / R^{(1.1)}(3)$	$y^{I}_{cons}/R^{(I)}$
2001	$5,47/^{37}/_{7}$	$3,76/^{51}/_{9}$	$8,31/^{7}/_{1}$	$5,00/^{26}/_{7}$
2002	$6,52/^{21}/_{5}$	$4,97/^{39}/_{7}$	$8,37/^{8}/_{1}$	$6,00/^{18}/_{4}$
2003	$6,97/^{18}/_{4}$	$5,57/^{25}/_{6}$	$8,44/^{7}/_{1}$	$6,50/^{10}/_{3}$
2004	$6,68/^{20}/_{4}$	$5,58/^{33}/_{5}$	$8,47/^{5}/_{1}$	$6,43/^{11}/_{4}$

* The top number is the value of II, the numbers under the 1st and 2nd slashes are SR ranks among the subjects of the Russian Federation, respectively, and among the regions of the Volga Federal District (VFD).

Comparative analysis of the dynamic of the 1st and 2nd blocks indicators identified the following "hot spots" (weaknesses) in the region on the category "quality of population":

- *deaths from infectious and parasitic diseases and tuberculosis (number of cases per 100 000 population):*
  - average of VFD ≈ 23,
  - Samara region. ≈ 40,
  - the best indicator ≈ 16 (Oryol region.);
- *number of people with disabilities (per 1 thousand population):*
  - average of VFD ≈ 60,0,
  - Samara region. ≈ 72,4,
  - the best indicator ≈ 19,5 (Chukotka);
- *number of people with congenital anomalies (per 1 thousand population)*
  - average of VFD ≈ 1,5,
  - Samara region. ≈ 2,1,
  - the best indicator ≈ 0,7 (Lipetsk and Sverdlovsk regions).

Therefore, these indicators should considered the main "bottlenecks" in the issue of improving the quality of the population and that SR it is namely the improvement of the situation determining the dynamics of these indicators (certain aspects of health care, the environment, etc.), is one of the priority tasks of socio-economic governance.

Finally, *on the 3rd step*, we turn to multiple-scenario calculations by determining on the "input" possible target values of performance indicators (components of the blocks) and determining on the "output", using the formulas obtained earlier for II, the values of integral indicators and the corresponding values Samara Region ranks. The final decision on specific actions in carrying out economic and social policies is made into account the evaluation and comparison of the expenditures required to implement one or another variant of the script. A set of possible scenarios presented in the so-called scenario map (see Table 3.7).

Analysis of the "scenario card" practically allows us to answer the question: "What is the value of the composite integral indicator of quality of the population of the Samara region (and thus its ranking among other subjects of the Russian Federation for this synthetic category), if it is possible to reach certain threshold values of performance indicators?"

Note that the analysis of other synthetic categories as well as of a single (consolidated) integral indicator QOL is conducted using exactly the same logics. Only in the latter case, in order to get to the performance indicators that constitute the blocks, *we would consistently need to go down two floors*, first to the level of integral indicators of the four basic synthetic categories and then one more floor to the blocks.

3.3.2 Moscow (Raw Data for 2002–2006)[10]

In this paragraph we are presents fragments of a large comprehensive study related to the implementation of the methodology (described in Chapter 2) in order to evaluate (using 2002–2006 data) quality of life of Muscovites and the level of socio-economic development of Moscow in comparison with other regions of the Central Federal District (CFD) and major cities and mega-cities of the world, as well as to identify the parameters of social and economic policy and institutional development which primarily affect the quality of life of Moscovites.

Therefore, creation of the information base for this research involves filling the hierarchical system of indicators (particular criteria) and integral indicators with a specific content, specified by the problems solved in this research and *operating on three levels*:

10 The text of the item based on the results of research commissioned by the Government of Moscow in 2008 (State Contract with the Institute of Economics of RAS number) DEPR/250-01-07 of 22.10.2007 on "Development strategy of Moscow until 2025 year"; CEMI was one of the subcontractors).

Table 3.7. Scenario card for the category “quality of the population” of Samara Region*

Current unified value of variable (see coding of variables in A2.6)	Change to a level						
	Variant 1	Variant 2	Variant 3	Variant 4	Variant 5	Variant 6	Variant 7
1 1.1.13 **0,670**	Chuvashia 0,738	Chuvashia 0,738	Chuvashia 0,738	Мордовия 0,760	Мордовия 0,760	Мордовия 0,760	Bashkortostan 0,835
2. 1.1.6 **0,948**	No change	No change	No change	No change	No change	No change	No change
3. 1.1.5 **0,446**	Average PFD 0,415	Average PFD 0,415	Average PFD 0,415	Sverdlovsk reg. 0,465	Sverdlovsk reg. 0,465	Sverdlovsk reg. 0,465	Chuvashia 0,540
4. 1.1.23+1.1.24 **0,489**	Average RF 0,514	Average RF 0,514	Average RF 0,514	Saratov reg. 0,618	Saratov reg. 0,618	Saratov reg. 0,618	Average VFD 0,644
5. 1.1.25 **0,536**	Average PFD 0,568	Average PFD 0,568	Average PFD 0,568	Tatarstan 0,704	Tatarstan 0,704	Tatarstan 0,704	Tatarstan 0,704
6. 1.1.26 **0,636**	Bashkortostan 0,673	Bashkortostan 0,673	Bashkortostan 0,673	Bashkortostan 0,673	Chuvashia 0,772	Chuvashia 0,772	Moscow 0,793
7. 1.1.27 **0,781**	No change	Saratov reg. 0,888	Saratov reg. 0,888	Saratov reg.0,888	Saratov reg. 0,888	Москва 0,944	Moscow 0,944
8. 1.1.28 **0,737**	No change	No change	Tatarstan 0,782	Tatarstan 0,782	Tatarstan 0,782	Mordovia 0,909	Mordovia 0,909
9 1.1.29 **0,558**	No change	No change	Chuvashia 0,611	Chuvashia 0,611	Chuvashia 0,611	Saratov reg. 0,621	Saratov reg. 0,621
10. 1.1.30 **0,565**	No change	No change	Chuvashia 0,634	Chuvashia 0,634	Chuvashia 0,634	Tatarstan 0,725	Tatarstan 0,725
11. 1.1.31 **0,334**	No change	No change	Volgograd reg. 0,385	Volgograd reg 0,385	Volgograd reg. 0,385	Mordovia 0,404	Mordovia 0,404
12. 1.3.6 **0,833**	No change	No change	Mordovia 0,845	Mordovia 0,845	Mordovia 0,845	Moscow	Moscow1

13. 1.4.2	No change	No change	No change	No change	No change	No change	No change
14. 1.3.2	No change	No change	Moscow region	Moscow region	Moscow region.	Mordovia	Mordovia
0,857			0,889	0,889	0,889	0,890	0,890
Consolidated Integral Indicator of QOL for the area changes from the actual value of 6.43 to:	6.62	6.68	6.83	7.09	7.24	7.5	7.7
Rank of region for quality of population will change from the value 11 to:	9	8	7	5	4	2	1

* The calculation is carried out from the “base” – 2004, that is under the “current unified values” we understand (both in the first column of the table, and variants) unified values 2004. divided by 10.

PFD – Privolzhsky Federal District (name of the territory).

RF – Russian Federation.

(a) *regional*, which is defined by a particular set of analysed indicators and indicators characterizing QOL in Moscow and the compared subjects of the Central Federal District;
(b) *cities inside RF*, which is formed by a specific set of analysed parameters and indicators characterizing QOL in Moscow and in each big city compared with it;
(c) *international cities*, which is defined by a particular set of analysed indicators characterizing QOL in Moscow and mega-cities of the world compared with it.

The structure and content of informational data for *interregional* analysis (level (a)) for each of the considered synthetic latent categories are described in Chapter 2.

A posteriori set of statistical indicators (particular criteria), which served as a base for building the meters (integral indicators) of synthetic categories of "*quality of life of the city population*" and "*level of economic development of the city*" (level (b)), are shown in Table 3.8.

Table 3.8. A posteriori sets of indicators for synthetic categories QOL and LED

Synthetic categories	Statistic indicators (particular criteria) of a posteriori set, characterizing given synthetic category
Quality of life of population (QOLP)	index of ethnic diversity the rate of natural population growth (per 1000 people)crude death rate (per 1000 of people) unemployment rate (%) analogue of GDP per capita (total volume of production in construction, industry, volume of paid services and retail turnover, all per capita) ratio of average wages and minimum subsistence level of working population housing supply (sq. m per capita) housing (sq. m. total area per capita) car supply (the number of cars in personal use per 1000 inhabitants) number of registered crimes (total number of registered crimes per 100.000 people)
Level of Economic Development (LED)	volume of industrial production per capita production volume in construction the per capita volume of total retail sales per capita volume of paid and household services per capita total investments per capita density of public transportation lines per square km area of the city (surface transport and metro – for cities with underground lines)

Comparative analysis of QOL and of the level of socioe-conomic development (LSED) in Moscow and 20 mega-cities in the world (level (c)) was carried out in the *unified* "phase" space of the following ten indicators:

- relative population density (city population density in relation to the country population density) (in 2007);
- number of owned cars per 1 thousand inhabitants (in 2006);
- cost of living index published by the agency "Mercer" (in 2007);
- housing, sq. m. per capita (US cities – in 2007, the others – in 2006);
- total number of registered crimes per 100 thousand people (US cities – in 2007, the others – in 2006);

- density of metro lines (km of tracks per 1 sq. km area). In Zurich – density of tram tracks (due to lack of subway) (in 2006);
- indicator of ethno-linguistic diversity;
- GDP per capita (in 2007, Asian cities, as well as Moscow and St. Petersburg – in 2006);
- unemployment rate (in 2006);
- relative wages (in arbitrary units relative to New York).

In this section of the book we are interested in how parameters of socio-economic policy and institutional development of the city affect the *quality of population, level of well-being and quality of social services* of Muscovites. The high value of institutions for the development of economy and society is now an established fact of economic theory (see e.g. [Polterovich, Popov, 2006]). Cross-country comparisons show [Rodric, Subramanian, Trebbi, 2005], that institutional factors affect the quality of life of the population more than a number of other seemingly more important factors (geography, foreign trade activity). And there is no doubt that various aspects of QOLP of a territory depend substantially on the parameters of its current socio-economic policy.

Thus, constructed and evaluated at an early *stage integral indicators of human potential (quality of population)* $y^{(1)}$, *level of well-being* $y^{(2)}$ *and quality of social services* $y^{(3)}$ (for explanation of their meaning and the method of construction, see Chap. 2) $y^{(3)}$are used as explained (resultant) variables and the parameters of socio-economic policy and institutional development of the area (in our case, the subject of the Russian Federation) are used as explanatory variables. A certain difficulty was the lack of official information on the parameters of institutional development of the regions of Russian Federation (in this respect, inter-country studies are in a better position, see. Paragraph 3.2). Relevant database had to be created, combining the data from disparate sources. A priori list of the parameters on socio-economic policy and institutional development in the region of the Russian Federation, analysed in this study, is given in Appendix 3.3. This list, with more than 40 indicators of institutional development and socio-economic policy, was formed of the following sections:
- conditions and characteristics of entrepreneurial activity (including the efficiency of the bureaucratic system, the level of protection of contracts and property rights, the availability of financial resources)
- information and communication technologies;
- criminality;
- foreign trade;
- budget performance (federal and regional);
- democracy (including the electoral code);
- demography, human potential;
- differentiation of the population by income.

As a result of econometric analysis (see 2.3.4) the following *a posteriori* set of indicators was selected (Table 3.9).

The indicator "*activeness of small business*" is built as a modified 1st principal component (see 2.3.6 of the book) of three indicators:
- number of small businesses per capita;
- average number of employees of small businesses per capita;
- percentage of products manufactured by small businesses in the GRP (in 2002–2004), and the ratio of the small business turnover to the GRP (in 2005–2006).

The index ranges from 0 to 10, and increases when the activity of small business in the region grows.

The aggregate "*physical security*" is built as a modified 1st principal component of three attributes:
- number of reported homicides and attempted murders per capita;
- number of registered facts of intentional infliction of serious bodily injury per capita;
- number of reported rapes and attempts on rape per capita.
- The index ranges from 0 to 10, and increases when the physical security of the regional population grows.
- *The Gini coefficient* for each year is unified by the formula (2.4) for $N = 1$.

Table 3.9. A posteriori set of indicators

Synthetic category of quality of life of population (explanatory variable)	Name of variable	Variable
Quality of population, $y^{(1)}$	SECUR.	Physical security for 2002–2006
	RESEAR.	Number of researchers with advanced degrees per 1000 people for 2002–2006
	ETHNO.	Ethno-linguistic diversity, 2002
	GINI	The Gini coefficient for 2002–2006
	NE	Growth in the share of budget expenditures on national economy in relation to the previous year (index) 2001–2006
	EDU	Growth in the share of budget expenditures on education in relation to the previous year (index) of 2001–2006
	HEALTH	Growth in the share of budget expenditures on health in relation to the previous year (index) of 2001–2006
Level of well-being, $y^{(2)}$	ASB	The activeness of small business for 2002–2006
	SECUR.	Physical security for 2002–2006
	UNEMP	The unemployment rate for 2002–2006
	NE	Growth in the share of budget expenditures on the national economy in relation to the previous year (index), 2001–2006
	EDU.	Growth in the share of budget expenditures on education in relation to the previous year (index) for 2001–2006
	HEALTH	Growth in the share of budget expenditures on health care and sports in relation to the previous year (index) for 2001–2006
	ln (dens)	The logarithm of the relative density of the population (2002–2006)

Table 3.9. (continued)

Synthetic category of quality of life of population (explanatory variable)	Name of variable	Variable
Quality of social services, $y^{(3)}$	FFSR	Proceeds from the Fund for Financial Support of Regions,% GRP
	SSF	Growth in spending the Social Security Fund (compared to the previous year)
	ln (ETHNO.)	The logarithm of the ethno-linguistic diversity
	ln GGR	The logarithms of the growth in the share of grants revenues in budget revenues (compared to previous year)
	GTRB	Growth in the share of tax revenues in budget revenues (compared to previous year)
	ln HCS	The logarithm of the growth of the share of expenditure on housing and communal services in the total expenditures of the budget (compared to previous year)
	NE	Growth in the share of budget expenditures on national economy (industry, energy and construction; compared to the previous year)

The optimum value is selected as the lowest value at which the uniform Gini coefficient included in the regression of the quality of the population with a positive sign (Table 3.10).

Table 3.10. Conventionally optimal values of the Gini coefficient

Years	2002	2003	2004	2005	2006
Optimal value of the Gini coefficient	0,368	0,376	0,390	0,387	0,391

The indicator "*ethno-linguistic diversity*" is calculated as follows. Let among N residents of the region for N_j people have a native language of the j $(j = 1, 2, \ldots, k$ where k is the total number of different languages represented in the region as native for the corresponding part of the population). Then, the indicator "ETHNO" of ethno-linguistic diversity is defined by formula

$$\text{ETHNO} = 1 - \sum_{j=1}^{k} \left(\frac{N_j}{N}\right)^2.$$

We can show that this indicator determines the probability of the event that two randomly selected residents of the region will have *different* native languages. It is built on the basis of census data of Russia in 2002 and is widely used in the literature as an institutional characteristic. In particular, it is published in the review of institutional indicators [Aron, 2000]. We have sufficient data to calculate this indicator only for 2002, however, because the ethno-linguistic diversity of the region is changing very slowly, it can be assumed that during 2003–2006 it remained unchanged.

And now let us comment on the situation in Moscow on some parameters of institutional development.

The activity of small businesses in Moscow is more than 2.5 times higher than the average for Russia (see Figures 3.2 and 3.4 in Appendix 3.4 indicator ASB). On this indicator, Moscow ranks first in Russia, a little ahead of St. Petersburg. According to Rosstat, about 20% of all employees of small companies in Russia work in Moscow. In addition, 20% of all small enterprises in the country are concentrated in Moscow. Their number is growing, and their annual turnover is one-third of the total annual total of all small enterprises in Russia.

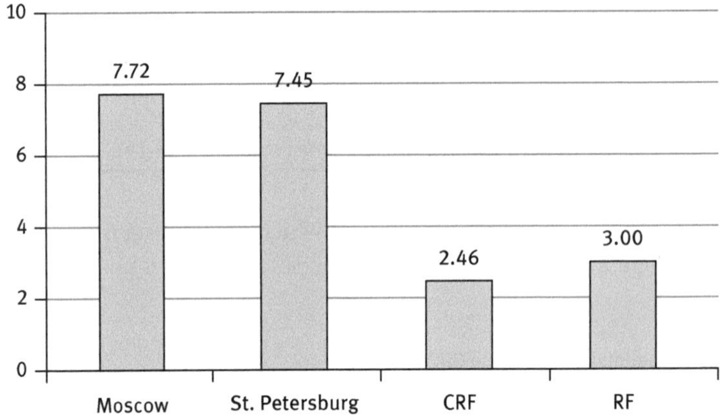

Figure 3.2. Indicator of the activity of small business in 2006

Innovative potential of Moscow's high. By the number of personal computers per capita Moscow is in first place in the Russian Federation. The public institutions of the capital are three times more likely to use computers with Internet access than the national average. Nearly half of all PhDs working in Russia lives in Moscow. More than a third of all scientists of the country (about 140 thousand people) work there. Unfortunately, their number is constantly decreasing: while the population of the capital as a whole has grown for 4 years to 56 thousand and amounted to 10.4 million people in 2006, for the period of 2002–2006 the number of researchers in Moscow decreased by more than 10 thousand. *Young people are leaving: the number of PhDs has remained unchanged, but the number of PhD candidates decreased.* This reflects the situation in Russia as a whole, where the "brain drain" is taking place.

Foreign Trade Activity in Moscow increased. Exports to non-CIS countries increased from 22.9 billion dollars in 2002 to 103 billion dollars in 2006 and in 2006 amounted to almost 40% of Russia's exports. Moscow is steadily leading among the regions of the Russian Federation by the volume of exports and imports.

Characteristics of social stratification. Ratio of funds and the Gini coefficient for Moscow have the highest values both on the Central Federal District and on

Russia. The magnitude of separation among the regions of Russia Moscow is followed by Tyumen are Samara region, St. Petersburg. In the Central Federal District Moscow is followed by Voronezh, Orel and Tambov regions. Moscow leads by a wide margin. According to Rosstat, the funds rate (the ratio of incomes of the richest 10% of the population to 10% of the income of the poorest of the population) in Moscow exceeds 40 times (Figure 3.3). This is very alarming, especially since it is known that the coefficient of funds calculated by Rosstat, is understated, as incomes of the richest part of the population is almost not considered.

Moscow is in second place in Russia after Tyumen on the value of the index of the "*independency of the regional budget from the federal*". Among the regions of the Central Federal District Moscow ranks first.

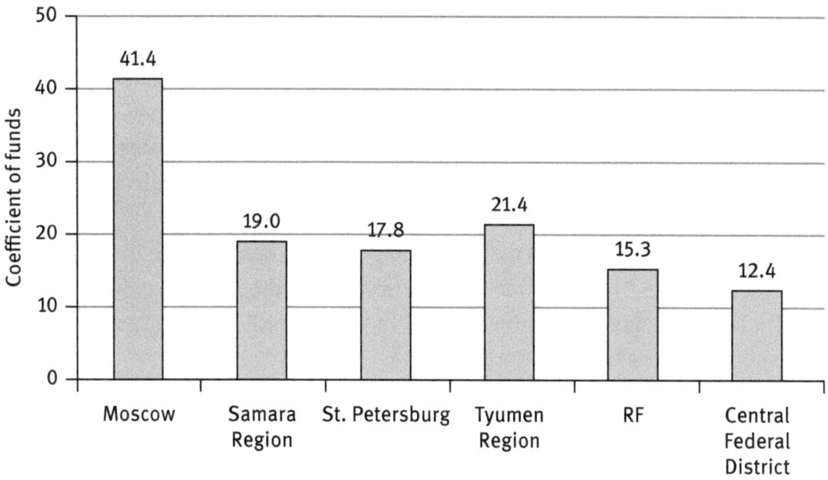

Figure 3.3. Relation of values of coefficient of funds of Moscow and following regions

Bureaucracy and corruption, protection of property rights do not meet the common position of leadership in Moscow. Here are some results of the survey of entrepreneurs conducted by "Opora Rossii" ("Support of Russia") together with the VTsIOM (All-Russian Center for Public Opinion) [Опора России, 2006]. On the indicator of illegal actions of officials in relation to entrepreneurs (see Appendix 3.4a indicator IAORE) Moscow occupies 51th place among the regions of the Russian Federation, and is close to Ingushetia and Dagestan. On the indicator of legal protection, the capital belongs to the "weak" regions of RF (according to the classification of "Opora Rossii") occupying 44th place on the index of "chances of the entrepreneur to defend its legitimate interests in court" (see Appendix 3.4a indicator DLIC).

Conditions of small businesses in Moscow are relatively favorable compared to other regions of RF. The study conducted by "Opora Rossii" considers Moscow among the regions with sustainable development of small business, where entrepreneurs often give a positive evaluation of the conditions of small business in the region compared to other regions. The value of the *index of institutional factors* in business development of Moscow is above average. Among the *conditions* of *small business development* were considered such factors as the size of the market, the value of consumer demand, the attitude to business by the authorities, the availability of resources (staff, production facilities, finance), as well as the level of security (protection from crime, extortion). The integral *index of institutional factors of business development* was calculated on the basis of indices obtained from the following indicators: attitude to small business by the authorities of the region, their impact, and support, legal protection, quality of the competitive environment and impact of big business, security, and availability of property and financial resources. Note that if on the first indicator Moscow leads among the regions of the Russian Federation, then on the second it is on the border with the disadvantaged regions. Among the regions of the Central Federal District Moscow is leading on the first indicator. On the *index of institutional factors* in the development of entrepreneurship Moscow ranked fourth after Belgorod, Bryansk and Lipetsk regions.

The results of econometric analysis of dependencies between integral indicators of quality of life and the parameters of socio-economic policy and institutional development

Thus, for each (*j*) of the three synthetic categories of quality of life in subject *i* of RF we have (see Table 3.9) on the one hand, the value of the corresponding integral indicator $y_{it}^{(j)}$ characterizing the *i* subject of RF in year *t*, and on the other, the value of the corresponding set of explanatory variables

$$\left(z_{it}^{(j.1)}, z_{it}^{(j.2)}, \ldots, z_{it}^{(j.k_j)}\right),$$

where $z_{it}^{(j.l)}$ is the value of the parameter *l* of socio-economic policy or institutional development (from the set corresponding to the synthetic category *j*, see Table 3.9) characterizing subject *i* of RF in the year $t(i = 1, 2, \ldots, 78; t = 1(2002), 2(2003), \ldots, 5(2006); j = 1, 2, 3; k_1 = 7, k_2 = 6; k_3 = 8)$.

Having conducted econometric analysis of the data or every $j(j = 1, 2, 3)$ and $t(t = 1, 2, \ldots, 5)$

$$\left(z_{it}^{(j.1)}, \ldots, z_{it}^{(j.k_j)}; y_{it}^{(j)}\right), \quad i = 1, 2, \ldots, 78 \tag{3.9}$$

we *must determine the exact appearance of the dependency* $y^{(j)}$ from $z^{(j.1)}, z^{(j.2)}, \ldots, z^{(j.k_j)}$, and in particular:

(a) determine which, from the a priori set of variables $z^{(j.1)}, \ldots, z^{(j.k_j)}$, have a statistically significant effect on $y^{(j)}$;

(b) find a general form of the sought dependency $y^{(j)}$ from $z^{(j.1)}, \ldots, z^{(j.k_j)}$or such transformations of the analysed variable which will bring the sought general form to a satisfactorily approximated linear function;

(c) after solving the problem (b), i.e. having determined the parametrical family $\{\varphi(z^{(j.1)}, \ldots, z^{(j.k_j)}; \Theta)\}$ within which we identified the sought regression dependency (where $\Theta = (\theta_1, \theta_2, \ldots, \theta_m)$ is a certain m-metric parameter), we need to obtain a consistent estimate $\hat{\Theta}$ (using the initial data (3.9)) of the parameters $\hat{\Theta}$;
(d) provide a meaningful interpretation of the resulting regression dependency

$$y^{(j)} = \varphi(z^{(j.1)}, \ldots, z^{(j.k_j)}; \hat{\Theta} + \varepsilon) \tag{3.10}$$

where ε is an accidental regression remainder, showing the influence on $y^{(j)}$ of the factors not accounted for in function φ.

Note 1. In the case then function φ is *linear* on explanatory variables $z^{(j.1)}, \ldots, z^{(j.k_j)}$, correlation (3.10) takes the following appearance:

$$y_{it}^{(j)} = \theta_1(t) + \theta_2(t) \cdot z_{it}^{(j.1)} + \cdots + \theta_{k_j+1}(t) \cdot z_{it}^{(j.k_j)} + \varepsilon_{it} \tag{3.10'}$$

So the problem (b) comes to estimation of the regression $\theta_1, \theta_2, \ldots, \theta_{k_j+1}$ (e.g. in this case the number of unknown parameters $m = k_j + 1$).

Note 2. Chronologically, the implementation of solutions of (a)–(c) is performed in interactive mode, and not strictly one after the other in the order they are listed.

The numerical experiment, including: a variation of the final composition of the explanatory variables, of the general form of the function φ, working with sharply outlying observations (Tyumen Region, Moscow and some other subjects of Russia), preliminary data partitioning (3.9) into the uniform (in some sense) subsamples, and struggling with multicolinearity etc., required the evaluation and analysis of several tens of version of models for each of the three synthetic categories.

As a result, we selected the following models of the sought dependencies.

1. Dependency of the quality of population ($y^{(1)}$) on the parameters of socio-economic policy and institutional development
This dependency has a linear form (3.10') and includes as explanatory variables, except for the free term $\theta_1(t)$, five indicators, namely:

$$y_{it}^{(1)} = \hat{\theta}_1(t) + \hat{\theta}_2(t) \cdot (\text{ETHNO})_{it} + \hat{\theta}_2(t) \cdot (\text{RESEAR.})_{it} + \hat{\theta}_3(t) \cdot (\text{SECUR.})_{it} +$$
$$+ \hat{\theta}_4(t) \cdot (\text{GINI})_{it} + \hat{\theta}_5(t) \cdot (\text{HEALTH})_{it} + \varepsilon_{it}. \tag{3.11}$$

Specific numerical values of the estimates , $\hat{\theta}_3(t)$ of their t-statistics and the coefficients of determination $\hat{R}^2(t)$ are given in Table 3.11.

We can see, first, that the resulting model is *quite stable over time*: its coefficients vary slightly (within the standard error of estimation) during the transition from one year to another. Furthermore, (3.11) and Table 3.11 show that the quality of the population is positively affected by increasing ethno-linguistic diversity indicators (ETNO), innovation (expressed in this case by the percentage of researchers with advanced degrees and the number of computers with Internet access in public institutions), physical security, and increasing the share of expenditure for health and sport (among total expenses of the regional budget), as well as the proximity of the characteristic of differentiation of the population by income (Gini coefficient) to its relatively-optimal value.

Table 3.11. Results of statistical estimation of the dependency of quality of population on the parameters of socio-economic policy and institutional development

Year*	Coefficients of explanatory variables**						$\hat{R}^2(t)$
	Free term	ETHNO.	RESEAR.	SECUR.	GINI	HEALTH	
2003	−2,00 (−1,76)	3,18(8,63)	0,63(4,81)	0,17(2,91)	2,0(3,06)	2,19(2,69)	0,60
2004	−1,11 (−1,09)	3,22(9,47)	0,63(5,30)	0,17(3,14)	1,55(2,96)	1,87(2,45)	0,64
2005	−0,84 (−0,84)	3,12(9,25)	0,68(5,72)	0,21(3,90)	1,47(2,90)	1,57(2,10)	0,65
2006	−0,99 (−0,96)	3,27(9,35)	0,67(5,41)	0,17(3.06)	1,55(2,95)	1,87(2,41)	0,64

* The starting point of calculations dates back to 2003 (and not 2002), since the index fell indicator of the growth in the share of the expenditures on health and sport in relation to the previous year was included in the set of explanatory variables

** In parentheses you can see the values of *t*-statistics, i.e. the ratio of the value of the coefficient estimate to the mean square error of estimation.

Let us now consider the situation in Moscow in each area of the city, certain explanatory variables of the model (3.11).

Physical security for the region. According to the officially registered number of murders and attempted murders (per 100 thousand people.). Moscow was in eleventh place in 2006 in Russia and the fourth in the CFA after Lipetsk, Voronezh and Kursk regions. Table 3.12 shows statistics of murders and attempted murders in Moscow in 2002–2006. In 2006, the situation improved slightly compared with the previous year. However, as the data shows, it remains serious. On average in Moscow there are three murders and attempted murders per day, according to Rosstat. For comparison, according to the US Department of Justice, published in “Crime in the United States”, in New York, with a population of 80% of the population of Moscow, the number of murders in the year did not exceed 600 in 2002–2006.

Table 3.12. The number of murders and attempted murders in Moscow

2002	2003	2004	2005	2006
1275	1342	1235	1331	1191

Innovation potential of the region. According to indirect characteristics of innovation potential, Moscow is in the first place in Russia. Table 3.13 provides data on the number of researchers with academic degrees in Moscow.

Table 3.13. Number of researchers with advanced degrees of Moscow

2002	2003	2004	2005	2006
46,784	45,950	44,305	43,929	43,642

On the number of computers in public institutions with access to the Internet Moscow also ranks first among the regions of Russia. The data shows a good positive trend (Table 3.14).

Table 3.14. The number of computers with Internet access in public institutions of Moscow per 100 people

2004	2005	2006
19	22	26

Ethno-linguistic diversity. In ethno-linguistic diversity Moscow takes 39th place among the regions of Russia. Among the regions of the Central Federal District it is the most ethnically diverse city, after the Yaroslavl region (see. Appendix 3, ETHNO). The Index of ethnic diversity in Moscow during the 2002 census was 0.22. Judging by the official statistics, no correction is required in this area. The issue of unregistered influx of migrants in Moscow, we leave outside the scope of this study.

Distribution by income. Indicators of income inequality in Moscow (Table 3.15) exceed all acceptable norms (see values of funds coefficient (CF) in the Table 3.5). Reducing stratification among Muscovites should remain one of the priorities of socio-economic policy of the government of Moscow.

Table 3.15. Coefficient of funds, Moscow

2002	2003	2004	2005	2006
51	51,8	44	38,6	41,4

The share of expenditures on health and sports in Moscow budget increases, which positively affects the quality of the population (Table 3.16).

Table 3.16. Spending on education, health and sports in Moscow's budget

Year	2002	2003	2004	2005	2006
Share of expenditures on health and sports	0,079	0,073	0,08	0,113	0,125
Growth of the share of expenditures on health and sports (index)	1,04	0,92	1,09	1,41	1,11

2. Dependency of the level of well-being on the parameters of socio-economic policy and institutional development.
   This dependency is also satisfactorily approximated by *a linear model* of the form (3.10'). In particular, the final version of the model includes, besides the free member $\theta_1(t)$, four indicators as explanatory variables, namely:

$$y_{it}^{(2)} = \hat{\theta}_1(t) + \hat{\theta}_2(t)\cdot(\mathrm{ASB})_{it} + \hat{\theta}_3(t)\cdot(\mathrm{SECUR})_{it}+$$

$$+\hat{\theta}_4(t)\cdot(\mathrm{UNEMP})_{it} + \hat{\theta}_5(t)\cdot(\mathrm{EDU})_{it-r} + \varepsilon_{it.} \tag{3.12}$$

Specific numerical values of estimates $\hat{\theta}_j(t)$, their t-statistics and the coefficients of determination $\hat{R}^2(t)$are given in Table 3.17.

Table 3.17. Results of statistical analysis of dependency of the level of well-being on the parameters of socio-economic policy and institutional development

Year	Coefficients of explanatory variables*					$\hat{R}^2(t)$
	Free term	ASB	SECUR.	UNEMP.	EDU.**	
2002	1,12(1,16)	0,21(3,26)	0,19(2,74)	−0,09(−3,31)	1,57(2,15)	0,51
2003	0,34(0,43)	0,28(5,06)	0,22(3,55)	−0,06(−2,98)	1,73(2,71)	0,61
2004	0,31(0,49)	0,25(5,37)	0,26(5,23)	−0,06(−3,72)	1,72(3,31)	0,70
2005	0,72(1,07)	0,24(4,60)	0,24(4,81)	−0,06(−3,46)	1,56(2,99)	0,68
2006	−0,59(−0,81)	0,30(5,06)	0,31(5,65)	−0,05(−2,66)	2,15(3,62)	0,69

* In parentheses are the values of t-statistics, i.e. the ratio of the estimate value to the mean square error of estimation.
** For $(\text{ОБР.})_{i.t-\tau}$ (translation in parenthesis: EDU.) we use the value of increasing the share of expenditures on education in the budget of the i- subject in the year $t-\tau$ where τ–lag of about three.

Analysis of Table 3.17 shows, as in the previous case, the relative resilience of this model's parameters (3.12) in time. We also see that the level of well-being of Muscovites is positively impacted by the increase of the activity indicators of small business (ASB), physical security (SECUR.) and the growth in the share of expenditures in the budget (compared to the previous year) for education, registered three years ($\tau = 3$) ahead of the current year, as well as the reduction of unemployment.

In the paragraph above we considered the situation in Moscow, associated with the level of physical security. We also noted a relatively good situation in Moscow with the institutions of small business. The state of affairs in Moscow in the area of expenditures on education (EDU) and unemployment (UNEMP.) cannot be considered alarming either. We can see from Table 3.18 that the share of expenditures on education is increasing, although this growth is slowing.

Table 3.18. Dynamics of the share of expenditures on education in Moscow's budget

Year	2002	2003	2004	2005	2006
Share of educational expenditures	0,086	0,09	0,10	0,12	0,13
Growth of the share of educational expenditures	1,14	1,04	1,07	1,21	1,09

Moscow, compared to other cities of the world, is in a clearly advantageous position on the level of unemployment. It is also the undisputed leader on this indicator among all subjects of the Russian Federation. In comparison, the unemployment rate in Moscow – 1.6%, in the nearest competitors – Singapore – 1.7%, in Beijing – 2.5%, in London – 6%, – 10.1% in Paris.

3. Dependence of the quality of social services on the parameters of socioeconomic policy and institutional development

   In the process of finding the model for this dependency we came to a model *linear in logarithms of analysed variables*, e.g.

$$\ln y_{it}^{(3)} = \hat{\theta}_1(t) + \hat{\theta}_2(t)\cdot\ln(\text{DENSITY})_{it} + \hat{\theta}_3(t)\cdot\ln(\text{ETHNO.})_{it} + \\ + \hat{\theta}_4(t)\cdot\ln(\text{HCS})_{it} + \hat{\theta}_5(t)\cdot\ln(\text{GGR})_{it} + \varepsilon_{it}. \quad (3.13)$$

Specific numerical values of estimates $\hat{\theta}_j(t)$of their t-statistics and the coefficients of determination $\hat{R}^2(t)$ are given in Table 3.19.

We can see that the results of the model evaluation according to 2005 and 2006 data practically do not differ. The first two explanatory variables (relative density of population and its ethno-linguistic diversity), although can be attributed to the parameters of institutional development, are not amenable to direct regulation. It should be noted, incidentally, that if the growth "ETHNO", as we have seen, had positive impact on the *quality of population*, its impact on the quality of social services proved just the opposite. Two variables from the scope of factors in our research, namely:

- increase in the share of expenditures on housing and communal services (HCS) in the total expenditures of the budget (in relation to previous year);
- increase in the share of grants revenue in total budget revenue (in relation to previous year) have a positive effect on the quality of social services, wherein the increase of the first variable (HCS) by 1% leads to the increased estimation $y^{(3)}$ by 0.11% and the increase of the second variable (GGR) by 1% raises the same estimation by 0.05% (according to their direct elasticity).

The dynamics of these variables as part of the consolidated budget of Moscow can be traced using the input data presented in Appendix 3.4.

Table 3.19. Results of statistical estimation of the dependency of the quality of social services on the parameters of the socio-economic and institutional development

Year*	Coefficients of explanatory variables**					$\hat{R}^2$
	Free term	ln (density)	ln (ETHNO.)	ln(HCS)	ln (GGR)***	
2005	0,95 (3,56)	0,06 (5,16)	−0,07 (−3,00)	0,11 (2,36)	0,05 (1,39)	0,48
2006	1,05 (1,87)	0,05 (4,90)	−0,08 (−3,47)	0,11 (2,38)	0,04 (0,31)	0,43

* The table shows the results of just over two years since the financial performance indicators of the regional budgets for the costs of housing and grants revenue in the budget up to 2005 have been extremely unstable and contradictory.

** In parentheses are the values of *t*-statistics, i.e. ratio of estimation value to the mean square error of estimation.

*** The variable "logarithm of growth in the share of grants in the income of regional budget (compared to the previous year)" is included in the model, although it has a relatively low importance.

Comparison of quality of life and of level of socio-economic development of Moscow and 20 other major megacities in the world. List of megacities and input statistics are given in Table 3.20. The last column of Table 3.20 contains the values of the indicator "Best Cities", published by the agency "Mercer". This indicator measures (using special methodology) on the conventional scale the degree of comfort of living in the city compared to New York (the latter is set to 100 points). Cities with the indicator "Best Cities" marked with an asterisk, were absent from the list of cities evaluated by "Mercer". The corresponding values are restored using suitable regression models of the indicator "Best Cities" using other parameters of the table.

We performed clustering of compared cities in the space of the first ten analysed indicators (using *k*-means method, see 12.4.9 in [Айвазян, Мхитарян; 2001]), and identified the model of ordered multiple choice (see 9.2.1 in [Айвазян, 2010]). The results of both approaches were close. Table 3.21 shows the results of clustering.

The fifth cluster, which includes Moscow, is characterized by relatively low level of GRP (per capita) and housing, nearly full employment, and a relatively low level of car own ship. It should be noted that *the comparison between Moscow and megacities of the world reveals a clear disparity between the cost of living in the city and the level of socio-economic development (also taking into account the level of QOLP of the city).*

Table 3.20. Initial data on the particular criteria of quality of life and key indicators of socio-economic development of mega-cities of the world (2006)

	Relative density of population	Ethnic diversity	GRP per capita	Cost of living	Average housing supply	Auto-mobile supply (per 1000 people)	Unem-ploy-ment level	Density of metro tracks	Number of registered crimes per 100,000 people	Wage index	Best cities
Moscow	1129	23	16777	134,4	19,7	243	1,6	264	2085	25,4	97,2*)
St. Petersburg	290	21	9000	103,0	22,4	228	2,4	163	2193	25,4	97,1*)
New York	337	41	61000	100,0	45,0	415	5,2	305	2291	100	100
Los Angeles	102	50	53300	87,1	44,0	551	4,8	22	4862	97	100*)
Chicago	152	57	51100	105,0	44,0	515	5,7	274	5652	94,7	100,4
Madrid	57	15	36900	92,1	30,0	290	12,0	522	3852	64,3	100,5
Milano	33	18	30100	104,4	38,0	610	6,0	456	3500	59,9	99,9
Rome	11	6	47000	97,6	35,0	590	11,0	31	1852	49,7	98,7*)
Barcelona	176	25	31800	89,2	32,0	414	11,0	1145	4292	66,6	100,6
Toronto	1241	54	40900	107,0	41,1	310	7,5	98	5102	80,4	105,4
Tokyo	17	4	63000	122,1	15,8	250	5,7	131	2427	87,4	102,3
Osaka	35	4	32000	108,4	18,8	260	7,2	620	2810	85	100,5
Beijing	6	7	10000	95,9	9,3	106	2,5	8	1914	10,9	94,7*)
Hong Kong	45	7	33800	119,4	7,1	45	4,9	159	1150	34,9	99,4*)
Sydney	136	22	39000	94,9	36,0	570	3,5	1	1900	79,6	106,5
Paris	217	25	46000	101,4	35,0	440	10,1	2474	6040	68,8	102,7
London	17	35	59400	126,3	31,9	333	6,0	256	4400	96	101,2
Berlin	17	21	22024	85,9	38,8	315	15,0	170	1700	82,1	105,2
Hamburg	10	25	33139	107,5	40,0	320	8,0	132	1820	80	103,6
Vienna	41	9	29000	96,9	36,0	360	3,8	158	320	81,2	107,7
Zurich	22	35	45000	107,6	37,0	380	2,6	1	300	124,2	108,1
Singapore	1	28	32000	100	20	103	1,7	103	745	38,9	102,5

Table 3.21. Results of clustering of mega-cities of the world in the space of the first 10 indicators in Table 3.20*

1st cluster	2nd cluster	3rd cluster	4th cluster	5th cluster
New York	Madrid	Los Angeles	Tokyo	Moscow
Milano	Barcelona	Chicago	Osaka	St. Petersburg
Rome	Paris	Toronto	Hong-Kong	Beijing
Sydney	Berlin	(101,93)	London	(96,33)
Hamburg	(102,25)		(100,85)	
Vienna				
Zurich				
(103,50)				

* In parenthesis are the values of the "Best Cities" indicator for the cities of the cluster.

3.4 Conclusions

1. Construction and trending of the II synthetic categories QOL, in addition to descriptive purposes (ratings of territories according to these synthetic categories, identifying leaders and outsiders, etc.), pursues the goals related to the assessment of socio-economic policies carried out on the analysed area as well as the identification of the so-called problem areas in its socio-economic development. In particular, the negative dynamics of II with respect to its previous values and the worsening of the situation of the analysed territory on the given II value compared to other areas points to the area of concern ("bottleneck") in socio-economic management of the territory. The specific form of II allows to identify those indicative of socio-economic indicators, which are key in determining II, namely to pinpoint priority areas of public life in terms of impacting these indicators by means of public policy. Of course, regional (federal) governments make their final choice or priorities based on the cost of the correction of these indicators (particular criteria) adjustments to the cost of certain indicators (partial criteria) and on the specific political and socio1economic situation.
2. Identifying and describing the existing relationships between the II of synthetic categories QOL, on the one hand, and the direct or indirect characteristics of socio-economic policies and the quality of institutions on the other plays an important role in building territorial (federal) support systems for the decisions of governments. The solution to this problem is achieved by means of *econometric analysis*, and allows to use as the results of the current socio-economic policy the impact not only on the indicators, (particular criteria) involved in the contraction, determining the II values (see point 1 of the "Conclusions"), but also on the parameters characterizing the policy and the quality of institutions.
3. The results of the problem solution, mentioned in point 2 of the "Conclusions", obtained for Russia using of cross-country econometric analysis of the data of 1995–2004 for 46 countries (see the list in Appendix 3.1), revealed the following priority areas for improvement in current socio-economic policies in Russia (as of 2009):
   - development of the knowledge economy and, in particular, the significant increase in government expenditures on research and development activities;
   - urgent actions to ensure significantly higher levels of physical security of society members and strengthening of the institution of private property;
   - implementation of institutional changes and reforms in the field of socio-economic policy aimed at significant reduction in the level of corruption in the country!

Priority is also assigned the task of *reducing income differentiation of population, improvement of environmental pollution and improvement of social responsibility of business leaders.*

4. Observation of the dynamics of the synthetic category "*quality of population*" in the inter-regional analysis aimed to investigate the quality of life of the Samara Region (3.3.1) is an example of the identification of key areas of concern in the socio-economic development of the region by identifying the statistical indicators (partial criteria), which are primarily responsible for the negative dynamics of the integral indicator. The problem areas in this example were *mortality from infectious and parasitic diseases and tuberculosis, prevalence of disability and people with congenital anomalies.*
5. The solution of the problem stated in point 2 of Conclusion in relation to the inter-Russian analysis with the emphasis on Moscow as the main object of the study, led to the following conclusions (as of 2008).
   A. Strengths (positive):
      - positive changes in quality of population (moving from 36th place in 2002 to the 5th place in 2006 among 78 subjects of the Russian Federation) and the undisputed leadership in this indicator among the subjects of the Central Federal District (next to Moscow on this indicator Belgorod region of Central Federal District occupies only the 35th place among 78 subjects of the RF);
      - sustainable leadership (over the past 3~4 years) in the quality of social services compared to all 78 subjects of RF;
      - sustainable leadership among all 78 subjects compared to the past five years on a single (consolidated) integral indicator QOL calculating QOL on the three synthetic categories: quality of population, level of well-being and quality of social services (but excluding the environmental component).
   B. Weaknesses (negative):
      - Moscow's relocation to a worthy position in the world in the level of material well-being of its population is not possible without a radical reduction of the following: funds coefficient characterizing the ratio of incomes of the richest 10% of population to the income of the poorest 10% of population of Moscow (now, according to various estimates, it ranges between 40 and 50, and Moscow takes the last place on this indicator among all the subjects of Russian Federation), the share of the poor and the depth of poverty (12th and 19th place respectively among all the subjects of Russian Federation), and also without a radical improvement of the provision of housing for its residents (the last place in Central Federal District);
      - Indicators, showing the alarming state of affairs in the social and demographic development of Moscow, include: coefficient of natural

population growth (24th place among the 78 subjects of RF), incidences of congenital anomalies (28th place), the number of disabled people per 1 thousand of population (71th place), the prevalence of drug and substance abuse (49th place), the prevalence of HIV-infected (62th place). Apparently, overcoming the negative trends in this area will be legitimately attributed to the most important strategic objectives in the development of Moscow. Among the measures aimed at the solution of this strategic task, we see not only the improvement of the health institutions, benefits for drugs and medical supplies for population, etc. In our opinion, the policy of full support (financial, informational, moral) of healthy lifestyle, including the development of mass sports, recreational physical education, construction of sports complexes, etc. should take precedence.
- Comparison between Moscow and 20 mega-cities of the world revealed a discrepancy between the cost living in the city, on the one hand, and the level of socio-economic development of the city and the quality of life of its population on the other: being the most expensive among the compared megacities (indicator value for the cost of living in Moscow's is the highest, see Table 3.20), Moscow is inferior to all of them (except for Beijing and St. Petersburg) by value "Best cities" characterizing quality of life in the city!

C. Possible strategic risks (threats):
- continuing trend of increasing differentiation of the population by income, characterized by the values of funds coefficient around 40–50, is a strategic danger forcing social tension to a critical level, fraught with social explosions;
- Moscow's outsider (low–transl.) position on a number of important socio-demographic characteristics (natural increase of population, prevalence of people with disabilities, people with congenital abnormalities, drug addiction, HIV infection) strategically may lead to critically lowering foundations of a healthy society, namely of human potential;
- there is a disparity between the quality and cost of living of Muscovites which could lead to the collapse of the entire financial system of the capital.

4 Microeconometric Analysis of Quality of Life and Living Standards

4.1 Types of consumer behaviour households and identification of key typological characteristics[1]

Following the *pragmatic* concept (see 1.2.4), while studying quality of life, we put at the forefront of the research the **interaction of the needs** (in the broadest sense of motivation Maslow's "hierarchy of needs" [Маслов, 1999]) **and the acual consumption**. Accordingly, the problem of studying the differentiation of needs, manifested in different types of consumer behaviour of households, and the related problem of identifying the key typological characteristics can be attributed to the *central* problems of not only the theory and practice of consumer choice but the whole subject of the analysis of the quality of life of population.

Note that since this book is talking about an *econometric approach*, it is assumed that the proposed methodology should be implemented using specific socio-economic data. Available information sources of the issue discussed in this section are basically standard, official data from sampling of household budget surveys (SHBS), conducted quarterly by Rosstat of Russian Federation (RF). With respect to content, *they cover, essentially, only the economic part* of the above-mentioned Maslow's "hierarchy of needs". Therefore, the results presented in this section should be attributed to *the analysis of economic well-being of the population*, although the methodology itself with the necessary supportive information (which should include the results of additional questionnaire-type surveys of population) can be used to analyse the quality of life, understood in the most general sense.

4.1.1 Theoretical basis of research, source approval (hypotheses) and their formalization

(1°). *First*, our study is based on the so-called **behavioural (statistical-genetic)** approach, whereby the needs of the population, as a socio-psychological and economic category, should be studied, modelled and predicted **by the analysis of the actual behaviour of consumers**. The actual consumption and the real needs **are formed in the household, the family** – a social unit where specific interests, motives and orientations of people are refracted.

---

1 This section contains the results obtained in the framework funded by the grant RHF №10-02-00633a "Econometric analysis of the typology of consumer behaviour of Russian households".

An alternative to the behavioural approach is a **normative** approach, the essence of which is reduced to an a priori (axiomatic) concept of needs, based on the postulates of physiology, psychology, health, architecture, etc. This approach underlines the calculations and design of the so-called statutory consumer budgets, claiming to define a set of vital needs of population. However, it conceals some controversy. Science can only formulate psychophysical requirements for the functioning of the human organism as a representative of a species. Thus, doctors, using the scientific method, can determine the number and the set of biologically active substances (proteins, fats, carbohydrates, salts, vitamins, etc.) required for the body and then normalize them for different population groups according to their gender, age, place of residence, the nature of work, employment, etc. Meanwhile, a person is a social phenomenon who functions as a member of society and has needs based not only on biological nature but is also connected with a whole complex of socio-economic factors, which are largely social. A human does not develop a need for specific nutrients (of which he may not even be aware) but for a specific set of foods. At the same time, the same combination of calories, proteins, fats and carbohydrates, vitamins, etc. could be realized by many different sets of foods varying by cost, nature, flavour and other qualities. Choosing the best, suitable, scientifically sound, rational, etc. from a plurality of possible combinations cannot be done using only scientific methods, from the standpoint of biological requirements of the organism. In this case, it is fundamentally important to take into account a wide variety of aspects of social conditions, and therefore, it is not surprising that the very task of identifying the needs transitions from the natural to the socio-economic sciences. The influence of environment and socio-economic factors has even greater impact on the formation of spiritual and intellectual needs. These needs are of a higher level and their role keeps increasing in modern society; they almost entirely depend on the development of society, culture, science, health, and social relations.

In addition, physiological norms underlying standard consumer budgets are defined *autonomously* and independently of each other for certain types of goods and services, whereas elements of the real structure of consumption are inseparable: the size and shape of the consumption of one type of goods and services affect the consumption of other goods. In other words, in reality, what we see are not separate needs but their *complexes realized in stereotyped consumer behaviour.*

Yet the truth is that we need a reasonable combination of behavioural and normative approaches. This, incidentally, concurs with the basic idea of Makarov [Макаров, 2010]. This basic idea "is to search and find in modern society the right balance between stiffness (in our case – the standard – S.A.) and flexibility (in our case – the freedom of choice of behaviour – S.A.), which provides stability of the system" (ibid. 11). Makarov states that "individual choice and collective choice are soft and hard components of the mechanism of distribution of wealth in human society" (42), and both of these elements, of course, must be present in such a mechanism. We are talking about standards in the broad sense (including social, energy-related, and environmental), far beyond food and other consumer products. Unfortunately, the very problem of the formation of standards is studied very poorly (Makarov calls it "a problem of the near future", ibid. 45). We note only that the standards should be differentiated according to the considered types of consumer behaviour (or, in Makarov's terms, "social clusters"). This, in particular, also follows from the well-known theory in the public sector of the economy, the "Tiebout hypothesis" [Tiebout, 1956].

(2^0). *Second*, we proceed from the assumption that the society has an objectively conditioned socio-psychological and economic differentiation of population's needs and the difference in consumption associated with it. The mechanism of this differentiation, functioning according to national particularities, socio-economic conditions, specifics of education and employment, and differences in culture and place in social production, although leading to ever greater individualization of consumption, acts in a way that **generates a relatively small (compared to the millions of households considered) number of types of uniform consumer behaviour**, covering the vast majority of consumption cells (households). Thus, we proceed from the objectively existing division of the totality of households into a number of similar (in terms of consumer behaviour) classes. Substantial (socio-economic) explanation of this fact is based on the concept of *group target structures of consumption* and a related concept of *targeted levels of consumption* for each of the considered set of needs, first introduced by V. V. Soptsov [Митоян, 1990]. In particular, the existence of the target level of expenditures (consumption) of the analysed set of goods (services) is established using an empirical test of two requirements:

- *density of distribution of households according to expenditure per capita on this set of goods (services) must have a local maximum at a flow rate of the target level;*
- *derivative of the regression function of income per capita according to the per capita expenditure on this set of goods (services) should abruptly increase when the flow value is equal to the target level.*

The sufficiency of these conditions for the existence of the target level of expenditure per capita is determined by the fact that the first and second requirements may simultaneously be carried out only at the level of per capita expenditures, corresponding to the *target* level of consumption of this type of goods, but on other levels these conditions are not linked functionally or statistically. To understand this, we need to consider in more detail the content side of the above-mentioned requirements.

The first requirement, linking the target level of per capita expenditure on this group of goods with a maximum density of distribution of households according to per capita expenditures for the same group of products, has a fairly transparent meaningful interpretation. Indeed, the target level of consumption is determined by (and also defines) the existence of a centripetal force that has both a subjective and a socio-economic nature. This force causes the family not only to strive for a certain level of consumption but also to dwell on it for a certain period of time. This obvious mechanism results in an increased concentration of families in the neighbourhood of the point of per capita expenditure, corresponding to the target level of consumption of this type of goods. Having multiple target levels of household expenditure of families on the goods of the same commodity group (commodity complex) determines the appropriate number of local maxima of the density of distribution of households by per capita expenditures on this group of goods.

Meaningful interpretation of the second requirement of the existence of the target level of consumption is much less obvious. In this connection, we should consider this matter in more detail. Phenomenological research methodology of household consumption, which did not include the mechanism of formation of the consumer behaviour of households, has led to the concept of monotony and limited value of the function of expenditure/income dependency. This approach completely excludes the target aspect of consumption.

However, it is obvious that the family that has reached the target level of consumption of goods of a commodity group, for a period of time, does not increase its per capita expenditures for this group and channels increased per capita funds to expand its consumption of other commodity groups. This leads to the fact that in considering the regression dependency of per capita household expenditure on goods of a certain type (*Y*) from per capita income of the family (*X*), the regression line *Y* on *X*, having reached the next target level cost Y1 in the process of monotonic increase and then followed by further increase in per capita income, will still remain a time constant (Figure 4.1).

Levels of per capita expenditures Γ_1 and Γ_2 determine the upper limits of per capita expenditures, above which there are no families associated with the corresponding (Y_1 and Y_2) target levels of consumption of this type of goods.

Note the important fact that the function shown in Figure 4.1, i.e. the function of dependency of per capita expenditures on the goods of this complex on per capita income, is ambiguous. However, when constructing the regression of per capita expenditures of sample households on the selected goods over per capita income, this important feature does not manifest itself.

If levels of Y_1 and Y_2 correspond to the local maxima of the sample household distribution density on per capita expenditure on goods of the same group, then we have the sufficient condition that these levels (Y_1 and Y_2) are the target. Then the whole range of per capita expenditures of sample households for goods of the group is divided into three so-called consumer layers.

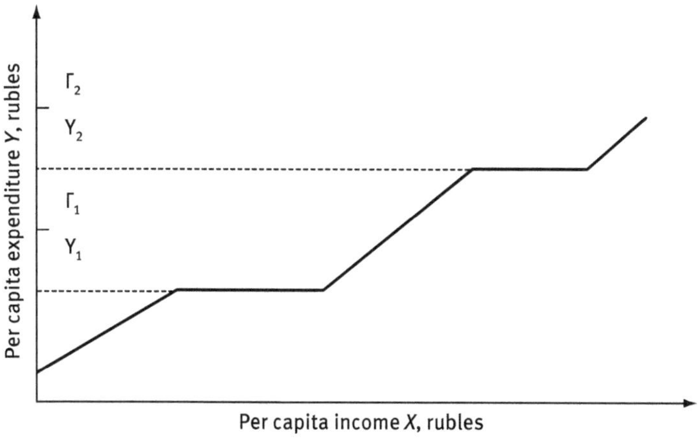

Figure 4.1. Regression of per capita expenditures on goods of one set to per capita family income as part of the study of consumption patterns

Y1 – first target level of per capita expenditures;	Γ_1 – the boundary of the first consumer layer;
Y2 – second target level of per capita expenditures;	Γ_2 – the boundary of the second consumer layer

The first target level corresponds to per capita expenditures of lower than Γ_1, the second to per capita expenditures larger than Γ_1 but lower than Γ_2, and the third to expenditures larger than Γ_2. Thus, any intersection (a combination in which each complex of goods is represented by one layer) of consumer layers of complexes of goods determines the consumer class.

(3^0). *Third*, we rely on a provision conforming to which an actually folding pattern of consumption is the result of the desire of a multitude of households to optimize their satisfaction of needs and to make their consumption pattern most appropriate, and the optimality criterion (the concept of "the most appropriate" patterns of household consumption) is separate for each type (layer) of consumer behaviour.

Note that this position is essentially a basic concept also in the consumer choice theory, which uses the so-called objective preference functions (or consumption functions) as a tool to study the problem of consumption structure and forecast. However, the researchers operating in this area tended to overlook the need for a preliminary phase of research associated with the identification of the existing typological structure of consumer behaviour. In fact, the construction of preference functions (using budget data) is legitimate *only when done separately for each type of consumer behaviour of households.*

(4^0). And finally, *fourth*, while evaluating the quality of life, we consider two aspects of causal attribution simultaneously: external circumstances and disposition of the analysed factor. At the same time, we remain on the positions of cognitive psychology just as proponents of *the expansion of human giftedness concept*. In our case, we are talking about the interaction of two *multidimensional* phenomena, one of which (*disposition of the analysed factor*) is **the behaviour of the household in consumption area** and the other (*external circumstances*) includes **determinant factors that impact the formation of people's needs.**

Determinants are many and varied. On the one hand, they refer directly to the consumer, acting in the sphere of consumption. They are gender, age and family status, structure of population, social status and labour specifics of the people employed in social production and education level and preparedness to perform work of various qualifications. All of these factors to some extent determine social activity of the population in all spheres of life. On the other hand, determinants characterize the external conditions of consumer life and describe the consumption domain. They include climate; national-ethnic features; type of settlement (town, village, the capital, farm, etc.); conditions of distribution of material and spiritual values established in the community and reflected in the level and income inequality, in the use of public consumption funds, and the scope of the national wealth in the consumption sphere.

Of course, the above list of determinant factors is not intended to be exhaustive. In addition, we did not deliberately included factors describing the state of the productive forces, which indirectly affects the overall level of consumption. Ultimately, we will only be interested in the task of formalized description of dependencies that allow, using the values of the specific (if possible, concise) set of determinant factors, characterizing a household (HH), to predict to which type of consumer behaviour we can most likely refer to this HH. In addition, the identification of this set of determinant factors (which we shall henceforth call typological signs) is also included in this research.

* * *

Note that the statistical-genetic (behavioural) approach to identifying and explaining the structure of needs does not contradict the idea of the possibility of influence on the formation of the population's needs of the society and, accordingly, does not rule out the regulatory role of the state in this process. On the contrary, the knowledge of the system of population preferences and its formal description allow to more clearly identify the boundaries and possibilities of exogenous influence on the actual development of the structure of demand.

From postulates (2°)–(4°), it is easy to conclude that the mathematical tool, the most suitable for model-formal description of the mechanism of differentiation of family consumption, identifying the main types of needs and consumer behaviour, definition of the main typological signs is *multivariate statistical analysis*, and particularly the sections such as cluster analysis, parameter estimation of mixtures of multivariate distributions, and reducing the dimensions of the tested feature space.

Below is the language of some of the general concepts and basic assumptions of this research in terms of multivariate statistical analysis.

(A) The hypothesis of consumer unit: The actual consumption, as well as the real needs, is formed in the family (household) as the primary social unit in which interests, motivation and orientation of a person are refracted.

Based on this assumption, we come to a necessity to analyse consumer behaviour and the relevant determinants (factors of life) of the scope of families as a whole and each of them individually. In particular, let O_i be a surveyed consumer unit (family) i; $i = 1, 2, ..., n$, where n is the total number of analysed families. For brevity, we denote $O = \{O_i,\ i = 1, 2, \ldots, n\}$ as the analysed scope of families.

On examining living conditions (determinants) of the family O_i, in fact, we will fix the values of a quantity (p) of indicators $x_i^{(1)}, x_i^{(2)}, \ldots, x_i^{(p)}$ of the situation of family i. Among these p indicators (attributes) can be both quantitative (income, total number of family members, size of living area, age of family members, etc.) and qualitative or categorized (quality of housing, etc.) classification (profession of family members, the industry in which the head of the family works, the landlord, etc.). Thus, with each family O_i, we can associate a p-dimensional vector (or p-dimensional observation):

$$X_i = \begin{bmatrix} x_i^{(1)} \\ x_i^{(2)} \\ \vdots \\ x_i^{(p)} \end{bmatrix},$$

which is considered in the corresponding determinant space $\mathbf{X}$ (i.e. $X \in \mathbf{X},\ i = 1, 2, \ldots, n$).

Similarly, in the study of consumer behaviour (i.e. actually prevailing consumption patterns) of family O_i, we fix a certain number (m) of indicators $y_i^{(1)}, y_i^{(2)}, \ldots, y_i^{(m)}$ of consumer behaviour, where $y_i^{(\nu)}$ $(\nu = 1, 2, \ldots, m)$ denotes the specific (i.e. calculated on the average per family member) amount of ν type of goods (goods or services, including savings) consumed by the i family during the base period (year) and expressed in physical or monetary units.

Thus, each family O_i is associated with X_i m-dimensional vector (m-dimensional observation):

$$Y_i = \begin{bmatrix} y_i^{(1)} \\ y_i^{(2)} \\ \vdots \\ y_i^{(m)} \end{bmatrix},$$

which is considered in the corresponding space of consumer behaviour $\mathbf{Y}$ (i.e. $Y \in \mathbf{Y},\ i = 1, 2, \ldots, n$).

To achieve the planned algorithmic scheme of the study, we need to make the spaces **X** and **Y** *metric*, or, in other words, to incorporate metrics into them (i.e. to determine the method of calculating the distance between any two elements in both space **X** and **Y**).

To arrive at the necessary initial mathematical concepts with which the given socio-economic categories and then setting their tasks are formalized, we need to clarify the wording of several abstracts.

(B) The hypothesis of the stratified nature of the behavioural space Y: It postulates the existence of a relatively small number of N types $S_{\mathbf{Y}}^{(1)}, S_{\mathbf{Y}}^{(2)}, \ldots, S_{\mathbf{Y}}^{(N)}$ of consumer behaviour, such that the differences in consumption patterns of families of the same type $S_{\mathbf{Y}}^{(k)}$ are random (i.e. due to the influence of many random, uncontrollable or unaccountable factors) and insignificant compared to the differences in consumer behaviour of families from different types, $S_{\mathbf{Y}}^{(k_1)}$ and $S_{\mathbf{Y}}^{(k_2)}$.

The geometric interpretation of hypothesis (B) means that there is a metric $\rho_{\mathbf{Y}}(O_i, O_j)$ in the space **Y**, taking into account the nature of the relationship of the individual components of consumer behaviour $y^{(1)}$ and the proportion of their influence on the differentiation patterns of consumption that the whole given set of "multidimensional points" of **O** families in this space naturally decomposes into a relatively small number of point "clots", or clusters $S_{\mathbf{Y}}^{(1)}, S_{\mathbf{Y}}^{(2)}, \ldots, S_{\mathbf{Y}}^{(N)}$, which are "at a distance" (in the sense of the metric $\rho_{\mathbf{Y}}$), but do not break up into equally spaced parts.

(B⁰) A reinforced hypothesis about the stratified nature of space: It differs from hypothesis (B) by an additional assumption, according to which the random spread of multidimensional points, i.e. of families of consumer behaviour $Y_i(k)$, belonging to a (k-type) $S_{\mathbf{Y}}^{(k)}$, is subject to an *m-dimensional normal law of probability distribution.* In other words, if $Y(k)$ is a random vector of consumer behaviour of a family, randomly extracted from a homogeneous set $S_{\mathbf{Y}}^{(k)}$ of families of k-type, then its probability density is given by a ratio:

$$f_k(Y) = \frac{1}{(2\pi)^{\frac{m}{2}} |\Sigma(k)|^{\frac{1}{2}}} e^{-\frac{1}{2}(Y-\mathbf{a}(k))^{\mathrm{T}} \Sigma^{-1}(k)(Y-\mathbf{a}(k))}. \tag{4.1}$$

In this ratio, vector

$$\mathbf{a}(k) = \begin{pmatrix} a^{(1)}(k) \\ a^{(2)}(k) \\ \vdots \\ a^{(m)}(k) \end{pmatrix} \tag{4.2}$$

sets the average (and at the same time the most characteristic) structure of all possible consumer behaviours of the k-type households, and the matrix

$$\Sigma(k) = \begin{pmatrix} \sigma_{11}(k), & \sigma_{12}(k), & \dots, & \sigma_{1m}(k) \\ \sigma_{21}(k), & \sigma_{22}(k), & \dots, & \sigma_{2m}(k) \\ \dots & \dots & \dots & \dots \\ \sigma_{m1}(k), & \sigma_{m2}(k), & \dots, & \sigma_{mm}(k) \end{pmatrix} \tag{4.3}$$

determined by the so-called covariance of the components of consumer behaviour of k-type families, i.e.:

$$\begin{aligned} \sigma_{qe}(k) &= \sigma_{lq}(k) = \operatorname{cov}\,(y^{(l)}(k), y^{(q)}(k)) = \\ &= \mathbf{E}\{(y^{(l)}(k) - a_{(l)}(k)) \cdot (y^{(q)}(k) - a_{(q)}(k))\}. \end{aligned} \tag{4.4}$$

At the same time, the distribution of the random vector $Y(k)$ is assumed to be *non-degenerate.*

Sample (empirical) analogues of theoretical values are $y_0^{(l)}(k)$ and $\sigma_{ql}(k)$, respectively:

$$\hat{a}^{(l)}(k) = \frac{1}{n_k} \sum_{i=1}^{n_k} y_i^{(l)}(k), \quad (l = 1, 2, \dots, m) \tag{4.2'}$$

$$\hat{\sigma}_{ql}(k) = \frac{1}{n_k} \sum_{i=1}^{n_k} (y_i^{(l)}(k) - \hat{a}^{(l)}(k))(y_i^{(q)}(k) - \hat{a}^{(q)}(k)) \quad (q, l = 1, 2, \dots, m), \tag{4.4'}$$

where n_k is the total number of households recorded in the k-type of consumer behaviour and $y_i^{(l)}(k)$ is the value of the l component of consumer behaviour registered in the i-th k-family type.

(C) The hypothesis of optimality: This is the short term for the assumption according to which for every (k) type of consumer family behaviour we postulated the existence of its own optimality (or rationality) criterion, the mathematical expression of which is naturally presented in the form of the so-called preference function or target functions of consumption $u^{(k)}(Y)$. This function allows to numerically measure the degree of preference (rationality) of each fixed "consumer behaviour" Y, and the rules for calculating the numerical value of this degree of preference (i.e. the specific form of the consumption function) is the same only for the families belonging to the same type, and changes when we transition from one type to another.

(D) Statistical manifestation of optimality hypothesis: It is a natural development of the optimality hypothesis (**C**). It postulates that the consumption structure Y, actually developed in a randomly selected family of k-type, is the result of objective laws (usually unconscious or only intuitively conscious by the family), equivalent to the evaluation and comparison of the different options for Y in terms of their degree of rationality. At the same time, a real measure of awareness of each individual family and the distorting effect of a set of random factors prevent it from reaching the exact optimal structure $Y_{\text{opt}}(k)$, but the objective trend of aspirations of a multitude of homogeneous (i.e. one type) families towards the optimization of their needs (under certain restrictions by income) is expressed in the fact that, as a result of the law of large numbers, the modal, i.e. the most commonly observed (in a broad class of cases it is the same as the statistical average) actual structure of consumption of households, k-type $Y_{\text{mod}}(k)$ is exactly optimal, i.e.

$$Y_{\text{opt}}(k) = Y_{\text{mod}}(k)$$

or is the same as

$$u^{(k)}(Y_{\text{mod}}(k)) = \max_{Y} u^{(k)}(Y). \tag{4.5}$$

with restrictions

$$\sum_{l=1}^{m} p^{(l)} y^{(l)}(k) = s(k) \tag{4.6}$$

$$y^{(j_q)} = \tilde{y}^{(j_q)}, \quad q = 1, 2, \ldots, L, \tag{4.6'}$$

where $p^{(1)}, p^{(2)}, \ldots, p^{(m)}$ are the retail price per unit of goods, respectively, $y^{(1)}, y^{(2)}, \ldots, y^{(m)}$ and $s(k)$ is the average value of the per capita average income of families (households) of the k-type of consumer behaviour (if specific quantities $y^{(1)}, y^{(2)}, \ldots, y^{(m)}$ of goods consumed by the family are expressed in monetary units, then, obviously, $p^{(1)} = p^{(2)} = \ldots = p^{(m)} = 1$).

Equation (4.6) shows the limitations of *resource type*, while equation $(4.6')$ shows the a priori restrictions on certain types of goods that are *rationed* or *regulated*. The relevance and obviousness of rationing and regulation of certain goods even in a market economy is clearly stated, for example, in [Makarov, 2010, sections 9.11–9.13]. However, they also mention (163) that "Mathematical economics has made a small contribution to the theory of rationing (see e.g. D.H. Howard *Rationing, quantity constraints and consumption theory*//Econometrica. 1977 Vol. 45. N 2.) There it is emphasized that the regulations distort the market equilibrium. Much more important is the question of how standards are generated by society."

(E) The hypothesis of the existence of typological signs/determinants of the trust level $(1-\beta)$: It is postulated that in the initial vector of factors of family life we can distinguish a sub-vector

$$\tilde{X} = \begin{pmatrix} x^{(j_1)} \\ x^{(j_2)} \\ \vdots \\ x^{(j_{p'})} \end{pmatrix} = \begin{pmatrix} \tilde{x}^{(1)} \\ x^{\sim(2)} \\ \vdots \\ \tilde{x}^{(p')} \end{pmatrix}, \quad p' < p$$

and choose metric such as $\rho\tilde{X}(O_i, O_j)$ in the corresponding space **X**, and in this metric space, we can find non-overlapping domains $S_{\tilde{X}}^{(1)}, S_{\tilde{X}}^{(2)}, \ldots, S_{\tilde{X}}^{(N)}$ for which the following statement is true: for any type of consumer behaviour $S_Y^{(k)}$, there is the corresponding range of values of typological signs $S_{\tilde{X}}^{t(k)}$, such that[2]

$$P\{Y \in S_Y^{(k)} | X \in S_{\tilde{X}}^{t(k)}\} \geq 1-\beta, \quad k = 1, 2, \ldots, N. \tag{4.7}$$

It is pertinent to note that the fulfilment of equation (4.7) allows to obtain an estimate of the accuracy of the method of determining the type of household consumption of the value of its typological sign. Namely, after the establishment of correspondences of type $k \leftrightarrow t(k)$ between each of consumer type $S_Y^{(k)}$ and the range of values of corresponding typological signs, we will determine the type of consumer behaviour for randomly selected family O^* using the following rule: "from

$$\tilde{X}(O^*) \in S_{\tilde{X}}^{t(k)}$$

it follows that

$$Y(O^*) \in S_Y^{(k)}\text{"},$$

then from (4.7) we easily find that the number of incorrectly classified families cannot exceed β.

4.1.2 General methodological logistics of the research

Solving the problem of the typology of consumer behaviour and identifying the main typological factors can now be represented as the following *four phases* of the general methodological logistics of the study, including the formulation of mathematical problems that must be solved at the same time.

---

2 Characters here $P\{A\}$ and $P\{A|B\}$ here and below denote, respectively, the probability of a random event A and the conditional probability of a random event A, provided that there is (was) the event B. Note that in the statement of hypothesis (**E**), the value must be clearly positive and sufficiently small.

Step 1. Collection and primary processing of raw data

As noted above, the source of information for solving the problem under discussion is the primary data (logs and diaries) from *sampling of household budget surveys* (SHBS) conducted quarterly by Rosstat RF and its regional agencies.

From the description of hypothesis (**A**) (see 4.1.1), we can see that the analysed objects (i.e. households – HH) act as multidimensional observations, or points, in two multidimensional feature space. On fixing as the coordinates of these points (HH) the values (or levels) of typological variables (i.e. the determinant factors), we see them in the space of the state , i.e. in the space of which the coordinates are the main indicators of family life. However, on fixing as the coordinates of the same points the indicator values of their consumer behaviour, we see them in the space of behaviour. Obviously, given the appropriate choice of the metric in spaces **X** and **Y**, the geometric proximity of the two points in space **X** will mean similarity of living conditions of the two respective families, and the geometric proximity of the points in space **Y** will show the similarity of their consumer behaviour.

Accordingly, at step 1, taking into account the specific characteristics of the original data $\{X_i, Y_i\}$, $i = 1, 2, \ldots, n$, it is necessary to solve the following two problems.

Problem 1. Selection of metric in space **Y**.

In other words, we are talking about the definition of a rule (algorithm), under which we have to calculate the "distance" between any two sets of consumer goods Y_i and Y_j. The success of this study largely depends on the successful solution of this problem. At the same time, we must not ignore the effect of essentially multivariate set of given behavioural traits $Y^{\mathrm{T}} = (y^{(1)}, y^{(2)}, \ldots, y^{(m)})$. Let's illustrate this idea with an example, which we artificially simplified (for clarity) by reducing the dimension of the given vector of consumer behaviour to two (i.e. let's give $Y^{\mathrm{T}} = (y^{(1)}, y^{(2)})$, where $y^{(1)}$ and $y^{(2)}$ are the implied specific volumes of consumer goods categories "Food" and "Non-food", respectively). The values $(y^{(1)}, y^{(2)})$ were recorded for each family in two different populations, and the families in "population I" differed significantly from the families in "population II" on a number of important factors of their life (income, geographic location, socio-demographic structure). Registration results are presented (on the conventional scale) in Figure 4.2, where the "crosses" are the geometric representation of population I and the "zeroes" of population II. Then, $\hat{a}^{(1)}(I)$ and $\hat{a}^{(1)}(II)$ are sample average expenditures of families I and II, respectively, for the category "Food", and $\hat{a}^{(2)}(I)$ and $\hat{a}^{(2)}(II)$ are average expenditures of families I and II, respectively, for the item "Non-food products and other services".

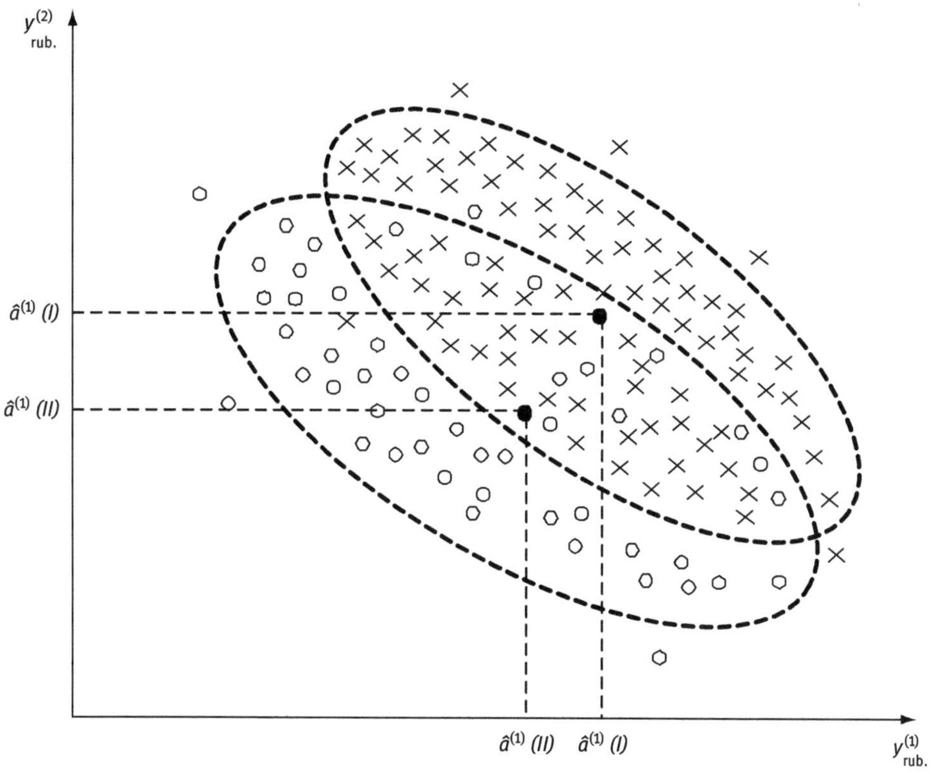

Figure 4.2. The geometric representation of the results of registration of expenditures on food $y^{(1)}$ and non-food products and services $y^{(2)}$ for families of two populations

The statistical analysis showed that the component-wise difference of the means $\Delta_1 = |\hat{a}^{(1)}(I) - \hat{a}^{(1)}(II)|$ and $\Delta_2 = |\hat{a}^{(2)}(I) - \hat{a}^{(2)}(II)|$ as well as the total component-wise difference of the means $\Delta_1 + \Delta_2$ is statistically significantly different from zero. In other words, if we focus on component differences without regard to their relationship, then it is not possible to establish a distinction in consumer behaviour of the two family populations. However, there is a difference and it is detected by using a metric that takes into account the nature of the relationships of the analysed components. In particular, based on the equity of the reinforced hypothesis of the stratified nature of the **Y** consumer behaviour space (hypothesis ($\mathbf{B}^o$)), make a natural assumption that the nature of the relationships between the components $y^{(l)}(k)$ and $y^{(q)}(k)$ (i.e. the covariance matrix $\Sigma(k)$) remains the same during the transition from one to another type of consumption, i.e.:

$$\Sigma(1) = \Sigma(2) = \ldots = \Sigma(N) = \Sigma .$$

Then, the most natural metric, as we know (see, for example, [Айвазян, 2010], A3.3), is the distance of Mahalanobis type:

$$\rho_Y(O_i, O_j) = [(Y_i - Y_j)^{\mathrm{T}} \Sigma^{-1} (Y_i - Y_j)]^{1/2}. \tag{4.8}$$

This is the metric we will use in space **Y**, especially because of its particular cases, corresponding to the *diagonal* covariance matrices of type

$$\Sigma = \begin{pmatrix} \sigma_{11} & & & 0 \\ & \sigma_{22} & & \\ & 0 & \ddots & \\ & & & \sigma_{mm} \end{pmatrix}$$

(which corresponds to mutually correlated components $y^{(1)}, y^{(2)}, \ldots, y^{(m)}$ with the dispersions $\sigma_{11}, \sigma_{22}, \ldots, \sigma_{mm}$) and

$$\Sigma = \begin{pmatrix} \sigma & & & 0 \\ & \sigma & & \\ & 0 & \ddots & \\ & & & \sigma \end{pmatrix}$$

(which corresponds to mutually non-correlated components $y^{(1)}, y^{(2)}, \ldots, y^{(m)}$ with common variance σ) and, as is easily seen, it can be reduced to the weighted Euclidean metric

$$\tilde{\rho}_Y(O_i, O_j) = \left[\sum_{l=1}^{m} \frac{1}{\sigma_{ll}} \left(y_i^{(l)} - y_j^{(l)} \right)^2 \right]^{1/2}$$

and the usual Euclidean metric

$$\tilde{\tilde{\rho}}_Y(O_i, O_j) = \left[\sum_{l=1}^{m} \left(y_i^{(l)} - y_j^{(l)} \right)^2 \right]^{1/2}.$$

Problem 2. Reducing the dimensionality of space **Y.**
It refers to the transition from the m-dimensional vector Y for initial consumer behaviour characteristics to the vector $\tilde{Y}$ of the substantially smaller dimension $\tilde{m}$ ($\tilde{m} \ll m$), the components of which, $y^{(l)}$, have to be formed as derived values from the initial characteristics $y^{(1)}, y^{(2)}, \ldots, y^{(m)}$ (including $\tilde{y}^{(l)}$ may repeat individual component of the vector Y), and at the same time be the most informative in terms of discovering the nature of differentiation of family consumer behaviour. The description of the methods for dimension reduction can be found, for example, in [Applied Statistics: Classification and dimension reduction, 1989]. In this problem, the formal methods of dimension reduction need to be combined with the methods of aggregation of the initial indicators, which are based on the content (socio-economic) analysis.

Step 2. Identification and modelling of the main types of consumer behaviour
The hypothesis that there are natural, objectively determined types of consumption, i.e. a number of family types with one class of families characterized by a relatively similar type of consumer behaviour, geometrically means disintegration of the set of points (families), studied in the behavioural space, into a corresponding number of bunches or clusters of points. Having identified these classes or clots using the methods of multivariate statistical analysis (method of cluster analysis, taxonomy), we obtain the basic types of consumer behaviour. And only after that we can constructively implement the method of constructing the objective preference functions, which develops and modifies the method proposed by V. A. Volkonsky [Волконский, 1973].

Accordingly, the realization of the objectives of phase 2 requires a solution to the following two problems.

Problem 3. Dividing a given population of households into a number (unknown to us) of classes so that the families belonging to the same class are characterized by a relatively similar consumer behaviour.
Mathematically, the problem is formulated as follows. There is a plurality of points $\{Y_i\}_{i=1,n}$ in a multidimensional space $\mathbf{Y}$ and the metric ρ_Y is defined in this space (using equation (4.8)). Starting from hypothesis (**B**) (or (**B°**)) of the stratified structure of population $\{Y_i\}_{i=1,n}$, we need to find a partition of this population S_Y into an unknown number N of disjoint classes $S_Y^{(1)}, S_Y^{(2)}, \ldots, S_Y^{(N)}$, which (the partition), in a sense, would reproduce in the most accurate way the stratification structure. Obviously, such a formulation of the problem needs to be clarified; in particular, we need to explain in what way we are going to reproduce the desired stratified structure of the analysed population of families in the most accurate way.

Here are two versions of clarifying the statement of problem 3, depending on which hypothesis, (**B**) or (**B°**), will be used in this study.

If hypothesis (B) on the stratified nature of the behavioural space $\mathbf{Y}$ is true, problem 3 is solved by the method of cluster analysis, for example, using *the procedure of "average N" with an unknown number of classes N* (see, e.g. [Айвазян, Мхитарян, 2001]). Ways to overcome the difficulties associated with the unknown number of classes and the search for *stable* (in some sense) types of partition are described in [Типология потребления, 1978, 59-68] (see also [Айвазян, 1996]).

If hypothesis ($\mathbf{B^0}$) on the stratified nature of the behavioural space $\mathbf{Y}$is true, problem 3 is solved by using the procedures for estimating the parameters of normal distributions, followed by classification of observations $Y_1, Y_2, \ldots, Y_n$ in accordance with the Bayes optimal decision rule. Indeed, hypothesis ($\mathbf{B^0}$) rules that the law of probability distribution of the whole given population $\mathbf{O}$ of multidimensional random points/families Y is a mixture of normal laws of probability distributions, each of which refers only to the families of one type of consumption. In other words, if $f(Y)$ is probability distribution density of a multidimensional random variable describing consumption patterns randomly extracted from family **O**, then, in accordance with ($\mathbf{B^o}$)

$$f(Y) = \sum_{k=1}^{N} \pi_k f_k(Y), \tag{4.9}$$

where $f_k(Y)$ is the density of the m-dimensional (non-degenerate) normal law (see (4.1)), which (density) determines the laws of the random variation of the structure of k-type family consumption, and π_k is the proportion (specific share) of k-type family consumer behaviour throughout the given family population **O**.

Obviously, in this case, the concept of a homogeneous class (type) of consumer behaviour ("strata") is formalized with the concept of multivariate normal general population, and the task of the best partitioning of the given population **O** into disjoint classes $S_Y^{(1)}, S_Y^{(2)}, \ldots, S_Y^{(N)}$ is reduced to the best statistical estimation of parameters N, π_k, $\mathrm{a}(k)$ and $\Sigma(k) (k = 1, 2, \ldots, N)$ in equation (4.9). Indeed, knowing the parameters N, π_k, $\mathrm{a}(k)$ and $\Sigma(k)$ and taking into account (4.9), we have comprehensive information on the nature of each individual family consumer type and on their specific representation in the total population of families **O**. So in accordance with Bayes optimal classification rule, the i-th household (observation Y_i) will be assigned to the class k_0, for which

$$\pi_{k0} \cdot f_{k0}(Y_i) = \max_{1 \le k \le N} \pi_k f_k(Y_i),$$

where the density of the m-dimensional normal distribution $f_k(Y)$ is defined by equation (4.1) (see the description of hypothesis ($\mathbf{B^o}$)).

Overcoming the challenges associated with an unknown number of classes N and multi-variance of analysed distributions is described in Прикладная статистика: классификация и снижение размерности, 1989; Айвазян, 1997].

Problem 4. Building the target functions of consumer preferences on the basis of family budget statistics.

The research methodology adopted for this study, as repeatedly noted above, is aimed mainly at (a) the identification of the nature of differentiation of consumer behaviour of population, (b) the development of analysis and prediction methods for this differentiation, and (c) the development of models that allow for partial management of consumer demand structure. If only the first two goals are undertaken, then to justify the use of the suggested methods in this book, it would be enough to rely on the assumptions in hypotheses (**A**), (**B**) (or ($\mathbf{B^o}$)), and (**E**). However, if we consider the desirability of moving on the path of socio-economic modelling, which allows to produce reasonable positive measures for the partial management of consumer demand, then one of the tools here is the use of retail prices, and, in particular, understanding the laws of their impact on demand. As it is well known [Слуцкий, 1963; Волконский, 1973, 1974], the most complete description of the patterns linking consumption (demand) volumes $y^{(l)}$ of various benefits to their retail prices $p^{(l)}$ and the income s can be obtained using target functions of consumer preferences $u^{(k)}(Y)$ $(k = 1, 2, \ldots, N)$, which determine (up to a monotonic transformation) the numerical value of measure of optimality of each fixed set of benefits Y for families of k-type of consumer behaviour (see the formulation of hypothesis (**C**)). However, despite the presence of interesting and profound theoretical and methodological research on the analysis of the consumption functions [Frish, 1959; Слуцкий, 1963; Волконский, 1973, 1974], the attempts of their particular construction on the basis of statistical data have not been successful. Among the reasons for this situation, we can note the inadequacy and poor realism of key assumptions, which served as a basis of the given methods (see, e.g. [Frish, 1959], which postulates the "independence on preference": the usefulness of one benefit does not depend on the amount of consumption of the other). The most promising in terms of application is the approach of V.A. Volkonsky, based on the so-called hypothesis of homogeneity [Volkonsky, 1973]. However, his attempts to statistically implement it on the **total** population of consumers, as expected, were unsuccessful for one simple reason: the homogeneity hypothesis suggests essentially the fulfilment of our hypotheses (**A**), (**B**), (**C**) and (**D**), but it is easily seen that these assumptions may be realistic only within a homogeneous consumer behaviour population of families. In the study [Volkonsky, 1973], this fact was not explicitly stated, and the remark, contained there, that the described "... models can in principle be applied also to specific classes of families somehow picked out from the main population" (p. 622) somehow leads away from the correct understanding of the valuable result formulated in the study: "*cannot* be applied ..." (Emphasis added – S.A.), but *must* be applied only to certain classes of households, and not "somehow picked out from the main population" but picked out according to a certain criteria of homogeneity of their consumer behaviour. Thus, we arrive at the fact that within the framework of solving the general problem of identifying a typology of consumption and constructing consumer types as one of its aspects we can get the implementation of, what we think, is one of the most effective method of constructing the objective functions of consumer behavior based on data on family budgets.

We will briefly describe this method, relying largely on the work of [Volkonsky, 1973] and making some modifications arising from the comments just made.

We proceed from the assumption that hypotheses (**A**), (**B⁰**), (**C**) and (**D**) are correct. As initial statistics, we will consider family budgets $\{Y_i(k)\}_{i=\overline{1,n_k}}$ of k-type of consumer behaviour $(k = 1, 2, \ldots, N)$, so that task 3 (the problem of partitioning of the whole analysed population of families O into classes on the basis of similarities in consumer behaviour of one class of families) is considered solved. Let's assume Y(k)(m) is the general I couldn't insert the real formula but Y should be capital and in bold, (k) is the power of Y and should be higher on the top of it and italic. (m) is fine as is..

Simultaneous execution of hypotheses ($\mathbf{B^0}$) and (**D**) guaranties, first of all, optimal statistical average of consumer behaviour of the families of each fixed (k) type, i.e. accuracy of the equation

$$u^{(k)}(\mathbf{a}(k)) = u^{(k)}(Y_{\text{opt}}(k)) = \max_{Y \in \mathbf{Y}^{(k)}(m)} u^{(k)}(Y), \tag{4.5'}$$

and secondly, a normal character of the random scattering consumer behaviour vectors $Y(k)$ not only in the totality of population of k family type (i.e. not only in the m-dimensional space Y) but in a narrower population of families $\mathbf{Y}^{(k)}(m-1)$ selected from $\mathbf{Y}^{(k)}(m)$, conforming to resource constraints (4.6). This means in particular that

$$u^{(k)}(\mathbf{a}_{\text{cond}}(k)) = u^{(k)}(Y_{\text{opt.cond}}(k)) = \max_{Y \in \mathbf{Y}^{(k)}(m-1)} u^{(k)}(Y),$$

where

$$\mathbf{a}_{\text{cond}}(k) = \mathbf{E}\left(Y(k) | \sum_{l=1}^{m} p^{(l)} y^{(l)} = s\right)$$

is conditional expectation of a random vector $f\,Y(k)$, calculated under condition (4.6), and $Y_{\text{опт.усл}}\,(k)$ the solution of the extreme problem such that

$$\begin{cases} u^{(k)}(Y) \to \max_{Y \in \mathbf{Y}^{(k)}(m)} \\ \sum\limits_{l=1}^{m} p^{(l)} y^{(l)} = s \\ y^{(j_l)} = \tilde{y}^{(j_l)}, \quad l = 1, 2, \ldots, L \end{cases}$$

or, which is the same as

$$u^{(k)}(Y) \to \max_{\mathbf{Y}^{(k)}(m-L-1)}.$$

Note that the population $\mathbf{Y}^{(k)}(m)$ is a subset of an m-dimensional space $\mathbf{Y}$, while the whole population $\mathbf{Y}^{(k)}(m-L-1)$ is a subset in a corresponding $(m-L-1)$-dimensional space.

In this situation, i.e. if hypotheses (**A**), (**B⁰**), (**C**) and (**D**) are accurate, the theoretical result proved in [Волконский, 1963] (see lemma on p. 632) can be used as the basis of a method of construction of quadratic approximation for the unknown function $u^{(k)}(Y)$ on the statistics of family budgets. This result as applied to our study can be formulated as follows: if the random variation of the structure $Y(k)$ of the consumer behaviour of families of k-type is adequately described by a non-degenerate m-dimensional normal law (see hypothesis ($\mathbf{B^0}$), formula (4.1)), and at the same time the structure $Y_{\text{mod}}(k)$, statistically more often observed in families of this type (in a symmetrical distribution, which includes normal, it is the same as the statistically average structure $Y_0(k)$), is also optimal in the sense of (4.5), then in normal conditions and in the sense of (4.5') the desired target function $u^{(k)}(Y)$ on the hyper-plane (i.e. while limited) $\sum_{l=1}^{m} p^{(l)} y^{(l)}(k) = s$ can be replaced up to a monotonic transformation by the density function $f_k(Y)$, and in particular

$$u^{(k)}(Y) = c_k(s) \cdot \psi_k(Y) + d_k(s),$$

where the function

$$\psi_k(Y) = -\frac{1}{2}(Y - \mathbf{a}(k))^{\mathrm{T}} \Sigma^{-1}(k)(Y - \mathbf{a}(k)),$$

and the constants (for fixed income s) $c_k(s)$ and $d_k(s)$ are functions of s.

Hence, in particular, it follows directly that, on the whole, population $\mathbf{Y}^{(k)}(m)$ (i.e. in the absence of restrictions on income) of the target function of the consumer behaviour of k-type households is a function of two variables $\psi_k(Y)$ and s, i.e.:

$$u^{(k)}(Y) = F(\psi_k(Y), s), \tag{4.10}$$

while $F(\psi_k, s)$ is a function, increasing for argument ψ_k.

To obtain a quadratic approximation of function $u^{(k)}(Y)$, let's expand it into the Taylor series in the neighbourhood of the point $\mathbf{a}(k)$ up to a member of the second order inclusive, using expression (4.10), the rule for composite function differentiating, and the fact that

$$\frac{\partial \psi_k}{\partial y^{(l)}} | Y_0(k) = 0, \quad \frac{\partial^2 s}{\partial y^{(l)} \partial y^{(q)}} = 0 \ (l, \ q = 1, \ 2, \ \ldots, \ m).$$

The result will be (up to a constant term and a constant positive factor):

$$u^{(k)}(Y) = P^{\mathrm{T}}(Y - \mathbf{a}(k)) - A_k[P^{\mathrm{T}}(Y - \mathbf{a}(k))]^2 - B_k(Y - \mathbf{a}(k))^{\mathrm{T}} \Sigma^{-1}(k)(Y - \mathbf{a}(k)). \tag{4.11}$$

In this equation, $P^{\mathrm{T}} = (p^{(1)}, p^{(2)}, \ldots, p^{(m)})$ is vector of retail prices[3] and the constants A_k and B_k must be determined with some additional considerations.

To determine the unknown constants A_k and B_k, we can use a combination of several approaches.

(a) Information on constants A_k and B_k contained in the postulate of "independences of preference" [Frish, 1959]. Mathematically, this postulate in relation to the benefits $y^{(l)}$ and $y^{(q)}$ is expressed by the following equation:

$$\frac{\partial^2 u^{(k)}(Y)}{\partial y^{(l)} \partial y^{(q)}} = 0,$$

which should express the independence of utility of consumption of the benefit unit $y^{(l)}$ from the consumption of the benefit $y^{(q)}$ (but not the independence of the

3 If a specific number $y^{(1)}, y^{(2)}, \ldots, y^{(m)}$ of goods consumed by the family is expressed in monetary terms, it is clear that we must have $p^{(1)} = p^{(2)} = \ldots = p^{(m)} = 1$.

demand on benefit $y^{(l)}$ from the purchase volume of the benefit $y^{(q)}$, as the demand for all goods is interdependent through restrictions on income). The use of this postulate for all components (even with sufficient aggregation) of vector Y in order to build the consumption function $u^{(k)}(Y)$ [Frish, 1959] seems unjustified to us, because in this way, this assumption seems clearly unrealistic.

However, it is quite possible to select in a special way one or more individual pairs of the components

$$(y^{(i_1)}, y^{(j_1)}),\ (y^{(i_2)}, y^{(j_2)}), \cdots, (y^{(i_T)}, y^{(j_T)})$$

of vector Y (or pairs of suitably segregated initial components), such that for each of them the hypothesis of "independence of preferences" would be approximately accurate, i.e.

$$\frac{\partial^2 u^{(k)}(Y)}{\partial y^{(i_t)} \partial y^{(j_t)}} = 0, \quad t = 1,\ 2,\ \ldots,\ \mathrm{T}. \tag{4.12}$$

Using $u^{(k)}(Y)$ in the form (4.11), considering the matrix $\Gamma(k) = (\gamma_{lq}(k))_{l,q=\overline{1,m}} = \Sigma^{-1}(k)$ and assuming for definiteness that the components of vector Y are expressed in monetary units, we obtain from (4.12) the following system of equations:

$$A_k + B_k \gamma_{i_t j_t}(k) = 0, \quad t = 1,\ 2,\ \ldots,\ \mathrm{T}.$$

By solving this system using the method of least squares, i.e. minimizing the sum

$$\sum_{t=1}^{\mathrm{T}} (A_k + B_k \gamma_{i_t j_t})^2$$

we can estimate only the equation $\tau = A_k/B_k$. Its estimation is

$$\hat{\tau} = -\frac{\sum_{t=1}^{\mathrm{T}} \gamma_{i_t j_t}(k)}{\mathrm{T}}.$$

(b) Information on constants A and B contained in the coefficient of elasticity of consumption by income. Under the elasticity $e_l(k)$ l of the benefit $y^{(l)}$ by income (for k-type families), we understand a simple (not logarithmic) derivative of $y^{(l)}$ by s, that is

$$e_l(k) = \frac{\partial y^{(l)}(k)}{\partial s}$$

the actual value of which can be obtained, for example, from fairly statistically available demand functions (Engel curves) [Volkonsky, 1973].

Simple calculations associated with differentiating the optimized consumption function lead to the following system of equations for constants A and B:

$$\varepsilon(k) = (U^{-1}(A_k, B_k)\cdot 1_m \cdot 1_m^{'})^{-1}(U^{-1}(A_k, B_k)\cdot 1_m), \tag{4.13}$$

where $\varepsilon(k) = (e_1(k), e_2(k), \ldots, e_m(k))^{\mathrm{T}}$ is a column vector of elasticities,

$$U(A_k, B_k) = -A_k \begin{pmatrix} 1 & 1 & 1 & 1 \\ 1 & 1 & \ldots & 1 \\ \ldots & \ldots & \ldots & \ldots \\ 1 & 1 & \ldots & 1 \end{pmatrix} - B_k \Gamma(k)$$

is a matrix of size $m \times m$, and $1_m = (1, 1, \ldots, 1)^{\mathrm{T}}$ is an m-dimensional column vector of ones. Unfortunately, system (4.13) is difficult for finding the numerical solution and requires iterative procedures.

An example of the empirical study in which this method is implemented is presented in the work [Колосницын, Макарчук, 1982].

Step 3. Selection of the most informative determinant factors (typological variables) and stratification of households in the space of these variables

Obviously, it would be incorrect to rely on the fact that the range of possible values of each of the typological signs will not intersect for the families of different types of consumer behaviour. In other words, the values of each of the typological signs individually and their set of population are subject to some uncontrollable spread when analysing families within each type of consumption, i.e. they are characterized by some law of probability distribution. It is natural to consider the determinant factors or their sets as the most informative, the difference in the distribution of which is the greatest during the transition from one class of consumer behaviour to another. This idea is the basis of the selection method of the most informative typological signs. Finally, having already selected a small number of the most informative determinant factors, we can try again to partition the analysed family population into classes of clots but this time in the space of the selected typological signs. So, the result of this partitioning will depend not only essentially on the content of the group of the most informative typological signs but also on how we calculate the distance between two points/families in this space and, in particular, on the weight of the typological signs participating in this distance. Therefore, we will try to pick the weight so that the result of partitioning of families into classes in the space of the most informative determinant factors in a sense is the least different from the partition of the same points/families that we obtained in the space of behaviour (step 2).

Step 3 of our logical plan, essentially, allows us to transition from the *consumption typology to consumer typology* and to find correspondence of these two structures. The result of this stage is the identification of the structure of the population by type of consumers determining the specific weight of each typological group. From the above, it follows that the content and the objectives of step 3 require solving the following two main problems.

Problem 5. The selection of the most informative (typological) variables in **X** *space of determinant factors.*

This problem is similar to task 2, which provided for the selection of the most informative factor in terms of identifying the differentiation of family consumer behaviour signs in the behaviour space **Y**. However, task 5 has two specific significant differences. First, p-dimensional signs analysed in space **X** are of mixed nature, i.e. we have quantitative, qualitative and classification signs included in the components of vector X. Ultimately, this difference in principle does not affect the overall statement of the problem, but creates only some additional complexity in the technical implementation of those elements of the logical solution of problem 2, which we may be needed to solve problem 5. The second difference is fundamental and, specifically, allows us to modify the general statement of the problem to make it possible to use a much more efficient method of selection of the most informative features than those used in task 2. It is a method based on the presence of the so-called training samples: in contrast to the conditions of problem 2, in problem 5 we are aware of the partitioning of these analysed objects (families) into homogeneous classes (types of use) $S_Y^{(1)}, S_Y^{(2)}, \ldots, S_Y^{(N)}$ before the analysis of the analysed system of signs X. Therefore, analysing the behaviour of various combinations of the components of X in different classes, it would be natural to name those most informative combination of these component behaviour changes dramatically during the transition from one class $(S_Y^{(k_1)})$ to another $(S_Y^{(k_2)})$.

Let us illustrate this by describing a mathematical formulation of the problem. Let us have an original p-dimensional sign $X = (x^{(1)}, x^{(2)}, \ldots, x^{(p)})^{\mathrm{T}}$ of a mixed nature. In order to standardize the methods of algorithmic processing of all components simultaneously of vector X, let's give to each component $x^{(l)}\,(l = 1, 2, \ldots, p)$ a finite set of its possible values (grades) $x^{(l)}(1), x^{(l)}(2), \ldots, x^{(l)}(m_l)$. So, if $x^{(l)}$ is a quantitative variable, then $x^{(l)}(\nu)$ is the middle of the ν–th grouping interval, into which the entire range of possible values of the random variable $x^{(l)}$ is divided; if $x^{(l)}$ is a sign of quality, then x is the number of graduation, determining the level (degree of manifestation) of a given property or quality; and if j is a classification (nominal) sign, then $x^{(l)}(\nu) = \nu$ is the number of the class to which a particular object may be assigned. It is easy to calculate that the total number M of possible values $X_1^0, X_2^0, \ldots, X_M^0$ of a multidimensional sign X is equal to the product $m_1 \cdot m_2 \ldots m_p$. For a formalized description of differentiation of an occasional sign X, observed during the transition from one family class $(S_Y^{(k_1)})$ to another $(S_Y^{(k_2)})$, we need to introduce a quantitative measure for distinguishing between the classes in terms of this sign's behaviour.

Content and experimental analysis of populations considered in the **X** space has led us in the long run to the most convenient and informative measure differentiation between the general household populations with numbers k_1 and k_2 to the so-called variation distance, given by the following equation:

$$\Delta(k_1, k_2; X) = \frac{1}{2}\sum_{i=1}^{M} |P_{k_1}(X_i^0) - P_{k_2}(X_i^0)| \tag{4.14}$$

where $P_j(X_i^0) = P\{X = X_i^0 | j\}$ is the probability of occurrence of value X_i^0 in j-th general population of families and the summing is performed for all $m_1 \cdot m_2 \ldots m_p$ possible values of the analysed characteristic X.

It is easy to check that the "distance" (4.14) is arranged in such a way that it is always non-negative and not greater than one, and $\Delta(k_1, k_2) = 0$ if and only if $P_{k_1}(X) \equiv P_{k_2}(X)$, i.e. when the laws of probability of distribution of sign X in classes numbered k_1 and k_2 are identically the same.

Obviously, in our empirical analysis, we will have to use the sample analogues of values involved in (4.14), i.e.

$$\hat{\Delta}(k_1,\ k_2;\ X) = \frac{1}{2}\sum_{i=1}^{M} |\hat{P}_{k_1}(X_i^0) - \hat{P}_{k_2}(X_i^0)|, \tag{4.14'}$$

where $\hat{P}_k(X)$ is the relative frequency of the occurrence (or fraction) among k-th population families of the kind of families, which have the value of the determinant factor that equals X, and the sum is carried out only by all values of X that have been registered at least once in corresponding populations.

Then for each fixed dimension $p' = 1, 2, \ldots, p-1$ and each possible set of components $X(p') = (x^{(j_1)}, x^{(j_2)}, \ldots, x^{(jp')})^T$ (the number of different sets of a given dimension p' equals the number of combinations of p elements of p', i.e. $C_p^{p'}$), we calculate the values $\hat{\Delta}(k_1,\ k_2;\ X(p'))(k_1,\ k_2 = 1, 2, \ldots, N;\ k_1 \neq k_2)$ and determine the most informative set $\tilde{X}(p')$ of predetermined dimension $\eta_0\ (\theta)$ from the condition

$$\hat{\Delta}_{av}(\tilde{X}(p')) = \max_{X(p')} \hat{\Delta}_{av}(X(p')) \tag{4.15}$$

or from the condition

$$\hat{\Delta}_{min}(\tilde{X}(p')) = \max_{X(p')} \hat{\Delta}_{min}(X(p')). \tag{4.15'}$$

In equations (4.15) and (4.15')

$$\hat{\Delta}_{min}(X) = \min_{(k_1;\, k_2)} \hat{\Delta}(k_1,\ k_2; X),$$

and

$$\hat{\Delta}_{av}(X) = \sum_{\substack{k_1, k_2 = 1 \\ k_1 \neq k_2}}^{N} \omega_{k_1 k_2} \cdot \hat{\Delta}(k_1,\ k_2; X),$$

where the "weight" of the pairs $(k_1, k_2)$of (k_1, k_2)classes are calculated by the formula

$$\omega_{k_1 k_2} = \frac{n_{k_1} + n_{k_2}}{\sum\limits_{i,j=1\, i \neq j}^{N} (n_i + n_j)} = \frac{n_{k_1} + n_{k_2}}{(N-1) \cdot n}$$

(here, as before, n_k is the number of families classified as a result of solving problem 3 to the k consumption type).

We propose to select the dimension of the most informative sign $\tilde{X}(p') \ (p' = 1, 2, \ldots, p)$ among all the signs satisfying criterion (4.15) or (4.15'), using substantive socio-economic considerations. In addition, it is useful to trace the changes in the characteristic $\hat{\Delta}_{\mathrm{av}}(\tilde{X}(p'))$ or $\hat{\Delta}_{\min}(\tilde{X}(p')$ depending on p' and to select for possible values of the dimension p', primarily those for which there are "outstanding" in terms of value leaps in differences

$$\hat{\Delta}(\tilde{X}(p')) - \tilde{\Delta}(\tilde{X}(p'-1)).$$

Thus, by solving problem 5, we have found some p'-dimensional signs $\tilde{X} = (\tilde{x}^{(1)}, \tilde{x}^{(2)}, \ldots, \tilde{x}^{(p')})^{\mathrm{T}}$, the components of which are either selected from a number of initial determinant sign X or are some of its functions. This sign $\tilde{X}$ can be considered as an experimental approach to some, not a priori known to us, typological signs, the existence of which is postulated by hypothesis (**E**). It will help us in the long run to determine (and to predict) the type of consumer behaviour of a family values using the values of its determinant factors.

Phase 3 is completed by the solution of the following problem.

Problem 6. Stratification of households in the space typological variables.
The result of clustering of the analysed family population $\mathbf{O} = \{O_i\}_{i=\overline{1,n}}$ into disjoint classes $S_{\tilde{X}}^{(1)}, S_{\tilde{X}}^{(2)}, \ldots, S_{\tilde{X}}^{(N)}$ in space $\tilde{X}$ depends not only on the composition of the components of vector $\tilde{X}$ of typological variables but also on how we calculate the distance $\rho\tilde{X}(O_i, O_j)$ between two points/families in this space.

We will introduce the distance between the objects specified by the multidimensional variables of mixed nature. We understand by the gradation of variables $\tilde{x}^{(l)}$: for quantitative variables, it will be one of the grouping intervals, used to cluster the range of variation of the sign; for a qualitative sign, it will be one of the categories that define the quality of the analysed object; for a classification sign, it will denote one of the homogeneous groups (classes) into which the analysed population of objects is divided. Let m_l be the total number of possible gradations according to $\tilde{x}^{(l)}$. Then each "observation", as a result of the measurement of the variables $\tilde{x}^{(l)}$ on the object O_i, will be conveniently represented by the binary vector $\tilde{X}_i^{(l)} = (\tilde{x}_i^{(l.1)}, \tilde{x}_i^{(l.2)}, \ldots, \tilde{x}_i^{(l.m_l)})^{\mathrm{T}}$, where all components except one will equal zero. At the same time, the number of the single component $\nu^{(l)}(O_i)$ of vector $\tilde{x}_i^{(l)}$ is determined by the number of that gradation of the variables, in which the object O_i was included. We must set a measuring procedure to compute the measure of variation $\delta^{(l)}(\nu_1, \nu_2)$ between $\nu_1^{(l)}$–th and $\nu_2^{(l)}$–th gradations of the variable $\tilde{x}^{(l)}$ ($\nu_1^{(l)}, \nu_2^{(l)} == 1, 2, \ldots, m_l$). When the grouping intervals and quality categories for quality and quantity variables are arranged *in order*, it is customary to use the following equation to denote this rate of variation:

$$\delta^{(l)}(\nu_1^{(l)}, \nu_2^{(l)}) = c_l \cdot |\nu_1^{(l)} - \nu_2^{(l)}|, \tag{4.16}$$

where the value c_l is a positive, "weight coefficient", which reflects the relative measure of importance of the variables $\tilde{x}^{(l)}$. In case of classification signs, we generally calculate this measure of variation by expert calculation. In the complete absence of meaningful considerations on the basis of which we could commensurate distances between different graduations/classes in the case of classification components $\tilde{x}^{(l)}$, we can propose, for example:

$$\delta^{(l)}(\nu_1, \nu_2) = \begin{cases} 0, & \text{if } \nu_1 = \nu_2 \\ c_l, & \text{if } \nu_1 \neq \nu_2 \end{cases}. \tag{4.16'}$$

Then, the distance between the objects O_i and O_j, which are specified by multivariate observations $\tilde{X}_i$ and $\tilde{X}_j$, respectively, is customary to set with the following equation:

$$\rho \tilde{X}^{(C)}(O_i, O_j) = \sum_{l=1}^{p'} \delta^{(l)}(\nu^{(l)}(O_i), \nu^{(l)}(O_j)), \tag{4.17}$$

where the "weights" c_l, determining the relative value of separate components $\tilde{x}^{(l)}$ ($c_l \geq 0$, $\sum_{l=1}^{p'} c_l = 1$), are the desired quantities and should be "matched" in the best, so to speak, way, and the measure of variation of gradations $\delta^{(l)}(\nu^{(l)}(O_i), \nu^{(l)}(O_j))$, which include objects O_i and O_j on the variable $\tilde{x}^{(l)}$, is calculated by means of (4.16), or by (4.16').

We need to find a weight vector $\tilde{C} = (\tilde{c}_1, \tilde{c}_2, \ldots, \tilde{c}_p)'$ for which the distance between the partitions S_Y and $S_{\tilde{X}}(\tilde{C})$ is the smallest, i.e.

$$d(S_Y, S_{\tilde{X}}(\tilde{C})) = \min_C d(S_Y, S_{\tilde{X}}(C)). \tag{4.18}$$

In equation (4.18), the partition S_Y is the result of problem 3, i.e. classification structure of decomposition of the solution of problem 3, i.e. classification structure S_Y

describes stratification of the analysed population of families O into the types of uniform consumer behaviour $S_Y^{(1)}, S_Y^{(2)}, \ldots, S_Y^{(N)}$; partition $S_{\tilde{X}}(C)$ is the result of the application of classification algorithms in space $\mathbf{X}$ of binary variables $\tilde{X}^{(l\cdot q)}$ with metric $\rho\tilde{X}^{(C)}$ defined by means of (4.16). A description of some classification algorithms of this type is given in Appendix 4.1.

Note. There are various ways to measure the distance between partitions (see, e.g. [Айвазян, Бежаева, Староверов, 1974; Миркин, 1976]). Here are two methods we used in the problem of typology of consumption. Let $S^{(1)}$ and $S^{(2)}$ be two different partition classes of the same set of n objects.

Distance Kemeny-Snell $d_{\text{k-c}}(S^{(1)}, S^{(2)})$ is determined using the following formula:

$$d_{\text{k-c}}(S^{(1)}, S^{(2)}) = \frac{1}{n(n-1)} \sum_{i=1}^{n} \sum_{j=1}^{n} |s_{ij}^{(1)} - s_{ij}^{(2)}|,$$

where $s_{ij}^{(l)} = 1$ if objects O_i and O_j are in the same class of l-th partition, and $s_{ij}^{(l)} = 0$ if the objects are in different classes of the l-th partition ($l = 1, 2$). It is easy to show that the values of the distance $d_{\text{k-c}}(S^{(1)}, S^{(2)})$ can vary from zero (when the partitions $S^{(1)}$ and $S^{(2)}$ coincide) to one (when one of the considered partitions n singleton classes, while the other unites all the objects into a single class).

Tanimoto distance $d_{\text{T}}(S^{(1)}, S^{(2)})$ (is basically equivalent to the distance Kemeny-Snell but in some cases is more convenient for computations) is defined by the following formula:

$$d_{\text{T}}(S^{(1)}, S^{(2)}) = \frac{1}{n} \sum_{i=1}^{n} \frac{n^{(1)}(O_i) + n^{(2)}(O_i) - 2n^{(1.2)}(O_i)}{n^{(1)}(O_i) + n^{(2)}(O_i) - n^{(1.2)}(O_i)},$$

where $n^{(l)}(O_i)$ is the number of objects belonging to a class that contains an object O_i in the l-th partition and $n^{(1.2)}(O_i)$ is the number of objects and at the same time members of both of the aforementioned classes. It can be shown that the distance Tanimoto can vary from 0 to $1-\frac{1}{n}$, and it reaches its extreme values at the same partitions as the distance Kemeny-Snell.

An optimization problem (4.18) and the appropriate procedure for finding extremum of function $d(C) = d(S_Y, S_{\tilde{X}}(C))$ of p' is rather laborious. However, in this particular case, we can use the approaches that will significantly simplify this procedure and reduce the need of computer time to quite acceptable boundaries. We will describe one such approach.

Let's consider identified vectors of most informative signs of the increasing dimension from 1 to p', found in the solution of problem 5, i.e.

$$\tilde{X}(1) = (\tilde{x}^{(1)}(1)),\ \tilde{X}(2) = (\tilde{x}^{(1)}(2),\ \tilde{x}^{(2)}(2))^{\mathrm{T}},\ \ldots$$

$$\ldots, \tilde{X}(p') = (\tilde{x}^{(1)}(p'),\ \tilde{x}^{(2)}(p'), \ldots, \tilde{x}^{(p')}(p'))^{\mathrm{T}}. \tag{4.19}$$

Let's fix the components $\tilde{x}^{(1)}(p'),\ \tilde{x}^{(2)}(p'), \ldots, \tilde{x}^{(p')}(p')$ included in the "resulting" (i.e. finally selected) most informative p'-dimensional sign $\tilde{X}(p')$, and track how many times each of them occurs in a component of a sequence of vectors (4.19). Let μ_l be the total number of "occurrences" of the variable $\tilde{x}^{(l)}(p')$ in the composition of the most informative sets $\tilde{X}(1),\ \tilde{X}(2), \ldots, \tilde{X}(p')$. Obviously $1 \le \mu_l \le p'$. Let us take as a zero approximation $C_0 = (c_1^{(0)},\ c_2^{(0)}, \ldots, c_{p'}^{(0)})$ for the desired point of the extremum $\tilde{C}$ value:

$$c_l^{(0)} = \frac{\mu_l}{\sum_{l=1}^{p'} \mu_l}$$

Let's choose the step size δ $(0,05, \le \delta \le 0,10)$ and on a part of the hyperplane $\sum C_l = 1, C_l \ge 0\ (l = 1, 2, \ldots, p')$ let's construct a "grid" with the nodes of the following form:

$$\begin{cases} C(t_1,\ t_2, \ldots, t_{p'}) = (c_1^{(0)} + t_1\delta,\ c_2^{(0)} + t_2\delta, \ldots, c_{p'}^{(0)} + t_{p'}\cdot\delta), \text{ where} \\ t_i = 0,\ \pm 1,\ \pm 2, \ldots;\ i = 1, 2, \ldots, p' \\ \sum_{i=1}^{p'} t_i = 0. \end{cases} \tag{4.20}$$

Further search for the minimum point (4.20) of the function *d(C)* is carried out by means of the directed enumeration of the values of the function at the nodes of the "grid" of the form (4.20). The search terminates at a point where the transition from any "neighbouring"[4] grid point does not decrease the value of the analysed function.

We can see from the described solution of the problem of determining the best metric such as (4.17) in the space $\tilde{X}$ that we also solve the problem of constructing the partition $S_{\tilde{X}}(\tilde{C}) = \{S_{\tilde{X}}^{(1)}, S_{\tilde{X}}^{(2)}, \ldots, S_{\tilde{X}}^{(N)}\}$, the least differing (in the sense of (4.18)) from the partition S_Y of the analysed population of families O into the types of consumption.

Step 4. Determination of the type of consumer behaviour of the household by the values of its typological variables

This step is devoted to, in essence, solving one problem, namely:

4 Points $C(t_1', t_2', \ldots, t_{p'}',)$ and $C(t_1'', t_2'', \ldots, t''_{p'})$ are called neighbouring only when $\max_{1 \le i \le p'} |t_i' - t_i''| = 1$.

Problem 7. Achieving the best (in terms of the hypothesis (E)) equation of the partitions of the analysed population of households into classes in spaces **X** *and* **Y**.

We proceed from the understanding that we have already solved problems 1–6. Let's define the substitution:

$$\Pi^* = \begin{pmatrix} 1 & 2 & \dots & N \\ t^*_{(1)} & t^*_{(2)} & \dots & t^*_{(N)} \end{pmatrix},$$

which shows a correspondence between the classes $S_Y^{(1)}, \dots, S_Y^{(N)}$ of the partition $S\tilde{X}(\tilde{W})$ (see the description of hypothesis (**E**)) in such a way that it meets the following requirements:

$$\min_{1 \le k \le N} \mathbf{P}\{\mathbf{Y} \in S_Y^{(k)} | \mathbf{X} \in S_{\mathbf{X}}^{(t*(k))}\} = \\ = \max_{\Pi} \min_{1 \le k \le N} \mathbf{P}\{Y \in S_Y^{(k)} | \mathbf{X} \in S_{\mathbf{X}}^{(t(k))}\}. \tag{4.21}$$

In this relation, the empirical analogue (statistical evaluation) of the conditional probability $P\{Y \in S_Y^{(k)} | \mathbf{X} \in S_{\mathbf{X}}^{(t(k))}\}$ is the percentage of those families among all families of class $S_{\mathbf{X}}^{t(k)}$ that entered into k-th type of consumer behaviour during the partitioning of the whole population of families **O** into classes in the space **Y** and the maximum on the right side is taken over all possible substitutions of Π. Thereafter, to simplify the record, let's renumber the classes of the partition $S_{\mathbf{X}}(C)$ so that $t*(k) = k$. Knowing the substitution of Π∗ satisfying criterion (4.21), we automatically get the level of trust 1−β of typological determinant signs, see hypothesis (**E**)), which, considering renumbering of the classes, will be defined by the following equation:

$$1-\beta = \min_{1 \le k \le N} \mathbf{P}\{Y \in S_Y^{(k)} | \mathbf{X} \in S_{\tilde{X}}^{(k)}\}$$

Note. Formation of classes $S_{\tilde{X}}^{(1)}, S_{\tilde{X}}^{(2)}, \dots, S_{\tilde{X}}^{(N)}$ could be carried out without the search for metric $\rho\tilde{X}^{(C)}$ and the partition $S_{\tilde{X}}(C)$ in the space X, but using the usual Bayesian approach. Namely, the range of values $S_{\tilde{X}}^{(k_0)}$, getting into which "signals" the referring of the appropriate family to consumption type $S_{\tilde{X}}^{(k_0)}$, is formed from all of these **X** values for which

$$\mathbf{P}\{\mathbf{X}|S_Y^{(k_0)}\} = \max_{1 \le k \le N} \mathbf{P}\{\mathbf{X}|S_Y^{(k)}\} \tag{4.22}$$

As shown by the analysis and existing experience of experimental calculations, the results of the classification of households by type of consumption based on the values of their typological signs **X** using both these approaches are practically the same. However, the latter approach, i.e. the approach based on Bayesian criterion (4.22), is less convenient from the practical point of view.

Step 4 concludes the proposed methodological approach to the research of the typology of consumption.

4.1.3 An approach to forecasts of consumption macrostructure

Each of the separate steps 2–4 of the above methodology is dedicated to solving a problem, which, in itself, is of *independent interest*: whether it is about identifying the main types of consumer behaviour of households *(step 2)*, the definition of the main factors that affect the selection of the household of a certain type of consumer behaviour *(step 3)*, describing the functions of consumers' preferences, according to which households of each type of consumer behaviour form their budget (*step 2*), or identifying the type of consumer behaviour of household values using characteristics of the household and environmental conditions such as its composition, average per capita income, place of residence, etc. (*step 4*). Realized as a complex, this methodology may serve as the basis for solving a number of urgent problems of the general theory and practice of consumption, demand and consumer choice.

As an example of one of these problems, we describe the general approach, based on this methodology, to a short-term (1–3 years) *forecast of a macrostructure of consumption on the basis of the* ***micro****economic data*. The basis of this approach is an empirically verified statement, according to which **the forecast of population structure in the space of determinant factors (which influence formation of human needs) is a much simpler task than the immediate forecast of behavioural characteristics of population.** On this basis, the following logical microeconometric research methodology is proposed.

1. *According to the initial data from SHBS for the current year, we implement steps 1–4 of the above methodology* (task 4 of constructing objective functions of consumer preferences can be excluded in this case from step 2).
2. *With the help of well-known methods of forecasting socio-demographic structure of population, the employment structure and the structure of the geographic dispersal, we forecast (for 1–3 years) specific weights of the classes in the space of typological signs uncovered in step 3.*
3. *Using the results of steps 2 and 4 on the values of typological variables, we determine types of consumer behaviour.*
4. *Using the known average number of benefits, consumed by households of each type of consumer behaviour, and forecast values of specific weight of these types, we estimate the total number of each type of consumed benefit.*

The disadvantages of this approach should include the fact that *it ignores the possibility of a new (not previously identified) type of consumer behaviour emerging over time.* However, in the forecast range of 1–3 years, the probability of the emergence of new types of consumer behaviour is low (and in case of their specific weight is negligible).

There are, of course, alternative approaches to forecasting consumption macrostructures. One of them (which operates with the same initial **micro**-data from SHBS) is based on *the direct analysis of the dynamics of consumer behaviour* without linking with the dynamics of the classes in the space of typological variable. The second approach can be based on the use of **macro**-data of trade statistics, which in this case are the tools of modern econometric analysis of multivariate time series (see, e.g. [Hamilton, 1994]). However, we reserve their description (as well as their criticism) outside the scope of this book.

4.1.4 Typology of consumption and income differentiation in russian society of 1996 (results of econometric analysis)

The research specifics related to the transitional period of Russian economy. The study was conducted in 1997, according to the 1996 data. Its specifics were related to, on the one hand, socio-economic transformations in Russian society at that time and, on the other, the emerging fundamental obstacles in obtaining representative and reliable information on income and consumption of population. Let us dwell on these specifics.

1. The first specific feature is the fact that the transition to a market economy practically excluded from the goal statement of the so-called problem of *scarce or rationed* goods but included market-based mechanisms in the formation of consumer prices. Moreover, it has opened the possibility of direct statistical evaluation of elasticity of consumption prices; in other words, it eliminated the *Soviet specifics* of the problem (and thus, to some extent simplified its solution). At the same time, the transition to a market economy has created the *specifics of the transition period*, which primarily included criminalization of the economy and development of a large sector of shadow economy (whose share in the total GDP by various expert estimates ranged from 40% to 50%), actually "wiping out" the middle class of Russian society, and in an unprecedented polarization and super-differentiation of population incomes. "*Wiping out" of Russian middle class and super-differentiation of population income* were the result of a peculiar way of reforms from the above and, in particular, of that specific manner of privatization of financial, raw material, industrial and agricultural resources of Russia implemented during the Perestroika period. As a result, the majority of those people that would form the Russian middle class, i.e. those who sincerely and actively sought to integrate into the new market reality and in reality had the necessary human and professional potential, faced the *lack of societal demand* and were driven by the society, horribly distorted by the new system of values, in the outsider strata. Only a tiny proportion of these people managed to join the elite, oligarchic, or regional corporate groups with property and income comparable to the largest international corporations and with

powerful means of influence on the national or regional level. However, it was the middle class and its socio-economic interests and consumer needs that provided a solid foundation for the stable functioning of the social institutions and the economy, supporting the foundations of preservation of the social order. It is exactly these factors that cause inefficiency of traditional statistical methods of population survey in order to estimate its income and consumption.

2. Let us explain the nature of fundamental difficulties with information support of the problem in the transitional period, which is exactly its second feature. Traditional analysis of income and consumption of the population (including the approach we proposed in [Consumer Typology, 1978]) uses as an information base the results of household budget surveys regularly conducted by Rosstat RF. However, the reliability and representativeness of the data was always doubtful primarily because of the way of selection of families, denial by a number of planned household to be surveyed, and the methodology of filling out questionnaire notebooks. In the post-Soviet period, these weaknesses were sharply exacerbated by the fact that a large number of Russian households (estimated range is up to 25% of the total population; see, for example, [Волков, 1997]) *is generally beyond the population framework from which the households are selected for budget network of the Rosstat RF.* The following groups are not represented in the sample of Rosstat, from which the initial statistics are formed: first, the entire sections of the poorest population (childless families, refugees and internally displaced persons, vagrants, persons without citizenship status in prison and recently released), and second, the *rich and super-rich part* of the society, largely consisting of people directly or indirectly related to the shadow economy and questionable (for legitimacy and economic decisions) processes of privatization of national riches.

Information base and the underlying assumptions. Given the specificity of the above transitional period of the Russian society, we had to create the information base, necessary to solve our problem, in a non-traditional way. Ultimately, the base was composed of three parts: 1. initial statistical data obtained as a result of a special sample *survey* of household budgets of the Russian Federation (referred to hereinafter as *actual statistics* – AS); 2. expert estimated values of certain parameters of the analysed general model: the specific weight of the rich and super-rich layers of the population, some of the structural parameters of their consumer behaviour (referred to hereinafter as *expert estimated model parameters* – EEMP); and 3. initial statistical data obtained from the *simulation model of Monte Carlo or bootstrap method* for expert estimated parameters of the corresponding distributions (referred to hereinafter as *simulated statistics* – SS). Let us describe in detail each of these components of information base.

1. *The actual statistical data* was initially represented by the results of special cross-sectional sample survey of budgets of Russian families in the cities and regions of Moscow in September–October 1996 in cities and regions of St. Petersburg, Saratov, Nizhny Novgorod, Perm and Novosibirsk. We examined 483 families selected using the *multi-stage stratified random method.* Unfortunately, of the four planned signs on

which we were supposed to carry out a stratified sampling (the territory, the level of urbanization, the scope of the employment of head of household and the monthly per capita income), we were able to complete only the first two. Later, the analysis was repeated on an extended sample. The expansion was accomplished by the addition of the data [RLMS, 1996], Round VI, which was the result of a special sampling survey of households by the Russian Federation in the framework of the project "The Russian Longitudinal Monitoring Survey". RLMS sample was designed as follows. For reasons of low population density and (or) difficult access (expensive) to survey objects, during the planning of sampling, the following areas were eliminated (in whole or in part): Tuva, Yakutia and the Chechen-Ingush autonomous republics, Krasnoyarsk Territory, Taimyr, Evenksk, Chukotka and Yamalo Nenets Autonomous Districts, Kamchatka, Sakhalin, and the Kaliningrad and Tyumen regions.[5] This reduced the analysed part of the Russian population by about 4.4%, i.e. by 6.5 million people. From the rest of the population (approximately 142 million people), using the same multistage stratified random sampling (stratification, just like in our sampling, was conducted on the basis of location and the level of urbanization of the population), we formed a one-stage sample of 3781 households, including 237 from Moscow, 192 from the Moscow region, 111 from St. Petersburg, 107 from Vladivostok, 100 from Saratov, 98 from Nizhny Novgorod, 96 from Smolensk, 89 from Kazan, 87 from Krasnoyarsk and 83 from Tomsk. (For more information on the organization of the survey, the content of questionnaires etc., see [RLMS, 1996].)

Thus, as AS we had a sample of the volume $n_1 = 4264$ in which as a component of the vector Y we identified specific (i.e. monthly per family member) expenses for the following products and services:

- $y^{(1)}$ – food (including alcohol and tobacco);
- $y^{(2)}$ – clothing and footwear;
- $y^{(3)}$ – industrial goods (including durable goods);
- $y^{(4)}$ – culture, leisure and sport;
- $y^{(5)}$ – health, health products and hygiene;
- $y^{(6)}$ – household services;
- $y^{(7)}$ – money savings of all kinds (including purchase of currency, deposits and securities);
- $y^{(8)}$ – real estate and luxury goods.

From the rather long list of signs $x^{(1)}, x^{(2)}, \ldots, x^{(p)}$ that characterize the family (household) and the conditions of its life, we have chosen only the following for this study:

- $x^{(1)}$ – *per capita income* ("low", "below average", "average", "high" and "very high");

5 When planning the sample in the RLMS project, we used administrative-territorial division of Russia as of 1989.

- $x^{(2)}$ – *sphere of employment of head of household* ("industry", "construction", "services", "agriculture" and "other");
- $x^{(3)}$ – *total number of family members* ("one", "couple", "three" and "more than three");
- $x^{(4)}$ – *type of location* ("metropolis", "one million-city", "medium and small town", "urban village" and "rural areas");
- $x^{(5)}$ – *quality of housing* ("bad", "satisfactory", "good" and "excellent").

Unfortunately, because of the above-described specific features of transitional period of the Russian society, it was not *possible to obtain a complete representative sample of families* in their stratification on the basis of per capita income, and even on the employment of head of household: participation in the shadow economy, in criminalized privatization of state property as well as owning expensive stocks, real estate or bank accounts abroad, for obvious reasons, is not advertised.

2. *Expert estimated* values of some *parameters of the model* allow to "patch" the described shortage in information supply. In fact, we use expert statistical information for that purpose. Thus, the difference between *the expert estimated real (true)* value of privatized state property and the statistics on its *face value (actual value realized in the course of privatization)* allows to evaluate the total amount of hidden income. Similarly, during the identification of the distribution model of the Russian population by per capita income (see below), the parameters of the first three components (stratas) of the mixture (i.e. the values q_j, m_j and Δ_j for $j = 1, 2, 3$) are evaluated *statistically*, while in the evaluation of the parameters of the two last components (i.e. the strata of the rich and super-rich) expert opinions are used substantially, including the above-mentioned amount of hidden income and expert estimated proportion of super-rich in the general population of the RF [Известия, 1995] Finally, the presence and identification of those three types of consumer behaviour, which are formed from the rich and super-rich strata, are also set on the basis of a combination of *expert estimations* (mainly pertaining to the averaged structures of family budgets of each type), *statistics* (of the average income of these types of families) and *initial assumptions*, based on which we carried out the specification of the model.

3. *Simulation statistics* are generated using *the Monte Carlo statistical method* and *the bootstrap method*. With Monte Carlo method, we simulated the samples that represent the types of consumer behaviour, emerging from the rich and super-rich population strata. For this purpose, it is enough to use the general view of the law of probability distribution, which governs the simulated observations, and the values of its parameters (the latter are defined with the above-mentioned expert-statistical method). Bootstrap method allows to replicate our existing AS in an amount that will provide the desired ratio of specific weights of the SS and AS (in this case, the volumes of data correlate nearly as 4:96).

Speaking of the *original model assumptions* on the basis of which *the specification of the analysed model* is accomplished, we need to add to the previously formulated hypotheses (**A**), (**B°**), (**C**), (**D**) and (**E**) (see point 4.1.1) a hypothesis (**F**) of the constancy of correlations and coefficients of variation in behavioural traits of various (consumer behaviour) classes of households, namely:

(**F**) *we postulate a special structure of the covariance matrix* $\Sigma(j)$, *in particular:*

$$\sigma_{lm}(j) = v^{(l)} v^{(m)} r_{lm}\, a^{(l)}(j)\, a^{(m)}(j), \tag{4.23}$$

where $v^{(s)} = \sqrt{\mathbf{D}y^{(s)}/(\mathbf{E}y^{(s)})^2}$ *(coefficient of variation of* $y^{(s)}$*)* and $r_{lm} = r(y^{(l)}, y^{(m)})$ *(coefficient of correlation between* $y^{(l)}$ and $y^{(m)}$*) are independent of j, i.e. remain the same for all types of consumer behaviour.*

Hypothesis (**F**) passes the statistical test for AS and allows to reduce the number of unknown parameters of the model presented in the vectors $\mathbf{a}(1), \mathbf{a}(2), \ldots, \mathbf{a}(k)$ and matrices $\Sigma(1), \Sigma(2), \ldots, \Sigma(N)$ with $44N$ to $8N + 36$ (where N is the number of different types of consumer behaviour).

Identifying types of consumer behaviour. So now we have a source of expert statistical information about the structure and volume of consumer expenditures of n Russian families in the form of a sample:

$$Y_1, Y_2, \ldots, Y_n. \tag{4.24}$$

This sample combines the version of AS replicated using the bootstrap method and modelled using the Monte Carlo method IS. The total volume of sample n (4.24) was 30,000 data, including $n_1 = 28, 800$ (including replication) AS and $n_2 = 1\,200$ SS.

Based on the assumptions set forth above, the following can be given the following mathematical formulation of the problem of identification of the main types of consumer behaviour.

There is a sample (4.24) *of the volume n of the entire population of families. Its law of probability distribution is described by the Gaussian N mixture (N is unknown), i.e. by density function* $f(Y|\Theta)$ *of the following form:*

$$f(Y|\Theta) = \sum_{j=1}^{N} \pi_j \varphi(Y|a(j); \Sigma(j)) \tag{4.25}$$

where $\varphi(Y|a(j); \Sigma(j))$ *is an eight-density of the normal distribution with the parameters* $a(j)$ and $\Sigma(j)$ *and the unknown vector parameter* Θ *includes all the unknown parameters of the right side of* (4.25), *i.e.:*

$$\Theta = (k; \pi_1, \ldots, \pi_{N-1}; a(1), \ldots, a(k); \Sigma(1), \ldots, \Sigma(N)).$$

We are required to use observation (4.24) to build the best (in some sense) estimates $\hat{\Theta}$ for the parameters Θ, as well as offer such classification rule $\delta(\hat{\Theta})$ that would allow to classify our observations on N types of consumer behaviour with the lowest probability of misclassification.

At the same time, because, in accordance with hypothesis (**B°**), each (j-th) type of consumer behaviour is identified as a population obeying the Gaussian distribution with the parameters $(\mathbf{a}(j);\ \Sigma(j))$, we are talking about the best rule of referring each of the n available observations Y_i to one of N Gaussian general populations. This problem is known in the multivariate statistical analysis as a problem of cluster analysis ("classification without learning"), based on the decomposition of a mixture of Gaussian distributions.

The approach to the solution of the first part of this problem – estimation of the unknown parameters Θ with a further simplifying assumption (4.23) of the structure of the covariance matrices $\Sigma(1), \Sigma(2), \ldots, \Sigma(N)$ – is given in [Aivazian, 1996]. Having defined the evaluations, we can define the best classification rule $\delta(\hat{\Theta})$ in the form of Bayesian classification procedure, namely, observation $\xi^{(l)}$ $(l = 1, 2, \ldots, 6)$ refers to the type of consumer behaviour "j_0" when

$$\hat{\pi}_{j_0}\varphi(Y_i|a(j_0);\ \Sigma(j_0)) = \max_{1\le j\le N} \hat{\pi}_j\varphi(Y_i|\hat{a}(j);\ \Sigma(j)) \tag{4.26}$$

It is this technique that has been used in the analysis of family budgets of the Russian Federation, that is, using a sample (4.24) as the source statistics.

We will explain briefly the general way of the practical implementation of the proposed approach.

1. According to the AS (i.e. the sampling of the volume $n_1 = 4264$), we will solve the problem of splitting the mixture (4.25) in which the covariance matrices $\Sigma(j)$ have a special structure (4.23), and the number of components of the mixture N' is less than the required total number of types of consumer behaviour N, as in this sample super-rich are not represented and the rich is under-represented. In the "output" of this step, we assess

$$\begin{aligned}&\hat{N}';\ \ \hat{\pi}'_1, \hat{\pi}'_2, \ldots, \hat{\pi}'_{\hat{N}'};\\ &\hat{\boldsymbol{a}}(1),\ \hat{\boldsymbol{a}}(2), \ldots, \hat{\boldsymbol{a}}(\hat{N}');\\ &\hat{v}^{(1)}, \ldots, \hat{v}^{(8)};\ \ \hat{r}_{lm}\ \ (\text{for}\quad l,\ m = 1,\ 2, \ldots, 8).\end{aligned}$$

2. Having assessed vector $Y_0 = (y_0^{(1)}, y_0^{(2)}, \ldots, y_0^{(8)})^{\mathrm{T}}$ of the general (total) volume of per capita consumption on the analysed expenditure items and by considering the unknown parameter values $\pi_{\hat{N}'+1}, \pi_{\hat{N}'+2}, \ldots, \pi_{\mathrm{N}}$ and $\hat{a}\,(\hat{N}'+1),\ \ a\,(\hat{N}'+2), \ldots, \mathrm{a(N)}$ using macroeconomic data, taking into account the need to recalculate specific weights $\hat{\pi}_1', \ldots, \hat{\pi}_{N'}'$ (that is transitioning to $\hat{\pi}_j = \hat{\pi}_j'(1 - \pi_{\hat{N}'+1} - \cdots - \pi_N)$), $j = 1,\ 2, \ldots, \hat{N}'$, we obtain the following equations connecting these unknown values:

$$\begin{cases} \sum_{j=1N^{\hat{\imath}}} \hat{\pi}_j + \pi_{N^{\hat{\imath}}+1} + \ldots + \pi_N = 1, \\ (1-\pi_{N^{\hat{\imath}}+1} - \cdots - \pi_N) \sum_{j=1N^{\hat{\imath}}} \pi'^{\hat{\imath}}_j \, \hat{a}(j) + \sum_{j=N^{\hat{\imath}}+1}^{N} \pi_j \mathbf{a}(j) = Y_0. \end{cases} \tag{4.27}$$

Using the fact that for some expenditure items expert estimation of the components $a^{(l)}(j)$ for the rich and especially super-rich appears to be quite transparent, and also the fact that at a high enough level of monthly per capita income (for example, with the average value of about 25,000 dollars) and expert assessment of the specific weight of the corresponding type of consumer behaviour is not too blurred, using equations (4.27) we get the assessments for k; $\pi_{N^{\hat{\imath}}+1}, \ldots, \pi_N$; $\mathbf{a}(N^{\hat{\imath}}+1), \ldots, \mathbf{a}(N)$; $\Sigma(N^{\hat{\imath}}+1), \ldots, \Sigma(N)$. When assessing the total number of type k we used cluster analysis of available data on the "rich", and in the evaluation of covariance $\Sigma(j)$ matrices (for $j \geq N^{\hat{\imath}}+1$), we used their special structure.

3. The amount of AS replication is due to the necessity of at least a minimum representation in the total sample (4.24) of the data on the super-rich families. In our analysis, as an assessment of the proportion of the type of consumer behaviour, formed mainly from the super-rich and wealthy part of the population, we took the value $\hat{\pi}_N = 0.0005$, which determined the amount of the total sample (4.24) $n = 30{,}000$ (the proportional representation of the super-rich families consisted of 15 units).

4. We repeat the statistical analysis of the mixture (4.25), but the sample (4.24) has the volume $n = 30{,}000$.

The results of this particular statistical analysis are shown in Table 4.1.

Note. In fact, about 15% of the analysed families were *only formally* (i.e. as a result of a formal application of classification rule (4.26)) assigned to one of the types of consumer behaviour shown in Table 4.1. Each of these families behaves as an outlier, i.e. in itself represents a certain (other than the eight types listed in Table 4.1) type of consumer behaviour. For their formal identification, we should use a *somewhat modified* Bayesian classification procedure (4.26), adapted to the case of an unknown number of required classes (types of consumer behaviour). Namely, in addition to rule (4.26), we formulated the following condition:

observation Y_i will not be attributed to either of the classes $(\hat{\pi}_j, \hat{a}(j), \Sigma(j))$, $j = 1, 2, \ldots, N$, if

$$\max_{1 \leq j \leq N} \hat{\pi}_j \varphi(Y_i | \hat{a}(j), \Sigma(j)) \leq \delta_0, \tag{4.26'}$$

where δ_0 is an a priori given threshold measure of likelihood of observation.

However, in view of the low representation of each of the classes, defined by such outlying observations (the share of each of them ranged from 0.00003 to 0.00007), we limited our analysis by the classes listed in Table 4.1.

Table 4.1. Consumption structure of the main types of consumer behaviour of the families of Russian Federation (September–October 1996)

Number of type of consumer behaviour	Specific weight of type, %	Expenditures on (% from the total)								Brief characteristics of the type in typological sign variables (average per capita income, thousand rubles)
		Food stuffs	Clothes and shoes	Other merchandise	Leisure, vacation, sports	Health, health products	Services, bills, dues	Savings	Real estate, luxury items	
1	34.20	61.8	14.6	6.0	3.21	6.6	7.8	–	–	Unemployed with occasional wages, low-paid, non-working retirees (less than 400)
2	39.42	48.6	21.2	11.2	3.5	3.0	12.5	–	–	The staff of the public sector and of slow income firms and companies (from 370 to 1200)
3	6.93	42.4	20.5	15.0	4.0	2.0	14.1	2.0	–	The majority of white collar employees (from 370 to 1000)
4	11.80	36.5	20.1	12.1	4.8	0.8	15.3	8.4	2.0	Ordinary employees of successful companies, enterprises, joint-stock companies (from 1000 to 7000)
5	3.65	30.6	16.4	20.5	9.9	1.0	13.0	6.6	2.0	Thriving group of the white collar workers (including artists) occupations (1000 to 7000)
6	3.3	32.8	10.2	10.4	5.0	2.0	9.1	19.2	11.3	The owners, major shareholders and key employees of successful companies, and ordinary strata of state bureaucracy elite, the average staff of the shadow economy (from 6000 to 30,000)
7	0.62	28.6	8.4	10.6	7.2	1.4	12.8	23.0	8.0	Some of the key employees of successful businesses, creative elite (5000 to 30,000)
8	0.05	18.0	7.8	6.2	6.8	0.5	10.1	25.7	24.9	The upper layers of state bureaucracy and mafia elite (more than 30,000)
Total	100.0									

Income as the most important typological variable. Let us now turn to task 5 from the list, which was given in connection with the description of the general problem of typology of consumption (see point 4.1.2). In particular, let's try to determine how informative is the average per capita family income in the problem of classifying families to a particular type of consumer behaviour. In this regard, let us briefly concentrate on the analysis and modelling of the distribution of the Russian population by per capita income (the results of this study are presented in more detail in the next section).

As is known, the values of a log-normally distributed random variable are formed and influenced by a very large number of mutually independent factors, and the impact of each individual factor is *multiplicative* in nature (i.e. a random growth caused by the influence of each factor is proportional to that already achieved by the time value of a given variable), "*slightly and evenly with equal probability of the sign*" (for more on such mechanisms of formation of random variables, see point 4.2.1). As applied to the problem of determining the distribution of households by per capita income, this means that the conditions of the genesis of log-normal observations are provided *only within a relatively homogeneous* (*by source of income generation, territorial and socio-professional signs*) population of families. So, within each (*j*-th) of the homogeneous in this sense population, the distribution of the population by per capita income will be described by a log-normal law, whose probability density function has the following form:

$$f(x; \mu_j, \sigma_j^2) = \frac{1}{\sqrt{2\pi}\sigma_j x} \exp\left\{ -\frac{(\ln x - \mu_j)^2}{2\sigma_j^2} \right\},$$

where μ_j and σ_j^2 are the parameters of the law, which determine the distribution of households by income from a single homogeneous population.

Note that this result has been repeatedly confirmed by the real statistics from experiments of different countries *with stable functioning economy*.

The specifics of the transitional period brought about changes in the underlying assumptions and, without cancelling the entire structure of the model, provided correction of the initial predispositions of a log-normal distribution of income.

In particular, the process of wiping out of the middle class and super-differentiation of income destroyed *a continuous spectrum of strata* and thus the normal character of distribution of value μ. Instead, we had (at that time) a *discrete spectrum of strata*, and therefore, a *discrete mixture of log-normal distributions* as a law of distribution $f_\xi(x)$ of all families of the society by per capita income (the mechanism of formation of a mixture of log-normal distributions is discussed in detail in point 4.2.2).

Therefore, while studying the influence of the factor *income*, we should offer *a concrete way of stratification of the Russian population*, in which the distribution of

population by average per capita income *within each stratum* is really subordinated to the log-normal law. The following describes the version of this stratification. It is based on accounting for differences in the following:
- *Sources of income generation*: federal budget, legal private enterprise, financial market, shadow economy (including the "machinations" with privatization);
- *Socio-professional characteristics*: unemployed of pre-retirement age, unemployed pensioners, employees and workers of different levels and domains, creative professions (including) liberal occupations, state official and mafia elites.

We did not explicitly consider a rather important (in terms of income inequality) *territorial sign*. However, in our model, it is involved as one of the random sources of differentiation operating within each stratum.

Thus, as a result of expert statistical analysis based on the partition of the total population of the Russian Federation on such strata in which the conditions of the formation of the log-normal distribution of income are observed, we proposed the following stratification of heads of families.
- Stratum 1 (*specific weight* q_1). Unemployed of pre-retirement age, with occasional (casual) earnings. Non-working pensioners and scholars. Low-paid wage earners.
  Stratum 2 (*specific weight* q_2). The majority of public sector employees and creative (including liberal arts) professions. Lower and middle-ranking employees of companies and enterprises of low-income industries.
- Stratum 3 (*specific weight* q_3). The rank and file employees of the state apparatus, firms, enterprises and joint stock companies of relatively prosperous sectors of the economy (fuel and energy complex, ferrous and non-ferrous metallurgy, financial market, profitable areas of trade, etc.).
- Stratum 4 (*specific weight* q_4). The owners, major shareholders and key employees of prosperous sectors of the economy (see above), ordinary employees of state elite; the base (middle and upper-middle) staff of the shadow economy and mafias; "top" intellectual (including artistic) elite.
- Stratum 5 (*specific weight* q_5). The highest positions in business, state bureaucracy and mafia elites.

Note that the *share of the hidden* (relative to the amount of declared) income is present in all of the five of these populations. However, according to various expert estimates, it varies from relatively low values in the strata 1–3 (at 20–30%) to 200–300% in stratum 4 and almost 2000% in stratum 5.

In accordance with the above stratification of the population of the Russian Federation and the general structure of the discrete model of mixture, the density function $g(z)$ of the probability distribution of the value of the *logarithm* of per capita income (i.e. on $z = \ln x$) can be represented in the following form:

$$g(z) = \sum_{j=1}^{5} q_j \varphi(z|\mu_j; \sigma_j^2),$$
$$\text{where } \varphi(z \mid \mu_j; \sigma_j^2) = \frac{1}{\sqrt{2\pi}\sigma_j} \exp -\frac{(z-\mu_j)^2}{2\sigma_j^2}\} \tag{4.28}$$

is a density of the normal distribution with the average value μ_j and the variance σ_j^2. The parameters q_j, μ_j and σ_j^2 should be evaluated on a sample or by an expert.

Expert-statistical analysis of the model and some practical conclusions. According to the above initial statistics, using the same expert-statistical approach that was used to identify the main types of consumer behaviour, the mixture (4.28) was analysed. This problem differs from the previous one by two simplifying reasons: firstly, only *one-dimensional* (rather than an eight-dimensional) mixture of normal distributions was analysed, and, secondly, the number of components of the mixture is known in advance (in this case $N = 5$). A detailed description of the analysis is given in [Айвазян, 1997]. Please note that in determining the parameters μ_j and σ_j^2 of the Gaussian population ($j = 1, 2, \ldots, 5$), we used previously known equations connecting the average values of (m_j) and the variance (Δ_o^2) of the log-normally distributed random variable (in this case of per capita income of j-th income group) with the average values $(\mu_j)$and variances (σ_j^2) of its logarithm, namely:

$$\begin{cases} m_j = \exp\left\{\mu_j + \frac{1}{2}\sigma_j^2\right\} \\ \Delta_j^2 = m_j^2(\exp\{\sigma_j^2\}-1), \end{cases} \quad j = 1, 2, \ldots, 5.$$

The results of the mixture (4.28) analysis and certain conclusions made on the basis of this analysis are presented in Tables 4.2 and 4.3.

All data of Goskomstat, which is used in this paper are taken from [I24] (see section “Information sources” in the References). From Table 4.3, it can be seen that the main difference with the official data of State Statistical Committee of the Russian Federation are concentrated, as one would expect on the “tails” of distributions. The reason is the lack of basic statistical Roskomstat data on representatives of “very poor” and “very rich” sectors of the population, as well as the propensity of some representatives of the “business” to concealing income.

Table 4.2. Main characteristics of components of mixed distribution of Russian families according to monthly per capita income (September 1996)

Number of strata (homogeneous group by income), j	Specific weight of the income group, q_j	Average value of per capita monthly income in group (thousand rub./number of min. salaries), m_j	Average quadratic deviation in group (thousand rub./number of min. salaries), Δ_j	Total specific (i.e. for one person) income of group (thousand rub./number of min. salaries), $q_j \cdot m_j$	Share of total income, $\tilde{q}_j$	Parameters μ_j and σ_j for corresponding normal distributions, $\varphi(z\vert\mu;\ \sigma_j^2)$
1	0.3900	350/4.67	101/1.35	136.5/1.82	0.113	$\mu_1 = 5,818$ $\sigma_1^2 = 0,0802$ $\sigma_1 = 0,2832$
2	0.4000	700/9.33	210/2.80	280.0/3.73	0.232	$\mu_2 = 6,508$ $\sigma_2^2 = 0,0862$ $\sigma_2 = 0,2936$
3	0.1700	1640/21.87	525/7.00	278.8/3.72	0.231	$\mu_3 = 7,344$ $\sigma_3^2 = 0,0975$ $\sigma_3 = 0,3122$
4	0.0397	12,000/160.00	5100/68.00	476.4/6.35	0.394	$\mu_4 = 9,310$ $\sigma_4^2 = 0,1660$ $\sigma_4 = 0,4075$
5	0.0003	120,000/1600.00	45,000/600.00	36.0/0.48	0.030	$\mu_5 = 11,629$ $\sigma_5^2 = 0,1316$ $\sigma_5 = 0,3627$
Sum $\sum$	1.0000	–	–	1207.7/16.10*	1.0000	–

* This amount is 54% higher than the official amount in September according to Goskomstat (see [Goskomstat, 1996, 176]): according to this source, the general income of the population in September 1996 amounted to 116 trillion rubles or 784 thousand rubles average per capita. This is due to undeclared (hidden) income of mainly the fourth and fifth income groups.

Table 4.3. Distribution of Russian population by per capita income (as of September 1996)

Average per capita income in the range (in thousand roubles)	In millions people		In %	
	Goskomstat data	model (4.28)	Goskomstat data	model (4.28)
till 400	35.5	44.5	24,00	30,07
400.1–600	33.2	32.8	22,40	22,17
600.1–800	25.4	24.2	17,20	16,34
800.1–1000	17.4	13.4	11,70	9,08
1000.1–1200	11.6	7.7	7,90	5,20
1200.1–1600	12.9	10.5	8,70	7,10
1600.1–2000	5.9	5.5	4,00	3,7
2000.1–25000 over 25000	6.1	$9,4\begin{cases} 9,2 \\ 0,2 \end{cases}$	4.1	$6,34\begin{cases} 6,22 \\ 0,12 \end{cases}$

Table 4.4. Joint (two-dimensional) distribution of Russian families belong to a particular income group (x) and by type of consumer behaviour (y)

Income groups – gradation *i* on *x*	**Types of consumer behaviour – gradation *j* on *y***								$p_{i.}$
	1	**2**	**3**	**4**	**5**	**6**	**7**	**8**	
1	0.3318	0.0479	0.0103	0	0	0	0	0	0.3900
2	0.0102	0.3348	0.0550	0	0	0	0	0	0.4000
3	0	0.0115	0.0040	0.1180	0.0365	0	0	0	0.1700
4	0	0	0	0	0	0.0333	0.0062	0.0002	0.0397
5	0	0	0	0		0	0	0.0003	0.0003
$p_{.j}$	0.3420	0.3942	0.0693	0.1180	0.0365	0.0333	0.0062	0.0005	1.0000

Although not too conspicuous (see Table 4.3), the difference between Goskomstat and model distribution series leads to *qualitatively different conclusions about the characteristics of income differentiation*. This applies to the Lorenz curves showing the degree of inequality in income distribution between population groups with different levels of material wealth [Aivazian, 1997], and to the evaluation of specific weights of the richest 20% and 10% groups in the total monetary income of the population. So, according to the official statistics (compare [I. 24], 106–107), these specific weights for Russia in 1995, were, respectively, 46.9% and 31.0%, while according to model (4.28) (however, as of September 1996) the same indicators are measured by the numbers of 65.2% and 56.6%. For comparison, it may be noted that among the developed countries such high values of these indicators do not exist (their values vary around the levels, respectively, around 40% and 25%), and the search for analogues of the Russian situation leads us to Brazil in 1989 (67.3% and 51.3%), South Africa in 1993 (63.3% and 47.3%) and Kyrgyzstan in 1992 (57.0% and 40.3%).

Table 4.4 shows the joint (two-dimensional) distribution of Russian families belonging to a particular income group (x) and by type of consumer behaviour (y), i.e. the values of probabilities $p_{ij} = \mathbf{P}\{x = i;\, y = j\}, p_{i.} = \mathbf{P}\{x = i\}$ and $p_{.j} = \mathbf{P}\{y = j\}$.

Analysis of Table 4.4 allows us to make some conclusions about the informativeness of obtained typological factors and their role in the stratification of the Russian population by source of income and socio-professional characteristics, adopted in the model mix (4.28) of the distribution of population by average per capita income (see the above-mentioned strata 1–5).

The decisive characteristic of the reliability of the forecast of type of consumer behaviour (j) on the values of typological signs $\tilde{X}$ is the conditional probability $\mathbf{P}\{y = j|\tilde{X}\}$.

Table 4.4 shows that if $\tilde{X}$ denotes the signs that form corresponding formed strata (populations) 1–5, then all types of consumer behaviour, except for the third, fifth and seventh are correctly restored by belonging to one or another stratum with a high probability. Indeed:

$$\mathbf{P}\{y = 1|x = 1\} = \frac{\mathbf{P}\{y = 1; x = 1\}}{\mathbf{P}\{x = 1\}} = \frac{p_{11}}{p_1} = 0,851;$$

$$\mathbf{P}\{y = 2|x = 2\} = \frac{\mathbf{P}\{y = 2; x = 2\}}{\mathbf{P}\{x = 2\}} = \frac{p_{22}}{p_2} = 0.837;$$

$$\mathbf{P}\{y = 4|x = 3\} = \frac{\mathbf{P}\{y = 4; x = 3\}}{\mathbf{P}\{x = 3\}} = \frac{p_{34}}{p_3} = 0.694;$$

$$\mathbf{P}\{y = 6|x = 4\} = \frac{\mathbf{P}\{y = 6; x = 4\}}{\mathbf{P}\{x = 4\}} = \frac{p_{46}}{p_4} = 0.839;$$

$$\mathbf{P}\{y = 8|x = 5\} = \frac{\mathbf{P}\{y = 8; x = 5\}}{\mathbf{P}\{x = 5\}} = \frac{p_{58}}{p_5} = 1.$$

From the analysis of the list of typological variables reflected in the last column of Table 4.1, and the variables that influenced the strata formed in the mixture model, we can conclude that in the latter only the sign of the professional area of the head of the family (compared with a set of typological signs) is missing. Adding (during the formation of strata) a nominal characteristic defining on *the binary scale*, the area of occupational activity of the head of the family ("intellectual-other") will lead to the splitting of populations (strata) 2, 3 and 4 (2^a,2^b), (3^a,3^b) and (4^a,4^b) and will increase the total number of strata to eight. Accordingly, constructed in this way, the typological sign $\tilde{X}$ will have the same number of gradations as a sign y ("type of consumer behaviour") and the contingency table formed (by analogy with Table 4.4) for nominal signs $(\tilde{X}; y)$ will allow us to calculate conditional probabilities $\mathbf{P}\{y = j|\tilde{X} = i\}$ $(i, j = 1, 2, \ldots, 8)$. It is easy to see that the values of the characteristics of the predictive power of the constructed models, namely, values of conditional probabilities $P^{(j)} = \max_{1 \le i \le 8} \mathbf{P}\{y = j|\tilde{X} = i\}$, for all values of j will turn out high enough (in our case all of them exceed 0.7)

Conclusions. The methodology of the analysis of consumer behaviour and the household income described in point 4.1 is, essentially, development and adaptation of the approach to the transitional period that was first proposed in [Typology of

Consumption, 1978]. The following moments are fundamentally new in this modified version of the approach:

- *including the process of expert evaluation of some parameters of the model* in the research process, providing, in particular, the use of a certain type of "connection" of statistical information of macro- and micro-levels on the volume and structure of consumption and expenditures of the population;
- supplementing existing real statistical data on income and consumption of households with especially modelled *simulation data*, electronically generated based on the existing AS (bootstrap method) or expert estimated values of the parameters of their distribution (Monte Carlo method);
- improving mathematical-statistical methods of splitting the mixtures of distributions of type (4.25) and (4.28). In the past while "fumbling" the total number of the components of such mixtures we used heuristic methods of the type "*N*-average with an unknown number of classes" with added preliminary projection of the analysed data to a plane of the first two principal components in the multidimensional case. Now we can offer a modern method based on the criteria of adequacy of the model (type ICOMP – and Akaike criteria) and projection pursuit [Aivazian, 1996].

Apparently, it is interesting to use this approach in some other (more detailed) versions of the structuring of family consumption, for example, with the special allocation of items of expenditure such as "alcohol consumption" and "expenditures on education and training". Another rather interesting result of this approach is the implementation of the related statistical methodology for restoring *the functions of consumer preferences* (separately for each identified type of consumer behaviour).

4.2 Analysis and modelling of the distribution relations in society

As we know, the total amount of funds of consumption of the country, accounting for a significant portion of the national income, is in turn subdivided into *personal funds* (payroll and business income) and *public funds* (payments to the elderly, disabled and students and free or discounted goods and services for population in areas such as education, health, social welfare, culture and information, sports, leisure and entertainment and housing).

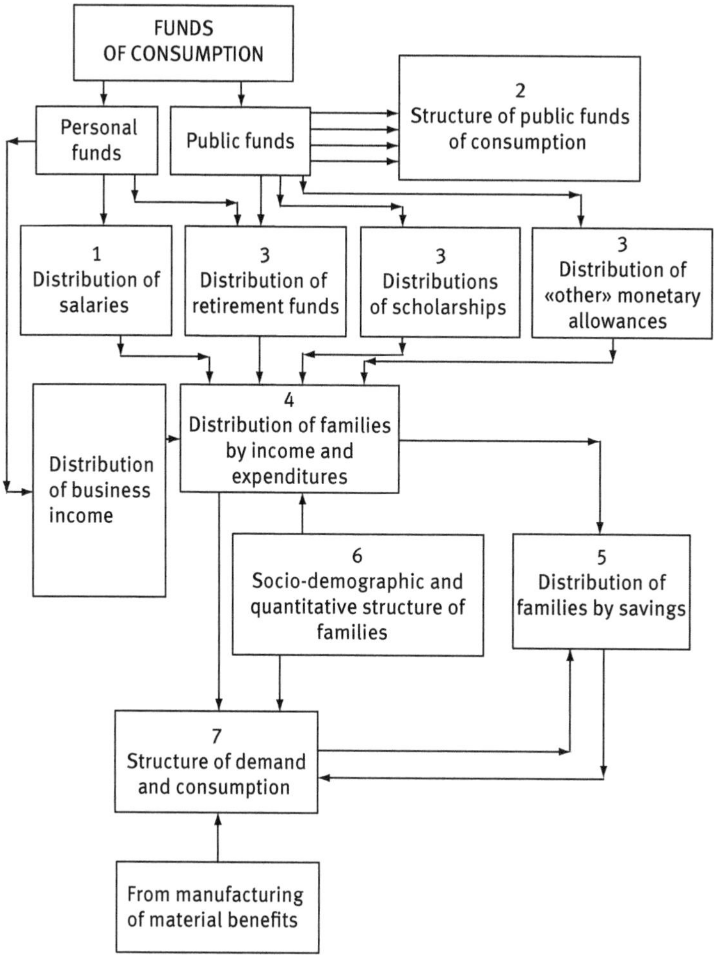

Figure 4.3. General picture of connections of the blocks of the system of distributive relationships in the society

In the whole socio-economic system that determines the nature of distributional relationships in the society, in accordance with objectively existing economic laws (caused in turn by the main goals of the society, adopted a system of values, etc.), we have a number of interrelated local mechanisms of individual socio-economic systems, such as (see Figure 4.3) (1) blocks of distribution of employees by salary and entrepreneurs by the volume entrepreneurial income; (2) blocks of distribution of public funds of consumption (the share of public funds allocated for health and culture, on state aid of the family, transportation, etc.); (3) blocks of distribution of the retired employees by pension size, scholarship winners by scholarships for families by additional ("other", i.e. not included in wages, pensions and grants) monetary allowances; (4) blocks of distribution of families by the average per capita income; (5) blocks of distribution of families by average per capita savings; (6) blocks of socio-demographic and quantitative structure of families and (7) blocks of structure of demand for different types of goods and services and their consumption.

This section is devoted to probabilistic statistical description of the functioning of blocks 1 ("salary"), 4 and 5. This is an attempt to penetrate the essence of *the very mechanism of formation of the analysed distributions* and not to be limited by the regular (based on the logic of "black box") statistical photograph of the considered phenomenon. Let us explain this idea. During the probabilistic and statistical modelling, and in particular at the stage of a priori information about the real nature of the mechanism of converting input parameters into the output (resulting) ones, *some part of this mechanism remains hidden from the researcher* (precisely this part, using common cybernetic terminology, is called a "black box"). The more professional knowledge of the mechanism of the investigated phenomenon the researcher demonstrates, the lower is the proportion of the "black box" in the general logics of modelling and the better operational and more accurate is the constructed model. Probabilistic statistical modelling, based entirely on the logic of "black box", allows the researcher to obtain a sort of instant statistical picture of the analysed phenomenon, which is generally unsuitable, for example, for forecasting purposes. In contrast, modelling, based on a deep professional analysis of nature of the phenomenon, can largely theoretically justify a general view of the constructed model, which, in particular, allows its legitimate use in the forecast calculation.

4.2.1 Population distribution model by salary size

Log-normal law of distribution of employed population by size of wages is widely used in the practice of analysis and construction of distributive relations of society. However, in general, the law is used as *a suitable formal approximation* of the unknown true distribution, which at best meets the requirements of the corresponding statistical goodness-of-fit tests.

Below we offer a variant of a rigorous justification of genesis of **specifically log-normal distribution** under the observance of some quite common and natural conditions. This justification is based on three facts measurable by experimental verification (see below):

(i) *within each homogeneous (on some non-negatively valued characteristic* θ *of* ***the degree of total giftedness of the individual****) population of workers, the distribution of their wages* $\eta(\theta)$ *is subject to log-normal law, i.e.* $\ln\eta(\theta)\in N(\mu(\theta);\sigma^2)$

(ii) the wages $\eta_0(\theta)$ of an individual, *characterized by the general level of giftedness* θ*, with the exclusion of the influence of random factors (i.e. in the idealized pattern in which wages are* ***uniquely*** *determined by the value* θ*) is determined by the law*

$$\eta_0(\theta) = c\,\theta^{\lambda},$$

where c and λ *and are positive constants;*

(iii) *characteristics of the general giftedness* θ, *like most biological and anthropological characteristics of the person, is subject to (in a general population of workers) a log-normal law, i.e.* $\ln\theta \in N(\mu_0; \sigma_0^2)$.

Note. Introduced in the 70s of the last century, *characteristic of general giftedness* θ [Aivazian, 1976] can be interpreted, in view of the common modern terminology in knowledge economy research, also as *a measure of the level of human capital of the individual.*

Assertion. If conditions (i)–(ii)–(iii) of the wages $\eta(\theta)$ of workers is subject to the log-normal distribution law with the parameters,

$$\begin{aligned} \mathbf{E}\,(\ln\eta(\theta)) &= \ln c + \lambda\mu_0 = \bar{\mu}_0 \\ \mathbf{D}\,(\ln\eta(\theta)) &= \sigma^2 + \lambda^2 \cdot \sigma_0^2 = \Delta^2 \end{aligned}$$

(therefore $\mathbf{E}\eta(\theta) = \bar{\mu}_0 e^{\frac{1}{2}\Delta^2}$ and $\mathbf{D}\eta(\theta) = \bar{\mu}_0^2 \cdot (e^{\Delta^2} - 1)\theta)$.
Proof

1. First, we will prove the log-normal law of distribution of the salaries of workers with a given (the same) value of the generalized indicator of giftedness . The values of the wages $\eta(\theta)$ of the given population of employees are influenced by a large number of mutually independent factors, and the impact ξ_i of each individual (i-th) factor has a *multiplicative* character, equally insignificant and equal probability of the sign,

$$\begin{cases} \eta_1 = \eta_0 + \xi_1 \cdot \eta_0 \\ \eta_2 = \eta_1 + \xi_2 \cdot \eta_1 \\ \dots\dots\dots\dots\dots\dots \\ \eta_N = \eta_{N-1} + \xi_N \cdot \eta_{N-1} \end{cases}, \tag{4.29}$$

where, according to (ii), $\eta_0 = \eta_0(\theta) = c\,\theta^{\lambda}$. From (4.29) it follows

$$\sum_{i=0}^{N-1} \left(\frac{\Delta\eta_i}{\eta_i} \right) = \xi_1 + \xi_2 + \ldots + \xi_N \tag{4.30}$$

where $\Delta\eta_i = \eta_{i+1} - \eta_i$. But the right side of (4.30) is the result of the *additive* action of many random factors that, under the above assumptions, should lead, as we know, to the normal distribution of this amount. At the same time, given a fairly large number of random addends (i.e. considering $N \to \infty$) and the relative insignificance of their impact (i.e. considering $\Delta\eta_i \to 0$), we can go from the sum on the left side (4.30) to the integral

$$\int_{\eta_0}^{\eta} \frac{d\eta}{\eta} = ln\eta - ln\eta_0 = ln\eta - ln\,(c\theta^{\lambda}).$$

We came to the conclusion that the logarithm of the examined value (reduced by a constant value $ln\,\eta_0$) is subject to the normal law with a mean value of zero, i.e.

$$F_\eta(y) = P\{\eta < y\} = P\{ln\eta < \ln y\} = P\{\ln\eta - \ln\eta_0 < \ln y - \ln\eta_0\} =$$
$$= \frac{1}{\sqrt{2\pi}\cdot\sigma} \int_0^{\ln y - \ln\eta_0} e^{-\frac{(t-\ln\eta_0)^2}{2\sigma^2}} dt,$$

where by the differentiation of the left and right side of this equation by x we obtain

$$f_\eta(y) = \frac{1}{\sqrt{2\pi}\cdot\sigma y} e^{\frac{-(\ln y - ln\eta_0)^2}{2\sigma^2}}.$$

And this means that workers with a given (fixed) level of total giftedness are distributed according to the size of their wage based upon the log-normal law, or what is the same:

$$\ln\eta(\theta) \in N(\mu(\theta); \sigma_2) \text{ (with a fixed } \theta), \tag{4.31}$$

where $\mu(\theta) = \ln\eta_0(\theta) = \ln c + \lambda\ln\theta.$

2. We can now prove that the **total** population of workers is normally distributed by the value $\ln\eta(\theta)$. Since we are interested in the distribution of workers by the size of the logarithm of their wages, the whole (total) population of workers constitutes the general population, which is a continuous **mixture** of populations, where the employees are characterized by their fixed level of general giftedness θ, given (μ_0, σ_0^2)-*normal* weighting function, as in accordance with (iii) $\ln\theta$ in the totality of population of workers normally distributed with parameters $\mu_0 = \mathbf{E}(\ln\theta)$ and $\sigma_0^2 = \mathbf{D}(\ln\theta)$. Therefore, the density function $f_\xi(x)$ of a random variable $\xi(\theta) = \ln\eta(\theta) = \ln c + \lambda\ln\theta$ can be represented in the following form:

$$f_\xi(x) = \int_{-\infty}^{+\infty} \frac{1}{\sqrt{2\pi}\sigma} e^{-\frac{(x-\mu(\theta))^2}{2\sigma^2}} \cdot \frac{1}{\sqrt{2\pi\bar{\sigma}_0^2}} e^{\frac{(\mu(\theta)-\bar{\mu}_0)^2}{2\bar{\sigma}_0^2}} d\mu(\theta), \tag{4.32}$$

where, considering (ii) and (iii), $\bar{\mu}_0 = \mathbf{E}\xi(\theta) = \ln c + \lambda\cdot\mathbf{E}(\ln\theta) = \ln c + \lambda\mu_0$ and $\bar{\sigma}_0^2 = \mathbf{D}\xi(\theta) = \mathbf{D}(\lambda\cdot\ln\theta) = \lambda^2\sigma_0^2$.

The result of the right-hand side of integration (4.32) can be obtained using the composition formula (convolution) for normally distributed addends (see, e.g. [Aivazian, Mkhitarian, 2001, Paragraph 4.4]) or using identical transformations of the integrand. In the latter case, let's transform the exponent for number *e*:

$$-\frac{1}{2}\left[\frac{x^2}{\sigma^2}-2x\frac{\mu}{\sigma^2}+\frac{\mu^2}{\sigma^2}+\frac{\mu^2}{\bar{\sigma}_0^2}-2\bar{\mu}_0\frac{\mu}{\bar{\sigma}^2}+\frac{\mu^2}{\sigma^2}\right]=$$

$$=-\frac{1}{2}\left[\frac{\mu^2(\bar{\sigma}_0^2+\sigma^2)}{\sigma^2\bar{\sigma}_0^2}-2\mu\left(\frac{x}{\sigma^2}+\frac{\bar{\mu}_0}{\bar{\sigma}_0^2}\right)+\frac{x^2}{\sigma^2}+\frac{\bar{\mu}_0^2}{\bar{\sigma}_0^2}\right]=$$

$$=-\frac{1}{2\frac{\sigma^2\bar{\sigma}_0^2}{\sigma^2+\bar{\sigma}_0^2}}[(\mu-M)^2+Z],$$

where $M=\frac{\bar{\sigma}_0^2x+\sigma^2\bar{\mu}_0}{\sigma^2+\bar{\sigma}_0^2}$, and $Z=\frac{\sigma^2\bar{\mu}_0^2}{\sigma^2+\bar{\sigma}_0^2}+\frac{\bar{\sigma}_0^2x_2}{\sigma^2+\bar{\sigma}_0^2}-\left(\frac{\sigma^2\bar{\mu}_0+\bar{\sigma}_0^2x}{\sigma^2+\bar{\sigma}_0^2}\right)$ do not depend on θ (i.e. on $\mu(\theta)$).

Back to (4.32), we have

$$f_\xi(x)=\frac{1}{\sqrt{2\pi}\frac{\sigma\bar{\sigma}_0}{\sqrt{\sigma^2+\bar{\sigma}_0^2}}\cdot\sqrt{2\pi}\sqrt{\sigma^2+\bar{\sigma}_0^2}}\int_{-\infty}^{\infty}e^{-\frac{1}{2\frac{\sigma^2\bar{\sigma}_0^2}{\sigma^2+\bar{\sigma}_0^2}}(\mu(\theta)-M)^2}d\mu(\theta)\times$$

$$\times e^{-\frac{1}{2\frac{\sigma^2\bar{\sigma}_0^2}{\sigma^2+\bar{\sigma}_0^2}}\left[\frac{\bar{\sigma}_0^2x^2(\sigma^2+\bar{\sigma}_0^2)}{(\sigma^2+\bar{\sigma}_0^2)^2}+\frac{\sigma^2\bar{\mu}_0^2(\sigma^2+\bar{\sigma}_0^2)}{(\sigma^2+\bar{\sigma}_0^2)^2}-\frac{\bar{\sigma}_0^4x^2+2\bar{\sigma}_0^2\sigma^2x\bar{\mu}_0+\sigma^4\bar{\mu}_0^2}{(\sigma^2+\bar{\sigma}_0^2)^2}\right]}=$$

$$=\frac{1}{\sqrt{2\pi}\sqrt{\sigma^2+\bar{\sigma}_0^2}}e^{-\frac{1}{2(\sigma^2+\bar{\sigma}_0^2)}(x-\bar{\mu}_0)^2}.$$

Therefore, we proved that when (i)–(ii)–(iii) are true, the whole population of workers is distributed upon the dimensions of the logarithm of their wages following normal law, i.e.

$$\xi(\theta)=\ln\eta(\theta)\in N(\ln c+\lambda\mu_0;\ \sigma^2+\lambda^2\sigma_0^2)\qquad(4.33)$$

where $\mu_0=\mathbf{E}(\ln\theta)$ and $\sigma_0^2=\mathbf{D}(\ln\theta)$. This means that the wages of workers $\eta(\theta)$ follows the log-normal law of distribution with

$$\begin{cases}\mathbf{E}\eta(\theta)=\bar{\mu}_0\,e^{\frac{\Delta^2}{2}}\\ \text{and } \mathbf{D}\eta(\theta)=\bar{\mu}_0^2(e^{\Delta^2}-1),\\ \text{where } \bar{\mu}_0=\ln c+\lambda\cdot E(\ln\theta)\text{ and }\Delta^2=\sigma^2+\lambda^2D(\ln\theta).\end{cases}\qquad(4.34)$$

The described model of formation (genesis) of distribution of workers on their wages allows us, besides everything else, to consider this process *in motion* and to use it for forecasting distributive relations. In particular, it is natural to consider a stationary (i.e. independence of time *t*) distribution of the random value on the limited time intervals *t* (at least within 15- to 30-year time intervals). Coefficient *c* (*t*) that reflects the degree of effective realization of natural giftedness of the given members of the society and, at the same time, the general dynamics of the welfare level of population has a tendency of growing. When it comes to coefficient $\lambda(t)$, together with the stationary behaving parameters $\sigma^2 = \mathbf{D}(\ln \eta(\theta))$ and $\sigma_0^2 = \mathbf{D}(\ln\theta)$, it determines the dynamics of differentiation of workers according to their wages.

We can supplement the offered construction of bio-anthropological interpretation of the model with a relevant experimental research, as a result of which we can make the values θ, c and λ included in it fairly tangible and fill them with a practical sense. So, by selecting for the research the employees of various categories of work complexity and subjecting them to a certain set of intellectual and other tests, using methods of multivariate statistical analysis (for example, within one of the models of factor analysis [Aivazian, Mkhitarian, 2001, Chapter 13]), we can determine the way of measuring value. Then, by comparing the obtained values θ, characterizing the overall giftedness of the analysed workers, with their wages, and using known methods of statistical research of dependencies, we can estimate the coefficients c and λ.

As for the results of experimental verification of the adequacy of the model (i.e. comparison of the main "output" features of the model with the existent ones), the following fact should be noted: the usual approach to modelling, as already noted, provides not just "a statistical picture" of the real situation, nor a formal experimental data fit into one of the theoretical models, but the mathematical description of **the mechanism of the phenomenon**, coming from more or undisputed (or confirmed by experiments) socio-economic initial premises. Therefore, one should not expect that the results arising from our model will *always* be well consistent with the experimental data. *The more logically the initial assumptions, "incorporated" in the base model and accepted by specialists as objective laws, are implemented (for example, in the form of an appropriate system of tariff conditions), the better is the coordination of the "model" and the experimental data.* In this sense, the results of the CEMI AS USSR successive (in time) comparison of model and experimental data on the distribution of workers' wages are symptomatic. We could clearly see in that work the two moments of extremely sharp discrepancy between model and real data. However, while these moments are moving to the past, there is a clear trend towards convergence of model and real data. A closer analysis shows that the moment of sharp discrepancy immediately followed a very significant distortion of the initial assumptions (i) and/or (ii). The first of these moments refers to the 60s of the last century and is connected with a sharp-willed directive for restructuring the existing wage scales, which led to a breach of the initial premises (ii). The second moment relates to the 90s of the last century, a period of radical socio-economic transformation of the Russian society, in the course of which premise (iii) was broken (model description of the specifics of this period, but in relation to the distribution of population by per capita expenditures, see 4.2.2). The fact that further on we have witnessed the convergence of model and real data, only shows that the objective economic laws gradually increase their impact on the character of distribution; they slowly "come to surface", and is steadily gaining various legal forms.

4.2.2 Analysis and modelling of regional population distribution by per capita expenditures[6]

Various metrics of poverty levels and indicators of population differentiation by income and expenditures are *key indicators of the quality of social policies*, and, in particular, are used to construct a targeted social assistance to low-income population, aimed at the maximum (within available funds) reduction of social tension in society.

Today, the Russian Federal authorities use [К методике, 1999; Великанова, Колмаков, Фролова, 1996; Великанова, Фролов, 1999] and other known researchers offer [Суворов, Ульянов, 1997; Ершов, Майер, 1998; Шевяков, Кирута, 1999] indicators and methods of their evaluation of selective budgetary statistics of households (even with the used equivalence scales and calibration of source data, based on the macroeconomic balance of revenues and ependitures of the population), which have drawbacks, resulting in serious distortion of the real values of these characteristics.[7]

We see several main reasons explaining this situation.

1. In the specific conditions of modern Russian economy, definition of indicators of poverty and the criteria by which a household should be classified as poor should be based on average per capita **expenditures** (*rather than income*, as is customary in most other studies). In support of this thesis, we will note that when considering expenditures instead of income:
   a) the problem of accounting or not accounting for late parts of salaries of the members of the households (HH) is eliminated;
   b) the issues of intentionally or unintentionally hidden part of the income, including income derived from informal sector, become immaterial;
   c) the range of factors that determine the level of welfare of HH, primarily due to the inclusion in this spectrum of *private farming* and *property components* (real estate, personal vehicles, jewellery, etc.), rental or sale of which can substantially maintain the level of welfare, legitimately expands.
2. The apparent failure of the statistical service (Goskomstat) of two-parameter log-normal model of population distribution region and throughout Russia per capita income; however, the main distortion that this model detects is on the

6 S. O. Kolenikov [Aivazian, Kolenikov, 2001] took part in the experimental computational part of the study, the results of which are described in this paragraph.

7 A number of studies [Aivazian, 1997; Suvorov, Ulianov, 1997; Sheviakov, Kiruta, 1999] showed that the indicator of income differentiation such as *funds coefficient* is lowered for not less than 2 times, and the estimates of "*the proportion of households with per capita income not exceeding the poverty level*", obtained by the methods described in all of the above-mentioned sources method, may vary between 1.5 and 2 times (proof of the latter fact is one of the results of this study, see below).

"tails" of the distribution, and the estimates are constructed for the above-mentioned characteristics.

3. Used by official statistical services, calibration of the model in which "tug" analysed the distribution is made in the direction of known socio-demographic structure of households and given (from the balances of income and expenses) value average per capita family income [Velikanova, Frolov, 1999] does not eliminate shift; it is improperly expected that the general form of the model density distribution and its fashion remain unchanged.
4. Proposed by other researchers [Ершов, Майер, 1998; Шевяков, Кирута, 1999], methods of approximation of the distribution under consideration and ways of "weighing" (calibration) initial observations also do not eliminate significant distortions of the real situation, as it does not allow us to estimate the proportion and structure of non-observed spectrum of the "rich" and "wealthy" households (the weighting gives a weight to an already existing observations but does not generate observations from a hidden part of the range).
5. Only the proportion of households with per capita income not exceeding the subsistence level is usually considered as the poverty level indicator [Определение основных показателей, 1999; Пилотные программы, 1999]. However, the choice of the poverty level indicator (or criteria by which a household should be classified as poor) should be made depending on the ultimate application goal of economic analysis. In particular, the formation of the targeted social assistance policies is to more correctly, from our point of view, focus on the characteristics of the "depth of poverty" type of indexes such as Foster, Greer, Thorbeck, which are a more sensitive measure of the level of social tension in society.
6. Modern Russian economic theory and practice have still not solved the problem of optimal distribution of funds allocated for social support of the poor, where optimality is understood in the sense of minimization of the analysed indicator of the level of social tension (see previous paragraph).

Objectives of the study, the results of which are presented in this point, were determined by the desire to overcome the shortcomings noted in points 1–6. In particular, we are talking about the development of the methodology of econometric analysis (on the basis of data on family budgets) of population distribution in Russia according to per capita expenditures and about constructing and statistically evaluating on the basis of this distribution the key characteristics of poverty and property differentiation of the population (various indicators of poverty rate, fund coefficient, etc.) and about the formulation and solution of the problem of optimal allocation of funds for targeted social support for the poor.

General formulations of the problems solved in the course of the study are conditioned by the above objectives. In an aggregated form, the issues at stake can be summarized in the form of the following three tasks (one *primary* and two *secondary*, applied):

Task 1 (auxiliary). *We need to build, theoretically substantiate and experimentally test the econometric model of population distribution interpreted in meaningful terms in the region of Russia according to the average total expenditures per capita, including development of methodologies for its (model) identification, based on official data from sampling of household budget surveys (SHBS) and some macroeconomic indicators of the balance of income and expenditures.*

The solution to this problem must take into account the specific conditions of the modern Russian economy, in particular, in modern Russian conditions, the dramatic role of *a certain source of distortion* of the law of distribution of per capita population income (expenditures) during the evaluation of this law using the results of the sample survey of household budgets. This source is the **conscious avoidance of a HH being surveyed that violates the intended sampling survey plan, and hence distorts the representativeness of the sample.**

It is possible to imagine some stratification of HH according to the probability rate of evasion from SHBS. In particular, in specific Russian conditions, it is necessary to provide for the existence of such category of HH, *whose representatives avoid SHBS* **with a probability of one** (the effect of *clipping* of the sample). Based on these specific conditions, it can be argued that this category combines HH, whose total per capita expenditures exceed a certain sufficiently high level, i.e. all "super-rich" HH.[8]

Of course, it is impossible to build such an econometric model of the distribution of the region's population by per capita income (or expenditures) that would have eliminated the possible distorting effect of this factor, without the wording and possible explanation and statistical testing of a number of additional working hypotheses and model assumptions. In our study, such *hypotheses* are the following:

- hypothesis H_1 of a general appearance of the analysed law of distribution of probabilities (l.d.p.);
- hypothesis H_2 of the shape of dependence of the probability of "loss" of the household from the network of officially statistically surveyed HH from its socio-economic and territorial characteristics and, in particular, from the total per capita household expenditures.

In addition, we rely in our study on two *model assumptions* H_3 and H_4:

- model assumption H_3 of the constancy of the coefficient of variation of per capita expenditures (in relation to a change in the socio-economic strata of the population for which it is calculated);

8 Note that the inclusion of another source of shift in statistical estimates of the law of distribution of population by average per capita income (expenditures), namely, the factor of deliberate distortions of the respondents' answers ("misreporting") for the purpose, for example, of concealment of informal sources of additional income, remains largely outside the scope of this study (some aspects of the influence of this factor on characteristics of interest to us are touched upon below in this point).

– model assumption H_4 on the general form l.d.p. per capita expenditures in the non-observed ("super-rich") strata of the population.

Hypothesis H_1 is based on the nature of transformations of the socio-economic structure of the Russian society (see below in this paragraph), and its statistical testing and use will allow to build, in terms of content, a relatively laconic and interpretable model of the distribution of population in the regions in Russia in terms of total per capita expenditures. Statistical testing and the use of *hypothesis* H_2 will allow to reduce the shift l.d.p. on income (expenditures), due to the influence of the factor "*refusal of HH to survey with a probability of less than one". Model assumptions* H_3 and H_4 perform a purely technical role. Their use will allow to take into account to some extent the distorting effect of the lack in statistically processed samples category "super-rich" HH.

Task 2 (additional, applied). *To consider a fairly wide class of poverty indicators based on the population distribution by total per capita expenditures and to formulate the problem of the best distribution of the sum S, allocated for social support of the poor, as a special optimization problem, in which indicators of poverty level from the above-mentioned class are considered as minimized criteria of social tension.*

Let's consider the following family as a class of indicators of poverty:

$$I\Big(w(x), f(x)\Big) = \int_0^{z_0} w(x) f(x)\, dx \tag{4.35}$$

where $f(x)$ is a function of density of family distribution on total per capita expenditures z_0, a so-called poverty line (value of poverty level), and a weighting function $w(x)$ is a continuous, differentiable, decreasing and convex downward on the interval $[0, z_0)$ function (these properties are determined by the natural assumption that during the transfer of any amount of money from the poor to less poor, the value of the indicator of poverty (4.35) will increase). Obviously, the family (4.35) includes (with a suitable choice of the weighting function $w(x)$) common factors such as poverty deficit ("poverty gap"), the index of the depth of poverty of Foster–Greer–Thorbeck ("Foster–Greer–Thorbeck index"), indicators of Dalton class ("Dalton Class Indicators") and, finally, the so-called measures with a discontinuous poverty line ("Poverty-Line-Discontinuous Measures" or "PLD measures") [Foster, Greer, Thorbeck, 1984; Hagenaars, 1987; Bourguignon, Fields, 1995].

Let S (the amount allocated for targeted social support for the poor) be smaller than the amount of money necessary for complete elimination of poverty. And let $\varphi(x|S)$ be a function that usually sets the rule of distribution of the sum S among population with per capita expenditures $x < z_0$ (for example, it may be a function of the density distribution of the sum S among the poor), and $\tilde{f}(x|\varphi, S)$ is the density of distribution of population according to total per capita expenditures, resulting *after* the distribution of social assistance in accordance with the rule $\varphi(x|S)$. In view of this, the value of the indicator of poverty level from (4.35) will also change, namely:

$$I\left(w(x), \tilde{f}(x|\varphi; S)\right) = \int_0^{z_0} w(x)\tilde{f}(x|\varphi; S)\, dx. \qquad (4.35')$$

Then, task 2 is reduced for determining the function $\varphi_0(x|S)$ in which the indicator of poverty level (4.35′) reaches its minimum (for given $w(x)$ and S), i.e.:

$$\varphi_0(x|S) = \arg\min_{\varphi} \int_0^{z_0} w(x)\tilde{f}(x|\varphi; S)\, dx. \qquad (4.36)$$

It should be emphasized that task 2 is considered in this work **in the context of the specific project to combat a relatively long-term poverty** [Брейтуэйт, 1999; Пилотные прграммы, 1999]. This results in the following two circumstances. First, the thesis of relatively high mobility of income groups [Богомолова, Тапилина, Ростовцев, 2000] is not valid within the category of the *long-term* (*permanent*) poverty thesis. Second, the main instruments of easing *such* poverty (alleviation of poverty) are *various forms of direct payments of subsidies to the needy*, and not incentives and motivation of social and labour activity of the needy, although the latter are certainly more efficient with respect to the group of *temporarily* poor, for example, among the unclaimed portion of the potential middle class.

Task 3 (auxiliary, applied). To calculate, based on solving the main problem, i.e. using the knowledge of the calibrated distribution law (on average per capita expenditures) of the population of Russia and its three regions, the numerical values of the basic characteristics of differentiation, that is, the coefficient of the funds and the Gini index, compare them with the official data of Goskomstat RF, and try to find countries/counterparts from the world practice.

Note that the factor of "truncation", and in particular, the loss of super-rich strata in the general econometric analysis of the distribution of the Russian population by per capita total expenditures (income), will have almost no impact on the indicators of poverty, representing the main interest in the solution of task 2, i.e. in addressing the problem of targeted social assistance to the poor families. After all, in calculating poverty indicators such as (4.35), *only the left "tail"* of the analysed distribution is used, while the operation of the model assumption H_4 may specify the behaviour of the distribution *on its right "tail"*. However, accounting for the super-rich strata leads to a significant increase in the values of various indicators of property differentiation of the population, of the Gini index, coefficient of funds (i.e. the **ratio of the total income of the richest 10% of the population to the total income of the poorest 10%**), etc. In turn, the characteristics of differentiation and *polarization* of population by expenditures (income) are the indicators of the level of social tension in society, that is why they are substantially related to the structure of the weighting function $w(x)$ in the poverty indicators of type (4.35): after all, in the task of reducing the level of long-term ("permanent") poverty, these indicators are interpreted *primarily as indicators of social tension in society.*

Literature review. First, let us discuss the familiar studies with the objectives closest to our main task. The model of distribution of the Russian population by total per capita expenditures, proposed in this project, is essentially a modification of the model of development and distribution of the population by *income*, first described in [Айвазян, 1997]. *The modification* consists in the introduction and evaluation of statistical "probability of evasion" of the household from the survey (see Hypothesis H_2), in replacing the income by *the expenditures*, in the formulation of the calibration algorithm, based on hypotheses H_2 and H_3 and on the knowledge of the average value of macro-characteristic of per capita expenditures, and in generating (via Monte Carlo simulation in accordance with the hypothesis) additional observations of the *unobserved* range of per capita expenditures.

Works [Айвазян, 1997; Ершов, Майер, 1998; Шевяков, Кирута, 1999; Суворов, Ульянова, 1997] contain a variety of reasons, confirming the validity of our criticisms 1–4 (see the beginning of point 4.2.2). In [Великанова, Колмаков, Фролов, 1996] we can see an approach that also relies on a mixture of log-normal distribution models; however, it is not equipped with the necessary tools to enable a competent econometric analysis of the mixture and offers no way to account for data hidden from direct observation. An approach based on polynomial approximation of the density of the analysed distribution law, described in [Yershov, Meyer, 1998], is too formal and does not allow to construct a phenomenological model of the analysed event, give a meaningful interpretation of the parameters of the model or take into account unobserved range expenditures.[9]

The main weakness of the approach described in [Суворов, Ульянова, 1997] is in the apparent inadequacy of the basic assumptions about the form of log-normal law of distribution of the population by income, although the authors considered three-parameter model (as opposed to the two-parameter model of "Goskomstat"). However, in the analysis of income instead of expenditures, the weakness of the basic assumptions about the "correctness" of the determination of the *modal* income on the results of the SHBS (which according to numerous experts, including the ones from "Goskomstat", has *a significant shift*) and also a formal approximation method of selection of the unknown parameters of the proposed model do not allow to fully approve the *model part* of that study. At the same time, *the economic analysis* of the income situation in Russia of the 1990s essentially, which helps to understand the mechanism of unobservable in SHBS right "tail" of the analysed distribution (essentially as a result of opportunities of certain narrow segments of the population in the sales and exploitation of parts of national wealth), in our view, deserves the most serious attention.

---

9 The calculations completed in [Aivazian, 1997] according to 1995–1996 showed that the "post-accounting" for "the rich" who partially avoided the surveys, and "the super-rich" who completely avoided the surveys, increases the Gini index from 0.376 to 0.53 and the funds coefficient from 12.9 to 22.8.

Let us dig more deeply into the critical analysis of the approach of A. Y. Sheviakova and A. Y. Kiruta [Sheviakov, Kiruta, 1999] and into its fundamental difference from the one we have adopted. This is due to the fact that in this work we have undertaken the most serious, in our opinion, attempt to describe *the actual* distribution of the population per capita income for the region, based on data from SHBS and "the balance of income and expenditure of population". This effort is based on a non-parametric approach to estimate, and includes in particular a certain procedure of shift elimination, the characteristic of the SHBS data, as well as the description of the procedure of aggregating regional data based on regional deflators and equivalence scales. We will note the most significant, in our view, shortcomings of this approach.

a) A weighing method ("calibration") of the available observations of SHBS, described in [Sheviakov, Kiruta, 1999] (without a reference to [Deville, Särndal, Sautory, 1993] where it was first proposed), essentially ignores all the population located to the right of the maximum of the observed values. In other words, the "right tail" of the distribution is completely ignored and therefore *the factor of the censorship of initial statistics is not considered.* In our model, the "tail" is reconstructed based on hypothesis H_4.
b) As a direct consequence of the above-mentioned drawback, we have a fundamentally erroneous conclusion of the authors [Sheviakov, Kiruta, 1999] that the "excessive economic inequality is entirely due to excessive poverty". Having ignored the right "tail" of the distribution, the authors could not have really come to a different conclusion.
c) An attractive at first sight "non-parametric" character of the approach, in fact, has two significant drawbacks. First, the estimate of the distribution law of the population by average per capita income obtained through this method *is a purely formal approximation* of the analysed unknown law and *cannot be meaningfully interpreted.* Second, the model obtained in this way is not suitable for solving the *forecast* problems.
d) While considering estimates of poverty, property differentiation and other indicators of well-being in the conditions of Russia's transitional period, it is better to appeal to *expenditures rather than to income* of population. This eliminates the problem of delayed payment of wages, hidden income, etc.

Now let us turn to the works most closely associated with the solution of task 2. The first thing to note implemented are the project [Braithwaite, 1999] and pilot programs [Pilot Program, 1999] conducted on the initiative and with the financial support of the World Bank. They made a legitimate attempt to measure poverty level based on the *reassessment of real per capita income of the household* (which in [Braithwaite, 1999] are called *potential consumer expenditures*). However, each of the specific ways of such reassessment proposed in these studies has significant disadvantages (we presented the results of the analysis of these methods in the preparation of the report [Национальная оценка, 1999]). In addition, in these papers, the sole criterion of poverty is the share of poor people (i.e. the criterion (4.35) with $w(x) \equiv 1$) and they do not attempt to determine the *optimal* method of distribution of the sum allocated for social support of low-income families (i.e. task 2 is not solved).

A quite full review of poverty indicators is given in the work of [Корчагина, Овчарова, Турунцев, 1999]. They discusses, in particular, one of the special cases, criterion (4.35), the so-called index of Foster–Greer–Thorbeck, and show the results of calculations of the index by the data of quarterly budget statistics of households in 1996. However, in this study, the index is calculated on the basis of *income* distribution, and most importantly, it is not associated with the task of optimization of targeted social support for the poor population.

Thus, in the *Russian* economic theory and practice, as far as we know, task 2 was neither considered nor completed. However, various aspects of this problem were researched in a number of works of foreign authors; however, they are also based on poverty indicators calculated on the basis of **income** distribution [Atkinson, 1987; Kanbur, 1987; Foster, Shorroks, 1988; Bourguignon, Fields, 1990, 1995; Ravallion, 1994]. In particular, in [Bourguignon, Fields, 1990], they proved that in "profitable" versions of indicators of poverty (4.35) with a weighting function $w(x)$ of the following appearance:

$$w(x) = \left(\frac{z_0 - x}{z_0}\right)^{\alpha}, \quad 0 \le x < z_0, \quad \alpha > 1 \tag{4.37}$$

(which conforms to the Foster–Greer–Thorbeck index) we consider optimal in the sense of (4.36) the so-called *pure strategy of pulling the poorest to the threshold value* $\bar{z}_0 < z_0$, where the threshold value is determined from the following condition:

$$N\left(\int_0^{\bar{z}_0} f(x)\,dx\right)\left(\int_0^{\bar{z}_0} (\bar{z}_0 - x) f(x)\,dx\right) = S \tag{4.38}$$

where N is the total population. This strategy ("allocation of p-type" in terms of co-authors) is that after determining a threshold value $\bar{z}_0$ every member of the society with a per capita income $x < \bar{z}_0$ receives an allowance of $\bar{z}_0 - x$. In [Bourguignon, Fields, 1990], it is proved that the mixed strategy ("allocation of mixed-type"), in which the portion of the sum S goes to the "pull-up" to the level $\bar{z}_0$ of the poorest, and the rest of S goes to the "pull-up" to the level z_0 of the richest of the poor (while, of course, on the right side of (4.38), by which the value $\bar{z}_0$ is determined, is a certain sum $S_1 < S$), can be optimal only under the condition $w(z_0) = \delta > 0$ (which corresponds, in terminology of [Bourguignon, Fields, 1995], to the poverty indicators with "a discontinued poverty line", "Poverty-Line-Discontinuous Measures", see above).

In connection with task 3, we will only mention here that the calculation of *measures of polarization* by per capita income introduced by Joan Maria Esteban and Debrey Ray [Esteban, Ray, 1994] relies heavily on the knowledge of these "tail" elements of the corresponding distribution. At the same time, in a number of studies, that measure is effectively used (along with, for example, a coefficient of funds and the Gini index) as *an indicator of social tension* in the society and a factor that is in a causal relationship with the level of crime [Fajnzulber, Lederman, Loayza, 1999]. This actually explains the attention that we give to the indicators of population differentiation by expenditures.

Discussion of basic hypotheses and model assumptions of the study. Task 1 stated above is based on theoretical grounds and/or experimental statistical testing of the hypotheses and model assumptions $H_1 - H_4$. Let's discuss them in detail.

Hypothesis H$_1$ *states that the distribution of Russian households by per capita total monetary expenditures may actually be adequately described with a mixture of log-normal laws.* This hypothesis is amenable to statistical testing using one of the consent criteria (its statistical verification on 1996 data, used, however, for average per capita income, was carried out in [Aivazian, 1997]).

In theoretical terms, this hypothesis is based on three postulates:

(a) the distribution of the population by per capita expenditures ξ within a strata homogeneous by the structure of income sources and territorial, socio-professional and demographic characteristics obeys the log-normal law with the parameters $a = \mathbf{E}(\ln\xi(a))$ and $\sigma^2(a) = \mathbf{D}(\ln\xi(a))$;

(b) if we present the society consisting of a continuous (by an average value of expenditure logarithms a) spectrum of such strata, then for some natural appearance of the mixed (weighting) function $q(a)$ the distribution of *the total population* by per capita expenditures will again obey the log-normal law;

(c) if violation of the continuity of spectrum of components of the society of various strata (i.e. with a substantial "washout" or elimination of certain strata, for example, of the so-called middle class) occurs, or in violation of the monotonic decrease of mixing function $q(a)$ during the distancing of the argument a from the total average of expenditure logarithms a_0, **the general log-normal distribution mentioned in (b) is transformed into a mixture of log-normal laws.**

Let us discuss each of these postulates.

Postulate (*a*) is very common in the studies of modelling of distributive relations in the society and *is based on a multiplicative effect on the expenditures (income, wages) of the factors* within the population of a homogeneous socio-economic strata. The mechanism for generating log-normal distribution in these situations is described in detail in the literature.

Postulate (*b*) is a simple consequence of the fact that with the normal distribution law of the different strata of value $a = \mathbf{E}(\ln \xi)$ (i.e. provided that mixing function $q(a)$ is described by $(a_0; \Delta^2)$ – normal density) the mixture of normal distributions of the logarithms of incomes $(a_0; \Delta^2)$ of various strata such as

$$\varphi(z) = \int_{-\infty}^{\infty} \frac{1}{\sqrt{2\pi}\,\sigma(a)} e^{-\frac{(z-a)^2}{2\sigma^2(a)}} q(a)\, da$$

represents (when $\sigma^2(a) = \sigma^2 = const$) a composition of normal laws and therefore again will become a normal distribution with parameters $a_0 = \mathbf{E}(\ln\xi)$ and $\sigma_0^2 = \sigma^2 + \Delta^2$. This fact was first noted and explained in [Айвазян, Рабкина, Римашевская, 1967].

[Monitoring Economic Conditions in the Russian Federation, 1997] *Postulate* (*c*) obviously holds in a degenerate situation when the **mixing function** $q(a)$ **is defined only in a finite set of discrete points** $a_1, a_2, \ldots, a_k$. The real situation in the Russian economy of the 1990s is, of course, more complicated. However, it was certainly characterized by a significant transformation of the mixing function $q(a)$. The specifics of the transitional period, without cancelling the overall design model of formation of the distribution of population by average per capita expenditures described in (a) and (b), brought about changes in the nature of the function $q(a)$.

Hypothesis $\mathbf{H_2}$ states that the probability of evasion of a household from an official budget survey is a certain type of function of the number of its socio-economic and territorial characteristics. This hypothesis is also amenable to statistical testing (the necessary statistics were taken from the results of RLMS [Monitoring Economic Conditions in the Russia Federation, 1996, 1997], and some additional information from Goskomstat RF). The hypothesis seems natural and was prompted during the discussion of the problem of E. B. Frolova, who heads the Department of Statistics of Standard of Living of the Russian Federal State Statistics Service. Obviously, this assumption is based on the experience of statisticians/registrars of SHBS and RLMS. Specification of hypothesis H_2 and the method of its econometric analysis are described below.

Model assumption $\mathbf{H_3}$ of the constancy of the coefficient of variation of per capita household expenditures, i.e. its independence from the number of a socio-economic stratum within which it is calculated. This model assumption is also amenable to statistical testing using a criterion of homogeneity of variances. As part of the theoretically and empirically proven position that population of the j-th stratum with the homogeneous geographical, socio-demographic and professional characteristics is distributed on per capita income and expenditures $\xi(j)$ in accordance with the log-normal law with parameters $a(j) = \mathbf{E}(\ln\xi(j))$ and $\sigma^2(j) = \mathbf{D}(\ln\xi(j))$ (see, e.g. [Aivazian, 1976]), hypothesis H_3 is equivalent to the following affirmation:

$$H_3' : \ \mathbf{D}\Big(\ln \xi(j)\Big) = \sigma^2 = \text{const.}$$

The equivalence of hypotheses H_3 and H_3' follows from the known equation, which is true within the log-normal general population:

$$\frac{[\mathbf{D}(\xi(j))]^{\frac{1}{2}}}{\mathbf{E}\xi(j)} = (e^{\sigma^2} - 1)^{\frac{1}{2}}.$$

Model assumption H4 states that the distribution of the population by per capita total expenditures x *in the range of statistically unobservable values* of these expenditures (i.e. where x_i is the value of average total expenditures in the i-th statistically surveyed household, and n is the total number of statistically surveyed households) can be approximated by a three-parameter log-normal law with a parameter of the shift equal to $x_0 = \max_{1 \le i \le n} \{x_i\}$ and the parameter $D(\ln\xi(k)) = \sigma^2$ independent of the number of stratum k and estimated by observations related to statistically surveyed strata of the population (see hypothesis H_3).

Statement H4 is not a statistical hypothesis, since it cannot be directly statistically verified using a particular statistical criterion (the required statistics are not available). Therefore, statement H_4 should be regarded an *initial model assumption*, the a priori validity of which can be justified only by suitable theoretical considerations and a posteriori by a comparison of the *actual* values of the main "output" characteristics with the corresponding values *obtained on the basis of the proposed model*. As theoretical considerations mentioned above, we can cite the following facts and evaluation of the specialists.

We can consider as one of the most significant consequences of the rapid collapse of the Soviet Union and its socio-economic system the formation of a narrow layer of "elite" from the highest ranks of party-bureaucratic and business elite, complemented by the most "advanced" representatives of organized crime, which, using specific methods of privatization, obtained the opportunity for overt and covert sales (on the international and domestic markets) of some appropriated items of national wealth. Various calculations of experts (see, e.g. [Suvorov, Ulianov, 1997]) show that **"commodity intervention" on the markets of the national wealth in the amount of 0.2–0.3% of its physical volume (per year) is equivalent to an additional increase in the gross income of the population in the 10–20%.** It is clear that the overwhelming share of the growth of gross income falls exactly on the **stratum of the "elite", which in view of the homogeneity of its social positions and the level of power can be classified as a specific socio-economic stratum.** Therefore, the law of distribution of the population by expenditures, mentioned in assumption H_4, refers to the population of this particular stratum.

Note that in normal circumstances we describe the distribution of the population by income (expenditures), with the income (expenditures) of this population exceeding a fixed level x_0, using *the Pareto law*. However, it is justified only in cases where the probability density function of **the total** population *decreases monotonically for all* $x \ge x_0$ (which, as a rule, takes place in a well-functioning economy). In case of the Russian situation in the 1990s, the above-described specifics of formation of the super-rich strata allows the existence of a local maximum of the density function to the right of point x_0.

Using hypotheses $H_1 - H_2$ and assumptions H_3 and H_4 allows to build an informal (i.e. interpreted in terms of content) distribution model of the Russian population by per capita total expenditures and to design, based on this model, a methodology of statistical evaluation of poverty indicators and differentiation, based on the data from household budget surveys and some macroeconomic indicators of the balance of population income and expenditures.

Key variables used in the study and their information sources: Following the study [Определение основных показателей, 1999], we define (in relation to the selected tact of time – *a quarter*) the cumulative monetary expenditure of HH as the sum of the following addends:

- $\xi^{(1)}$*quarterly consumer expenditures*, as the sum of the costs of food, alcohol, non-food products for personal consumption of HH and for services of a personal nature;
- $\xi^{(2)}$ *expenditures of intermediate consumption* (expenses of HH on personal subsidiary plots);
- $\xi^{(3)}$ quarterly average gross of fixed capital of HH (purchase of land and real estate, precious metal products, costs of construction and renovation of housing);
- $\xi^{(4)}$ *quarterly sum of all paid taxes and other obligatory payments* (including alimony, debt, club and society dues);
- $\xi^{(5)}$ *sum of balances of cash on hand and the growth of organized* savings (including purchase of currency and securities and bank deposits);
- $\xi^{(6)}$ cost estimation of quarterly consumption of products produced in personal subsidiary plots.

Thus, the aggregate (total) per capita monthly expenditures ξ inadvertently extracted from the general population of households are

$$\xi = \frac{1}{3m_\xi}\sum_{l=1}^{6}\xi^{(l)},$$

where m_ξ is the number of *conventional consumer units* in the sample households and the values $\xi^{(l)}$ $(l = 1, 2, \ldots, 6)$ are defined above (additional division of the total sum of quarterly expenditures by 3 reduces the analysed random variable ξ to a more familiar factor for Russians, that is, monthly basis).

Note on the equivalence scales. The exact determination of the number m_ξ depends on the choice of the particular system of calculating *equivalence scales*. It is well known (see, e.g. [Equivalence Scales, Well-being, Inequality and Poverty]) that there are different approaches to the construction of equivalence scales. The Russian state statistical services, for example, calculate the value m_ξ based on the condition that children and family members of retirement age are estimated at, respectively, 0.9 and 0.6 consumer units. Equivalence scale OECD (Organization of Economic Cooperation and Development) is built on the principle of per capita income adjustment accounting for resource economy in the analysed income cell when *jointly* used. In accordance with this principle, the head of HH is given the weight of 1.0, other adults 0.7, and every child 0.5. *Justification of the choice of a particular equivalence scale within the specificity of modern Russian society is a separate serious problem, which is outside of the scope of our study.* In it, we determined that m_ξ simply equals the number of members of the analysed household.

The observed values $x_i, x_i^{(1)}, x_i^{(2)}, \ldots, x_i^{(6)}$ of random variables, respectively, $\xi, \xi^{(1)}, \xi^{(2)}, \ldots, \xi^{(6)}$ are the results of a statistical survey of the i-th household, obtained by standard SHBS Goskomstat and RLMS (fifth, sixth, seventh and eighth rounds). *A macro-characteristic of total per capita expenditures of population on the region* (μ^{macro}) is derived from the quarterly "Balance of income and expenditures of population" of Goskomstat RF [Методологические положения по статистике, 1996]. The value μ^{macro} has the same meaning and the same structure as the variable ξ. However, values μ^{macro} and its components are calculated not with the sample statistics SHBS but with macro-regional *trade statistics, tax services, banking information and securities market.*

The relative frequency (share) $p(x)$ *of households with average and total expenditures* x, *evaded (or refused) from a statistical survey within the observation period,* is assessed according to Goskomstat RF and RLMS (rounds 5–8). We also studied the parameters of the socio-demographic structure of families of the region (the regional average: the number of family members, share of children, the proportion of retirees, etc.).

Let's discuss in more detail RLMS data and SHBS, which were used as information source of our research.

1. *Data RLMS, fifth, sixth, seventh and eighth rounds* [Monitoring Economic Conditions in the Russian Federation, 1997]. These data on expenditures account for quite a broad range of categories, though the time ranges that include the expenditures of each category are different. For example, expenditures on food (about 60 positions) are calculated on a weekly window; fuel expenses, services (about 10), rent, club fees and insurance premiums, as well as savings and loans on a window of 30 days, and the cost of non-food products and durable goods (about 10 broad categories) on 3 months. RLMS maintains the records of food production at the personal subsidiary plots (for the previous year), together with the cost of its maintenance (facilities and equipment, fertilizers, seeds and purchase of seedlings, livestock and poultry, etc.). All these data are converted so that the resulting expenditures correspond to one calendar month. This "purified" data is published in the derivatives data files of RLMS.

Of course, the obtained results should be interpreted taking into account the quality of the source data. For example, the level of family welfare, as measured by the volume of consumption, should include the depreciation of consumer durables, real estate and vehicles; however, to our knowledge, this observation is only a *theoretical* possibility of the development of budget surveys, which is extremely difficult to implement.

2. The results of the sample budget survey by Goskomstat of households in three regions of the RF (second quarter of 1998), obtained in the course of the joint project

of the working group CEMI and Goskomstat. According to the Goskomstat procedure [Определение основных показателей, 1999], the sample was formed on the principle of representative types of households on the basis of micro-census of 1994. Quarterly budget survey consists of a household filling a diary of expenditures for two weeks twice within the quarter and of an intermediate monthly survey. Based on the collected initial data on the expenditures, Goskomstat produces the following aggregate indicators, which will be used in the work: monetary expenditures (denras – the sum of actual expenses incurred by members of HH during the reference period – includes consumer expenditures and expenses not related to consumption); consumer spending (potras – part of the income for purchases of consumer goods and services); final consumption expenditures of households (konpot) and disposable HH resources (rasres – the sum of money, denres, and natural product, natdox, which were in possession of HH during the reference period, i.e. monetary expenditures, savings and natural products of HH deferred to the end of the accounting period). Budget surveys were also supplemented by a special questionnaire, for the research of the quality of life [Айвазян, Герасимов, 1998].

Description of the model and meaningful interpretation of its parameters. Let us use ξ (thousand rub.) to denote the average annual expenditure of randomly selected representative of the Russian population and ξ_j (thousand rub.) – the average per capita expenditure of the individual randomly selected from the population j-th homogeneous socio-economic stratum. Then, according to hypotheses H_1 and H_4, the distribution density of the random variable ξ is described by the model of a mixture of log-normal law such that

$$f(x|\Theta) = \sum_{j=1}^{k} q_j \frac{1}{\sqrt{2\pi}\,\sigma_j x} e^{\frac{(\ln x - a_j)^2}{2\sigma_j^2}} + q_{k+1} \frac{1}{\sqrt{2\pi}\,\sigma_{k+1}\cdot(x-x_0)} e^{\frac{(\ln(x-x_0)-a_{k+1})^2}{2\sigma_{k+1}{}^2}}$$

where $\Theta = (k;\ q_1, \ldots, q_{k+1};\ a_1, \ldots, a_{k+1};\ x_0;\ \sigma_1^2, \ldots, \sigma_{k+1}^2)$ are the parameters of the model, which could have the following meaningful interpretation:

$k+1$ is the number of components of the mixture (each component is interpreted as a stratum of the population homogeneous by its socio-economic characteristics);

q_j $(j = 1, 2, \ldots, k+1)$ is the a priori probability of the appearance of observations, representing the j-th component of the mixture (the specific weight of the j-th homogeneous stratum in the entire population of the region);

x_0 is a threshold value of per capita expenditures, separating *statistically available* range of change in expenditures $(x \le x_0)$ from *statistically inaccessible range* $(x > x_0)$;

$a_j = \mathbf{E}(\ln \xi_j)$ $(j = 1,\ 2, \ldots, k+1)$ is the theoretical average values of logarithms of per capita expenditures (averaging is produced by the entire population of the j-th stratum);

$\sigma_j^2 = \mathbf{D}(\ln \xi_j)$ $(j = 1,\ 2, \ldots, k+1)$ are the variances of the logarithms of average expenditures, calculated based on population of the j-th stratum.

It is assumed that the entire population of $(k+1)$-th (the richest) stratum has per capita expenditures that exceed a threshold value x_0, and *completely avoids sampling household budget surveys*. The remaining households are available for statistical surveys, although they may evade (refuse) them with a probability $p(x)$, where $p(x)$ monotonically increases by per capita consumption function (see hypothesis H_2).

Econometric analysis to identify model (4.39) implies, in particular, estimation of parameters Θ according to data from SHBS and "Balance of income and expenditure of population".

Methodology of the econometric analysis of the model. Econometric analysis of the model implies, in particular (after the stage of its specifications, see below in this paragraph), *statistical estimation of parameters*. The peculiarity of this case lies in certain shortcomings of the necessary information support (i.e. data from SHBS), expressed in the evasion of the part of households slated for sampling survey from SHBS, including a complete absence in the sample of representatives of the "right tail" of the distribution $f(x|\Theta)$ (in econometric literature, such problems are often called *problems of selective sampling* or *the sample selection problem*, see, e.g. [Вербик, 2008]). You can find more sources on the problems of information support of micro-econometric analysis of the level and way of life of the population in Paragraph 4.4 below. Here, we touch on these issues only in connection with the problem of statistical estimation of parameters Θ of the function $f(x|\Theta)$ defined by equation (4.39).

1. The probability p of evasion of a household from surveying as a function of some of its characteristics. In this section, we will focus on three characteristics as variables that determine the probability of evasion p:

$z^{(1)} = \ln \xi$ is a logarithm (natural) of the total per capita expenditures of HH;

$z^{(2)}$ is a characteristic of the place of residence of HH (with gradations *city, metropolitan areas, countryside* and *townships*), which is represented in the model by three dummy variables $z^{(2.j)}$, $j = 1, 2, 3$;

$z^{(3)}$ is a characteristic of the level of education of head of household (with gradations of *below high school, high school, industrial and technical schools, technical* and *higher education*), which is represented in the model by four dummy variables $z^{(2.l)}$, $l = 1, 2, 3, 4$.

Dependence of probability of evasion p from $Z = (1, z^{(1)}, z^{(2.1)}, z^{(2.2)}, z^{(2.3)}, z^{(3.1)}, z^{(3.2)}, z^{(3.3)}, z^{(3.4)})^T$ was analysed within *logit-model* of the following type:

$$p(Z) = P\{\eta_i = 1 \mid Z\} = \frac{e^{\beta^T Z}}{1 + e^{\beta^T Z}} \tag{4.40}$$

where

$$\eta_i = \begin{cases} 1, & \text{if } i\text{-th HH evaded survey,} \\ 0, & \text{if not,} \end{cases}$$

and $\beta = (\beta_0, \beta_1, \beta_{21}, \beta_{22}, \beta_{31}, \ldots, \beta_{34})^T$ is the column vector of the desired (subject to statistical evaluation) model parameters (4.40).[10] At the same time, characteristics $z^{(2)}$ and $z^{(3)}$ are introduced into the model in the form of *dummy variables*, so that after the econometric analysis of the model (4.40) the resulting function $p(Z)$ gives us *a whole set of models* that describe the relationship between the probability of evasion from the survey p and the logarithm of total per capita expenditures $z = z^{(1)}$ at different combinations of gradations of variables $z^{(2)}$ and $z^{(3)}$. It will be convenient to define the elements of this set of functions with the help of

$$p_{kl}(z) = p(z \mid z_k^{(2)}, z_l^{(3)}) = P\{\eta_i = 0 \mid z^{(1)} = z,\ z^{(2)} = z_k^{(2)},\ z^{(3)} = z_l^{(3)}\}, \qquad (4.40')$$

$$k = 1, 2, 3, 4; \qquad l = 1, 2, 3, 4, 5$$

(obviously, the total number of such functions will equal 20).

In Appendix 4.2, you can observe the results of estimating the parameters β of function (4.40) according to RLMS data (rounds 5–8). Calculations confirmed a statistically significant monotonically increasing dependence of the probability p from $z^{(1)}$ with any combination of gradations of the associated variables $z^{(2)}$ and $z^{(3)}$. You can also see the results of econometric analysis of the simplification (paired) version of model (4.40), in which we analysed the dependence p only from $z = z^{(1)} = \ln \xi$:

$$p(z) = P\left\{\eta_i = 1 \mid z^{(1)} = z\right\} = \frac{e^{\beta z}}{1 + e^{\beta z}}. \qquad (4.40'')$$

2. Calibration (weighting) of available observations. Analysis of the dependences (4.40) and (4.40″) may be interesting in itself. However, in our study, functions (4.40′) and (4.40″) are used later to calibrate existing observations and distribution of the region (country) by total per capita expenditures, evaluated according to budget data: if the source data contains statistics for each (i-th) household surveyed in addition to the value of its total per capita expenditures and the "values" of associated variables $z_{k_i}^{(2)}, z_{l_i}^{(3)}$, then we can use for calibration function (4.40′) (where k_i and l_i are the numbers of gradations recorded, respectively, for the variables $z^{(2)}$

10 In the *original* formulation of the problem, an a priori set of explanatory variables $Z = (1, z^{(1)}, z^{(2)}, \ldots, z^{(p)})^T$ in the logit model (4.40) was much wider: in addition to the variables $z^{(1)}$(of logarithm of total per capita expenditures HH) $z^{(2)}$ (characteristics of residence of the family) and $z^{(3)}$ (level of education of the head of household) were also included six variables that characterize socio-demographic composition of HH (the number of children of all ages, retired men, retired women, separately men and women of working age), as well as the demographic characteristic of head of household. However, a subsequent statistical analysis of the coefficients βin their significant difference from zero, supplemented by the analysis of significant differences of the derived functions (4.40) for different sets of explanatory variables, led us ultimately to precisely this version of the model.

and $z^{(3)}$ in the i-th observation); if we have *only the value of the total per capita expenditures*, we have to be limited to the so-called rough calibration using function (4.40″). Taking this into account for future reference in order to simplify the notation, we shall denote functions (4.40′) and (4.40″) using $p(z)$, in the case of calibration of *logarithmized* observed values of expenditures, and using $p(x)$, if we have in mind calibration of the *original* observations (measured in thousand rub.).

Let $f(x)$ be a density function of distribution of the region's population by per capita expenditures. Then, n is the total number of statistically surveyed residents of the region and x^* is a set value of per capita expenditures, and the number of observations $\nu(x^*)$ belonging to a small (of width Δ) neighbourhood of the point x^* under the condition that no one evades the survey will look as follows:

$$\nu(x^*) \approx nf(x^*)\Delta. \tag{4.41}$$

Real (observed in the sample of volume n) number of observations $\tilde{\nu}(x^*)$, calculated based on the known probabilities of evading survey $p(x)$, will be

$$\tilde{\nu}(x^*) \approx nf(x^*)\left(1 - p(x^*)\right)\Delta \tag{4.42}$$

From (4.41) and (4.42) it follows that

$$\nu(x^*) = \tilde{\nu}(x^*) \cdot \frac{1}{1 - p(x^*)}$$

In particular, choosing as points x^* the values of per capita expenditures x_i $(i = 1, 2, \ldots, n)$, observed in the sample, and taking sufficiently small values Δ, we will have

$$\tilde{\nu}(x_i) = 1$$

$$\nu(x_i) = \frac{1}{1 - p(x_i)}.$$

And this means that if according to the available sample

Observed values x	x_1	x_2	$\ldots$	x_n
Weights of observed values	$\frac{1}{n}$	$\frac{1}{n}$	$\ldots$	$\frac{1}{n}$

(4.43)

we want to evaluate the *true* density function $f(x)$, then we have to transition to a weighed (calibrated) sample

Observed values x	x_1	x_2	...	x_n
Weights of observed values	ω_1	ω_2	...	ω_n

(4.43')

where the weights ω_i are determined by the formula

$$\omega_i = \frac{\frac{1}{1-p(x_i)}}{\sum_{j=1}^{n}\left(\frac{1}{1-p(x_j)}\right)}$$

Note that the greater the weigh ω_i is, the greater the probability of survey evading $p(x_i)$, and $\sum_{i=1}^{n} \omega_i = 1$.

3. Estimation of the parameters of the statistically observable components of the mixture. At this stage, we are trying to solve the problem of estimating the parameters $k, \tilde{q}, \ldots, \tilde{q}_k, a_1, \ldots, a_k, \sigma_1^2, \ldots, \sigma_k^2$ in a mixture of log-normal distributions of type

$$\tilde{f}(x) = \sum_{j=1}^{k} \tilde{q}_j \frac{1}{\sqrt{2\pi}\,\sigma_j x} e^{-\frac{(\ln x - a_j)^2}{2\sigma_j^2}} \tag{4.39a}$$

of sample (4.43). The problem is reduced to estimation of the same parameters in a mixture of *normal* distributions of the form

$$\tilde{\varphi}(z) = \sum_{j=1}^{k} \tilde{q}_j \frac{1}{\sqrt{2\pi}\,\sigma_j} e^{-\frac{(z-a_j)^2}{2\sigma_j^2}}$$

using the sampling

Observed values z	z_1	z_2	...	z_n
Weights of the observed values	ω_1	ω_2	...	ω_n

(4.43'')

where $z_i = \ln x_i \ (i = 1, 2, \ldots, n)$.

The results from solving this problem using the initial statistics of the eighth round of RLMS, as well as the data from household budget surveys of Komi Republic, Volgograd and Omsk regions (for the second quarter of 1998), are shown in the next section. The problem is solved by the method described in [Aivazian, 1996], using the results of [Dempster, Laird, Rubin, 1977] and [Rudzkis, Radavicius, 1995], with the software implemented in the packages "Classmaster" and STATA (for a brief description of the methods and algorithms, see Appendix 4.3).

4. Estimation of the unobserved component of the mixture and of the entire distribution as a whole. Let the specific weight of an unobserved ($\hat{k}+1$-th) component of the mixture be equal to $q_{\hat{k}+1}$, and the mean value of logarithms of per capita expenditures of this stratum be $a_{\hat{k}+1}$. Then, the overall average per capita expenditures μ for the *entire* population of the region calculated using model (4.39) accounting for the values obtained at the previous stage $\hat{k}$; $\hat{\bar{q}}_1, \ldots, \hat{\bar{q}}\hat{k}$; $\hat{a}_1, \ldots, \hat{a}\hat{k}$; $\hat{\sigma}_1^2, \ldots, \hat{\sigma}\hat{k}^2$ will be determined by the following formula:

$$\mu = \int_0^\infty x\left(\sum_{j=1}^{\hat{k}} \hat{q}_j \frac{1}{\sqrt{2\pi}\,\hat{\sigma}_j x}\, e^{-\frac{(\ln x-\hat{a}_j)^2}{2\hat{\sigma}_j^2}} + q_{\hat{k}+1}\frac{1}{\sqrt{2\pi}\,\sigma_{\hat{k}+1}(x-x_0)}\, e^{-\frac{(\ln(x-x_0)-a_{\hat{k}+1})^2}{2\sigma_{\hat{k}+1}^2}}\right)dx,$$

where

$$\hat{q}_j = \hat{\bar{q}}_j(1-q_{\hat{k}+1}),\ \ j=1,2,\ldots,\hat{k}.$$

Taking into consideration the properties of the log-normal distribution, we obtained:

$$\mu = \sum_{j=1}^{\hat{k}} \hat{q}_j\, e^{\frac{1}{2}\hat{\sigma}_j^2+\hat{a}_j} + q_{\hat{k}+1}\left(x_0 + e^{\frac{1}{2}\sigma_{\hat{k}+1}^2+a_{\hat{k}+1}}\right). \tag{4.44}$$

The value μ, defined by formula (4.44), depends on the unknown values $q_{\hat{k}+1}, a_{\hat{k}+1}$, and also on x_0 and $\sigma_{\hat{k}+1}^2$. By construction, the value x_0 shall be equal to the maximum observed value of per capita expenditures, i.e.

$$x_0 = \max_{1\le i\le n}\{x_i\}.$$

If hypothesis H'_3 is true (see above), an overall estimation $\hat{\sigma}^2$ of the value σ^2 is calculated using the following formula:

$$\hat{\sigma}^2 = \sum_{j=1}^{\hat{k}} \hat{q}_j\,\hat{\sigma}_j^2$$

and the value $\sigma_{\hat{k}+1}^2$ shall be equal to $\hat{\sigma}^2$.

Then, in the coordinate plane $(q_{\hat{k}+1}, a_{\hat{k}+1})$, the level line is calculated from the condition:

$$\mu(q_{\hat{k}+1}, a_{\hat{k}+1}) = \mu^{macro} \tag{4.45}$$

where $\mu(q_{\hat{k}+1}, a_{\hat{k}+1})$ is calculated using formula (4.44) with $x_0 = \max_{1\le i\le n}\{x_i\}$ and $\sigma_{\hat{k}+1}^2 = \hat{\sigma}^2$, and μ^{macro} is an average value of per capita expenditures, obtained from

"*Balances of income and expenditures of population*" for the analysed region and the corresponding tact of time.

The exact choice of the *point* $(\hat{q}_{\hat{k}+1}, \hat{a}_{k+1})$ on line (4.45) requires additional conditions or expert data. During the actual creation of the level line (4.45), it is useful to take into account the following considerations:

a) from the general considerations it is evident that

$$q_{\hat{k}+1} << \min_{1\le j\le \hat{k}} \{q_j\}$$

(the symbol << means "considerably smaller"; in our case, we should state that q_{k+1} is approximately *one order smaller* than the value $\min_{1\le j\le \hat{k}} \{q_j\}$).

b) the level line (4.45) is created in the form of a table, in which the "input" is a grid of values $q_{\hat{k}+1}$ and the "output" are the corresponding values $a_{\hat{k}+1}$ calculated from condition (4.45) accounting for (4.44). It is appropriate to choose granularity of the value grid $q_{\hat{k}+1}$, for example, as follows at $\min_{1\le j\le \hat{k}} \{q_j\} = m\cdot 10^{-2}, \quad 1\le m\le 9$:

$$q_{\hat{k}+1} = \begin{cases} \nu\cdot 10^{-2}, & \nu = m-1,\ m-2, \ldots, 1; \\ \nu\cdot 10^{-3}, & \nu = 9,\ 8, \ldots, 1; \\ \nu\cdot 10^{-4}, & \nu = 9,\ 8, \ldots, 1. \end{cases}$$

c) starting from (4.44), we can obtain the following upper bound value of the share of special weight q_5 of the non-observed strata:

$$q_5 < \tfrac{1}{x_0}\left(\mu^{macro} - \sum_{j=1}^{\hat{k}} \hat{q}_j e^{\hat{a}_j + \frac{\sigma_j^2}{2}}\right).$$

Indicators of levels of poverty and social tension in the problem of targeted social support for low-income families. If we restrict the class of weighting functions $w(x)$, involved in the expression of poverty indicators (4.35), by the functions of form (4.37), use the results of [Bourguignon, Fields, 1990] on the optimal allocation of a financial support to low-income families in this case and use the results of estimation of the density function $f(x)$ in terms of total per capita expenditures, then we can formulate the following optimal rule of targeted social assistance to the low-income families:

using the defined ("input") parameters of the problem, that is, the total population $\boldsymbol{N}$ of the analysed region, the poverty line z_0, the sum $\boldsymbol{S}$ allocated to target social support of the population, the density function $f(x)$ describing the distribution of the population in the region in terms of total per capita expenditures, and the value $\alpha > 1$, narrowing down the Foster–Greer–Torbekka index used in the problem $I_\alpha^{(z_0)}(f) = \int_0^{z_0} \left(\frac{z_0 - x}{z_0}\right)^\alpha f(x)\,dx$, we will define the threshold value $\tilde{z}_0$ from equation (4.38), i.e.:

$$N \cdot I_0^{(\bar{z}_0)}(f) \cdot \bar{z}_0 \cdot I_1^{(\bar{z}_0)}(f) = S; \qquad (4.36')$$

every resident of the region with total per capita expenditures $x < \bar{z}_0$ receives an allowance in the amount of $\bar{z}_0 - x$.

Clearly, a change in the appearance of the weighting functions $w(x)$ can lead to the following modification in formulation of the optimal allocation rules of the targeted social assistance.

We calculated and analysed *the proportion of the poor* (i.e. indices of Foster–-Greer–Torbekka $I_\alpha^{(z_0)}(f)$ when $\alpha = 0$) and *indicators of social tension* (i.e. indices of Foster–Greer–Torbekka $I_\alpha^{(z_0)}(f)$ for $\alpha = 2$) for each of the above-mentioned regions of Russia in three different ways: (1) through direct non-parametric estimations; (2) using estimates based on log-normal distribution model of population on per capita total expenditures and (3) using the estimates based on a model of log-normal distribution mixture. The results of these calculations and of the corresponding analysis are given in the following part.

The results of the econometric analysis of the model. Formulated at the beginning of point 4.2.2, tasks 1, 2 and 3 and the methodology of this study described earlier led to the following **general logistics of the econometric analysis:**

1. We conduct a non-parametric analysis of the distribution of population by total per capita expenditures on the basis of the data of Goskomstat of Russia for the second quarter of 1998 in the Komi Republic, Volgograd and Omsk regions (for each region separately), as well as according to the database of RLMS (eight round, data for November 1998) for Russia as a whole; on the "output" of this phase of research are histograms and basic numerical characteristics of the corresponding distributions (see Appendix 4.4).
2. On the basis of panel data from RLMS (five to eight rounds) on the evasion of households to participate in a survey[11] within multiple and paired logit models, we evaluated the functions $p_{kl}(x)$ (see, eq. (4.40) and (4.40$'$)) and $p(x)$ (see formula (4.40$''$)), describing *the dependence of the probability p of evasions of HH from surveying from on the value x of its total per capita expenditures* for various combinations of gradations of variables that determine the character of the country of residence of this HH, and the educational level of the head of HH (see Appendix 4.2).
3. In accordance with the above methodology we *calibrate the source of statistical data*: rough (i.e. using paired logit-model (4.40$''$)) for regional data and more refined (i.e. using multivariate logit model (4.40)) for nationwide data from

11 This data is obtained from the employees of the Institute of Sociology (RAS) P. Kozyreva and E. Artamonova.

RLMS, 8th round; with the goal of elimination of the shift in the assessment analysed l.d.p. caused by the effect of "survey *evasion*" within the statistical range of the observed values of average per capita expenditures HH (see introduction to Paragraph 4.2).

4. We repeat step 1, i.e. again we perform *a non-parametric analysis of the analysed distributions*, but this time **based on calibrated data**; the results are compared with the results of step 1.
5. In accordance with the above-described methodology, we *estimate the parameters in models of mixture of distributions* (three regional and nationwide) for calibrated data within the observed statistical range of the values of average total household expenditures (see Appendix 4.3).
6. In accordance with the methodology described above, we assessed the unobserved component mixture in each of the four analysed models (three regional and nationwide), with the goal of elimination of the shift in the estimation of analysed probability distribution caused by the effect of "truncation" outside of statistically observable range of values of total per capita expenditure of HH.
7. On the basis of distributions received in steps 4 and 6, we calculate and analyse *indicators of poverty, and the differentiation of expenditures* for the three regions under consideration (as of the second quarter of 1998) and for Russia as a whole (as of November 1998).

   We present below the results of this statistical analysis as in A4.1–A4.5 and Tables A4.2–A4.12 of Appendices 4.2 and 4.4, which show the following:

   a) *The distribution of the population by per capita expenditures of any region within or across the country could be described during the second half of the 90s of the last century neither by a log-normal law nor a by a two-component* mixture of log-normal laws. Indeed, from the content of the fifth and sixth columns of Tables A4.5–A4.8, we can see that the hypothesis of a log-normal probability distribution must be rejected in all cases with an error of the first kind, having no less than three zeroes before the first significant digit, and the hypothesis subordination of the analysed distributions to the law of probability distribution, described by a *two-component* mixture of log-normal laws, is rejected with a significance level not exceeding (except for Omsk region) 0.085. The apparent consistency of the hypothesis with the Omsk region data, however, cannot serve as a sufficient justification for its acceptance: all quality criteria for model identification (the values of the log likelihood function, Akaike and Bozdogan criteria, and finally, the value of the statistic χ^2 of the corresponding consent criterion) are significantly better for the three-component mixture model also in the case of Omsk region.
   b) *Statistical algorithms for estimating the parameters in the model of the mix of normal distributions, programmed in "Classmaster" and STATA packages, tend to derive for estimation the numbers of components of the mixture* $\hat{k} = 3$ or $\hat{k} = 4$, *i.e. indicate that the population of the region and the country as a*

whole can be represented (in a statistically surveyed range of per capita expenditures) as a mixture of three or four socio-economic strata, although this is not necessarily expressed in the form of, respectively, three or four local maxima of corresponding density function. Volgograd region was the only exception to this rule. While in all other cases (i.e. in Russia as a whole, in the Komi Republic and the Omsk region) the increase in the number of components in the mixture, starting from $k = 5$, leads to a sharp deterioration of the values of all analysed quality criteria of model identification. Data analysis for Volgograd region I produced an estimate of the number of components in the mixture $\hat{k} = 5$ as the best.

c) *In comparison with a similar picture in 1996 [Aivazian, 1997], the stratification of the population into expenditure strata for 1998 is much less clearly expressed, which corresponds to the above-mentioned tendency to a gradual return of the analysed distribution to the log-normal appearance as the economy leaves the transitional period*; the graphs from [Aivazian, 1997] describing probability distribution of Russia's population on per capita income as of fall 1996 demonstrate the presence of local maxima of the density distribution. As we have seen (point 4.1.4), classification procedures based on this model allowed a certain population stratification and its socio-economic analysis of each identified stratum separately. Identification of strata and holding a similar socio-economic analysis *of the data of 1998* turned out much more problematic: it required the use of additional information.

d) *In this work, we used two simultaneous methodologies:* ***the calibration method*** *of observation (as a means of levelling the effect of "**survey evasion**") and a "non-statistical"* ***method of "imputation" of an unobservable component of the mixture*** *(as a means of levelling the effect of "truncation"). These methods help to explain a substantial part of about a 40% difference that exists between the real and the official (registered by state statistical agencies) income (expenditures) of the population.* The estimate of the unobserved (latent) component of the mixture is constructed up to the "line level" of the parameters $q_{\hat{k}+1}$ and $a_{\hat{k}+1}$ with certain restrictions on the ranges of variations of these parameters (see equations in point 4 of the "Methodology of econometric analysis of the model"). An example of a line level (created according to RLMS data, eighth round) is shown in Figure 4.4. It appears, however, that from the point of view of the distribution characteristics that we are interested in (poverty indicators, the average value and differentiation and polarization characteristics of the population in terms of per capita expenditure) *the specification of selecting a certain point* $(\hat{\mu}^{(\hat{k}+1)}, \hat{q}_{\hat{k}+1})$ on the line level is almost irrelevant. Poverty indicators depend essentially only on the "left tail" of the distribution, and characteristics of differentiation and polarization, of course, will increase substantially when incorporated into a model of latent strata, but their values are practically independent from exactly where we choose the point $(\hat{\mu}^{(\hat{k}+1)}, \hat{q}_{\hat{k}+1})$ on the line level. We will only note that according to the estimations of different studies the share of hidden income (expenditures) is from 25% to 40% (see point 4.1.4). Note that the proposed methodology for analysing calibration of observations increases the official expenditures by 2–3%, and the introduction of a latent strata into the model by 20–30% (according to the RLMS data, the eighth round, this increment was as follows: $(1211-913)/913 = 0.326 = 32.6\%$).

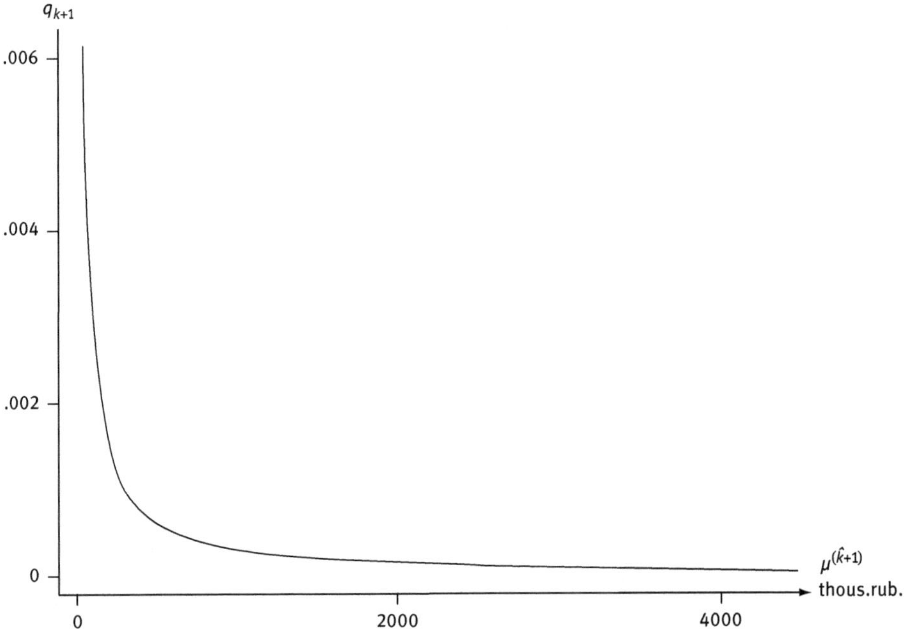

Figure 4.4. The relationship between specific weight and the average of the per capita expenditures used in determining the parameters of distribution of latent strata of population

e) *Estimations of indicators of poverty and social tension and differentiation of expenditures*. Tables 4.5 and 4.6 show the values of the indicators of poverty, social tension and differentiation of expenditures; in terms of a family $I_\alpha^{(z_0)}(f)$ of indices of Foster–Greer–Torbekka, the indices given in Table 4.5 are the indices $I_0^{(z_0)}(f)$ of the proportion of the poor and $I_2^{(z_0)}(f)$ is *an indicator of social tension due to the presence of a layer of the poor*. Table 4.5 includes official Goskomstat RF data (column 4), the data taken from [Pilot Program, 1999] (column 5, only for two regions included in the pilot program), direct non-parametric estimations of the parameters calculated from the above calibrated RLMS data and three analysed regions (columns 8 and 9), as well as estimates based on the log-normal model (column 6 and 10) and the model of the mix of log-normal distributions (columns 7 and 11). Table 4.6 contains information on the results of each of the two stages of calibration of the analysed distribution (first phase – weighing of the available observations, based on the knowledge of the function $p(x)$, see (4.43′); second phase – introduction into the model and estimation of the unobservable component of the mixture based on the knowledge of the average per capita expenditures μ^{macro}, derived from macro-statistics) and also the values of the basic characteristics of differentiation by expenditures: the Gini index and funds coefficient[12]; for each region, official statistical sources do not provide values of these characteristics, so in our work they are calculated as non-parametric estimations on the available sample data; the result is compared with the values of the same characteristics calculated using mixture model (4.39) proposed in this chapter.

The analysis of the data in Tables 4.5 and 4.6 leads to the following conclusions:
1. There is a substantial variation in the values of the analysed parameters between regions and (for each fixed region) between the methods of evaluation. In our view, the most accurate way to calculate the value of these characteristics is a non-parametric method of direct evaluation based on calibrated data (columns 8 and 9). And although it does not really change the ratings of the regions on these indicators, it provides significantly higher values of poverty indicators (for Komi Republic and Volgograd region it doubled!) than the official statistics. At the same time, the estimates derived from model of a mix yield in all cases the results significantly closer to direct non-parametric estimates than the official statistics, or a log-normal model.
2. Although the specific weight of the “non-observed” super-rich stratum included in the model is relatively small (in most cases it is measured in tenths and even hundredths of a percent), the impact of this stratum on the values of the key characteristics of differentiation and polarization of the population is significant. For example, the Gini index, according to official statistics, in November 1998, had the value of 0.380 for the country. For the straight non-parametric estimation of initial data RLMS (eighth round), it equaled 0.478. Having taken into account the latent stratum, its value increases to 0.599! The values of the Gini index by regions increase roughly by the same rate (with the exception of the Omsk region). The range of values of the decile ratio vacillates from 1.5 to 3 times. They are calculated using the shifted (unrepresentative) sample (column 8) and the model accounting for the latent stratum of the population (4.39) (column 9).

12 Let me remind you that the coefficient of funds in the Russian state statistics is the ratio of total income (expenses) of 10% of the richest population to total income (expenses) of 10% of the poorest population.

Table 4.5. Indicators of poverty and social tension (Russia, November 1998) *

Region #	Name of region	Poverty level, thousand rub.	Share of the poor					Foster–Greer–Torbekka poverty indices (social tension)		
			Goskomstat official data	Pilot Programs estimation	Estimation based on log-normal model	Estimation based on mixture model	Direct non-parametrical estimation	Direct non-parametrical estimation	Estimation based on log-normal model	Estimation based on mixture model
1	2	3*	4*	5	6	7	8	9	10	11
1	Russian	0.636	28.4	–	52.5	52.8	53.9	0.137	0.132	0.136
2	Komi Republic	0.466	20.6	26.7	53.8	56.2	56.7	0.127	0.130	0.128
3	Volgograd Region	0.368	31.5	49.2	62.0	62.7	63.0	0.177	0.177	0.177
4	Omsk Region	0.372	25.2	–	42.6	43.2	44.2	0.082	0.089	0.081

* * The data for columns 3 and 4 are taken from [socio-economic situation in Russia, 1998; Short-term socio-economic indicators, 1999] and [Social status and standard of living of the Russian population, 1999].

Table 4.6. The results of calibratio of distributions and the main indicators of differentiation by expenditures (Russia – as of November 1998, regions – as of the second quarter of 1998)

i	Region, information base, number of surveyed individuals	Average expenditures (thousand rub.)			Gini index		Decile ratio	
		Initial (by sample)	After calibration (increase, %)	Using latent stratum (increase, %)	Goskomstat data* or initial sample**	Proposed model (5)	Goskomstat data * or initial sample **	Proposed model (5)
1	**Russia,RLMS-8, n = 11,397**	0.913	0.932(2)	1.211(32,6)	0.380*0.478**	0.599	13.5*	45.8
2	Komi Republic, SHBS, n = 1089	0.633	0.686(8)	1.159(83.1)	0.395**	0.667	15.6**	43.7
3	Volgograd region,SHBS, n = 1263	0.412	0.433(5)	0.641(55,6)	0.389**	0.590	14.0**	32.0
4	Omsk region,SHBS, n = 1244	0.611	0.641(5)	0.699(14,4)	0.357**	0.442	10.5**	14.8

How do we explain the obtained differences and to what extent can we trust the results obtained on the basis of model (4.39)?

Table 4.5 shows the **characteristics of poverty and social tension**, the calculation of which is based on the "left tail" of the distribution of population expenditures. Therefore, as expected, the differences in the calculations made within an ordinary log-normal model and based on the mixture model (4.39) (compare columns 6 and 7, as well as 10 and 11) *are quite small, though systematic* (all values obtained from mixture model (4.39) are higher than the corresponding "log-normal" values by 1–2%). It is difficult to explain the spectacular differences between the "log-normal" characteristics and the official ones of the poverty level (compare columns 6 and 4): in fact, in accordance with Goskomstat methodology, such calculations would have to be made *using exactly the log-normal distribution model!* However, first, we used the RLMS data (eighth round), while the Goskomstat used the routine SHBS data of the fourth quarter of 1998, and second, the method for calculating the parameters of a log-normal approximation of distribution used by Goskomstat differs from the standard in mathematics and statistical practice.

Table 4.6 shows **the characteristics of population differentiation by expenditures**, the calculation of which is based simultaneously on two "tails" of the distribution – the left and right. And since one of the main salient features of mixture model (4.39) consists in a significant modification of the "right tail" of the distribution (by introducing and indirectly estimating the latent strata of the super-rich), the difference between the values of the characteristics of differentiation obtained by using model (4.39) (see columns 7 and 9) and the corresponding "Goskomstat" data (see columns 6 and 8) is, of course, quite impressive. Note that the results obtained by model (4.39) for the value of the Gini index and the funds coefficient may seem at first overly large. In support of this view, we can make the following observations. In model (4.39), the overwhelming proportion of the difference that exists between the μ^{macro}, estimated by macro-statistics average value of per capita expenditures, and the value μ, obtained using SHHBS, can be explained by disregarding a latent "super-rich" stratum (compare numbers in parentheses in columns 4 and 5 in Table 4.6). If this share is too high, then the values of the Gini and funds coefficients, calculated by model (4.39), will turn out elevated.[13] At the same time, in all the other approaches used to date, *this difference is fully compensated by calibration of distribution within the statistically examined range of variation of the values of per capita income (expenses)*, which means a complete disregard for the existence of

13 A comparison of the obtained Gini index values (for Russia) from model (4.39), equal to 0.610, with the values of this indicator in developed European countries and the United States (for these countries, the value of the Gini index varies in the neighbourhood of 0.3 and does not exceed 0.36) also suggests the idea of overestimation. However, for countries known for their high economic inequality (Brazil, Mexico, South Africa), this indicator is close to 0.6.

statistically unobservable population strata. Perhaps the truth lies somewhere between these two approaches? The answer to this question to some extent is contained in the following part.

The analysis of the stability of estimation of differentiation characteristics with respect to the systematic distortion of input data (the influence of the "misreporting" factor). Exaggerating the role of the non-observed strata in explaining the difference $\mu^{macro} - \mu$ in model (4.39) can occur, in particular, because of the systematic shift of sample survey data, due to the "misreporting" factor (see footnote 8). In other words, we are talking about deliberate lowering of their income and expenses by the respondents involved in the statistical sample surveys of HH.

In order to study the effects of the "misreporting" factor on the values of the Gini and funds coefficients derived from model (4.39) using data RLMS (eighth round) of Russia, we carried out the following calculations.

Let the distorting effect of the "misreporting" factor be measured by the value

$$\lambda = \frac{\mu_{truth} - \mu_{cal}}{\mu_{cal}} \times 100\%$$

where μ_{cal} (thousand rub.) is the average value of per capita household expenditures, estimated by sampling (within the observed range) and calibrated by the method described above in point 4.2.2 (in this case $\mu_{cal} = 0,932$, see column 4 in Table 4.6) and μ_{truth} (thousand rub.) – an unknown true average value of per capita expenditures represented in the sample of HH.

Obviously, our previous calculations (the results are presented in Tables 4.5 and 4.6, as well as in Tables A4.5–A4.8 of Appendix 4.3) give $\lambda = 0$, i.e. by assuming that $\mu_{cal} \approx \mu_{truth}$. The total excess of expenditures μ^{macro} over the calibrated expenditures μ_{cal} in the observable part of the range of the analysed sign is approximately 30% ((1.211 – 0.932)/0.932 = 0.299). The world practice of sample surveys of HH indicates that the mismatch of sampling and macroeconomic within several (usually less than 10) percent is a common phenomenon. In other words, it can be assumed that *some of the difference* $\mu^{macro} - \mu$ *still accounts for the available monitored range of the expenditure spectrum.*

Table 4.7. Analysis of sensitivity of Gini index and funds coefficient towards the influence of the "misreporting" factor

Degree of distortion of data, λ		**0%**	**5%**	**10%**	**15%**
Values of the characteristic of differentiation, obtained from model (5) accounting for the "misreporting" factor	Gini index	0.599	0.592	0.569	0.554
	Funds coefficient	45.8	42.5	39.5	35.6

The results of calculations of the Gini index and funds coefficient, performed within model (4.39) for various values of , are given in Table 4.7.

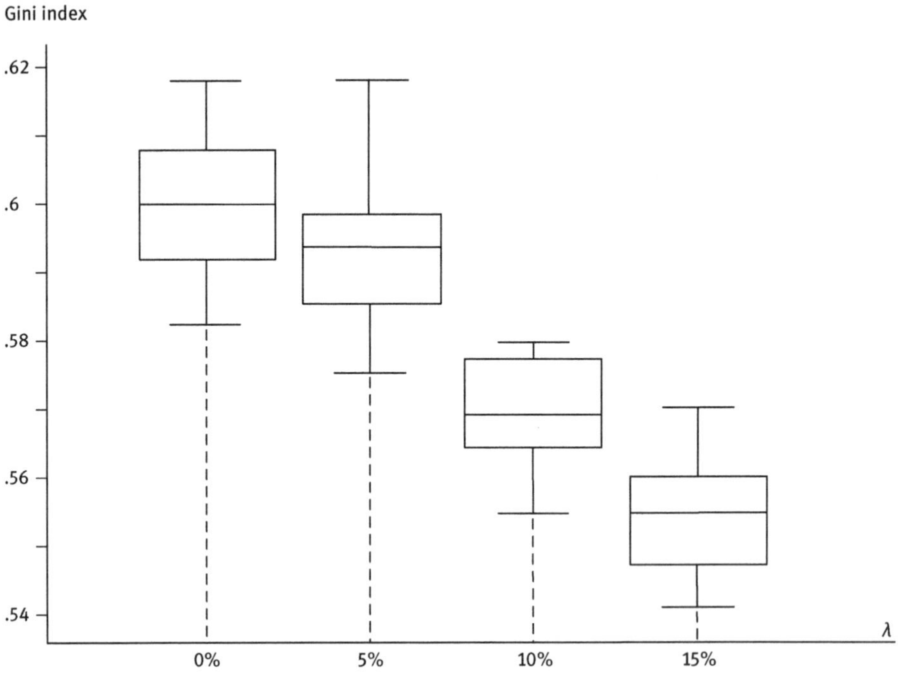

Figure 4.5. Confidence intervals for Gini index at various values of

Calculations were made based on data generated using *Monte Carlo simulation observation subordinate to distribution* (4.39), in which the parameters σ^2, a_j, q_j $(j = 1, 2, \ldots, k)$, a_{k+1} *and* q_{k+1} were estimated by the method described earlier in this paragraph and in Appendix 4.3 for sample RLMS data (eighth round)[14] **increased by** λ%λ (each). For each value, we simulated 20 samples; the volume of each was 400.000. This volume of each sample was selected so that each latent stratum with the weight q_{k+1}≤0, 1% would have a sufficient number of representatives (about 100) in the sample. Presence of 20 samples for each value of λλ allowed to calculate the approximate scattered true values of estimated characteristics of differentiation, i.e. their interval estimates with a level of confidence no less than 0.95. Examples of such interval estimates are shown in Figure 4.5.

The results obtained show sufficient stability of the estimates based on model (4.39). Taking into account the comments made above and assuming quite realistic values of 10–15%, *we can conclude that the values of the Gini and funds coefficients would most plausibly equal approximately to 0.55–0.57 and 36–39, respectively.*

14 In the econometric literature, this method of generating data is often referred to as **a parametric bootstrap simulation.**

4.2.3 Modelling the mechanism for determining the distribution of families of average per capita savings

Let ω be a vector parameter that defines the type of family (its per capita income, number of people, socio-demographic structure), and $\xi_\omega(t)$ be the average per capita value of savings, fixed in a randomly selected family ω and of "age" (the duration of the existence of the family in this type) t, $h_\omega(x;t)$ *and* $H_\omega(x;t)$, respectively, the density function and the distribution function of the random variable $\xi_\omega(t)$. We are interested in the limit (on $t\to\infty$) *stationary* distribution of family types $\omega(H_\omega(x)=H_\omega(x;\infty))$ and all families $(h(x)=\int h_\omega(x)\,dP(\omega))$[15] on per capita income variables and simultaneously in the dependence of basic value characteristics of the first family of distributions on the parameter ω.

As a model "entry", let's introduce the following characteristics of consumption in the families of the type $\omega:\ \alpha(\omega)$ where $\alpha(\omega)$ is the specific value (i.e. attributable to one family member per unit of time) of the "slow-moving" part of family income (in other words, this specific value of family funds, remaining after the necessary systematic expenditures on food, rent, etc.); and $\lambda_\omega(x)$ is a characteristic of the intensity of spending of the slow-moving part of income in families with x rubles of average per capita savings, i.e. during the time Δt such expenditure takes place (purchase of durable goods – cars, country houses, expensive luxury items, where etc.). Then probability $\lambda_\omega(x)\cdot\Delta t$; $\Psi_\omega(z|x)$ is a distribution function of the above-mentioned expenses occurring with the intensity $\lambda_\omega(x)$.

The initial assumptions of the model are as follows:

(A) The increasing (in time) process of monetary expenditures of the slow-moving part of the income is described by the jumping additive Markov process $Z_\omega(t|x)$ with the probability of a jump $\lambda_\omega(x)$ and the distribution function of the jump value $\Psi_\omega(z|x)$.

B) Specifications $\alpha(\omega),\lambda_\omega(x)$ and $\Psi_\omega(z|x)$ satisfy the conditions that ensure the existence of a *stationary* distribution $H_\omega(x)$, which is independent of the distribution $H_\omega(x;0)$ of the random variable $\xi_\omega(0)$.[16]

Affirmation. If the assumptions (A) and (B) are valid, then the distribution functions of the family per capita savings $H_\omega(x;t)$ and $H_\omega(x)$ are defined as the solutions of the following equations, respectively:

15 The experimental calculations have shown that even the minimum possible duration of the existence of the family in this type is sufficient for the convergence of the distribution $H_\omega(x;t)$ to its stationary form. By $P(\omega)$ we denote a distribution function, in general, of a *multidimensional* random value $\omega$and the integration is performed in the sense of Stieltjes integral.

16 The essence of this assumption, roughly speaking, is a natural assumption that the monetary receipts and expenditures in sustainable types of families have a sustained, steady character, eliminating the possibility (with unlimited tracking over time) of unlimited savings or unlimited "dive" into debt.

$$\frac{\partial H_\omega(x;t)}{\partial t} = -\alpha(\omega)\frac{\partial H_\omega(x,t)}{\partial x} + \\ + \int_x^\infty \lambda_\omega(u)\left[1 - \Psi_\omega\left((u-x)|x\right)\right]\frac{\partial H_\omega(u,t)}{\partial u}du$$

and

$$\alpha(\omega)\frac{dH_\omega(x)}{dx} = \int_x^\infty \lambda_\omega(u)\left[1 - \Psi_\omega\left((u-x)|x\right)\right]\frac{dH_\omega(u)}{du}du.$$

Corollary 1. If the function $\Psi$is such that it allows the expansion

$$1-\Psi_\omega\left((u-x)|x\right) = \varphi_\omega(u)g_\omega(x)$$

then

$$h_\omega(x) = \frac{c_1}{\varphi_\omega(x)}\exp\left\{-\frac{1}{\alpha(\omega)}\int_{c_2}^{x}\lambda_\omega(u)\,du\right\} \quad (4.46)$$

where c_1 and c_2 are certain constants.

Corollary 2. If $\lambda_\omega(x) = k(\omega)\,x$ and $\Psi_\omega(z|x) = z|x$ (i.e. the cost of one expenditure of the slow-moving part of the income is equally distributed within the interval of current savings x), then

$$H_\omega(x;t) = \begin{cases} 1-e^{-\frac{k(\omega)x^2}{2\alpha(\omega)}}\cdot e^{\frac{k(\omega)(x-\alpha(\omega)t)^2}{2\alpha(\omega)}}[1-H_\omega(x;0)], & \text{if } x > \alpha(\omega)t, \\ 1-e^{-\frac{k(\omega)x^2}{2\alpha(\omega)}}, & \text{if } x \le \alpha(\omega)\,t \end{cases}$$

and therefore

$$H_\omega(x) = 1-\exp\left\{-\frac{R(\omega)x^2}{2\alpha(\omega)}\right\}$$

(known in technology as the Weibull distribution).

Corollary 3. Let us consider a simplified case in which the type of family is determined only by the value of its per capita income ω, and let us focus on the distribution of per capita value of savings of families whose income exceeds a certain predetermined level ω_0. If the level ω_0 is high enough (let's say not less than the average income, these very families present major interest to us from the point of view of savings), then the distribution of these households by per capita income η, as it is known, is well approximated by the Pareto law, i.e. by the distribution function of the form $F(\omega) = P(\eta < \omega) = 1-(\omega_0/\omega)^{\nu}$. In this case, the transition from the distribution families of the given income group $H_\omega(x)$, defined by (4.46), to the distribution of *all* families (with incomes in excess of ω_0) by per capita savings yields (under additional assumptions $\alpha(\omega) = \alpha_0 \cdot \omega$, $\lambda_\omega(x) = \lambda(x)$ and $\varphi_\omega(x) = \varphi(x)$)

$$h(x) = \frac{c}{\varphi(x) \int_{\omega_0}^{x} \lambda(u)\, du} \Gamma_\Theta(\nu) \qquad (4.46a)$$

where $\Gamma_\Theta(\nu) = \int_0^\Theta e^{-t} t^{\nu-1} dt$ is an incomplete gamma-function, and $\Theta = \left(\int_{\omega_0}^{x} \lambda(u)\, du / \alpha_0 \omega \right)$.

The described model of the distribution of savings of the population can be taken as an object of *econometric analysis*. Special sample surveys of the population, "tuned" to this analysis, will allow statistical estimate of the "input" parameters of the model $(\alpha(\omega), \lambda_\omega(x), \Psi_\omega(z))$, to check the validity of the initial assumptions (*A*) and (*B*) and to conduct an experimental analysis of the results of its application.

4.3 Problems of information support of microeconometric analysis of the level and lifestyle of population

As we know, *econometric analysis is based on the results of economic measurement*. And if **the accuracy** and/or **representation** of these data are, for some reason, called into question, the results of the econometric analysis may also cause doubt. The main source of data of **micro**-econometric analysis in Russia is the national statistical office's quarterly sample surveys of household budgets (SHBS) carried out by state statistical services. They cover almost 50,000 households of the Russian Federation, distributed by regions (subjects of the RF) approximately proportionally to their population. Plans of sample surveys are drawn up in each region in accordance with the requirements of the *regional* representativeness of the sample and are implemented by regional state statistical services.[17]

17 For more about SHBS, see, e.g. [*Instrumentarii*, 2002]. Along with SHHBS, there are other sources. This is primarily the result of the National Population Census (which, unfortunately, is rarely performed, which is why their results quickly become obsolete). The Russia Longitudinal Monitoring Survey (RLMS) takes place annually, since the 90s of the last century. The latter, although far richer in content than SHHBS, covers only about 4000 households in Russia, and sampling is planned in a way that does not ensure representativeness of each region (for more on RLMS, see [RLMS, 1996]).

In this section we will consider two tasks of analysis and correction of the results of sampling household budget surveys. *The first task* stems from the fact that in practice the results of SHHBS form a sample that, due to objective circumstances, *always deviates from a representative (random) sample* of the general population analysed in the region. We call this case **the task of analysis and correction of sample survey data under the conditions of sampled selectivity** (see explanation of this term below). *The second task* is generated by the aggregation defects that occur in the calculation of the average annual values of certain derivatives of these SHHBS characteristics of the respective quarterly averages. In other words, we are talking **about the analysis and elimination of defects of temporal aggregation of the average values** of the analysed indicators.

4.3.1 Data analysis and correction of selected sample surveys in the sample selection conditions

The main problems of the quality of information sources delivered by the results of SHHBS are related to the practical **impossibility of implementation of a plan of sample survey**, due to two factors:

(i) Among each stratum of population planned for survey, **a certain percentage** of household refuse surveying;

(ii) *Households with average monthly income exceeding a certain high enough threshold do not get in the sample, i.e. refuse surveying with the probability* ***that equals one***.

These two facts lead to a **shift** based on sampling of results of econometric analysis in relation to the actual situation that characterizes **the entirety** of the analysed population. Accordingly, we are facing the task of a certain correction of SHBS data aimed at eliminating (or at least the maximum possible reduction) of this shift.

Both of these features of the SHBS data fall within the general problem known in econometrics and applied statistics as the sample selection problem (see, e.g. [Verbeek, 2008], 7.4 and 10.7). The essence of the problem is that certain objective conditions in the course of collecting the necessary sample data hinder the formation of *a random (representative)* sample, and this leads to the shift of statistical inferences based on this sample. In particular, the rule (necessary for the formation of a random sample) is violated, under which each of the elements of the analysed final general population of volume N falls into the sample with equal probability $1/N$.

Obviously, some correction of such non-representative aiming at its approximation to the *random* one is only possible if **there is some additional information**, such as the parameters of the distribution of certain characteristics of the elements of the analysed general population, directly or indirectly associated with the analysed characteristics (in our case – with per capita incomes and household expenditures).

It should be noted that the fact (i) is more or less inherent in SHBS practice in other countries, while the circumstance (ii), meaning, in particular, the complete lack in the sample of the "right tail" of analysed the distribution of per capita expenditures (or income) can be attributed to *the Russian specifics*.

In the context of *only one factor* (i) we can obtain a satisfactory solution to the problem of correcting the existing survey results in order to eliminate (or at least minimize) the shift obtained on the basis of their using *the methods of weighing the available survey data* (see, e.g. [Deming, 1943; Kalton, Flores-Gervantes, 2003]). Among these methods, we select the most effective methods of family CALMAR (the methods for CALibration on MARgins, see [Zieschang, 1990; Deville, Särndal, 1992; Deville, Särndal, Sautory, 1993]). They are based on the implementation of the following ideas: the weights ω_i for the *i*-th element of the sampling are selected in such a way that the weighted *sample* characteristics of the indicators, whose values are known for the **general population** (let's call these characteristics "control"), do not differ from the latter, subject to the least possible deviation (in some sense) of these weights from baseline.

However, with the *simultaneous action of both factors* (i) and (ii) the approach is incompetent, especially in solving problems of estimating various characteristics of **differentiation** of the distribution of population by average per capita income or expenditures, such as the funds or the Gini coefficients. In this case we propose a different approach.

You can see this approach below described in terms of a *parametric* formulation of the problem, i.e. in a situation where some general appearance of the analysed population distribution of per capita income or expenditure is postulated (in the given version a *log-normal* form of the distribution is postulated). Perhaps this approach could also be used for *a semi-parametric case*, i.e. to situations where only a general view of a "right tail" of the analysed distribution is postulated (which may be, for example, the Pareto distribution or a three-parameter, shifted log-normal distribution). Since we used the above-mentioned method of family CALMAR of weighing the available observations in the iterative procedure below, we will give a brief overview of this family of methods, adhering generally to the work [Deville, Särndal, Sautory, 1993].

Family of methods CALMAR of weighing the elements of a sample obtained under the conditions of a sample selection

Let us have observations

$$y_1, y_2, \ldots, y_n \tag{4.47}$$

a random variable η, while sample (4.47) was formed in violation of the conditions to ensure its representativeness (randomness). This means that when we use *the original* weights d_i assigned to the observations y_i, we cannot guarantee the solvency of statistical inferences based on sample (4.47). So, generally speaking, the limit (at $n \to \infty$) ratios (of probability) **will not** be observed:

$$\lim_{n\to\infty}\left(\sum_{i=1}^{n} d_i y_i^m\right) \to \mathbf{E}\eta^m, \quad m = 1, 2, \ldots,$$

The general formulation of the problem can be presented as follows: it is necessary to fix the weights $d = (d, d_2, \ldots, d_n)^T$ *so (i.e. it is required to find the new weights* $W = (w_1, w_2, \ldots, w_n)^T$*) the sampled first moment of the analysed random variable* η (i.e. $\bar{y}(W) = \sum_{i=1}^{n} w_i y_i$) *is as close as possible to (or exactly equal to) its* ***theoretical*** *first moment* $\mathbf{E}\eta$ *while the* ***new*** *weights (W) would be in a sense, as closely as possible to the* ***original*** *(d).*

Additional information used to solve this problem by methods CALMAR is as follows: the formation of sample (4.47), along with the values of the analysed random variable η, we register the values $X_i = (x_i^{(1)}, x_i^{(2)}, \ldots, x_i^{(k)})^T$, $i = 1, 2, \ldots, n$, of a range of supplementary (we'll call them "*control*") random variables $\xi = (\xi^{(1)}, \xi^{(2)}, \ldots, \xi^{(k)})^T$ on each element of the analysed population; theoretical average values of these variables $X_0 = \mathbf{E}\xi = (\mathbf{E}\xi^{(1)}, \mathbf{E}\xi^{(2)}, \ldots\ldots, \mathbf{E}\xi^{(k)})^T$ are presented in *macroeconomic* data.

In our case, average per capita income (expenditures) of the household is used as the analysed random variable η and as control variables we can use, for example, the same average per capita expenditures, proportion of children (under 15 years), retirees, etc. Obviously, the *theoretical (general)* averages of these control variables may be obtained from the macroeconomic statistics.

Mathematical formulation of the problem requires the introduction of the measure of distance $\rho(d; W)$ between the initial (*d*) and *sought* (*W*) weights. It is determined by the following equation:

$$\rho(d;\, W) = \sum_{i=1}^{n} d_i \cdot G\left(\frac{w_i}{d_i}\right) \tag{4.48}$$

where the function $G(z)$ must have the following properties:

$G(z)$ is positive and strictly convex;

$G(1) = G'(1) = 0$;

$G''(1) = 1$.

Then conditional optimization problem to minimize (in *W*) the distance $\rho(d; W)$ provided that $\sum_{i=1}^{n} w_i X_i = X_0$ is reduced to the following problem:

$$\Sigma d_i \cdot G\left(\frac{w_i}{d_i}\right) - \lambda^T (\Sigma w_i X_i - X_0) \rightarrow \min_W \quad (4.49)$$

where $\lambda = (\lambda_1 \lambda_2 \ldots \lambda_k)^T$ is the vector of Lagrange multipliers.

The solution of problem (4.49) is given in the following form:

$$w_i = d_i \cdot F(X_i^T \lambda) \quad (4.50)$$

where $F(u) = g^{-1}(u)$ (the inverse of the function $g(z) = G'(z)$), and the vector of Lagrange multipliers λ is determined by numerical methods from the system of equations:

$$\sum_{i=1}^{n} d_i F(X_i^T \lambda) \cdot X_i = X_0 \quad (4.51)$$

(details of the corresponding computational procedure are described in Section 11 of the work [Deville, Särndal, Sautory, 1993]).

Note that the an inverse function of $g(z)$ is provided by the properties of the function $G(z)$, while $F(0) = F'(0) = 1$.

Here are the options for functions $G(z)$ used in weighing observations by methods CALMAR:

$G(z) = \frac{1}{2}(z-1)^2;\ F(u) = 1 + u;$

$G(z) = z\log z - z + 1;\ F(u) = \exp(u);$

$G(z) = \frac{1}{8}\left[(z-a_{\min})\log\frac{x-a_{\min}}{1-a_{\min}} + (a_{\max}-z)\log\frac{a_{\max}-z}{a_{\max}-1}\right],$

where $a_{\min} and a_{\max}$ are certain constants, such as $z \in [a_{\min}, a_{\max}]$, while $a_{\min} < 1$ and $a_{\max} > 1$, and

$$b = \frac{a_{\max}-a_{\min}}{(1-a_{\min})(a_{\max}-1)};\ F(u) = \frac{a_{\min}(a_{\max}-1) + a_{\max}(1-a_{\min})\exp(bu)}{a_{\max}-1 + (1-a_{\min})\exp(bu)}$$

Now we can proceed to the description of the proposed procedure for solving **task 1** (see point 4.2.2), i.e. to the procedure of the analysis and correction of SHBS data to assess the distribution of the region's population by average per capita income (or expenditures).

Iterative procedure of simultaneous adjustment of weights *W* and estimation of the parameters of the analysed distribution

So, we have observations $(y_i;\ m_i)$ $(i = 1, 2, \ldots, n_0)$, where $y_i = \ln s_i$, s_i is the value of average per capita income (expenditures) in the i-th surveyed household in the region, and m_i is the number of members of the i-th household. At that time, *the statistically available range of values s is limited on the top by s_0* value (generally unknown). Assuming a log-normal distribution of the total population by per capita average expenditures s (i.e. in the **full** range of its possible values), we have the task of estimating the parameters a, σ^2 and $y_0 = \ln s_0$ of the distribution density function of the logarithm of per capita income $\eta = \ln s$ among *truncated* on the right population of the region:

$$f_{cond}(y|a;\ \sigma^2;\ y_0) = \frac{\frac{1}{\sqrt{2\pi}\sigma}\, e^{-\frac{1}{2\sigma^2}(y-a)^2}}{\int\limits_{-\infty}^{y_0} \frac{1}{\sqrt{2\pi}\sigma}\, e^{-\frac{1}{2\sigma^2}(t-a)^2}\, dt} \tag{4.52}$$

According to available observations from this "truncated" population:

$$Y = \{(y_1,\ y_1, \ldots, y_1), (y_2,\ y_2, \ldots, y_2)\ldots(y_{n_0},\ y_{n_0}, \ldots, y_{n_0})\} \tag{4.53}$$

where the multiplicity of value y_i in the sample equals $m_i (i = 1,\ 2, \ldots, n_0)$.

Direct application of *the maximum likelihood method* to the data y_i with weights $d_i = \frac{m_i}{n}$ $(i = 1, 2, \ldots, n_0;\ n = \sum_{i=1}^{n_0} m_i)$ does not provide consistent estimates for the unknown parameters due to the non-representativeness of sample (4.53). Therefore, *a preliminary* "*reweighting*" of the observations (4.53) using, for example, the procedure CALMAR[18] and given macro-values (per capita) control variables $z^{(3)}\beta$ However, we know the values of control variables X_0 for the **entire** general population, but to adjust the weights with the method CALMAR with a *truncated* distribution (4.52), we must use the corresponding values calculated *only for a truncated* (on top by expenditure) population of households. The share $q(y_0)$ of the "truncated" population is determined by the following equation:

$$q(y_0) = \int\limits_{y_0}^{\infty} \frac{1}{\sqrt{2\pi}\sigma}\, e^{-\frac{1}{2\sigma^2}(y-a)^2}\, dy \tag{4.54}$$

i.e. it itself depends on the unknown parameters $(a; \sigma^2; y_0)$. Therefore, we propose the following **iterative procedure of simultaneous adjustment of weights** $W = (w_1, w_2, \ldots, w_n)^T$ **and of estimating the parameters** $(a; \sigma^2; y_0)$ **of function (4.52).**

The logarithmic likelihood function of estimations (4.53), subject to distribution (4.52), has the following form:

18 If you have the data needed to build the "bounce function", the phase of reweighting observations can be carried out using this function (see 4.2.2).

$$l\left(Y|a;\sigma^2;y_0\right)=\ln\left(\prod_{l=1}^{n}f_{yc}\left(y_l|a;\sigma^2;y_0\right)\right)=$$
$$=-n\cdot\left[\ln\left(\int_{-\infty}^{y_0}e^{-\frac{1}{2\sigma^2}(t-a)^2}dt\right)+\frac{1}{2\sigma^2}\sum_{i=1}^{n_0}d_i(y_i-a)^2\right],\qquad(4.55)$$

where $n=\sum_{i=1}^{n_0}m_i$, and values y_l form n_0 series of identical elements y_i in sample (4.53), the length of each series m_i and the corresponding weight $d_i=\frac{m_i}{n}$.

The main idea of the proposed iterative procedure is as follows. Each (ν-th) iteration consists of two parts. In the first part, using the above-described method CALMAR and set values of control variables, we estimate the weights $W(\nu)$. In the second part of the iteration, using the *weighted* observations (4.53), we solve the problem of maximizing (for a; σ^2 *and* y_0) the likelihood function (4.55); using the obtained maximum likelihood estimations $\hat{a}_{ml}^{(\nu)}$, $\hat{\sigma}_{ml}^2(\nu)$ *and* $\hat{y}_{ml}(\nu)$, we estimate the share (4.54) population covered by the sample:

$$\hat{q}_0^{(\nu)}(y_0)=\int_{\hat{y}_{ml}(\nu)}^{\infty}\frac{1}{\sqrt{2\pi}\hat{\sigma}_{ml}(\nu)}e^{-\frac{1}{2\hat{\sigma}_{ml}^2(\nu)}(y-\hat{a}_{ml}(\nu))^2}dy.\qquad(4.56)$$

The procedure is iterative in nature, because before we start using it, we do not know what proportion of the population was not covered by the sample, and therefore do not know by how much the values of control variables X_0 in the procedure CALMAR need to be truncated.

In the first part of the **zero iteration** in the CALMAR procedure in determining the weights $W(0)=(w_1(0),w_2(0),\ldots,w_n(0))^{\mathrm{T}}$, we use not-truncated values of control variables X_0. Since underestimated expenditures constructed from sample (4.53) can be explained by **two** factors (refusal to be surveyed by some proportion of the relatively wealthy part of households from the *surveyed* range of expenditures and the complete absence in the sample of households with average per capita expenditures in excess of the value $s_{\max}$), but we ignored the second factor at zero iteration, the weights $w_i(0)$ assigned in this iteration to relatively wealthy households of the *statistically surveyed range of* η *variation* will be somewhat **elevated**, and therefore, the share $\hat{q}_0^{(0)}$ of the non-sampled rich population will be **lowered.** On the next (first) iteration in determining weights $W(1)$ we will use as control variables $X_0(1)$ already *truncated* values $X_0(1)=X_0\cdot(1-\hat{q}_0^{(0)})$. Accordingly, the value of the weights $w_i(1)$ of relatively more affluent households of the surveyed range η will decrease. That means that as a result of the second part of the iteration the value of the share $\hat{q}_0^{(1)}$ of non-sampled population will somewhat increase, etc. Empirically we established a very rapid convergence of this iteration process, i.e. while $\nu\to\infty$ $\hat{y}_0(\nu)\to\hat{y}_0,\hat{a}(\nu)\to\hat{a}$, $\hat{\sigma}^2(\nu)\to\hat{\sigma}^2$ *and* $\hat{q}_0^{(\nu)}(y_0)\to\hat{q}_0$.

Note 1. In the work [Deville, Särndal, Sautory, 1993], the authors proposed several options to select the function of distance $G(z)$ between the used (d_i) and sought weights (w_i). They also describe computational algorithms and programs CALMAR to implement the method. However, as already mentioned, the problem is solved under the assumption that the observations are presented in the sample (households) of the **total** range of possible values of the analysed random variable. Therefore, in our formulation of the problem if we limit CALMAR data correction by weighing only, it will lead to inadequate (concentrated only in the surveyed, truncated on the top range of the analysed random variable) distribution law.

Note 2. In the work of [Sheviakov, Kiruta, 1999] they also considered the problem of estimating the problem of distribution of the population (households) in the region by per capita income (expenditures) based on unrepresentative data SHHBS. They essentially offer the option (2°) (see above) of weighing method CALMAR, but without the reference to the primary source. In that study the authors also ignore the complete absence of observations in the "right tail" of the analysed distribution.

Note 3. When implementing the second part of the proposed iterative procedure, it is useful to consider the following. On iteration zero when calculating the maximum likelihood estimates of parameters $y_0,\ a\ and\ \sigma^2$ you can use as zero approximation the following values:

$$\hat{y}_0 = \ln s_{\max}(n),\quad where\ s_{\max}(n) = \max\{s_1, s_2, \ldots, s_n\};$$

$$\hat{a}_0 = \sum_{i=1}^{n} w_i(0)\cdot \ln s_i;$$

$$\hat{\sigma}_0^2 = \sum_{i=1}^{n} w_i(0)(\ln s_i - \hat{a}_0)^2.$$

Then maximization (on $y_0,\ a\ and\ \sigma^2$) of log-likelihood function (4.55) is reduced to the minimization of the expression

$$\int_{-\infty}^{y_0} e^{-\frac{1}{2\sigma^2}(t-a)}\, dt$$

in the vicinity of the point $(\hat{y}_0,\ \hat{a}_0,\ \hat{\sigma}_0^2)$, which may be accomplished, for example, with the method of grids.

On any other $(\nu+1)$-th iteration we can use the same scheme with the replacement of variables $\hat{y}_0,\ \hat{a}_0\ and\ \hat{\sigma}_0^2$ by the values $\hat{y}_{0ml}(\nu),\ \hat{a}_{ml}(\nu)\ and\ \hat{\sigma}_{ml}^2(\nu)$, respectively.

4.3.2 Analysis and elimination of defects in time aggregation average values

In this task, we are talking about a common practice in the statistical analysis of the *aggregation defects* that occur in the calculation of the average **annual** values of certain derivatives of SHHBS characteristics of the respective quarterly or monthly averages (see, e.g. [Scott et al., 2000]).[19] For example, in accordance with the currently accepted methodology, the assessment $\bar{q}(x_0)$ of the annual characteristic of the share of the poor $q(x_0)$ (i.e. the people whose per capita monthly income is below the "poverty line" x_0) is defined as the simple average of the four values of the corresponding quarterly characteristics $q_i(x_0)$ $(i = 1, 2, 3, 4)$, i.e.:

$$\bar{q}(x_0) = \frac{1}{4}\sum_{i=1}^{4} q_i(x_0).^{20}$$

We will show that with this method of calculation of the annual characteristics of poverty, **we will always have a positive** shift (assuming that the true share of the poor $q(x_0) < 0.5$), i.e.:

$$\bar{q}(x_0) = q(x_0) + \Delta\Big(q(x_0)\Big),$$

where $\Delta(q(x_0))$ is a positive (at $q(x_0) < 0.5$) systematic error, depending on the annual value of the true proportion of the poor $q(x_0)$, which in turn is determined by the parameters of distribution of the permanent component of per capita income.

Mathematical formulation of the problem and its solution

In accordance with the "permanent income hypothesis" of M. Friedman [Фридман, 1957], we assume that the average per capita monthly income ξ_i, characterizing a randomly taken individual in the i-th quarter of the year, is made up of the permanent (annual) component ξ and a variable (quarterly) component δ_i, i.e.:

$\xi_i = \xi + \delta_i \; i = 1, \ldots, 4,$

where quarterly components δ_i, being (as also is ξ) random variables, are independent of ξ and of each other and have mean values $E\delta_i$ equal to zero, and a constant (at i) dispersion σ^2.

Let's introduce the density function (f) and the corresponding distribution functions (F) for the law of probability distribution of the random variable ξ

($f_\xi(x)$ and $F_\xi(x)$) and for the random variables δ_i ($f_i(y)$ and $F_i(y)$). Then, using the known formulas, we can calculate the following probabilities:

– the probability $p_1(x_0)$ of a poor individual (annualized) as a result of the i-th quarterly survey not to fall into the category of the poor:

$$p_1(x_0) = P\{\xi + \delta_i \geq x_0;\ \xi < x_0\} = \int_0^{x_0} \Big(1 - F_i(x_0 - x)\Big) f_\xi(x)\, dx\ ;$$

19 Elena Borisovna Frolova drew my attention to this task.

20 Here and below we assume that quarterly income and the poverty lines are deflated according to the quarterly rate of inflation.

- the probability $p_2(x_0)$ of non-poor individual (annualized) as a result of the i-th quarterly survey to fall into the category of the poor:

$$p_2(x_0) = P\{\xi + \delta_i < x_0;\ \xi \geq x_0\} = \int_{x_0}^{\infty} F_i(x_0 - x) f_\xi(x)\, dx.$$

So $p_1(x_0)$ is the share (relative to the total population) of poor population at an annual rate, wrongly attributed, according to the results of the i-th quarterly survey, to the non-poor.

Similarly, $p_2(x_0)$ is the share (relative to the total population of non-poor population at an annual rate, wrongly attributed, according to the results of the i-th quarterly survey, to the poor).

From the construction follows the formula linking the proportion of the poor in the annual $(q(x_0))$and quarterly $(\bar{q}(x_0))$ basis:

$$q(x_0) = \bar{q}(x_0) - \left(p_2(x_0) - p_1(x_0)\right). \tag{4.57}$$

Using the symmetry (relative to zero) l.d.p. of random variables δ_i (i.e. $F_i(y) = 1 - F_i(-y)$), and the fact that

$$f_\xi(x_0 + z) \geq f_\xi(x) \text{ for any } x \in [0, x_0] \text{ and } z \in [0, Z_0 - x_0],$$

where Z_0 is the greater of the roots of the equation $f(z) = f(x_0)$, it can be shown that the value of the *difference* $p_2(x_0) - p_1(x_0)$ is always positive.

Setting (or estimating) distributions of the random variables ξ and δ_i (note that all δ_i, $i = 1, 2, 3, 4$ are identically distributed and mutually independent!), we can estimate the exact value of the shift in the assessment $q(x_0)$ of quarterly data. From the same distribution δ_i, in particular, follows that in this model the quarterly share of poor $q_i(x_0)$ remained virtually unchanged during the year (including deflated poverty lines and income).

Numerical example

Let the "poverty line" x_0 be set by the value $x_0 = 3.5$, and let (for ease of calculation) the random variables ξ and δ_i be *categorical* variables with a finite number of values, and having the following symmetric l.d.p. (see Tables 4.8 and 4.9).

Table 4.8. The law of probability distribution of random variable ξ

x_i^0	$x_1^0=1$	$x_2^0=2$	$x_3^0=3$	$x_4^0=4$	$x_5^0=5$	$x_6^0=6$	$x_7^0=7$	$x_8^0=8$	$x_9^0=9$
$P\{\xi=x_i^0\}$	0.05	0.08	0.12	0.15	0.20	0.15	0.12	0.08	0.05

Table 4.9. The law of probability distribution of random variables δ_i

z_j^0	$z_1^0=-3$	$z_2^0=-2$	$z_3^0=-1$	$z_4^0=0$	$z_5^0=1$	$z_6^0=2$	$z_7^0=3$
$P\{\delta_i=z_j^0\}$	0.08	0.12	0.16	0.28	0.16	0.12	0.08

True annual characteristic $q(x_0)$ of the share of the poor is calculated directly from the set law of probability distribution in Table 4.8 of a random variable ξ:

$$q(x_0)=P\{\xi<x_0\}=P\{\xi<3.5\}=P\{\xi=1\}+P\{\xi=2\}+P\{\xi=3\}=0.25. \quad (4.58)$$

Using the mutual dependence of the random variables ξ and δ (and therefore, $P\{\xi=i,\ \delta=j\}=P\{\xi=i\}\cdot P\{\delta=j\}$) and the laws of probability distribution of these random variables set in Tables 4.8 and 4.9, we will calculate $p_1(x)$ and $p_2(x)$:

$$\begin{aligned} p_1(x_0)&=P\{\xi<3.5;\ \xi+\delta\geq 3.5\}=\sum_{i=1}^{3}P\{\xi=i\}\times\sum_{j\geq 4-i}^{3}P\{\delta=j\}=\\ &=0.05\cdot 0.08+0.08(0.08+0.12)+0.12(0.08+0.12+0.16)=\mathbf{0.0632} \end{aligned} \quad (4.59)$$

$$\begin{aligned} p_2(x_0)&=P\{\xi\geq 3.5;\ \xi+\delta<3.5\}=\sum_{i=4}^{9}P\{\xi=i\}\times\sum_{j\leq 3-i}^{-3}P\{\delta=j\}=\\ &=0.15\cdot(0.08+0.12+0.16)+0.20\cdot(0.08+0.12)+0.15\cdot 0.08=\mathbf{0.106}. \end{aligned} \quad (4.60)$$

Placing the obtained values (4.58)–(4.59)–(4.60) in ratio (4.57), we see that the average quarterly estimate of the annual poverty level is about 4% higher than its true value.

Note 1. Having assessed using the available statistics dispersion σ^2 of the quarterly variations in income around its permanent level, and setting, for example, a $(0,\sigma^2)$-normal general appearance of a random variable $\delta_i$and a log-normal distribution value of permanent (annual) income, it is possible, using the scheme described above, to estimate the actual value of the shift.

Note 2. Using the same scheme, we can carry out the research of the problem of aggregation of monthly and quarterly data in the task of estimating the number of other characteristics of the analysed distribution, such as the dispersion and some indicators of differentiation (see, for example, performing similar tasks as well as some other approaches to their solution in [Scott, Strode, 2000]).

4.4 Conclusions

1. The proposed concept of analysis of the problem of typology of household consumer behaviour is based on the hypothesis of the existence of the so-called targeted group consumption patterns and on the associated concept of targeted levels of consumption. This explains the fact that the mechanism of the objectively existing differentiation of needs of the population, operating according to national circumstances, socio-economic conditions, specifics of growing up and labour capacity, differences in culture level and place in the society lead (despite all the greater individualization of consumption) **to a relatively small** (compared with the total number of households) **number of types of uniform consumer behaviour**, covering the vast majority of households.
2. **The behavioural approach** to the identification and description of the existing types of consumer behaviour is an alternative to *normative* and is based on **the empirical analysis on the actual behaviour of consumers**. Mathematical tools used to implement this approach (including not only identification and description of the main types of consumer behaviour but also identification of those descriptive characteristics of the household, the so-called determinant factors, that mainly condition its assignment to a particular type of consumer behaviour) includes advanced methods of *cluster and discriminant analysis*. The specific implementation of the behavioural approach to Russian data in 1996 allowed to identify and analyse eight types of consumer behaviour, covering more than 85% of considered households (see Table 4.1) by the main determinant factor, the average per capita family income (see Table 4.2).
3. In the overall socio-economic system, which determines the nature of distributive relations in society, in accordance with objectively acting economic laws in the society (which in turn are caused by the main goals of the society, the adopted a system of values, etc.), we have a number of operating interrelated local mechanisms of individual socio-economic systems, such as (see Figure 4.3) blocks of distribution of workers according to their wages and entrepreneurs – according to business income; block of the structure of social consumption funds (the share of public funds allocated for health care and culture, for state family assistance, transport and etc.); blocks of retirees distribution by pension size, scholarship recipients – by scholarship size, families – by supplementary (“other”, i.e. not part of the wages, pensions or scholarships) cash benefits; block of distribution of families by the size of average per capita income; block of distribution of families by the size of average per capita savings; the block of socio-demographic and quantitative family structures and the block of structure of demand by population for various types of goods and services and their consumption.

4. *The bio-anthropological model of the distribution of employees by salary size*, considered in 4.2.1, in contrast to the common formal adaptations of existing experimental data into one of the theoretical models, describes the **genesis, i.e. the very mechanism of formation of the analysed distribution,** implemented within the initial assumptions operating under objective economic laws (their formalization is presented by conditions (i)–(ii)–(iii) in 4.2.1).
   The results of the successive (in time) comparison of model and experimental data of the distribution of workers' wages, conducted by CEMI AS USSR, seem to be symptomatic in this sense. They clearly delineated two points of extremely sharp discrepancy in model and real data. However, with the removal of these moments in the past, there is a clear trend towards convergence of model and real data. A closer analysis showed that the moment of sharp disagreement immediately followed a very significant distortion of the initial assumptions (i) and/or (ii).
   The first of these moments refers to the 60s of the last century and is connected with a sharp-willed directive for restructuring of the wage schedules existing at that time, which led to a breach of the initial assumption (ii). The second point relates to the more recent 90s years, to the period of radical socio-economic transformations in the Russian society, during which assumption (iii) was broken (see more on model description of the specifics of this period, but in relation to the distribution of population by per capita expenditures in 4.2.2). The fact that further on we have observed the convergence of model and real data, reveals only that the objective economic laws gradually increase their impact on distribution types; they increasingly "come to the surface," steadily gaining various legal forms.
5. The specificity of the transition period in Russia caused some changes in the nature of "mixing functions" q(a) without cancelling a general logarithmically normal model of distribution for the average per capita income (expenditures). The phenomenon of appearance of a *discrete* mixture of log-normal distributions (instead of a special type of the *continuous* mixture, existing in the steady regime of stable economy, which again yields log-normal distribution) is due to the fact that in the transitional period there are changes in the structure of demand for labour and human capital. In view of these changes, the professionals who are not in demand are forced to switch to other, usually less profitable, sources of income. In particular, these changes applied to the "Soviet middle-class" professionals. Combined with low mobility, characteristic of the Russian labour market, these structural changes have caused "wiping-out" of individual groups of workers. At the same time, the appropriation of significant flows of income has given rise to completely new "very rich" population groups. So, a few of quite well-defined (on the revenue scale) groups has emerged, which led to a discrete mixture of distributions (log-normal within each group), hence the natural attempt of approximation of the sought distribu-

tion using a discrete log-normal mixture. Note that during the transition of the Russian economy and its approach to normal steady state the “mixing” function $q(a)$ will also return to its normal appearance, and therefore, the desired distribution of the population by expenditures will become more like a regular two-parameter log-normal distribution. Our preliminary calculations and comparison of the results of 1996 and 1998 confirm this trend.

6. In the specific conditions of modern Russian economy, definition of poverty and property differentiation indicators, the criteria by which a household should be included in the category of the poor, should be based on the average per capita *expenditures* (*rather than income*, as is the case in most other studies). When considering the expenditures instead of income:
   - the problem of “accounting – not accounting” for the delays in salary payments to members of HH is removed;
     the relevance of the issues related to intentionally or unintentionally hidden parts of income, including income obtained in the informal sector of the economy decreases;
   - the set of factors determining the well-being of HH primarily due to the inclusion of personal subsidiary plots and property components (real estate, personal vehicles, jewellery, etc.), rental or sale of which could substantially maintain the level of well-being.[21]
7. Statistical analysis of the model proposed in 4.2.2 provides:
   - An estimation of the parameters $\Theta = (k;\ q_1, q_2, \dots, q_k;\ a_1, a_2, \dots, a_k;\ \sigma_1^2, \sigma_2^2, \dots, \sigma_k^2)$ of the final discrete mixture of distributions (4.39a) in the *observed* range of per capita expenditures by using corresponding statistical procedures;
   - Special *calibration* of the obtained distribution based in the probability of evasions from survey as a function of (4.40′) for the value of per capita expenditures introduced in the model: place of residence and level of education of head of household;
   - Revaluation (recovery) of the *unobserved* $(\hat{k}+1)$-th component of the analysed mixture of distributions (4.39) *using a repeated calibration* of the model based on “tightening” of the model average value to exogenously given (from macro-statistics) value of the average per capita expenditures.

21 When forming the value of the per capita consumption of the household, all consumption expenditures include expenditures on intermediate consumption (including sustaining personal subsidiary plots), gross savings of fixed capital (purchase of land, property, construction costs and housing repairs, etc.), the sum of all paid taxes and other mandatory payments, balances of cash and the growth of organized savings (including purchase of currency and securities, bank deposits, etc.), and finally, the valuation of consumed products, produced by private subsidiary plots. In the work [Aivazian, Kolenikov, 2001] the entire sum was divided by the number of members of HH, i.e. equivalence scales were not used. However, the introduction in the proposed model of any given equivalence scales *is a purely technical moment*, which will influence the main conclusions of the work.

8. In contrast to the methods used by the state statistical-based services, as well as the known approaches suggested by other researchers, the model of population distribution by total per capita expenditures and the methodology of its econometric analysis, described in 4.2.2, *allow to take into account the evaded surveying HH*. This is achieved by introducing and statistically evaluating the function describing the probability of survey evasion of HH from a range of their economic, social, territorial and demographic characteristics, as well as by a special assessment of latent strata of the population.
9. The efficiency of the proposed method of econometric analysis of the distribution of population by average total per capita expenditures is demonstrated on the solution of two virtually unrelated applied tasks:
   - estimation of the main indicators of the population poverty level within the issue of optimal organization of the targeted social assistance to low-income segments of the population (task 2);
   - estimation of the main indicators of population differentiation by welfare, which serve as indicators of social tension level (task 3).
10. Using the techniques proposed in this study *introduces significant changes only in the "right tail" of the analysed distribution*. Accordingly, the estimation of main characteristics of poverty (the proportion of the poor, the index of Foster--Greer–Torbekka) based on this assessment, almost entirely defined by the "left tail" of the distribution, as would be expected, are almost no different from the traditional. Therefore, the main result in task 2 should be considered the rule of optimal organization of targeted social assistance to "persistently poor" segments of the population formulated in 4.2.2 (the identification of the level $\tilde{z}_0$ of expenditure to which the very poor should be "pulled", and so on).
11. The most important specifications for using the proposed methodology *are associated with the statistical analysis and interpretation of various measures of population differentiation by level of well-being, such as the Gini index and decile ratio.* We have demonstrated a relative stability of the proposed estimates of these characteristics in relation to the realistic variation of the initial model assumptions and, in particular, the influence of "misreporting" factor. Our calculations have showed that the value of the Gini coefficient and decile ratio for Russia in November 1998 equalled, respectively, 0.55–0.57 and 36–39 instead of 0.38 and 13.5, as follows from the official statistics.
12. It is advisable to implement the methodologies, based on the proposed model, on the *regional level*. Only after bringing the regions to the "common denominator" using the appropriate deflators and coefficients that take into account regional differences of purchasing power of ruble, the composition of the consumer basket, calculation of the poverty level, etc., the results can be aggregated by region to obtain a nationwide total.
13. The model, described in 4.2.3, of the formation mechanism of the distribution of households by per capita average savings can be taken as an object

of *econometric analysis*. Special sample surveys, tuned to this analysis, will allow to statistically verify the validity of the adopted initial assumptions ((**A**) and (**B**)), to estimate the "input parameters" of the model $(\alpha(\omega), \lambda_\omega(x), \psi_\omega(z))$ and experimentally analyse the results of its practical application.

14. The problems of improvement of methods of obtaining reliable and representative results from sample surveys of the population, especially the results of sampling household budget surveys, are still very significant. These problems are related to, firstly, practical *impossibility of the realization of the plan of a representative sample survey (the sample selection problem)* and, in addition, to a widespread statistical practice defects of temporal aggregation of average values of the analysed parameters. In 4.2.2, 4.3.1 and 4.3.2 we described some approaches to solving these issues.

A2 Appendices to Chapter 2

A2.1 The structure and content of the information included in WCY (The World Competitiveness Yearbook)

Macroeconomic data of the countries included in the WCY, are divided into blocks, each of which in turn contains a number of sections. Since the structure of blocks and partition in 2001 underwent a major change, here are two options to describe WCY:

- *the structure and content of WCY until 2001 (for example, [WCY, 1999]);*
- *the structure and content of WCY after 2001 (for example, [WCY, 2004]).*

Full database of WCY has been supported for a number of years by Central Economics and Mathematics Institute (CEMI) of the Russian Academy of Sciences (as of July 2011, CEMI has a database of indicators from WCY for 1997–2011).

A2.1a The structure and content of the information included in WCY unit (e.g. [WCY, 1999])

The 1999 yearbook [WCY, 1999] contains the 288 macrovariabes (including 107 expert estimates) of 47 countries. They are divided into blocks and sections as follows.

Block 1: Domestic economy – 30 indices (including 4 variables evaluated by experts)

Sections in block 1

1.1. Value added	–9 (1);
1.2. Investments	–2 (0);
1.3. Savings	–2 (0);
1.4. Final consumption	–4 (0);
1.5. Economic sector's performance	–6 (0);
1.6. Cost of living	–4 (0);
1.7. Adaptiveness	–3 (3).

Block 2: Internationalization – 45 indices (including 15 variables evaluated by experts)

Sections in block 2

2.1. Current account balance	– 6 (0);
2.2. Exports of goods and services	– 10 (1);
2.3. Imports of goods and services	– 5 (0);
2.4. Exchange rate	– 3 (1);
2.5. Portfolio investments	– 2 (0);
2.6. Foreign direct investments	– 6 (0);
2.7. National protectionism	– 8 (8);
2.8. Openness	– 5 (5).

Block 3: Government – 48 indices (including 26 variables evaluated by experts)

Sections in block 3

3.1. National debt	– 8 (1);
3.2. Government expenditure	– 3 (1);
3.3. Fiscal policies	– 14 (3);
3.4. State efficiency	– 11 (11);
3.5. State involvement	– 7 (6);
3.6. Justice and security	– 5 (4).

Block 4: Finance – 27 indices (including 16 variables evaluated by experts)

Sections in block 4

4.1. Cost of capital	– 3 (1);
4.2. Availability of capital	– 8 (7);
4.3. Stock market's dynamism	– 5 (2);
4.4. Banking sector's efficiency	– 11 (6).

Block 5: Infrastructure – 32 indices (including 8 variables evaluated by experts)

Sections in block 5

5.1. Basic infrastructure	– 10 (5);
5.2. Technological infrastructure	– 13 (3);
5.3. Energy self-sufficiency	– 5 (0);
5.4. Environment	– 4 (0).

Block 6: Management – 36 indices (including 16 variables evaluated by experts)

Sections in block 6

6.1. Productivity	– 12 (0);
6.2. Labour costs	– 5 (0);
6.3. Corporate performance	– 5 (3);
6.4. Management efficiency	– 9 (8);
6.5. Corporate culture	– 5 (5).

Block 7: Science and technology – 26 indices (including 11 variables evaluated by experts)

Sections in block 7

7.1. Total expenditure on R&D	– 5 (0);
7.2. R&D personnel	– 6 (2);
7.3. Technology management	– 5 (5);
7.4. Scientific environment	– 5 (3);
7.5. Intellectual property	– 5 (1).

Block 8: People – 44 indices (including 11 variables evaluated by experts)

Sections in block 8

8.1. Population characteristics	– 5 (0);
8.2. Labour force characteristics	– 8 (2);
8.3. Employment	– 8 (0);
8.4. Educational structure	– 11 (3);
8.5. Quality of life	– 7 (1);
8.6. Attitudes and values	– 5 (5).

A2.1 b The structure and content of the information included in WCY since 2001 (for example, WCY [2004])

The 2004 yearbook [WCY, 2004] contains data on 323 macrovariables (including the 113 variables, evaluated by experts on 10-point scale, where 0 - is the worst situation, and 10 - is the best) of large groups of individuals from 52 countries and 8 regions (a list of countries and regions is given in Appendix 3.1). These data are divided into four blocks, and each block in turn is divided by sections. So every measure is encoded by three numbers: the number of block, partition number and serial number of the index in this section. However, some sections may be divided into sub-sections.

Block 1: Economic performance – 83 indices (including 4 variables evaluated by experts).
1.1. Domestic economy (sub-sections: volume, growth, prosperity, forecasts) – 33 (1);
1.2. International trade – 20 (0);
1.3. International investment – 17 (3);
1.4. Employment – 9 (0);
1.5. Prices – 4 (0).

Block 2: Government efficiency – 77 indices (including 44 variables evaluated by experts)
2.1. Public finance – 11 (1);
2.2. Fiscal policy – 14 (3);
2.3. Institutional framework (with sub-sections: the central bank, government effectiveness) – 17 (13);
2.4. Business legislation (sub-sections: transparency, competition and regulation, labour regulation, regulation of the capital market) – 22 (20);
2.5. Societal framework – 13 (7).

Block 3: Business efficiency – 69 indices (including 36 variables evaluated by experts)
3.1. Productivity – 11 (0);
3.2. Labour market – 20 (10);
3.3. Finance (sub-sections: Bank efficiency, stock market efficiency, finance management) – 22 (10);
3.4. Management practices – 10 (10);
3.5. Attitudes and values – 6 (6)

Block 4: Infrastructure – 94 indices (including 29 variables evaluated by experts)
4.1. Basic infrastructure – 24 (6);
4.2. Technological infrastructure – 18 (6);
4.3. Scientific infrastructure – 22 (5);
4.4. Health and environment – 17 (6);
4.5. Education – 13 (6).

A2.2 A priori and a posteriori synthetic sets of partial criteria category of "life quality" in a cross-country analysis (according to WCY 2009)*

No	Variable (criterion): name and unit of measure	Symbol	Code in WCY, 2009	Formula for unifying transformation
1	GDP per capita based on purchasing power parity, USD	$x^{(1)}$	1.1.22	(2.2)
2	Productivity, USD	$x^{(2)}$	3.1.04	(2.2)
3	Spendings on private consumption per capita, USD	$x^{(3)}$	1.1.23	(2.2)
4	The illiteracy rate among the population aged over 15 years, %	$x^{(4)}$	4.5.14	(2.3)
5	20% rate of funds, times**	$x^{(5)}$	$\frac{2.5.07}{2.5.06}$	(2.4)
6	CPI, %	$x^{(6)}$	1.5.01	(2.3)
7	Life expectancy (at birth), years	$x^{(7)}$	4.4.05	(2.2)
8	Infant mortality rate: average number of deaths under 1 year per 1000 births	$x^{(8)}$	4.4.07	(2.3)
9	CO_2 emission, metric tonnes per $1 million of GDP	$x^{(9)}$	4.4.17	(2.3)
10	Total expenditure on R&D, % of GDP	$x^{(10)}$	4.3.02	(2.4)

* For some indicators [WCY, 2009] are merely lagged (1–2 years) values.

** 20% rate of funds is calculated as the ratio of total income of the richest 20% of the population to the total income of the poorest 20% in the population.

A2.3 Original and standardized statistical data for the cross-country analysis (source: WCY, 2009)

A2.3a Original data

No	Country	Particular criteria									
		$x^{(1)}$	$x^{(2)}$	$x^{(3)}$	$x^{(4)}$	$x^{(5)}$	$x^{(6)}$	$x^{(7)}$	$x^{(8)}$	$x^{(9)}$	$x^{(10)}$
1	2	3	4	5	6	7	8	9	10	11	12
1	Argentina	13775	44.71	4662	2.4	17.87	8.1	75	17	698.6	0.51
2	Australia	34935	17.36	25063	1	7.00	4.4	82	6	521.6	2.09
3	Austria	37913	39.24	26229	1	4.40	3.2	80	4	225.5	2.56
4	Belgium	35183	48.08	25260	1	4.87	4.5	79	5	293	1.87
5	Brazil	10041	11.01	4983	9.5	21.82	5.6	72	20	305.1	1.1
6	Bulgaria	11923	12.06	4423	1.7	4.40	12.3	73	12	1501.8	0.48
7	Canada	35669	37.33	25161	1	5.54	2.4	81	6	421.4	1.89
8	Chile	14176	16.26	5992	3.5	15.79	8.7	78	9	410.3	0.68
9	China Mainland	5825	4.70	928	8.4	12.07	5.9	73	24	2109.4	1.49
10	Colombia	8716	9.59	3440	6.4	25.08	7	74	21	365.3	0.16
11	Croatia	16760	20.90	9433	1.3	4.77	6.1	76	6	422	0.81
12	Czech Republic	24688	27.44	10179	1	3.49	6.3	77	4	850	1.54
13	Denmark	35677	39.51	30495	1	4.31	3.4	79	4	201.4	2.55
14	Estonia	19638	21.82	9482	1	5.53	10.4	73	6	920.5	1.14
15	Finland	34750	43.12	26437	1	3.82	3.8	79	3	318.8	3.47
16	France	33693	50.14	25738	1	5.58	3.3	81	5	166.5	2.08
17	Germany	34907	41.26	25009	1	4.34	2.6	80	5	282.8	2.53
18	Greece	29280	39.93	22580	2.9	6.19	4.2	80	4	351.4	0.57
19	Hong Kong	43040	36.13	18702	1.7	14.72	4.3	82.3	2	220.7	0.81
20	Hungary	18868	25.31	10066	1.1	3.84	6.1	73	7	498.9	0.97
21	Indonesia	3888	4.09	1362	8.6	5.15	10.3	68	34	923.3	0.35
22	Ireland	42993	49.63	30262	1	5.68	4.2	80	4	201.9	1.31
23	Israel	26642	35.78	15613	2.9	7.88	4.6	81	5	435.6	4.68
24	Italy	30057	41.47	22776	1.1	6.46	3.3	81	4	240.4	1.13
25	Japan	33403	37.28	21747	1	3.37	1.4	83	4	278.2	3.4
26	Jordan	5088	11.65	2777	6.9	6.91	14.9	71	25	1233.2	0.75
27	Kazakhstan	11000	10.96	3575	1	4.67	17	64	29	2244.7	0.28
28	Lithuania	18266	21.91	9300	1	6.35	11.1	71	9	454.4	0.82
29	Luxembourg	79379	59.45	36014	1	4.63	3.4	80	4	263.3	1.63
30	Malaysia	13538	16.51	3613	8.1	12.34	5.4	72	12	984.8	0.57
31	Mexico	13986	14.37	6613	7.6	12.81	5.2	74	35	437.3	0.46
32	Netherlands	39215	45.74	24429	1	5.09	2.5	80	5	263.2	1.7
33	New Zealand	26985	29.50	16922	1	6.84	4	80	6	348.5	1.18
34	Norway	53481	57.23	36689	1	3.88	3.8	80	4	109.3	1.57
35	Peru	8374	7.36	2834	9.5	15.32	7.3	73	25	302.6	0.15
36	Philippines	3400	4.22	1287	6.6	11.00	9.3	68	32	565.7	0.1

(continued)

No	Country	Particular criteria									
		$x^{(1)}$	$x^{(2)}$	$x^{(3)}$	$x^{(4)}$	$x^{(5)}$	$x^{(6)}$	$x^{(7)}$	$x^{(8)}$	$x^{(9)}$	$x^{(10)}$
1	2	3	4	5	6	7	8	9	10	11	12
37	Poland	16747	21.39	8348	1	5.63	4.2	75	7	895.8	0.56
38	Portugal	22781	25.84	15273	5.1	7.91	2.6	79	4	288.7	1.18
39	Qatar	54779	33.86	12206	9.8	7.16	15.1	77	11	698.8	0.31
40	Romania	13271	15.90	6100	2.4	4.84	7.9	73	16	771.7	0.53
41	Russia	15524	17.95	5780	1	8.65	14.1	66	13	1604.2	1.12
42	Singapore	47646	36.36	15416	5.6	9.80	6.5	80	3	309.9	2.27
43	Slovak Republic	21302	25.57	9917	1	3.95	4.6	74	8	670.1	0.46
44	Slovenia	27941	31.39	13879	1	4.14	5.7	78	4	405.8	1.53
45	South Africa	9887	17.32	3443	12	17.77	10.6	51	69	1326.1	0.92
46	South Korea	25268	25.33	10142	2	4.75	4.7	79	5	500.6	3.01
47	Spain	31053	38.62	19865	2.6	6.00	4.1	81	4	265.8	1.2
48	Sweden	36180	40.48	24068	1	4.02	3.4	81	4	122.8	3.64
49	Switzerland	41068	36.59	37219	1	5.43	2.5	82	5	113.5	2.9
50	Taiwan	30254	29.70	10462	2.4	6.03	3.5	77.9	7	740	2.62
51	Thailand	8013	6.76	2251	5.9	7.78	5.4	72	8	1049.7	0.2
52	Turkey	13526	21.06	7161	11.3	7.28	10.4	73	26	452.3	0.58
53	Ukraine	7075	8.69	2353	1	4.08	25.3	67	24	2879.6	0.85
54	United Kingdom	35347	39.08	28011	1	7.21	3.7	79	6	220.4	1.76
55	USA	45017	47.78	32564	1	8.48	3.8	78	8	432.3	2.67
56	Venezuela	12197	14.40	7142	7	15.79	30.6	74	21	808.6	0.23

A2.3b Data used in the unification of the measurement scales

j	Particular criteria $x^{(j)}$ (code)	$x_{min}^{(j)}$	$x_{max}^{(j)}$	$x_{max}^{(j)}-x_{min}^{(j)}$	$x_{Opt}^{(j)}$ *	Formula for unifying transformation
1	2	3	4	5	6	7
1	$x^{(1)}$ (1.1.22)	3400.00	56000	52600.0	56000.0	(2.2)
2	$x^{(2)}$ (3.1.04)	4.09	59.45	55.36	59.45	(2.2)
3	$x^{(3)}$ (1.1.23)	928.0	37219.0	36291.0	37219.0	(2.2)
4	$x^{(4)}$ (4.5.14)	1.00	12.0	11.0	1.0	(2.3)
5	$x^{(5)} \left(\frac{2.5.07}{2.5.06}\right)$	3.37	18.50	15.13	5.00	(2.4)
6	$x^{(6)}$ (1.5.01)	1.40	19.0	17.6	1.4	(2.3)
7	$x^{(7)}$ (4.4.05)	60.00	83.0	23.0	83.0	(2.2)
8	$x^{(8)}$ (4.4.07)	2.00	37.0	35.0	2.0	(2.3)
9	$x^{(9)}$ (4.4.17)	109.30	2300.0	2190.7	109.3	(2.3)
10	$x^{(10)}$ (4.3.02)	0.10	4.68	4.58	2.3	(2.4)

* In the case of non-monotonic dependence of the QOL from $x^{(j)}$ the value $x_{opt}^{(j)}$ is determined as the average of the life quality values for the three best countries (according to the expert 10-point assessment).

A2.3c Unified initial statistics

i	Country	Unified particular criteria										$y_{i_{St.}}$
		$x^{(1)}$	$x^{(2)}$	$x^{(3)}$	$x^{(4)}$	$x^{(5)}$	$x^{(6)}$	$x^{(7)}$	$x^{(8)}$	$x^{(9)}$	$x^{(10)}$	
1	2	3	4	5	6	7	8	9	10	11	12	13
1	Argentina	1.97	2.40	8.97	8.73	0.47	6.19	6.52	5.71	7.31	2.48	4.58
2	Australia	6.00	6.35	6.65	10.00	8.52	8.30	9.57	8.86	8.12	9.12	9.28
3	Austria	6.56	7.34	6.97	10.00	9.55	8.98	8.70	9.43	9.47	8.91	9.64
4	Belgium	6.04	7.95	6.70	10.00	9.90	8.24	8.26	9.14	9.16	8.19	8.71
5	Brazil	1.26	1.25	1.12	2.27	0.30	7.61	5.22	4.86	9.11	4.96	4.79
6	Bulgaria	1.62	1.44	0.96	9.36	9.56	3.81	5.65	7.14	3.64	2.35	3.53
7	Canada	6.13	6.00	6.68	10.00	9.60	9.43	9.13	8.86	8.58	8.28	9.27
8	Chile	2.05	2.20	1.40	7.73	2.01	5.85	7.83	8.00	8.63	3.19	6.68
9	China Mainland	0.46	0.11	0.00	3.27	4.76	7.44	5.65	3.71	0.87	6.60	4.72
10	Colombia	1.01	0.99	0.69	5.09	0.00	6.82	6.09	4.57	8.83	1.01	4.82
11	Croatia	2.54	3.04	2.34	9.73	9.83	7.33	6.96	8.86	8.57	3.74	5.58
12	Czech Republic	4.05	4.22	2.55	10.00	8.88	7.22	7.39	9.43	6.62	6.81	7.59
13	Denmark	6.14	6.40	8.15	10.00	9.49	8.86	8.26	9.43	9.58	8.95	8.98
14	Estonia	3.09	3.20	2.36	10.00	9.60	4.89	5.65	8.86	6.30	5.13	6.20
15	Finland	5.96	7.05	7.03	10.00	9.13	8.64	8.26	9.71	9.04	5.08	8.12
16	France	5.76	8.32	6.84	10.00	9.57	8.92	9.13	9.14	9.74	9.08	8.17
17	Germany	5.99	6.71	6.64	10.00	9.51	9.32	8.70	9.14	9.21	9.03	8.89
18	Greece	4.92	6.47	5.97	8.27	9.12	8.41	8.70	9.43	8.89	2.73	6.18
19	Hong Kong	7.54	5.79	4.90	7.27	2.80	8.35	9.70	10.00	9.49	3.74	6.59
20	Hungary	2.94	3.83	2.52	9.91	9.14	7.33	5.65	8.57	8.22	4.41	5.25
21	Indonesia	0.09	0.00	0.12	3.09	9.89	4.94	3.48	0.86	6.28	1.81	4.26
22	Ireland	7.53	8.23	8.08	10.00	9.50	8.41	8.70	9.43	9.58	5.84	8.39
23	Israel	4.42	5.72	4.05	8.27	7.87	8.18	9.13	9.14	8.51	0.00	6.91
24	Italy	5.07	6.75	6.02	9.91	8.92	8.92	9.13	9.43	9.40	5.08	6.86
25	Japan	5.70	6.00	5.74	10.00	8.79	10.00	10.00	9.43	9.23	5.38	6.90
26	Jordan	0.32	1.37	0.51	4.64	8.58	2.33	4.78	3.43	4.87	3.49	5.66
27	Kazakh-stan	1.44	1.24	0.73	10.00	9.76	1.14	1.74	2.29	0.25	1.51	4.45
28	Lithuania	2.83	3.22	2.31	10.00	9.00	4.49	4.78	8.00	8.42	3.78	5.22
29	Luxem-bourg	10.0-0	10.0-0	9.67	10.00	9.73	8.86	8.70	9.43	9.30	7.18	9.21
30	Malaysia	1.93	2.24	0.74	3.55	4.56	7.73	5.22	7.14	6.00	2.73	7.43
31	Mexico	2.01	1.86	1.57	4.00	4.21	7.84	6.09	0.57	8.50	2.27	4.63
32	Nether-lands	6.81	7.52	6.48	10.00	9.93	9.38	8.70	9.14	9.30	7.48	8.91
33	New Zealand	4.48	4.59	4.41	10.00	8.63	8.52	8.70	8.86	8.91	5.29	8.91

(continued)

(continued)

i	Country	Unified particular criteria										$y_{i_{St.}}$
		$x^{(1)}$	$x^{(2)}$	$x^{(3)}$	$x^{(4)}$	$x^{(5)}$	$x^{(6)}$	$x^{(7)}$	$x^{(8)}$	$x^{(9)}$	$x^{(10)}$	
1	2	3	4	5	6	7	8	9	10	11	12	13
34	Norway	9.52	9.60	9.85	10.00	9.17	8.64	8.70	9.43	10.00	6.93	9.16
35	Peru	0.95	0.59	0.53	2.27	2.35	6.65	5.65	3.43	9.12	0.97	4.33
36	Philippines	0.00	0.02	0.10	4.91	5.56	5.51	3.48	1.43	7.92	0.76	4.81
37	Poland	2.54	3.13	2.04	10.00	9.54	8.41	6.52	8.57	6.41	2.69	4.26
38	Portugal	3.68	3.93	3.95	6.27	7.84	9.32	8.26	9.43	9.18	5.29	6.12
39	Qatar	9.77	5.38	3.11	2.00	8.40	2.22	7.39	7.43	7.31	1.64	8.13
40	Romania	1.88	2.13	1.43	8.73	9.88	6.31	5.65	6.00	6.98	2.56	2.68
41	Russia	2.30	2.50	1.34	10.00	7.30	2.78	2.61	6.86	3.18	5.04	2.89
42	Singapore	8.41	5.83	3.99	5.82	6.44	7.10	8.70	9.71	9.08	9.87	8.89
43	Slovak Republic	3.40	3.88	2.48	10.00	9.23	8.18	6.09	8.29	7.44	2.27	6.30
44	Slovenia	4.67	4.93	3.57	10.00	9.37	7.56	7.83	9.43	8.65	6.76	6.81
45	South Africa	1.23	2.39	0.69	0.00	0.54	4.77	0.00	0.00	4.45	4.20	5.08
46	South Korea	4.16	3.84	2.54	9.09	9.81	8.13	8.26	9.14	8.21	7.02	6.18
47	Spain	5.26	6.24	5.22	8.55	9.26	8.47	9.13	9.43	9.29	5.38	7.71
48	Sweden	6.23	6.57	6.38	10.00	9.28	8.86	9.13	9.43	9.94	4.37	9.08
49	Switzerland	7.16	5.87	10.00	10.00	9.68	9.38	9.57	9.14	9.98	7.48	9.71
50	Taiwan	5.11	4.63	2.63	8.73	9.24	8.81	7.78	8.57	7.12	8.66	5.84
51	Thailand	0.88	0.48	0.36	5.55	7.94	7.73	5.22	8.29	5.71	1.18	5.93
52	Turkey	1.93	3.07	1.72	0.64	8.31	4.89	5.65	3.14	8.43	2.77	5.00
53	Ukraine	0.70	0.83	0.39	10.00	9.32	1.14	3.04	3.71	0.00	3.91	2.73
54	United Kingdom	6.07	6.32	7.46	10.00	8.36	8.69	8.26	8.86	9.49	7.73	7.16
55	USA	7.91	7.89	8.72	10.00	7.42	8.64	7.83	8.29	8.53	8.45	8.53
56	Venezuela	1.67	1.86	1.71	4.55	2.01	0.00	6.09	4.57	6.81	1.30	2.44

A2.4 The results of the implementation of the procedures for formation posteriori set particular criteria QOL priori set of cross-country analysis

Table A2.4a. Correlation matrix of unified particular criteria*

	$\tilde{x}^{(1)}$	$\tilde{x}^{(2)}$	$\tilde{x}^{(3)}$	$\tilde{x}^{(4)}$	$\tilde{x}^{(5)}$	$\tilde{x}^{(6)}$	$\tilde{x}^{(7)}$	$\tilde{x}^{(8)}$	$\tilde{x}^{(9)}$	$\tilde{x}^{(10)}$
$\tilde{x}^{(1)}$	1	0.93	0.83	0.45	0.39	0.52	0.76	0.72	0.58	0.63
$\tilde{x}^{(2)}$		1	0.89	0.56	0.46	0.61	0.77	0.74	0.63	0.66
$\tilde{x}^{(3)}$			1	0.57	0.32	0.60	0.74	0.64	0.61	0.62
$\tilde{x}^{(4)}$				1	0.62	0.35	0.47	0.70	0.18	0.46
$\tilde{x}^{(5)}$					1	0.21	0.31	0.48	0.05	0.37
$\tilde{x}^{(6)}$						1	0.73	0.61	0.69	0.52
$\tilde{x}^{(7)}$							1	0.80	0.74	0.52
$\tilde{x}^{(8)}$								1	0.54	0.54
$\tilde{x}^{(9)}$									1	0.32
$\tilde{x}^{(10)}$										1

* The intersection of row $\tilde{x}^{(i)}$ and column $\tilde{x}^{(j)}$ gives the correlation coefficient between $\tilde{x}^{(i)}$ and $\tilde{x}^{(j)}$.

Table A2.4b. The covariance matrix of unified particular criteria*

	$\tilde{x}^{(1)}$	$\tilde{x}^{(2)}$	$\tilde{x}^{(3)}$	$\tilde{x}^{(4)}$	$\tilde{x}^{(5)}$	$\tilde{x}^{(6)}$	$\tilde{x}^{(7)}$	$\tilde{x}^{(8)}$	$\tilde{x}^{(9)}$	$\tilde{x}^{(10)}$
$\tilde{x}^{(1)}$	7.04	6.44	6.50	3.48	3.03	3.24	4.39	5.21	3.62	4.44
$\tilde{x}^{(2)}$		6.87	6.90	4.28	3.49	3.76	4.39	5.32	3.85	4.63
$\tilde{x}^{(3)}$			8.80	4.93	2.73	4.22	4.77	5.18	4.26	4.86
$\tilde{x}^{(4)}$				8.58	5.30	2.46	3.02	5.62	1.26	3.59
$\tilde{x}^{(5)}$					8.50	1.44	1.96	3.84	0.37	2.86
$\tilde{x}^{(6)}$						5.62	3.76	3.96	3.81	3.27
$\tilde{x}^{(7)}$							4.77	4.81	3.78	3.04
$\tilde{x}^{(8)}$								7.55	3.45	3.92
$\tilde{x}^{(9)}$									5.49	1.98
$\tilde{x}^{(10)}$										7.09

* The intersection of the row $\tilde{x}^{(i)}$ and column $\tilde{x}^{(j)}$ gives the correlation coefficient between $\tilde{x}^{(i)}$ and $\tilde{x}^{(j)}$.

Table A2.4c. Features of multicollinearity for the variables $\tilde{x}^{(1)}$–$\tilde{x}^{(10)}$ (values R_j^2)*

j	Symbol for the particular criterion: $\tilde{x}^{(j)}$	Symbol for the particular criterion: Codes in WCY [2009]	The name (meaning) of particular criterion	R_j^2
1	$\tilde{x}^{(1)}$	1.1.22	GDP per capita at PPP, USD	0.897
2	$\tilde{x}^{(2)}$	3.1.04	Productivity, USD	0.927
3	$\tilde{x}^{(3)}$	1.1.23	Spendings on private consumption per capita, USD	0.855
4	$\tilde{x}^{(4)}$	4.5.14	The share of the illiterate population, %	0.748
5	$\tilde{x}^{(5)}$	$\left(\frac{2.5.07}{2.5.06}\right)$	20% rate of funds, times	0.504
6	$\tilde{x}^{(6)}$	1.5.01	CPI, %,	0.660
7	$\tilde{x}^{(7)}$	4.4.05	Life expectancy, years	0.823
8	$\tilde{x}^{(8)}$	4.4.07	Infant mortality rate (Avg. number of cases)	0.825
9	$\tilde{x}^{(9)}$	4.4.17	CO_2 emissions, metric tonnes	0.694
10	$\tilde{x}^{(10)}$	4.3.02	Total expenditures on R&D, % of GDP	0.526

* R_j^2 — is the value of the corrected coefficient of determination between $\tilde{x}^{(j)}$ and all other nine particular criteria.

Table A2.4d. Progress of the step-by-step procedure of "serial connection" predictors applied to variable $\tilde{x}^{(1)}-\tilde{x}^{(10)}$– with an a priori excluded $\tilde{x}^{(1)}-\tilde{x}^{(2)}$

Nstep,k	The number of selected predictors,k	The set of selected predictors $\tilde{x}^{(j_1)}, \tilde{x}^{(j_2)}, ..., \tilde{x}^{(j_k)}$	The coefficient of determination $R_k^2 = R_{adj}^2(y_з; \tilde{x}^{(j_1)}, ..., \tilde{x}^{(j_k)})$
1	1	$\tilde{x}^{(3)}$	0.569
2	2	$\tilde{x}^{(3)}$ и $\tilde{x}^{(7)}$	0.629
3	3	$\tilde{x}^{(3)}$, $\tilde{x}^{(7)}$ и $\tilde{x}^{(10)}$	0.675*
4	4		
–	–	Further increase in the number of predictors increased R_k^2 at each step, less than 0.01. and in the long run (i.e. with $k = 8$) to only 0.708.	
–	–		
–	–		
8	8		

* The coefficients of the multiple regression model of $y_{iз}$ on $\tilde{x}_i^{(0)} \equiv 1$, $\tilde{x}^{(3)}$, $\tilde{x}^{(7)}$ and $\tilde{x}^{(10)}$, respectively: 2.418; 0.228; 0.308 and 0.216.

A2.5 The results of the calculated QOL index and ranking countries according to WCY [2009]

i	Country	Values of QOL index			Ranks of the countries on QOL index		
		$\hat{y}_i$(w.l.)	y_{iSt}	$\hat{y}_i$(lear.)	$\hat{y}_i$(w.l.)	y_{iSt}	$\hat{y}_i$(lear.)
1	Argentina	5.134	4.58	6.13	37	47	27
2	Australia	8.023	9.28	9.29	13	3	3
3	Austria	8.451	9.64	8.88	6	2	8
4	Belgium	8.234	8.71	8.28	11	13	12
5	Brazil	3.507	4.79	3.27	45	44	38
6	Bulgaria	4.320	3.53	2.48	40	52	45
7	Canada	8.094	9.27	8.76	12	4	10
8	Chile	4.686	6.68	4.11	39	27	32
9	China Mainland	3.015	4.72	3.68	50	45	36
10	Colombia	3.243	4.82	2.10	47	42	48
11	Croatia	5.960	5.58	4.22	31	37	31
12	Czech Republic	6.494	7.59	5.67	27	20	29
13	Denmark	8.399	8.98	9.13	7	8	5
14	Estonia	5.694	6.2	4.05	33	30	33
15	Finland	7.896	8.12	7.25	15	18	18
16	France	8.499	8.17	9.12	4	16	6
17	Germany	8.252	8.89	8.80	10	11	9
18	Greece	7.151	6.18	6.19	21	31	26
19	Hong Kong	6.934	6.59	6.68	24	28	21
20	Hungary	5.973	5.25	3.85	30	38	35
21	Indonesia	2.567	4.26	0.79	54	50	53
22	Ireland	8.479	8.39	8.18	5	15	13
23	Israel	6.362	6.91	4.65	29	23	30
24	Italy	7.700	6.86	7.32	18	25	17
25	Japan	7.828	6.9	7.78	17	24	16
26	Jordan	3.117	5.66	2.26	49	36	47
27	Kazakhstan	2.852	4.45	0.00	53	48	56
28	Korea	6.700	6.18	6.21	26	31	25
29	Lithuania	5.433	5.22	3.07	35	39	40
30	Luxembourg	9.339	9.21	9.30	1	5	2
31	Malaysia	3.945	7.43	2.30	43	21	46
32	Mexico	3.503	4.63	2.91	46	46	42
33	Netherlands	8.320	8.91	8.16	9	9	14
34	New Zealand	6.998	8.91	6.53	23	9	23
35	Norway	9.227	9.16	9.28	2	6	4
36	Peru	2.893	4.33	1.79	52	49	50
37	Philippines	2.542	4.81	0.39	55	43	54
38	Poland	5.692	4.26	3.49	34	50	37

(continued)

(continued)

i	Country	Values of QOL index			Ranks of the countries on QOL index		
		$\hat{y}_i$(w.l.)	y_{iSt}	$\hat{y}_i$(lear.)	$\hat{y}_i$(w.l.)	y_{iSt}	$\hat{y}_i$(lear.)
39	Portugal	6.443	6.12	6.12	28	33	28
40	Qatar	5.432	8.13	3.97	36	17	34
41	Romania	4.795	2.68	2.74	38	55	43
42	Russia	4.287	2.89	2.01	41	53	49
43	Singapore	7.377	8.89	8.07	20	11	15
44	Slovak Republic	5.859	6.3	3.27	32	29	39
45	Slovenia	7.034	6.81	6.29	22	26	24
46	South Africa	1.680	5.08	0.06	56	40	55
47	Spain	7.427	7.71	7.11	19	19	20
48	Sweden	7.862	9.08	7.19	16	7	19
49	Switzerland	8.726	9.71	10.00	3	1	1
50	Taiwan	6.853	5.84	6.60	25	35	22
51	Thailand	3.993	5.93	1.57	42	34	51
52	Turkey	3.690	5	2.93	44	41	41
53	Ukraine	3.132	2.73	1.45	48	54	52
54	United Kingdom	8.006	7.16	8.41	14	22	11
55	USA	8.378	8.53	8.94	8	14	7
56	Venezuela	2.967	2.44	2.61	51	56	44

Rank correlation coefficient: between rankings by y_3 and $\hat{y}$ (lear.) **0.8545**; and between the rankings for $y_{St.}$ and **0.8517.**

A2.6 A priori and a posteriori partial criteria for each of the categories of synthetic quality of life of the Russian Federation regions

No	The source code of the particular criterion	Synthetic category section index	Source*	Number of the variable in a posteriori set**	Formula for unifying transformation
1	2	3	4	5	6
		1. The quality of population			
		1.1. *Reproduction, demographic structure and physical health*			
1	1.1.1.	Percentage of the urban population	[I1]		
2	1.1.2.	Percentage of men in the general population	[I1]		
3	1.1.3.	Number of births per 1000 of people in population	[I1]		
4	1.1.4.	Number of deaths per 1000 of people in population	[I1]		
5	1.1.5.		[I4]	2.1	(2.2)
6	1.1.6.	The number of deaths under 1 year age per 1000 live births (infant mortality-ness)	[I4]	1.2	(2.3)
7	1.1.7.	Share (%) of people below working age	[I4]		
8	1.1.8.	Share (%) of persons in working age (1998)	[I4]		
9	1.1.9.	Share (%) of persons in retirement age	[I4]		
10	1.1.10.	Dependency ratio per 1000 people of working age in total	[I4]		
11	1.1.11.	Dependency ratio by persons below working age per 1000 people of working age	[I4]		
12	1.1.12.	Dependency ratio by persons older than working age per 1000 people of working age	[I4]		

(continued)

(continued)

No	The source code of the particular criterion	Synthetic category section index	Source*	Number of the variable in a posteriori set**	Formula for unifying transformation
1	2	3	4	5	6
13	1.1.13.	Life expectancy (at birth): the whole population – both sexes	[14]	1.1	(2.2)
14	1.1.14.	Life expectancy (at birth): the whole population – men	[14]		
15	1.1.15.	Life expectancy (at birth): the whole population – women	[14]		
16	1.1.16.	Life expectancy at birth: city – both sexes	[14]		
17	1.1.17.	Life expectancy at birth: city – men	[14]		
18	1.1.18.	Life expectancy at birth: city – women	[14]		
19	1.1.19.	Life expectancy at birth: village – both sexes	[14]		
20	1.1.20.	Life expectancy at birth: village – men	[14]		
21	1.1.21.	Life expectancy at birth: village – women	[14]		
22	1.1.22.	The number of illnesses reported in patients with newly diagnosed, per 1000 people	[14]		
23	1.1.23.	The number of deaths from infectious and parasitic diseases per 1000 people	[14]	1.3	(2.3)
24	1.1.24.	The number of deaths from tuberculosis per 1000 people	[14]		
25	1.1.25.	Number of deaths from cancer per 100.000 people	[14]	2.2	(2.3)
26	1.1.26.	The number of deaths from diseases of the cardiovascular system per 100.000 people	[14]	2.3	(2.3)

(continued)

No	The source code of the particular criterion	Synthetic category section index	Source*	Number of the variable in a posteriori set**	Formula for unifying transformation
1	2	3	4	5	6
27	1.1.27.	The number of deaths from respiratory diseases per 100.000 people	[I4]	1.4	(2.3)
28	1.1.28.	The number of deaths from diseases of the digestive system per 100.000 people	[I4]	1.5	(2.3)
29	1.1.29.	The number of deaths from accidents, injury and poisoning per 100.000 people	[I4]	1.6	(2.3)
30	1.1.30.	The number of persons with disabilities to 1000 people	[I4]	2.4	(2.3)
31	1.1.31.	Registered cases of congenital anomalies (malformations) per 1000 people	[I6], [I8]	1.7	(2.3)

* See "Information materials" in References.
** The number in the 1st rank in the variable number posteriori set defines the block number, and in the 2nd, the variable number in a given block.

		1.2. The ability to form and maintain a family			
32	1.2.1.	Registered in the year of marriages per 1000 people aged from 16 to 45 years old	[I1]		
33	1.2.2.	Divorces per 100 marriages	[I1]		
		1.3. The level of education and culture			
34	1.3.1.	The proportion of pupils at daytime general educational institutions in the total number of children aged 7–17 years, %	[I5]		
35	1.3.2.	The proportion of students in secondary specialized educational institutions as part of young people aged 16–29 years, %	[I5]	3.2	(2.2)

(continued)

(continued)

No	The source code of the particular criterion	Synthetic category section index	Source*	Number of the variable in a posteriori set**	Formula for unifying transformation
1	2	3	4	5	6
36	1.3.3.	The proportion of university students in the population aged 16–34 years, %	[I5]		
37	1.3.4.	Students in specialized secondary educational institutions per 1000 people employed in the economy	[I6], [I8]		
38	1.3.5.	College students by 1000 employment in the economy	[I6], [I8]		
39	1.3.6.	Percentage with a higher special education among the employed in the economy	[I6], [I8]	3.1	(2.2)
		1.4. The skill level of the population			
40	1.4.1.	"Present" labour productivity $\left(\frac{\text{thous.rub.}}{\text{people, year}}\right)$ (in unified for all regions of the standardized structure of employment)	[I1], [I5]	3.3	(2.2)
		2. The level of well-being			
		2.1. Population incomes and expenditures			
41	2.1.1.	Gross regional product (GRP) per capita, $\left(\frac{\text{thous.rub.}}{\text{people, year}}\right)$	[I1], [I3]	1.1	(2.2)
42	2.1.2.	CPI, %	[I1], [I3]		
43	2.1.3.	Purchasing power per capita income relative to the subsistence minimum sets	[I1], [I3]	1.2	(2.2)
44	2.1.4.	The share of the population with income below the subsistence minimum, %	[I1], [I3]	1.3	(2.3)

(continued)

No	The source code of the particular criterion	Synthetic category section index	Source*	Number of the variable in a posteriori set**	Formula for unifying transformation
1	2	3	4	5	6
45	2.1.5.	Ratio of funds	[I1], [I3]	1.4	(2.4)
46	2.1.6.	The share of expenditure on food in the consumption expenditure of the population, %	[I1], [I3]		
47	2.1.7.	The volume of retail trade turnover and paid services per capita (adjusted for purchasing power), thous. rub	[I1], [I3]	1.6	(2.2)
		2.2. Housing and property			
48	2.2.1.	Total area of housing per capita, sq. m.	[I3]	2.1	(2.2)
49	2.2.2.	The proportion of the total area of housing privately owned by citizens, %	[I3]		
50	2.2.3.	The share of housing housed in dilapidated buildings, %	[I3]	2.3	
51	2.2.4.	Commissioning of apartment houses (total area per capita)	[I3]	2.2	(2.3)
52	2.2.5.	The proportion of families who are registered for housing, %	[I3C]	1.5	
53	2.2.6.	Provision of the population by own cars (per 1000 people)	[I3]		(2.2)
		2.3. Social infrastructure capacity			
54	2.3.1.	The density of public roads paved (km/ 1000 m² area)	[I3]	2.4	(2.2)

(continued)

(continued)

No	The source code of the particular criterion	Synthetic category section index	Source*	Number of the variable in a posteriori set**	Formula for unifying transformation
1	2	3	4	5	6
55	2.3.2.	The level of teledensity (the weighted sum of the numbers of telephone sets per 100 households in urban and rural areas)	[13]		
		3. The quality of social services			
		3.1. Working conditions			
56	3.1.1.	The purchasing power of the average wage, subsistence of the minimum sets	[13]		
57	3.1.2.	Unemployment rate, %	[15]	1.1	(2.3)
58	3.1.3.	The level of the economic activity, %	[15]		
59	3.1.4.	The number of victims of fatal outcome or disability at one time or more per 1000 workers, people	[15], [18]	2.4	(2.3)
60	3.1.5.	Arrears of wages per employed in the economy in relation to the subsistence minimum, %	[15], [18]	1.4	(2.3)
61	3.1.6.	The proportion of workers who have at least one kind of benefit or compensation for work in adverse working conditions (in industry, construction, transport, communications), %	[15], [18]		
62	3.1.7.	The purchasing power of the average granted pension, (the ratio of the average pension to the subsistence minimum)	[15], [18]	4.1	(2.2)
63	3.1.8.	Percentage of industrial workers employed in hazardous work conditions	[15]	2.3	(2.3)

(continued)

No	The source code of the particular criterion	Synthetic category section index	Source*	Number of the variable in a posteriori set**	Formula for unifying transformation
1	2	3	4	5	6
64	3.1.9.	Load unemployed per vacancy	[I5]	1.2	(2.3)
65	3.1.10.	The share of long-term unemployed	[I5]	1.3	(2.3)
		3.2. Physical, material and social security of the members of society			
66	3.2.1.	The number of reported crimes per 100.000 people	[I15], [I8]		
67	3.2.2.	Deaths from homicide (per 100.000 people)	[I15], [I8]		
68	3.2.3.	The number of reported homicides and attempted murders per 100.000 people	[I15], [I8]	2.5	(2.3)
69	3.2.4.	The number of registered facts of intentional infliction of serious bodily injury per 100.000 people	[I15], [I8]	2.6	(2.3)
70	3.2.5.	The number of reported rapes and attempted rapes by 100.000 people	[I15], [I8]	2.7	(2.3)
71	3.2.6.	The number of robberies, robberies, thefts from apartments of citizens per 100.000 people	[I15], [I8]	2.8	(2.3)
72	3.2.7.	The number of deaths in road accidents per 100.000 people	[I15], [I8]		
73	3.2.8.	The total number of registered crimes per 100.000 people	[I15], [I8]		
74	3.2.9.	Total number of incidents of bullying per 100.000 people	[I15], [I8]		

(continued)

(continued)

No	The source code of the particular criterion	Synthetic category section index	Source*	Number of the variable in a posteriori set**	Formula for unifying transformation
1	2	3	4	5	6
75	3.2.10.	The depth of poverty	[I1], [I3]	1.5	(2.3)
76	3.2.11.	Ratio of population to the income of the poorest 10% of the population	[I1], [I3]	3.3	(2.3)
77	3.2.12.	The average value of social benefits (the total amount of social payments, divided by the number of them receiving)	[I6], [I8]	4.2	(2.2)
78	3.2.13.	The ratio of the number of seats in the centres of social services to the total number of those in need	[I6], [I8]	4.3	(2.2)
79	3.2.14.	Reaching those in need for social services (elderly and disabled home)	[I6], [I8]	4.4	(2.2)
80	3.3.1.	The number of registered illegal embezzlement per 100.000 people	[I6], [I8]	2.9	(2.3)
81	3.3.2.	The number of persons employed in the dispensary about substance abuse based on 100.000 people	[I15], [I8]	3.1	(2.3)
82	3.3.3.	The number of persons employed in the dispensary about addiction per 100.000 people	[I15], [I8]	3.1	(2.3)
83	3.3.4.	The number of persons employed in the dispensary about alcoholism and alcoholic psychosis per 100.000 people	[I15], [I8]	2.2	(2.3)

(continued)

No	The source code of the particular criterion	Synthetic category section index	Source*	Number of the variable in a posteriori set**	Formula for unifying transformation
1	2	3	4	5	6
		3.3. Characteristics of social pathology			
84	3.3.5.	Died from suicides per 100.000 people	[I15], [I8]	2.10	(2.3)
85	3.3.6.	The prevalence of social diseases (tuberculosis, syphilis, viral hepatitis)	[I15], [I8]	2.1	(2.3)
86	3.3.7.	The number of registered HIV-infected people (100.000 people)	[I15], [I8]	3.2	(2.3)
		3.4. Social and territorial mobility of the population			
87	3.4.1.	The immigration rate (10.000 people)	[I1]	1.6	(2.2)
		3.5. The socio-political health of society			
88	3.5.1.	Percentage of population who took part in the last elections of the federal or regional level	[I22]		
		4. The quality of the ecological niche			
		4.1. Air			
89	4.1.1.	The mass of pollutants released into the atmosphere from stationary sources, an average of 1 sq. M. km area of the region, million tonnes/year	[I7], [I8]	1.1	(2.3)
90	4.1.2.	The mass of pollutants released into the atmosphere from stationary sources in the average per capita, kg/year	[I7], [I8]		
91	4.1.3.	Percentage of release into the atmosphere of	[I7], [I8]		

(continued)

(continued)

No	The source code of the particular criterion	Synthetic category section index	Source*	Number of the variable in a posteriori set**	Formula for unifying transformation
1	2	3	4	5	6
		harmful substances in the total mass of pollutants emitted from stationary sources of their separation, %			
92	4.1.4.	The proportion of sulphur dioxide in the total mass of emissions of harmful substances into the atmosphere from stationary sources, %	[17], [18]		
93	4.1.5.	The proportion of carbon monoxide in the total mass emissions of harmful substances into the atmosphere from stationary sources, %	[17], [18]		
94	4.1.6.	The proportion of nitrogen oxides in the total mass of emissions of harmful substances into the atmosphere from stationary sources, %	[17], [18]		
95	4.1.7.	Sulphur dioxide emitted into the atmosphere from stationary sources in the average per capita kg/year	[17], [18]		
96	4.1.8.	Carbon monoxide released into the atmosphere from stationary sources in the average per capita kg/year	[17], [18]		

(continued)

No	The source code of the particular criterion	Synthetic category section index	Source*	Number of the variable in a posteriori set**	Formula for unifying transformation
1	2	3	4	5	6
97	4.1.9.	Nitrogen oxides released into the atmosphere from stationary sources in the average per capita kg/year	[17], [18]		
		4.2. Water			
98	4.2.1.	The proportion of water used in the total volume of water withdrawn from natural sources, %	[17], [18]		
99	4.2.2.	The proportion of fresh water used for drinking and sanitary needs in the total volume of water used, %			
100	4.2.3.	The proportion of fresh water used for production needs in the total volume of water used, %	[17], [18]		
101	4.2.4.	The proportion of fresh water used for irrigation, watering and agricultural water supply in the total volume of water used, %	[17], [18]		
102	4.2.5.	The ratio of the volume of wastewater discharged into surface waters to the volume of water taken from water bodies, %	[17], [18]		
103	4.2.6.	Share discharged into surface waters contaminated with sewage in the total volume of water withdrawn from water bodies in the region	[17], [18]	1.3	(2.3)

(continued)

(continued)

No	The source code of the particular criterion	Synthetic category section index	Source*	Number of the variable in a posteriori set**	Formula for unifying transformation
1	2	3	4	5	6
104	4.2.7.	The share of standard-treated water in the total volume of wastewater discharged into surface waters, %	[17], [18]		
105	4.2.8.	Fresh water used for drinking and sanitary needs, the average per capita m^3/year	[17], [18]		
106	4.2.9.	Discharged into surface water bodies of the polluted water per 1000 km^2 in the region, bln. m^3/year	[17], [18]	1.2	(2.3)
		4.3. Soil			
107	4.3.1.	Formed in enterprises per year of toxic waste production (TWP), the average 1000 tonnes per 1000 km^2	[17], [18]	1.4	(2.3)
108	4.3.2.	Specific total value of TWP in storage in the region at the end of the year, 1000 tonnes per 1000 km^2		1.5	(2.3)
		4.4. Biodiversity and natural ecosystems			
109	4.4.1.	The proportion of the area of nature reserves, hunting reserves and national parks in the region's total area	[17], [18]	3.1	(2.2)
110	4.4.2.	Nature reserves, hunting reserves and national parks in the average per capita, ha	[17], [18]		

(continued)

No	The source code of the particular criterion	Synthetic category section index	Source*	Number of the variable in a posteriori set**	Formula for unifying transformation
1	2	3	4	5	6
111	4.4.3.	The proportion of the area of nature reserves, hunting reserves and national parks in the regions of the Russian Federation, the total area of nature reserves, hunting reserves and national parks of Russia, %	[17], [18]		
		4.5. Status of natural ecosystems			
112	4.5.1.	The area sown and planted forests (minus the area of dead forest plantations), divided by the total forest area in the region	[17], [18]	2.1	(2.2)
113	4.5.2.	Area calves, entered into the category of valuable forest species, divided by the total forest area	[17], [18]	2.2	(2.2)
114	4.5.3.	The share of the burned area in the forest fires in the total forest area	[17], [18]	2.3	(2.3)
115	4.5.4.	The share of re-cultivated land (net of "violations" of the ground) in the total area of farmland	[17], [18]	4.1	(2.2)

A2.7 Initial statistics posterior set of indicators (in initial and unified form) in regions of Russia

A2.7a The quality of population, 2003. 1 part

		Life expectancy		Infant mortality		Rate of natural increase		Deaths from infections and tuberculosis		Mortality from tumours		Deaths from cardiovascular diseases		Deaths from respiratory diseases	
	Formula for unification	(2.2)		(2.3)		(2.2)		(2.3)		(2.3)		(2.3)		(2.3)	
1	2	3		4		5		6		7		8		9	
		x	$\tilde{x}$	x	$\tilde{x}$	x	$\tilde{x}$	x	$\tilde{x}$	x	$\tilde{x}$	x	$\tilde{x}$	x	$\tilde{x}$
1	Adygeya	67.5	9.0	10.8	7.8	−5.1	5.2	50.9	3.8	189.2	6.6	947.8	4.4	51.1	8.2
2	Altai	60.2	0.6	24	0.3	1.1	8.1	56.6	3.3	161	7.9	703.1	8.4	110.8	1.1
3	Altai Territory	66.1	7.7	11.3	7.3	−5.1	5.2	72.8	1.6	206.7	4.1	880.7	5.4	98.5	1.8
4	Amur Region	61.3	1.6	17.2	1.0	−4.2	5.6	85.8	0.8	167.4	7.6	821.6	6.6	75.1	4.8
5	Arkhangelsk Region	62.1	2.6	12.4	5.3	−7.1	4.2	40.6	6.3	188.9	6.6	1001.3	3.6	62.3	6.9
6	Astrakhan Region	65.4	6.4	12.6	4.8	−3.5	5.9	88.9	0.6	204.3	4.5	809.9	6.9	49	8.4
7	Bashkortostan	66.4	8.3	12.5	5.0	−3.1	6.1	27.6	8.6	149.1	8.6	818.8	6.7	78.4	4.3
8	Belgorod Region	68.3	9.4	9.4	8.9	−6.9	4.3	18.9	9.5	187.4	6.7	1075.2	2.9	51.1	8.2
9	Bryansk Region	64.6	5.6	13	4.1	−10	2.8	58.8	3.1	218.3	2.6	1238.1	1.5	67.5	6.2
10	Buryatia	61.3	1.6	14.3	2.3	−1.9	6.7	56.5	3.3	168.4	7.6	708.7	8.4	89.8	2.5
11	Chelyabinsk Region	64.8	5.7	12.7	4.5	−6.1	4.7	42.5	5.8	215.8	2.9	907.5	4.9	93.8	2.1
12	Chita Region	60.2	0.6	16.3	1.2	−3.3	6.0	67	2.3	163.2	7.8	817.8	6.7	123.5	0.4
13	Chukotka Autonomous Area	59.1	0.0	28	0.0	2.3	8.7	17.2	9.6	128.1	9.3	374.8	9.5	68.8	5.9
14	Chuvash	66.1	7.7	8.8	9.5	−5.2	5.1	29.2	8.3	142.8	8.9	767	7.8	145.1	0.0
15	Daghestan	72.3	10.0	14.5	2.1	9.9	7.7	30.2	8.0	70.5	10.0	300.8	9.7	69.6	5.6

16	Irkutsk Region	60.7	1.0	13.7	3.1	−4.6	5.4	83.7	1.0	180.3	7.1	797.5	7.2	96.1	2.0
17	Ivanovo Region	62.4	2.9	12.5	5.0	−13.3	1.2	41.3	6.1	242.3	0.9	1360.8	0.5	110	1.1
18	Jewish Autonomous Region	60.8	1.1	22.3	0.4	−4.1	5.6	113	0.0	179.3	7.2	889.1	5.3	51	8.2
19	Kabardino-Balkarian	68.7	9.5	14.9	1.7	0.1	7.7	45.3	4.7	109.2	9.6	611.7	8.8	44	8.9
20	Kaliningrad Region	61.7	2.2	14.9	1.7	−8.7	3.4	77.7	1.3	195.9	5.5	866.2	5.7	66.9	6.3
21	Kalmykia	66.2	7.9	13.9	2.8	1.5	8.3	67.7	2.2	142.5	8.9	440.9	9.4	57.7	7.4
22	Kaluga Region	64.6	5.5	11.8	6.3	−10.2	2.7	50.1	3.9	236	1.3	1150.4	2.2	76.8	4.5
23	Kamchatka Region	63.0	3.7	14.3	2.3	−1.4	6.9	45.8	4.5	155.5	8.4	583.5	8.9	41.8	9.5
24	Karachayevo-Circassian	68.1	9.3	12.8	4.4	−0.8	7.2	43.6	5.5	149.6	8.6	725.4	8.2	64.2	6.7
25	Karelia	60.8	1.1	8.1	9.8	−9.7	3.0	46.3	4.4	209.2	3.6	1085	2.8	74.2	5.0
26	Kemerovo Region	61.9	2.3	11.8	6.3	−7.8	3.9	90.3	0.5	210	3.5	950.1	4.4	89.8	2.5
27	Khabarovsk Territory	61.9	2.3	15.6	1.4	−5.5	5.0	67.5	2.2	193.1	6.0	889.7	5.2	69.5	5.6
28	Khakassia	60.8	1.1	18.4	0.9	−6	4.7	84.9	0.9	195.6	5.6	886.4	5.3	88.2	2.7
29	Kirov Region	63.7	4.4	13.2	3.9	−10.4	2.6	32.3	7.4	203.2	4.8	1146.7	2.2	118	0.6
30	Komi	61.7	2.1	9.4	8.9	−4.3	5.6	38.2	6.8	164.6	7.7	752.9	7.9	70.5	5.4
31	Kostroma Region	62.4	2.9	13.4	3.5	−12.2	1.8	27.1	8.7	219.2	2.5	1248.9	1.4	84.2	3.4
32	Krasnodar Territory	67.3	8.9	9.5	8.8	−5.4	5.0	51.8	3.7	204.6	4.4	928	4.7	49.2	8.4
33	Krasnoyarsk Territory	63.0	3.8	14.1	2.5	−4.7	5.4	69.7	1.9	191.4	6.3	742	8.0	78.6	4.2
34	Kurgan Region	63.9	4.7	14.5	2.1	−7.7	3.9	81	1.1	224.9	2.1	805.9	7.0	105.9	1.4
35	Kursk Region	65.4	6.4	12.5	5.0	−10.9	2.4	45.6	4.6	215.2	2.9	1291.3	1.0	113.3	0.9
36	Leningrad Region	61.7	2.2	9.3	9.0	−13.1	1.3	75.1	1.5	233.2	1.6	1264.6	1.3	81.4	3.8
37	Lipetsk Region	65.9	7.2	10.5	8.3	−9.6	3.0	26.2	8.9	198.7	5.3	1227.4	1.6	74.1	5.0
38	Magadan Region	63.0	3.8	11.9	6.1	−1.4	6.9	28.9	8.4	179.9	7.1	565.2	8.9	81.1	3.9
39	Mari El	64.0	5.0	10.8	7.8	−6	4.7	24.4	9.1	158.9	8.1	835.4	6.3	130.6	0.2
40	Mordovia	66.3	8.1	10.6	8.2	−8.8	3.4	26.3	8.9	207.1	4.0	1051.6	3.1	86.1	3.0
41	Moscow	69.6	9.8	11.5	6.9	−4.7	5.4	22.9	9.2	262.3	0.3	947	4.4	41.9	9.4
42	Moscow Region	65.3	6.1	11.5	6.9	−9.1	3.3	36.5	6.9	248.7	0.6	1053.1	3.1	57	7.5
43	Murmansk Region	63.2	4.0	8.9	9.4	−4	5.7	21.8	9.3	158.2	8.2	761.6	7.8	56.4	7.7
44	Nizhny Novgorod Region	64.0	5.0	13.5	3.3	−11	2.3	48.9	4.1	232.7	1.7	1280.5	1.1	70.8	5.4
45	North Ossetia – Alania	68.5	9.5	10.6	8.2	−1.3	7.0	52.1	3.7	154	8.4	797.3	7.2	35.9	10.0

(continued)

(continued)

		Life expectancy		Infant mortality		Rate of natural increase		Deaths from infections and tuberculosis		Mortality from tumours		Deaths from cardiovascular diseases		Deaths from respiratory diseases	
	Formula for unification	(2.2)		(2.3)		(2.2)		(2.3)		(2.3)		(2.3)		(2.3)	
1	2	3		4		5		6		7		8		9	
		x	$\tilde{x}$	x	$\tilde{x}$	x	$\tilde{x}$	x	$\tilde{x}$	x	$\tilde{x}$	x	$\tilde{x}$	x	$\tilde{x}$
46	Novgorod Region	60.8	1.1	12.8	4.4	−14.3	0.8	68.9	2.0	237.4	1.2	1460.6	0.1	85.1	3.2
47	Novosibirsk Region	65.8	7.2	11.9	6.1	−4.9	5.3	75.1	1.5	217	2.8	802.1	7.1	59.2	7.2
48	Omsk Region	66.0	7.6	11.9	6.1	−4.1	5.6	58.8	3.1	211.4	3.3	796.2	7.3	84.7	3.3
49	Orel Region	65.2	6.1	11.1	7.5	−9.9	2.9	16.5	9.6	222	2.3	1170	2.1	91.4	2.3
50	Orenburg Region	65.4	6.4	12.7	4.5	−4.4	5.5	39.3	6.7	202.1	5.0	863.8	5.7	70.6	5.4
51	Penza Region	65.7	7.0	15.1	1.6	−9.5	3.1	31.1	7.7	204.6	4.4	1126.4	2.4	69	5.8
52	Perm Region	62.3	2.8	13.9	2.8	−7	4.3	49.6	4.0	192.9	6.0	962.4	3.9	92	2.3
53	Primorye Territory	63.1	3.9	17.7	0.9	−5.1	5.2	95.2	0.4	196.3	5.5	823.3	6.6	69.4	5.6
54	Pskov Region	60.9	1.2	16.4	1.1	−15.9	0.0	50.6	3.9	227.8	1.9	1561.4	0.0	101.6	1.6
55	Rostov Region	66.5	8.4	14.2	2.4	−7.1	4.2	68.1	2.1	206	4.2	954.4	4.2	42.9	9.2
56	Ryazan Region	63.9	4.8	13.3	3.7	−12.9	1.4	43.1	5.6	251.2	0.5	1159	2.1	94.8	2.1
57	Sakha (Yakutia)	64.0	4.9	13.2	3.9	4.8	9.9	29.3	8.2	131.7	9.2	414.3	9.5	39.1	9.8
58	Sakhalin Region	61.4	1.8	12.5	5.0	−5.5	5.0	41.5	6.1	201.5	5.1	842.5	6.1	67.4	6.2
59	Samara Region	65.6	6.9	9	9.3	−6.6	4.4	42.9	5.7	190.4	6.4	853.7	5.9	55.8	7.8
60	Saratov Region	65.5	6.7	13.7	3.1	−7.9	3.8	40.9	6.2	208.1	3.8	964.2	3.9	42.3	9.3
61	Smolensk Region	62.3	2.8	11.5	6.9	−13.3	1.2	64.8	2.6	216.2	2.8	1335.6	0.7	83.1	3.5
62	St. Petersburg	66.5	8.4	8	9.9	−8	3.8	38.9	6.7	275.4	0.0	1028.2	3.4	49	8.4
63	Stavropol Territory	67.1	8.8	11.5	6.9	−4.7	5.4	44.6	5.0	194	5.9	902.2	5.0	47.3	8.6
64	Sverdlovsk Region	64.0	4.9	12.2	5.7	−6.9	4.3	54.9	3.4	207.8	3.9	935.2	4.6	78	4.3

65	Tambov Region	65.5	6.7	11.7	6.5	−11.5	2.1	39.8	6.6	242.3	0.9	1206.4	1.8	105.9	1.4
66	Tatarstan	67.8	9.2	10.8	7.8	−3.6	5.9	27.4	8.7	178.1	7.2	843.1	6.1	58.1	7.3
67	Tomsk Region	64.4	5.4	17.2	1.0	−3.9	5.7	41.2	6.2	202.6	4.9	649.7	8.6	56.1	7.8
68	Tula Region	63.0	3.8	12.1	5.8	−14.7	0.6	65.3	2.5	267.7	0.2	1284.4	1.1	120.4	0.5
69	Tuva	54.3	0.0	27.6	0.1	5.4	9.8	145.1	0.0	115	9.5	501.3	9.2	79.1	4.2
70	Tver Region	61.5	2.0	13.5	3.3	−14.9	0.5	45.4	4.7	247.1	0.7	1525.2	0.0	104.2	1.5
71	Tyumen Region	66.1	7.7	9.6	8.8	3	9.0	43.7	5.4	129.1	9.2	458	9.3	43.6	9.0
72	Udmurtian	64.3	5.3	13.1	4.0	−4.2	5.6	44	5.3	159	8.1	792.3	7.3	112.1	1.0
73	Ulyanovsk Region	65.3	6.1	12.2	5.7	−8.4	3.6	31.7	7.6	210.3	3.4	1062.8	3.0	65.5	6.6
74	Vladimir Region	63.2	4.0	9.3	9.0	−11.5	2.1	43.6	5.5	233.8	1.6	1299	1.0	76.6	4.6
75	Volgograd Region	66.4	8.3	14.7	1.9	−6.8	4.4	64.9	2.6	222.6	2.2	967.2	3.9	62.1	6.9
76	Vologda Region	62.5	2.9	12.6	4.8	−9.4	3.1	31.2	7.7	222.4	2.2	1094.2	2.7	83.6	3.5
77	Voronezh Region	66.2	7.9	10.9	7.7	−10.7	2.5	30.4	7.9	192.5	6.1	1000	3.6	68.1	6.1
78	Yaroslavl Region	63.5	4.2	10.9	7.7	−11.7	2.0	31.5	7.6	236.6	1.3	1148.2	2.2	84.9	3.2

A2.7a The quality of population, 2003. 2 part

		Mortality from diseases of the digestive system		Prevalence of unnatural death		Number of the disabled		Congenital anomalies		The proportion of specialists with higher education among the employed in the economy		The reduced productivity (GRP/employment in the economy)		The proportion of students in secondary and higher education among young people	
1	2	3		4		5		6		7		8		9	
	Formula for unification	(2.3)		(2.3)		(2.3)		(2.3)		(2.2)		(2.2)		(2.2)	
		x	$\tilde{x}$	x	$\tilde{x}$	x	$\tilde{x}$	x	$\tilde{x}$	x	$\tilde{x}$	x	$\tilde{x}$	x	$\tilde{x}$
1	Adygeya	55.4	5.1	184	8.1	83.1	3.8	1.3	7.2	23.7	7.8	65.0	0.1	19.7	5.4
2	Altai	33.5	9.5	405.7	0.3	93.5	1.8	2.3	1.7	25.9	8.9	94.8	2.5	17.2	2.5
3	Altai Territory	37.7	9.2	235.2	6.1	75.1	4.5	2.3	1.7	21.1	5.4	82.4	1.5	16.8	2.2
4	Amur Region	52.6	6.1	342.4	1.5	64.8	6.7	1.5	5.6	19.8	3.7	127.0	5.8	19.5	4.9
5	Arkhangelsk Region	55.5	5.0	311.3	2.8	78.2	4.2	3.3	0.5	17.5	1.1	171.9	7.7	17.9	3.0
6	Astrakhan Region	45	8.4	210.1	7.3	41.9	9.1	1.5	5.6	21.4	5.8	123.9	5.3	19.3	4.4
7	Bashkortostan	34.7	9.4	197.7	7.7	46.2	8.9	1.1	8.0	16.7	0.6	156.8	7.2	20.6	6.8
8	Belgorod Region	37.7	9.2	163.4	8.7	105.1	0.9	1.3	7.2	20.1	4.2	119.2	4.7	20.2	6.2
8	Bryansk Region	57.8	4.5	233	6.2	85.2	3.4	0.8	8.9	20.2	4.5	80.7	1.2	18.7	3.9
9	Buryatia	48.4	7.6	409.3	0.3	66.7	6.1	1.3	7.2	20.5	4.9	122.8	5.0	15.5	1.4
10	Chelyabinsk Region	55	5.3	267.2	4.3	54.8	8.1	2	2.8	21.7	6.2	141.0	6.7	21.1	7.2
11	Chita Region	53.1	6.0	401.8	0.4	67.3	6.0	2.1	2.3	19.0	2.5	118.2	4.6	16.0	1.6
12	Chukotka Autonomous Area	42.1	8.8	353.7	1.0	19.5	9.9	1	8.3	22.2	6.7	537.5	10.0	5.0	0.0

13	Chuvash	60.1	3.9	281.3	3.7	69	5.7	3.8	0.2	20.2	4.5	82.7	1.5	23.5	8.6
14	Daghestan	22.2	9.9	53.4	9.8	61.5	7.2	4.1	0.1	19.3	3.0	69.8	0.4	12.6	0.5
15	Irkutsk Region	78.3	1.1	350.8	1.1	70.7	5.3	1.6	4.4	22.5	7.1	152.4	7.1	20.6	6.7
16	Ivanovo Region	78.1	1.1	274.2	3.9	71.7	5.0	1.7	3.9	19.5	3.3	75.6	0.7	22.5	7.9
17	Jewish Autonomous Region	69.4	2.5	269.2	4.2	63.2	7.0	2.4	1.4	16.4	0.4	107.4	3.7	16.7	2.1
19	Kabardino-Balkarian	56.7	4.8	98.8	9.5	46.7	8.8	1.5	5.6	29.6	9.5	91.0	2.2	13.2	0.7
20	Kaliningrad Region	80.8	0.8	336.8	1.7	67.4	6.0	1.3	7.2	24.2	8.4	122.7	5.0	14.5	1.1
21	Kalmykia	54.6	5.6	175.8	8.3	61.9	7.2	1.4	6.5	25.6	8.9	81.8	1.4	19.6	5.2
22	Kaluga Region	66.1	3.1	221.8	6.6	86.5	3.1	1.2	7.7	21.9	6.4	111.2	3.9	17.4	2.6
23	Kamchatka Region	51.4	6.4	235.8	6.1	33.8	9.5	2	2.8	21.5	5.9	161.4	7.4	20.0	6.0
24	Karachayevo-Circassian	47.8	7.8	135.5	9.2	87	3.0	0.8	8.9	27.0	9.1	78.3	1.0	14.6	1.1
25	Karelia	55.5	5.0	336.3	1.8	103	1.1	2.3	1.7	18.4	1.6	136.1	6.4	18.3	3.4
26	Kemerovo Region	65.3	3.2	354	1.0	59.5	7.7	2.3	1.7	19.0	2.5	134.4	6.3	18.7	3.9
27	Khabarovsk Territory	82.5	0.6	310.5	2.8	58.9	7.8	2	2.8	27.3	9.2	168.0	7.6	24.5	9.0
28	Khakassia	71.3	2.1	352.4	1.1	66.3	6.3	2.7	0.9	20.7	5.2	119.4	4.7	19.7	5.5
29	Kirov Region	42.8	8.7	339.8	1.6	76.4	4.4	0.9	8.6	17.9	1.3	87.3	1.9	21.0	7.1
30	Komi	71.8	1.9	326.2	2.2	60.7	7.4	2	2.8	17.3	1.0	219.5	9.0	16.4	1.9
31	Kostroma Region	59.7	4.0	317.7	2.5	100.6	1.2	1	8.3	17.1	0.8	100.6	3.3	16.6	2.0
32	Krasnodar Territory	56.8	4.7	176.3	8.3	65.8	6.5	1.6	4.4	21.4	5.8	131.1	6.1	15.3	1.4
33	Krasnoyarsk Territory	77.7	1.2	273	4.0	67.6	5.9	1.6	4.4	22.2	6.7	198.0	8.8	20.5	6.6
34	Kurgan Region	51.4	6.4	301.5	3.2	56.8	8.0	1.5	5.6	16.3	0.4	85.0	1.8	18.0	3.1
35	Kursk Region	49.2	7.2	163.9	8.7	79	4.2	1.5	5.6	22.4	6.9	99.5	3.1	22.4	7.9
36	Leningrad Region	73	1.6	314.4	2.7	106.8	0.7	1.2	7.7	19.2	2.8	186.6	8.4	14.0	0.9
37	Lipetsk Region	47.9	7.8	214.7	7.0	93.3	1.8	0.7	9.7	20.1	4.3	176.4	8.0	17.1	2.4
38	Magadan Region	50	6.9	254.9	4.9	39.3	9.3	1	8.3	22.1	6.5	224.6	9.1	19.7	5.4
39	Mari El	54.7	5.5	322	2.4	87.3	2.9	1.6	4.4	23.4	7.6	75.6	0.7	20.1	6.1
40	Mordovia	42.1	8.8	184.5	8.1	89	2.2	1.8	3.6	23.2	7.5	92.2	2.2	23.9	8.9
41	Moscow	51.2	6.5	119.3	9.4	116.3	0.2	1.9	3.3	42.8	10.0	433.5	10.0	34.5	10.0
42	Moscow Region	65.1	3.3	246.8	5.3	85.0	3.4	0.8	8.9	24.8	8.6	173.5	7.8	23.1	8.3

(continued)

(continued)

		Mortality from diseases of the digestive system		Prevalence of unnatural death		Number of the disabled		Congenital anomalies		The proportion of specialists with higher education among the employed in the economy		The reduced productivity (GRP/employment in the economy)		The proportion of students in secondary and higher education among young people	
1	2	3		4		5		6		7		8		9	
	Formula for unification	(2.3)		(2.3)		(2.3)		(2.3)		(2.2)		(2.2)		(2.2)	
43	Murmansk Region	53.9	5.8	207.3	7.4	27.3	9.7	0.8	8.9	20.0	3.9	185.3	8.4	15.5	1.5
44	Nizhny Novgorod Region	69.4	2.5	240.7	5.6	102.6	1.1	1.4	6.5	20.0	4.0	128.0	5.9	23.2	8.4
45	North Ossetia – Alania	79.3	1.0	107.9	9.5	66.5	6.2	1.7	3.9	32.6	9.8	76.9	0.8	18.4	3.6
46	Novgorod Region	71.1	2.2	338.1	1.7	109.9	0.5	2.3	1.7	18.8	2.2	124.5	5.5	19.6	5.0
47	Novosibirsk Region	49.2	7.2	235.3	6.1	60.2	7.6	1.5	5.6	24.0	8.2	140.9	6.7	26.8	9.4
48	Omsk Region	45.7	8.2	215.8	6.9	70	5.4	2.1	2.3	20.0	4.0	134.0	6.3	21.6	7.5
49	Orel Region	56.4	4.8	251.1	5.1	87.1	3.0	1.5	5.6	22.7	7.2	111.5	4.0	24.2	8.9
50	Orenburg Region	52.5	6.1	252.8	5.0	87.9	2.7	3	0.7	17.9	1.3	122.7	5.0	20.0	5.9
51	Penza Region	48.2	7.7	257.4	4.8	65.5	6.6	1.2	7.7	21.5	5.9	72.9	0.5	20.2	6.2
52	Perm Region	71.8	1.9	337.9	1.7	83.8	3.7	1.9	3.3	16.7	0.6	172.5	7.8	18.3	3.4
53	Primorye Territory	68.4	2.7	254.8	4.9	57.5	7.9	2.5	1.1	23.6	7.8	125.9	5.7	19.4	4.5
54	Pskov Region	69.9	2.4	328.9	2.1	96.3	1.5	0.7	9.7	18.8	2.2	90.2	2.1	16.8	2.2
55	Rostov Region	48.3	7.7	150.3	9.0	88.2	2.6	2.5	1.1	23.7	7.9	97.5	2.8	22.2	7.8
56	Ryazan Region	64.7	3.3	241.1	5.5	110.1	0.5	1.6	4.4	22.5	7.0	124.5	5.5	20.2	6.2

57	Sakha (Yakutia)	44.8	8.4	238.5	5.8	44.3	9.0	1.6	4.4	20.0	3.9	283.0	9.5	19.7	5.6
58	Sakhalin Region	70.9	2.2	348.7	1.2	50.2	8.5	2.2	2.0	18.7	2.0	248.1	9.3	14.3	1.0
59	Samara Region	48.4	7.6	229.4	6.3	72.4	4.9	2.1	2.3	24.3	8.4	174.6	7.9	23.3	8.5
60	Saratov Region	49.8	7.0	217	6.9	51.1	8.4	2.3	1.7	23.9	8.1	111.6	4.0	21.7	7.5
61	Smolensk Region	65.6	3.2	290.2	3.5	94.9	1.6	3.2	0.5	19.6	3.4	107.7	3.7	18.7	3.9
62	St. Petersburg	59.6	4.0	167.7	8.6	128.7	0.0	1.5	5.6	37.2	10.0	183.0	8.3	26.7	9.4
63	Stavropol Territory	50.9	6.6	158	8.9	72.6	4.9	0.9	8.6	22.9	7.4	102.2	3.4	19.5	4.8
64	Sverdlovsk Region	62.5	3.6	309.7	2.9	74	4.7	0.7	9.7	18.8	2.2	151.4	7.0	21.9	7.6
65	Tambov Region	61.2	3.8	234.3	6.1	107	0.7	1.5	5.6	19.5	3.3	98.2	2.9	19.0	4.1
66	Tatarstan	49.8	7.0	189.1	8.0	61.2	7.3	1.5	5.6	20.6	5.0	185.4	8.4	23.0	8.3
67	Tomsk Region	59.7	4.0	218.3	6.8	50.9	8.5	2.5	1.1	28.1	9.3	217.3	9.0	28.1	9.6
68	Tula Region	78.9	1.0	262.2	4.6	88.2	2.6	1	8.3	18.1	1.5	100.3	3.3	17.6	2.7
69	Tuva	98	0.0	508.2	0.0	81.6	4.0	1.6	4.4	21.6	6.1	80.2	1.1	10.6	0.2
70	Tver Region	72.5	1.6	349.6	1.1	70.7	5.3	0.9	8.6	19.2	2.8	113.2	4.1	18.0	3.1
71	Tyumen Region	33.3	9.5	206.4	7.4	45	8.9	1.5	5.6	20.7	5.1	640.2	10.0	18.5	3.6
72	Udmurtian	82.4	0.6	281.8	3.7	62.2	7.1	1.3	7.2	18.8	2.2	125.1	5.6	22.9	8.2
73	Ulyanovsk Region	38.1	9.1	270.5	4.1	82.8	3.9	1.6	4.4	19.2	2.8	96.2	2.6	21.1	7.2
74	Vladimir Region	63.2	3.5	252.2	5.0	93.1	1.8	1.3	7.2	19.3	2.9	93.4	2.4	26.7	9.4
75	Volgograd Region	58.9	4.3	196.9	7.7	71.8	5.0	1.9	3.3	20.8	5.2	110.9	3.9	20.1	6.1
76	Vologda Region	96.5	0.0	298.4	3.4	100.4	1.3	1.8	3.6	17.0	0.8	183.7	8.3	18.2	3.3
77	Voronezh Region	47.2	8.0	149.7	9.0	89.2	2.2	0.8	8.9	24.2	8.4	98.6	3.0	25.4	9.2
78	Yaroslavl Region	84.5	0.4	311.1	2.8	88.1	2.6	2.1	2.3	20.4	4.8	156.8	7.2	19.4	4.6

A2.76 Level of welfare of the population, 2003

		GDP per capita		The ratio of per capita income and the minimum subsistence level		The share of the population with incomes below the subsistence minimum		Ratio of funds		The proportion of the total area of housing per capita		Total housing area per capita		Number of cars in personal use		The density of roads		The proportion of old and dilapidated housing		The volume of trade turnover and paid services per capita (adjusted)	
	Formula for unification	(2.2)		(2.2)		(2.3)		(2.4)		(2.2)		(2.2)		(2.2)		(2.2)		(2.3)		(2.2)	
1	2	3		4		5		6		7		8		9		10		11		12	
		x	$\tilde{x}$	x	$\tilde{x}$	x	$\tilde{x}$	x	$\tilde{x}$	x	$\tilde{x}$	x	$\tilde{x}$	x	$\tilde{x}$	x	$\tilde{x}$	x	$\tilde{x}$	x	$\tilde{x}$
1	Adygeya	25.1	0.5	153.7	2.7	34.9	2.4	9.3	9.8	22.7	8.6	0.1	2.5	151.9	7.0	207.0	8.6	1.2	8.6	24.2	2.3
2	Altai	40.5	1.6	134.2	1.4	37.6	1.3	8.3	9.5	15.2	0.0	0.1	2.0	137.8	5.2	31.0	2.5	5.5	2.1	17.2	0.5
3	Altai Territory	41.8	1.9	160.5	3.7	34.9	2.4	11.3	8.2	19.2	3.0	0.2	3.9	151.5	6.9	86.0	4.2	2.2	5.6	30.7	5.9
4	Amur Region	50.6	4.2	148.2	2.1	34.9	2.4	9.7	9.5	19.6	3.9	0.1	2.2	152.5	7.1	19.0	2.0	6.2	1.6	22.5	1.6
5	Arkhangelsk Region	71.0	7.8	197.7	6.6	23.5	6.7	10.2	9.1	22.5	8.4	0.1	0.9	106.0	1.6	12.6	1.4	7.4	1.0	29.7	5.0
6	Astrakhan Region	60.2	6.2	202.7	7.1	23.4	6.7	10.8	8.6	18.6	1.6	0.6	10.0	149.0	6.3	61.0	3.5	11.9	0.4	31.9	6.5
7	Bashkortostan	77.7	8.3	235.6	8.7	20.3	8.4	13.3	6.7	18.6	1.6	0.4	8.8	152.9	7.2	154.0	6.4	1.1	8.8	40.5	9.2
8	Belgorod Region	64.8	7.1	198.1	6.7	22.9	7.0	9.9	9.3	22.0	7.3	0.4	9.5	145.3	5.9	241.0	9.0	1.5	7.9	30.2	5.5
9	Bryansk Region	43.6	2.3	179.2	5.1	28.4	5.0	10.0	9.2	21.9	7.1	0.2	3.9	70.8	0.6	185.0	8.0	1.4	8.1	29.5	4.9
10	Buryatia	42.1	2.0	164.1	4.0	36.1	1.9	14.3	5.9	17.6	0.6	0.2	5.3	91.0	1.0	18.0	1.9	7.0	1.1	25.5	3.1
11	Chelyabinsk Region	66.7	7.3	105.8	0.5	25.9	5.6	10.9	8.5	19.8	4.4	0.2	6.2	149.0	6.3	101.0	4.7	1.6	7.7	30.8	6.1
12	Chita Region	43.2	2.3	175.3	4.9	44.1	0.8	11.3	8.2	18.2	1.0	0.1	1.2	131.9	4.0	23.0	2.2	3.6	3.9	26.0	3.4
13	Chukotka Autonomous Area	97.1	9.7	70.2	0.0	28.4	5.0	13.3	6.7	22.0	7.3	0.4	9.0	28.7	0.1	46.0	2.9	3.0	4.7	18.4	0.7
14	Chuvash	46.4	2.9	156.5	3.1	32.7	3.2	8.6	9.7	19.8	4.4	0.4	9.4	71.7	0.6	256.0	9.2	1.7	7.2	25.8	3.3
15	Daghestan	27.1	0.5	111.3	0.7	52.7	0.4	10.4	8.9	18.4	1.1	0.2	6.0	50.8	0.3	148.0	6.1	29.6	0.0	19.9	0.9
16	Irkutsk Region	61.0	6.5	198.6	6.7	32.2	3.4	14.9	5.5	19.3	3.2	0.1	1.1	151.1	6.8	16.0	1.7	5.3	2.3	30.2	5.5
17	Ivanovo Region	29.8	0.6	111.5	0.7	58.0	0.3	8.4	9.5	21.5	6.4	0.1	1.7	101.6	1.4	164.0	6.8	2.0	6.7	18.4	0.7
18	Jewish Autonomous Region	38.3	1.1	302.1	9.4	36.3	1.8	10.1	9.2	20.0	4.7	0.1	0.7	143.6	5.8	0.8	0.4	5.3	2.3	24.0	2.2
19	Kabardino-Balkarian	38.8	1.1	158.2	3.3	31.4	3.7	8.6	9.7	17.6	0.6	0.2	7.4	102.1	1.4	235.0	9.0	3.7	3.8	28.8	4.5

20	Kaliningrad Region	50.4	4.2	173.7	4.7	35.1	2.3	8.4	9.5	19.2	3.0	0.2	7.5	221.0	9.9	305.0	9.5	3.7	3.8	29.6	4.9
21	Kalmykia	52.7	4.7	120.2	0.9	54.6	0.4	11.8	7.8	19.8	4.4	0.2	4.4	134.0	4.5	37.0	2.7	1.3	8.3	12.5	0.1
22	Kaluga Region	48.1	3.5	161.7	3.9	33.2	3.1	8.6	9.7	22.2	7.7	0.2	6.4	136.2	4.9	175.0	7.4	5.9	1.8	29.8	5.0
23	Kamchatka Region	46.6	3.0	191.3	6.1	27.9	5.1	11.1	8.4	21.2	5.9	0.0	0.5	229.0	10.0	2.9	0.5	4.7	2.8	24.8	2.8
24	Karachayevo-Circassian	36.2	1.0	146.9	2.0	38.0	1.2	10.2	9.1	18.4	1.1	0.2	3.4	123.3	2.9	134.0	5.5	0.5	9.5	24.4	2.4
25	Karelia	53.7	5.1	204.2	7.3	19.0	9.1	9.4	9.7	21.8	6.9	0.1	2.2	189.6	8.9	39.0	2.7	3.9	3.6	31.9	6.5
26	Kemerovo Region	58.5	5.9	199.5	6.8	20.2	8.5	12.2	7.5	19.7	4.2	0.2	4.7	122.1	2.7	61.0	3.5	6.0	1.8	36.5	8.4
27	Khabarovsk Territory	59.0	5.9	203.7	7.3	25.4	5.8	12.2	7.5	19.5	3.6	0.1	1.5	157.7	7.6	6.3	0.8	2.1	6.1	30.9	6.1
28	Khakassia	48.4	3.8	176.9	5.0	29.0	4.6	10.5	8.8	18.9	2.2	0.2	5.0	176.8	8.3	41.0	2.8	4.9	2.6	24.1	2.2
29	Kirov Region	44.6	2.5	154.5	2.7	31.9	3.5	8.2	9.4	20.3	5.1	0.1	1.8	114.0	2.0	75.0	3.9	4.3	3.1	23.9	2.1
30	Komi	88.2	9.1	269.1	9.1	18.5	9.3	17.4	3.5	22.2	7.7	0.2	3.5	131.4	3.9	13.1	1.5	6.7	1.1	42.5	9.5
31	Kostroma Region	49.2	4.0	158.2	3.4	33.9	2.9	9.7	9.5	22.8	8.7	0.1	2.8	117.5	2.2	94.0	4.5	3.5	4.1	23.5	1.9
32	Krasnodar Territory	60.2	6.1	187.5	5.7	29.8	4.3	12.8	7.1	18.7	1.8	0.3	8.5	187.1	8.8	143.0	5.9	1.5	7.9	37.9	8.7
33	Krasnoyarsk Territory	81.4	8.6	216.0	8.1	24.7	6.1	14.3	5.9	20.0	4.7	0.2	5.5	131.4	3.9	5.5	0.7	3.4	4.3	34.4	7.6
34	Kurgan Region	32.2	0.8	138.4	1.6	41.3	0.9	13.0	6.9	19.1	2.8	0.1	1.6	150.6	6.6	92.0	4.4	6.2	1.6	22.9	1.6
35	Kursk Region	46.5	3.0	166.0	4.1	30.3	4.1	9.9	9.3	22.0	7.3	0.2	7.0	126.5	3.4	208.0	8.6	0.7	9.3	24.9	2.8
36	Leningrad Region	68.1	7.5	90.9	0.3	38.1	1.1	7.7	9.0	23.3	9.1	0.3	8.0	160.5	7.7	124.0	5.3	2.5	5.2	23.2	1.8
37	Lipetsk Region	81.7	8.6	214.0	8.1	22.6	7.2	11.6	8.0	21.9	7.1	0.3	8.4	153.6	7.3	217.0	8.8	0.9	9.1	34.6	7.7
38	Magadan Region	70.9	7.8	211.6	8.0	18.5	9.3	12.5	7.3	20.7	9.8	0.0	0.5	189.9	8.9	4.8	0.7	8.1	0.9	28.5	4.3
39	Mari El	40.9	1.7	125.4	1.1	48.5	0.6	10.9	8.5	20.2	5.0	0.2	6.1	97.0	1.1	142.0	5.9	2.0	6.7	20.0	0.9
40	Mordovia	48.6	3.8	150.0	2.3	36.5	1.7	9.2	9.8	21.1	5.8	0.2	4.5	96.7	1.1	167.0	6.9	1.6	7.7	21.0	1.1
41	Moscow	169.2	10.0	525.3	10.0	20.1	8.6	22.0	0.0	22.8	2.0	0.4	9.5	260.0	9.8	650.0	10.0	0.4	9.6	100.9	10.0
42	Moscow Region	60.3	6.2	184.7	5.4	25.1	5.9	10.7	8.7	23.3	9.1	0.6	10.0	200.5	9.4	352.0	9.6	1.1	8.8	35.3	8.0
43	Murmansk Region	59.4	6.0	345.3	9.6	22.3	7.4	11.7	7.9	22.3	7.9	0.0	0.3	141.5	5.6	17.0	1.8	1.9	6.9	33.8	7.2
44	Nizhny Novgorod Region	74.4	8.1	192.3	6.2	22.7	7.1	9.6	9.5	21.3	6.1	0.2	4.0	128.2	3.6	175.0	7.4	2.1	6.1	32.7	6.9
45	North Ossetia – Alania	40.7	1.6	170.3	4.4	23.6	6.6	11.1	8.4	24.9	9.6	0.2	5.4	123.4	3.0	286.0	9.4	2.0	6.7	29.9	5.2
46	Novgorod Region	58.8	5.9	206.9	7.7	28.7	4.8	10.7	8.7	23.7	9.2	0.2	3.6	119.7	2.4	159.0	6.6	3.5	4.1	30.7	6.0
47	Novosibirsk Region	52.5	4.6	159.0	3.5	36.6	1.7	11.3	8.2	18.6	1.6	0.2	6.4	103.5	7.9	55.0	3.3	2.7	5.0	38.3	8.8
48	Omsk Region	53.4	5.0	199.5	6.8	21.5	7.9	12.8	7.1	19.5	3.6	0.2	3.2	136.5	5.0	57.0	3.4	1.3	8.3	35.6	8.1
49	Orel Region	66.1	7.2	186.0	5.5	28.1	5.1	11.1	8.4	21.2	5.9	0.3	8.2	133.4	4.4	171.0	7.1	2.2	5.6	30.3	5.6
50	Orenburg Region	65.7	7.2	170.6	4.4	28.9	4.7	9.0	10.0	19.1	2.8	0.2	7.5	163.1	7.9	107.0	4.8	4.1	3.3	22.6	1.6
51	Penza Region	40.4	1.6	153.6	2.6	33.9	2.9	8.4	9.5	21.3	6.1	0.2	3.9	107.6	1.7	151.0	6.3	1.7	7.2	26.7	3.6
52	Perm Region	80.1	8.5	235.5	8.7	21.8	7.7	15.5	5.0	18.9	2.2	0.2	3.0	119.2	2.3	66.0	3.6	4.3	3.1	36.8	8.5
53	Primorye Territory	40.3	1.5	146.6	2.0	37.8	1.2	9.4	9.7	18.9	2.2	0.1	1.7	187.1	8.8	43.0	2.8	3.0	4.7	26.0	3.4
54	Pskov Region	47.9	3.4	132.2	1.4	25.8	5.6	9.1	9.9	24.5	9.5	0.2	3.1	121.9	2.7	180.0	7.8	3.3	4.5	34.9	7.9
55	Rostov Region	45.7	2.7	206.7	7.7	24.0	6.4	11.7	7.9	19.1	2.8	0.2	7.1	153.8	7.3	123.0	5.3	2.2	5.6	38.5	8.9

(continued)

(continued)

		GDP per capita		The ratio of per capita income and the minimum subsistence level		The share of the population with incomes below the subsistence minimum		Ratio of funds		The proportion of the total area of housing per capita		Total housing area per capita		Number of cars in personal use		The density of roads		The proportion of old and dilapidated housing		The volume of trade turnover and paid services per capita (adjusted)	
	Formula for unification	(2.2)		(2.2)		(2.3)		(2.4)		(2.2)		(2.2)		(2.2)		(2.2)		(2.3)		(2.2)	
1	2	3		4		5		6		7		8		9		10		11		12	
56	Ryazan Region	61.1	6.5	174.7	4.8	29.5	4.4	9.6	9.5	22.5	8.3	0.2	6.6	141.6	5.6	173.0	7.3	2.3	5.4	28.0	4.0
57	Sakha (Yakutia)	85.4	8.9	226.2	8.5	20.4	8.3	12.8	7.1	19.5	3.6	0.3	8.2	123.7	3.0	2.4	0.5	11.4	0.4	33.7	7.2
58	Sakhalin Region	65.9	7.2	303.0	9.4	26.0	5.5	12.0	7.7	21.4	6.3	0.1	1.0	202.0	9.4	9.9	1.1	7.9	0.9	27.9	3.9
59	Samara Region	88.8	9.2	293.1	9.1	23.0	8.0	17.7	3.4	20.0	4.7	0.3	8.0	183.3	8.6	141.0	5.8	1.6	7.7	52.5	10.0
60	Saratov Region	53.5	5.0	180.0	5.2	29.0	4.6	9.4	9.7	21.7	6.8	0.2	4.8	140.6	5.5	103.0	4.7	3.6	3.9	27.9	3.9
61	Smolensk Region	53.1	4.9	193.4	6.3	23.9	6.5	9.8	9.4	22.8	8.7	0.3	7.7	116.7	2.1	179.0	7.8	2.0	6.7	33.8	7.2
62	St. Petersburg	76.9	8.3	171.1	4.5	16.5	9.7	13.0	6.9	20.9	5.5	0.4	9.0	201.5	9.4	630.0	9.8	0.4	9.6	42.1	9.4
63	Stavropol Territory	44.7	2.6	157.8	3.3	34.4	2.7	10.2	9.1	19.1	2.8	0.2	7.2	172.6	8.2	113.0	5.0	1.1	8.8	33.4	7.1
64	Sverdlovsk Region	68.7	7.5	244.6	8.9	17.8	9.5	12.3	7.5	20.4	5.1	0.2	4.5	133.2	4.3	55.0	3.3	1.7	7.2	38.9	8.9
65	Tambov Region	47.6	3.3	206.2	7.6	22.5	7.2	11.9	7.8	21.6	6.6	0.2	6.8	148.5	6.2	160.0	6.6	3.3	4.5	33.9	7.3
66	Tatarstan	88.3	9.1	240.6	8.9	19.8	8.8	13.1	6.8	19.7	4.2	0.4	9.4	121.5	2.6	196.0	8.4	2.0	6.7	38.8	8.9
67	Tomsk Region	87.2	9.0	133.6	1.4	21.9	7.7	11.8	7.8	19.0	2.5	0.2	7.0	129.3	3.7	12.0	1.3	6.0	1.8	34.2	7.5
68	Tula Region	55.1	5.3	189.5	5.9	21.4	7.9	8.0	9.2	22.5	8.3	0.1	2.6	138.9	5.3	199.0	8.4	6.4	1.4	27.9	3.9
69	Tuva	25.2	0.5	125.3	1.1	48.7	0.6	10.1	9.2	12.6	0.0	0.1	0.8	97.0	1.1	15.0	1.6	12.7	0.3	13.3	0.2
70	Tver Region	52.4	4.5	146.5	2.0	39.0	1.1	7.7	9.0	24.5	9.5	0.2	6.0	137.9	5.2	179.0	7.8	4.1	3.3	25.6	3.2
71	Tyumen Region	299.6	10.0	342.0	9.6	13.1	10.0	20.6	1.1	18.3	1.0	0.4	8.9	193.5	9.1	7.1	0.9	6.4	1.4	48.0	9.9
72	Udmurtian	62.3	6.8	156.3	3.0	31.1	3.9	7.8	9.1	18.1	0.9	0.2	6.0	149.5	6.4	136.0	5.6	2.1	6.1	24.8	2.8
73	Ulyanovsk Region	48.5	3.8	153.1	2.6	35.5	2.1	11.2	8.3	20.8	5.4	0.1	2.5	130.4	3.8	119.0	5.2	0.9	9.1	31.3	6.3
74	Vladimir Region	51.9	4.4	155.8	3.0	31.5	3.7	8.0	9.2	22.4	8.1	0.2	5.1	124.0	3.1	192.0	8.2	2.3	5.4	21.7	1.3
75	Volgograd Region	53.3	5.0	208.4	7.8	22.3	7.4	10.3	9.0	19.4	3.4	0.2	3.2	135.7	4.8	78.0	4.0	1.9	6.9	35.3	8.0
76	Vologda Region	93.9	9.5	190.4	6.0	20.0	8.6	10.8	8.6	23.1	9.0	0.2	5.6	135.6	4.8	79.0	4.0	4.7	2.8	29.3	4.8
77	Voronezh Region	41.6	1.9	188.2	5.7	30.1	4.2	12.0	7.7	22.9	8.9	0.3	8.6	175.9	8.3	175.0	7.4	0.8	9.2	31.9	6.6
78	Yaroslavl Region	77.2	8.3	222.6	8.4	19.3	9.0	10.7	8.7	21.5	6.4	0.2	3.6	107.7	1.7	171.0	7.1	4.6	2.9	28.3	4.2

A2.7c The quality of social services, 2003, 1 part

		Migration		The index of the "depth of poverty"		Unemployment		The number of applications per vacancy		Long-term unemployment (more than 12 months'), %		Arrears of wages		The incidence of socially significant diseases (syphilis, hepatitis, tuberculosis)	
	Formula for unification	(2.2)		(2.3)		(2.3)		(2.3)		(2.3)		(2.3)		(2.3)	
1	2	3		4		5		6		7		8		9	
		x	$\tilde{x}$	x	$\tilde{x}$	x	$\tilde{x}$	x	$\tilde{x}$	x	$\tilde{x}$	x	$\tilde{x}$	x	$\tilde{x}$
1	Adygeya	15.9	8.4	6.7	1.6	15.8	0.9	3.7	2.6	3.5	1.4	55.4	0.8	7.98	9.3
2	Altai	−2.1	4.9	5.9	2.2	10.2	3.2	8.1	1.0	0.8	9.9	47.1	1.4	1.66	1.1
3	Altai Territory	−23.2	2.8	7.0	1.3	12.1	1.6	6.9	1.1	2.4	5.2	37.5	5.0	4.93	4.9
4	Amur Region	−31	2.0	5.2	3.6	10.1	3.3	6.7	1.1	3.4	1.6	41.2	3.4	0.87	0.6
5	Arkhangelsk Region	−39.5	1.5	2.9	8.4	9.9	4.0	3.4	3.4	3.3	1.8	29	8.3	7.01	8.0
6	Astrakhan Region	−0.6	5.3	3.6	6.7	10.1	3.3	3.6	2.8	2.2	6.1	41.6	3.2	3.77	3.4
7	Bashkortostan	6.6	7.5	3.5	7.1	8.2	5.8	2.7	4.6	2.0	7.4	33.2	7.0	6.14	6.5
8	Belgorod Region	77.4	9.8	3.7	6.6	8.2	5.8	2.1	5.9	2.1	7.2	28.2	8.6	8.21	9.5
9	Bryansk Region	−1	5.1	4.9	3.9	7.3	6.9	2.7	4.6	2.3	5.8	45.3	1.7	4.64	4.3
10	Buryatia	−35.4	1.8	7.2	1.2	16.8	0.8	2.5	5.2	2.4	4.4	40.5	3.7	0.89	0.6
11	Chelyabinsk Region	−6.2	4.5	4.0	5.8	5.9	8.4	3.2	3.8	2.5	4.0	25.1	9.2	5.61	5.5

(continued)

(continued)

		Migration		The index of the "depth of poverty"		Unemploy-ment		The number of applications per vacancy		Long-term unemploy-ment (more than 12 months'), %		Arrears of wages		The incidence of socially significant diseases (syphilis, hepatitis, tuberculosis)	
	Formula for uni-fication	(2.2)		(2.3)		(2.3)		(2.3)		(2.3)		(2.3)		(2.3)	
1	2	3		4		5		6		7		8		9	
12	Chita region	–42.6	1.3	8.6	0.9	15.3	0.9	39.2	0.0	2.4	5.2	47.4	1.4	1.77	1.2
13	Chukotka Autono-mous Area	–100	0.3	4.3	5.3	4.8	9.0	1.5	7.2	4.3	0.4	24.8	9.2	6.90	7.8
14	Chuvash	0.4	5.6	5.6	3.0	8.6	5.1	2.7	4.6	1.2	9.6	33.8	6.6	4.97	5.0
15	Daghestan	–19.3	3.3	13.0	0.4	20.1	0.5	71.5	0.0	1.8	8.4	60.1	0.6	7.44	8.7
16	Irkutsk Region	–19.5	3.3	6.1	2.0	11.7	1.8	3.7	2.6	2.8	3.3	38.2	4.6	1.50	1.0
17	Ivanovo Region	2.8	6.3	12.5	0.4	6.5	7.8	1.6	7.0	2.0	7.4	35.7	5.6	6.54	7.2
18	Jewish Autono-mous Region	–3.4	4.8	5.8	2.5	6.5	7.8	2.3	5.6	3.3	1.6	44.5	2.0	0.55	0.4
19	Kabardino-Balk-arian	-24.4	2.6	5.5	3.3	22	0.3	14.4	0.6	3.8	0.9	56.5	0.8	8.17	9.4
20	Kaliningrad Region	40.3	9.3	5.1	3.7	7.5	6.6	1.1	8.3	1.4	9.3	43.6	2.3	1.49	1.0
21	Kalmykia	–57	0.9	14.0	0.3	17.9	0.6	15.3	0.5	1.6	9.0	57.1	0.7	4.56	4.2
22	Kaluga Region	8.1	7.7	4.9	3.9	6.2	8.2	0.7	9.5	1.9	7.8	35.8	5.5	5.84	6.1
23	Kamchatka region	–75.6	0.5	3.6	6.7	11.4	2.0	3.6	2.8	4.8	0.1	44.3	2.0	6.13	6.4

24	Karachayevo-Circassian	−44.1	1.1	7.8	1.0	19	0.5	3.6	2.8	2.7	3.6	61.6	0.5	4.63	4.3
25	Karelia	4.7	7.2	2.2	9.5	8.5	5.3	2.9	4.2	2.4	5.3	34.1	6.5	3.50	2.9
26	Kemerovo Region	3.6	6.7	3.0	8.3	9.7	4.2	3.5	3.1	2.2	6.6	38.1	4.7	2.41	1.6
27	Khabarovsk Territory	4.9	7.2	3.5	7.2	6.1	8.3	4.7	1.9	3.0	2.7	41.4	3.3	1.80	1.2
28	Khakassia	13.4	8.2	4.6	4.7	10.5	2.7	5.4	1.6	1.9	7.9	42.8	2.6	2.61	1.9
29	Kirov Region	−27.3	2.2	4.5	4.8	7.4	6.7	2.7	4.6	1.3	9.5	20.5	9.6	5.74	5.9
30	Komi	−59.2	0.8	2.8	8.7	11.9	1.6	4.8	1.8	4.3	0.4	39.8	4.0	4.99	5.0
31	Kostroma Region	1.2	5.8	5.8	2.5	6.3	8.1	1.3	7.8	2.2	6.6	41.7	3.1	8.34	9.5
32	Krasnodar Territory	26.8	9.0	5.5	3.2	10.1	3.3	0.7	9.5	2.9	3.1	36.3	5.4	6.48	7.0
33	Krasnoyarsk Territory	−20.5	3.2	3.9	6.2	11.2	2.1	4.7	1.9	2.3	5.6	40.5	3.7	4.02	3.6
34	Kurgan Region	−43	1.2	8.9	0.8	9.6	4.3	4.2	2.1	2.3	5.9	35.5	5.8	6.49	7.0
35	Kursk Region	−27.5	2.2	4.8	4.2	8.6	5.1	2	6.1	2.2	6.1	33.6	6.7	5.76	5.9
36	Leningrad Region	87	9.9	5.6	2.8	8.7	5.0	0.8	9.2	2.8	3.3	18.3	9.7	6.31	6.6
37	Lipetsk Region	12.8	8.1	4.0	5.9	4.4	9.4	0.8	9.2	3.3	1.7	37.5	5.0	3.44	2.8
38	Magadan Region	−90	0.4	2.3	9.4	10.2	3.2	5.8	1.4	3.1	2.3	35.8	5.5	3.49	2.8
39	Mari El	−6.5	4.4	11.3	0.5	12.2	1.6	3.3	3.6	1.9	7.9	42.4	2.8	5.51	5.5
40	Mordovia	−26.4	2.3	6.6	1.7	7.4	6.7	2.4	5.4	2.4	4.5	42.5	2.7	7.53	8.9
41	Moscow	52	9.5	3.3	7.8	1.3	10.0	0.4	10.0	2.2	6.6	28	8.7	9.20	9.9
42	Moscow Region	99.2	10.0	3.6	6.9	4.3	9.5	1	8.6	3.0	2.8	26.9	9.0	6.58	7.2
43	Murmansk Region	−70.2	0.6	2.8	8.6	10	3.8	5.3	1.6	2.1	7.0	40.3	3.8	4.53	4.1
44	Nizhny Novgorod Region	6.1	7.4	3.0	8.2	6.1	8.3	0.8	9.2	1.8	8.3	31.7	7.8	6.36	6.7
45	North Ossetia – Alania	−24.1	2.6	4.4	5.0	10.1	3.3	3.6	2.8	3.6	1.1	72	0.3	7.71	9.0
46	Novgorod Region	4.6	7.1	4.6	4.6	5	8.9	1.3	7.8	2.2	6.7	33.3	6.9	3.59	3.1

(continued)

(continued)

		Migration		The index of the "depth of poverty"		Unemployment		The number of applications per vacancy		Long-term unemployment (more than 12 months'), %		Arrears of wages		The incidence of socially significant diseases (syphilis, hepatitis, tuberculosis)	
	Formula for unification	(2.2)		(2.3)		(2.3)		(2.3)		(2.3)		(2.3)		(2.3)	
1	2	3		4		5		6		7		8		9	
47	Novosibirsk Region	–8.9	4.1	6.5	1.7	11	2.3	1.2	8.1	1.6	8.9	61.2	0.5	2.68	2.1
48	Omsk Region	–41	1.4	3.4	7.7	9.5	4.4	2.3	5.6	3.2	2.1	40.1	3.9	3.37	2.7
49	Orel Region	1.9	6.0	5.5	3.2	7.8	6.4	1.8	6.6	1.5	9.1	40	3.9	7.10	8.1
50	Orenburg Region	–18	3.4	4.5	4.9	11.3	2.0	1	8.6	2.5	3.9	40.1	3.9	3.89	3.5
51	Penza Region	2.5	6.1	5.6	2.9	9.1	4.7	2.3	5.6	2.4	4.4	43.7	2.3	6.72	7.5
52	Perm Regoin	–11.3	3.8	3.7	6.4	6.9	7.3	3	4.1	2.4	4.6	35.2	6.1	2.61	1.9
53	Primorye Territory	–26.7	2.2	5.3	3.5	7.9	6.3	3.9	2.3	3.7	1.0	35.5	5.8	1.38	0.9
54	Pskov Region	18.3	8.5	4.0	5.8	8.1	6.1	3.3	3.6	2.1	7.1	39.1	4.3	4.36	3.8
55	Rostov Region	0.2	5.5	3.9	6.1	12.3	1.5	1.4	7.5	4.2	0.5	27.6	8.9	7.24	8.3
56	Ryazan Region	0.6	5.6	4.8	4.3	8.3	5.6	1.4	7.5	2.5	4.1	32.5	7.4	5.83	6.1
57	Sakha (Yakutia)	–44.6	1.1	2.6	9.1	9.4	4.5	2.2	5.7	3.2	2.2	42.4	2.8	2.74	2.2
58	Sakhalin Region	–73.2	0.5	3.5	7.2	9.1	4.7	1.5	7.2	3.8	0.9	32.8	7.3	2.00	1.3
59	Samara Region	9.8	7.9	4.2	5.5	4.4	9.4	1.9	6.4	2.1	7.0	31.1	7.9	6.85	7.8

60	Saratov Region	3.7	6.7	4.3	5.2	10.5	2.7	0.9	8.9	2.4	4.9	46.8	1.5	5.63	5.6
61	Smolensk Region	3.3	6.5	3.4	7.5	10.9	2.4	1.7	6.8	2.4	4.5	45.3	1.7	3.54	3.0
62	St. Petersburg	10.9	8.0	2.1	9.5	4.1	9.6	0.9	8.9	2.9	2.9	27.5	8.9	6.82	7.7
63	Stavropol Territory	25.2	8.9	6.1	2.0	10.3	3.0	1.4	7.5	1.8	8.5	35.4	5.9	8.19	9.4
64	Sverdlovsk Region	2.7	6.2	2.5	9.2	7.6	6.5	1.5	7.2	2.6	3.7	31.9	7.8	4.41	3.9
65	Tambov Region	−17.5	3.5	4.1	5.7	9.1	4.7	8.4	1.0	2.9	3.0	48.4	1.2	7.49	8.8
66	Tatarstan	24.1	8.9	3.4	7.7	6.7	7.6	2	6.1	1.9	7.7	33.3	6.9	7.29	8.4
67	Tomsk Region	−7.3	4.3	3.1	8.1	13.7	1.2	6.4	1.2	2.8	3.4	35	6.2	3.15	2.5
68	Tula Region	−4.1	4.7	2.8	8.6	5.3	8.7	1.8	6.6	3.3	1.8	20.6	9.5	5.48	5.4
69	Tuva	−22.5	2.9	9.7	0.7	20.7	0.4	42.7	0.0	3.3	1.6	65.3	0.4	0.13	0.3
70	Tver Region	−1.4	5.0	5.9	2.2	6.6	7.7	1.1	8.3	1.9	8.1	33.7	6.7	4.88	4.8
71	Tyumen Region	31.9	9.1	1.6	9.8	8.3	5.6	2.8	4.4	3.9	0.8	29.4	8.2	2.77	2.2
72	Udmurtian	−8.5	4.2	4.2	5.4	6.7	7.6	2.8	4.4	1.8	8.4	32.7	7.3	2.77	2.2
73	Ulyanovsk Region	−22.3	2.9	6.9	1.4	7.4	6.7	3.7	2.6	2.2	6.2	37.9	4.8	5.40	5.4
74	Vladimir Region	6.9	7.5	4.8	4.1	10.1	3.3	2.6	5.0	3.0	2.7	36.2	5.4	6.62	7.3
75	Volgograd Region	−10.9	3.9	3.4	7.7	10.9	2.4	1.1	8.3	2.3	5.4	28.9	8.3	4.74	4.5
76	Vologda Region	3.8	6.7	2.8	8.9	4.8	9.0	2.1	5.9	1.7	8.9	28.3	8.6	5.93	6.2
77	Voronezh Region	17.7	8.5	5.6	2.7	8.1	6.1	3.3	3.6	2.4	5.0	45.3	1.7	7.42	8.6
78	Yaroslavl Region	18.5	8.5	2.7	9.0	5.7	8.5	1.1	8.3	1.9	8.3	32.7	7.3	7.65	9.0

A2.7c The quality of social services, 2003, 2 part

		DWC in industry, construction and transport		Occupational injuries		The number of murders and attempted murders		Intentional infliction of grievous bodily harm		The number of rapes and attempted rapes		Robbery, robberies, thefts from apartments		Embezzlement and theft (economic crimes)		The prevalence of alcoholism and alcoholic psychosis	
	Formula for unification	(2.3)		(2.3)		(2.3)		(2.3)		(2.3)		(2.3)		(2.3)		(2.3)	
1	2	3		4		5		6		7		8		9		10	
		x	$\tilde{x}$	x	$\tilde{x}$	x	$\tilde{x}$	x	$\tilde{x}$	x	$\tilde{x}$	x	$\tilde{x}$	x	$\tilde{x}$	x	$\tilde{x}$
1	Adygeya	14.14	8.9	3.63	5.0	13.67	8.9	21.07	8.9	6.50	3.7	169.65	8.7	27.53	5.5	1702.72	4.1
2	Altai	34.63	2.1	6.77	0.9	29.54	2.2	67.95	1.4	18.7	0.0	196.06	8.0	55.17	1.4	2048.26	2.2
3	Altai Territory	26.83	4.6	5.50	1.4	21.52	4.4	43.89	3.9	6.79	3.1	475.45	1.7	34.39	3.5	1853.85	3.1
4	Amur Region	32.73	2.9	6.29	1.0	25.06	3.3	51.12	2.7	6.01	4.2	307.77	5.5	12.54	9.5	1464.16	6.4
5	Arkhangelsk Region	46.68	0.7	8.34	0.5	21.43	4.4	41.35	4.5	5.43	5.1	286.19	6.2	20.78	7.2	1478.19	6.2
6	Astrakhan Region	31.73	3.3	2.43	8.6	20.74	5.0	36.89	5.5	4.69	6.8	381.87	3.8	35.03	3.3	1584.48	5.3
7	Bashkortostan	25.99	4.8	1.78	9.5	14.79	8.4	34.93	6.0	4.39	7.3	172.20	8.6	39.37	2.7	1328.91	7.4
8	Belgorod Region	27.52	4.5	3.67	4.8	10.78	9.5	20.69	8.9	2.84	9.4	151.80	9.3	25.60	6.1	1364.90	7.2
9	Bryansk Region	24.52	5.0	3.59	5.1	15.28	8.0	28.07	7.4	3.58	8.8	320.07	5.1	32.95	4.1	2291.91	1.4
10	Buryatia	33.15	2.7	2.55	8.3	39.10	0.6	78.21	0.9	8.70	1.5	522.97	1.2	33.99	3.7	1009.02	8.9
11	Chelyabinsk Region	38.42	1.5	3.30	6.1	29.45	2.2	55.86	2.2	7.14	2.5	419.85	2.9	21.76	6.9	1576.38	5.5
12	Chita Region	31.47	3.4	2.84	7.3	50.25	0.1	88.75	0.5	10.5	0.6	540.69	1.1	47.63	1.8	1760.36	3.6
13	Chukotka Autonomous Area	45.27	0.9	4.48	3.4	22.94	3.9	95.60	0.4	9.56	1.1	158.70	9.1	32.50	4.1	3201.09	0.3
14	Chuvash	21.54	6.1	2.11	9.1	15.44	7.8	25.07	7.8	6.73	3.3	282.46	6.3	31.71	4.3	1880.09	3.0
15	Daghestan	9.95	9.6	1.77	9.5	9.30	9.7	5.29	9.8	3.67	8.7	29.40	10.0	9.03	9.8	603.90	9.7
16	Irkutsk Region	33.71	2.3	3.44	5.6	46.04	0.3	90.61	0.5	4.75	6.7	639.41	0.7	26.57	5.8	1434.87	6.7
17	Ivanovo Region	35.22	1.9	4.56	3.2	20.93	4.9	43.97	3.9	4.57	7.0	623.85	0.7	27.79	5.3	2765.42	0.6

18	Jewish Autonomous Region	28.73	4.2	4.95	2.2	37.85	0.7	80.97	0.8	10.5	0.6	664.74	0.6	17.84	8.4	783.39	9.5
19	Kabardino-Balkarian	12.89	9.1	1.87	9.4	8.11	9.9	11.33	9.6	4.11	7.7	92.55	10.0	19.31	7.8	831.71	9.4
20	Kaliningrad Region	23.60	5.4	3.06	6.5	17.54	6.8	39.61	4.8	4.94	6.2	388.53	3.6	67.92	1.0	1437.40	6.7
21	Kalmykia	6.87	9.8	1.75	9.5	31.25	1.6	34.00	6.2	9.62	1.1	168.27	8.8	39.75	2.7	1139.53	8.4
22	Kaluga Region	17.77	7.7	3.62	5.0	14.70	8.4	32.02	6.5	6.48	3.7	242.15	7.2	40.92	2.5	1507.86	6.0
23	Kamchatka Region	49.16	0.5	4.50	3.3	21.33	4.5	37.89	5.3	9.82	1.0	245.04	7.2	111.76	0.3	2597.62	0.8
24	Karachayevo-Circassian	10.10	9.5	2.04	9.2	23.08	3.8	18.05	9.1	4.11	7.7	212.12	7.8	25.52	6.1	1194.81	8.1
25	Karelia	46.90	0.7	7.63	0.7	24.44	3.5	48.60	3.0	5.48	5.0	386.47	3.6	23.77	6.5	2517.02	0.9
26	Kemerovo Region	60.07	0.0	7.24	0.7	29.80	2.1	71.22	1.2	5.13	5.8	322.49	5.0	16.94	8.6	1489.74	6.1
27	Khabarovsk Territory	37.40	1.6	5.67	1.2	41.17	0.5	72.48	1.1	9.23	1.3	671.25	0.5	89.67	0.5	1360.03	7.2
28	Khakassia	24.08	5.2	4.96	2.2	28.86	2.3	63.60	1.6	6.80	3.1	473.59	1.8	24.39	6.4	1597.28	5.0
29	Kirov Region	23.84	5.3	9.06	0.4	16.18	7.3	35.72	5.8	6.98	2.7	265.31	6.7	19.28	7.8	2020.40	2.3
30	Komi	54.74	0.1	8.45	0.5	32.64	1.3	60.05	1.9	10.2	0.7	477.66	1.7	148.1	0.0	1667.73	4.4
31	Kostroma Region	29.80	3.9	5.20	1.8	20.41	5.2	53.15	2.5	3.97	8.1	425.60	2.8	29.27	4.8	2372.91	1.2
32	Krasnodar Territory	16.53	8.2	2.93	7.0	12.61	9.2	21.61	8.7	3.91	8.3	145.51	9.5	16.99	8.6	1578.82	5.4
33	Krasnoyarsk Territory	37.50	1.6	3.95	4.0	26.63	2.7	55.86	2.2	5.01	6.1	351.26	4.4	35.08	3.3	1242.05	7.8
34	Kurgan Region	20.86	6.4	4.74	2.7	36.83	0.8	40.69	4.6	8.42	1.6	443.42	2.5	18.50	8.1	1597.75	5.0
35	Kursk Region	33.83	2.2	2.69	7.8	14.15	8.6	22.57	8.4	7.20	2.4	209.16	7.8	92.52	0.5	1718.37	3.9
36	Leningrad Region	22.41	5.8	4.70	2.8	29.46	2.2	45.09	3.6	5.41	5.1	392.87	3.4	13.56	9.3	2163.72	1.8
37	Lipetsk Region	33.91	2.2	2.66	7.9	12.44	9.2	26.37	7.6	3.90	8.3	234.19	7.4	32.78	4.1	2302.15	1.4
38	Magadan Region	44.66	1.0	4.94	2.2	29.43	2.2	48.31	3.1	5.00	6.1	349.83	4.4	20.35	7.3	4493.24	0.0
39	Mari El	13.70	9.0	5.31	1.6	24.99	3.3	35.07	5.9	8.01	1.9	400.44	3.2	17.48	8.5	1643.05	4.6
40	Mordovia	19.40	7.1	4.86	2.4	18.50	6.4	29.51	7.2	6.92	2.8	238.23	7.3	42.43	2.4	1383.59	7.1
41	Moscow	12.32	9.2	2.00	9.3	12.92	9.1	19.94	9.0	3.18	9.2	398.59	3.3	19.18	7.8	1004.94	8.9
42	Moscow Region	18.94	7.3	2.72	7.7	22.24	4.1	31.47	6.6	5.00	6.1	250.89	7.0	15.58	8.9	1784.86	3.5
43	Murmansk Region	55.64	0.1	3.28	6.2	17.97	6.6	40.46	4.7	2.26	9.7	313.78	5.3	14.16	9.1	1057.72	8.7
44	Nizhny Novgorod Region	20.80	6.5	2.87	7.2	15.07	8.2	31.08	6.7	4.72	6.7	462.30	2.2	40.53	2.6	2616.72	0.8
45	North Ossetia – Alania	18.78	7.3	3.46	5.5	15.53	7.8	10.59	9.6	2.82	9.4	152.88	9.3	27.34	5.6	648.14	9.7
46	Novgorod Region	19.25	7.1	4.66	2.9	25.17	3.3	41.61	4.4	5.82	4.4	304.87	5.6	37.71	2.9	2791.23	0.5
47	Novosibirsk Region	18.90	7.3	5.10	2.0	19.62	5.7	45.55	3.5	6.38	3.9	502.81	1.4	62.15	1.1	1488.73	6.1

(continued)

(continued)

		DWC in industry, construction and transport		Occupational injuries		The number of murders and attempted murders		Intentional infliction of grievous bodily harm		The number of rapes and attempted rapes		Robbery, robberies, thefts from apartments		Embezzlement and theft (economic crimes)		The prevalence of alcoholism and alcoholic psychosis	
	Formula for unification	(2.3)		(2.3)		(2.3)		(2.3)		(2.3)		(2.3)		(2.3)		(2.3)	
1	2	3		4		5		6		7		8		9		10	
48	Omsk Region	14.36	8.8	4.51	3.3	19.45	5.8	43.40	4.0	6.92	2.8	315.84	5.2	83.65	0.7	1201.51	8.0
49	Orel Region	19.65	7.0	4.01	3.9	18.50	6.4	29.97	7.0	3.98	8.1	301.64	5.7	49.63	1.6	1815.40	3.3
50	Orenburg Region	33.05	2.7	3.48	5.5	25.86	3.0	36.69	5.6	4.29	7.5	265.72	6.7	47.56	1.8	957.38	9.0
51	Penza Region	13.34	9.0	2.58	8.2	14.42	8.5	23.01	8.3	3.05	9.3	186.10	8.2	19.04	7.9	1669.01	4.4
52	Perm Region	31.43	3.4	4.58	3.1	31.51	1.5	84.61	0.6	11.6	0.2	781.29	0.3	26.90	5.7	2231.87	1.6
53	Primorye Territory	33.55	2.4	5.61	1.3	32.97	1.2	52.83	2.5	5.29	5.4	459.66	2.3	25.01	6.2	1946.62	2.7
54	Pskov Region	16.90	8.0	2.94	6.9	16.34	7.2	41.58	4.4	5.31	5.3	361.43	4.2	34.69	3.4	1797.11	3.4
55	Rostov Region	29.92	3.9	3.81	4.4	15.43	7.8	24.04	8.0	4.75	6.7	269.43	6.6	28.34	5.0	1376.12	7.1
56	Ryazan Region	15.80	8.4	3.62	5.0	17.85	6.7	29.36	7.2	3.54	8.8	167.85	8.8	20.51	7.2	1723.08	3.9
57	Sakha (Yakutia)	31.96	3.2	3.44	5.6	32.67	1.3	77.36	0.9	6.32	3.9	420.51	2.9	12.76	9.4	1974.76	2.5
58	Sakhalin Region	31.49	3.4	5.50	1.4	38.41	0.7	59.46	1.9	4.80	6.5	700.92	0.5	88.07	0.6	3473.36	0.1
59	Samara Region	26.10	4.8	2.99	6.7	25.04	3.3	29.32	7.2	5.45	5.0	333.10	4.7	30.63	4.5	1486.10	6.1
60	Saratov Region	24.14	5.1	2.66	7.9	18.24	6.5	27.81	7.4	4.75	6.7	269.27	6.6	33.79	3.8	1235.56	7.8
61	Smolensk Region	19.84	6.9	2.53	8.4	20.59	5.1	38.88	5.0	2.89	9.4	348.41	4.4	48.38	1.7	1894.26	2.9
62	St. Petersburg	18.42	7.5	3.07	6.5	17.84	6.7	21.18	8.8	2.26	9.7	334.93	4.7	8.89	9.8	904.74	9.2
63	Stavropol Territory	17.47	7.9	2.41	8.7	15.90	7.5	22.75	8.3	5.68	4.6	239.85	7.3	30.38	4.6	1104.26	8.5
64	Sverdlovsk Region	37.34	1.6	3.89	4.2	26.91	2.7	61.82	1.7	8.11	1.8	530.34	1.1	47.73	1.8	1069.08	8.7
65	Tambov Region	15.44	8.5	3.49	5.4	12.94	9.1	19.54	9.0	4.46	7.3	149.89	9.4	19.16	7.8	2289.42	1.4
66	Tatarstan	21.84	6.0	2.38	8.8	16.03	7.4	22.78	8.3	5.77	4.5	269.05	6.6	36.87	3.1	1193.93	8.1

67	Tomsk Region	31.63	3.3	3.34	6.0	19.84	5.5	42.66	4.2	6.42	3.8	449.93	2.4	59.49	1.1	1715.13	3.9
68	Tula Region	21.39	6.1	3.40	5.7	21.36	4.5	38.07	5.2	3.98	8.1	182.54	8.3	13.06	9.4	2102.94	2.0
69	Tuva	29.78	3.9	2.39	8.8	76.47	0.0	322.22	0.0	41.8	0.0	508.67	1.3	55.65	1.3	1648.82	4.6
70	Tver Region	21.31	6.2	4.23	3.6	25.57	3.1	45.85	3.4	5.02	6.1	335.43	4.7	20.88	7.2	1940.27	2.7
71	Tyumen Region	29.41	4.0	3.00	6.7	22.50	4.0	62.53	1.7	5.85	4.4	471.18	1.9	29.63	4.7	1680.16	4.3
72	Udmurtian	22.50	5.8	4.83	2.5	18.99	6.1	46.41	3.3	9.72	1.1	461.93	2.2	27.87	5.3	1598.59	5.0
73	Ulyanovsk Region	31.73	3.3	2.89	7.1	29.74	2.1	37.33	5.4	6.71	3.3	175.07	8.5	23.06	6.6	1335.28	7.3
74	Vladimir Region	17.39	7.9	3.91	4.1	20.83	5.0	34.13	6.1	5.49	4.9	382.28	3.8	28.02	5.2	1979.61	2.5
75	Volgograd Region	21.31	6.2	3.74	4.5	20.83	5.0	31.82	6.6	7.49	2.1	306.21	5.6	59.91	1.1	1558.69	5.6
76	Vologda Region	43.66	1.1	8.40	0.5	19.66	5.6	52.89	2.5	8.25	1.7	343.67	4.5	34.49	3.5	1347.33	7.3
77	Voronezh Region	15.47	8.5	2.72	7.7	16.24	7.2	25.00	7.8	3.64	8.7	306.60	5.6	22.74	6.7	1566.02	5.6
78	Yaroslavl Region	33.81	2.2	4.27	3.6	19.52	5.7	47.66	3.1	7.29	2.3	682.72	0.5	19.94	7.5	2087.70	2.0

A.2.7c The quality of social services, 2003, 3 part

		The incidence of death from suicide		Ratio of funds		The prevalence of substance abuse		Prevalence of HIV		The purchasing power of pensions		The total amount of social payments to poor people		Coverage of the elderly and disabled social care at home		Coverage of the elderly and disabled social service centres, social security	
	Formula for unification	(2.3)		(2.4)		(2.3)		(2.3)		(2.2)		(2.2)		(2.2)		(2.2)	
1	2	3		4		5		6		7		8		9		10	
		x	$\tilde{x}$	x	$\tilde{x}$	x	$\tilde{x}$	x	$\tilde{x}$	x	$\tilde{x}$	x	$\tilde{x}$	x	$\tilde{x}$	x	$\tilde{x}$
1	Adygeya	24.21	8.1	9.3	7.7	198.95	3.8	24.93	7.0	122.17	8.0	18.68	1.7	81.50	2.1	0.25	1.0
2	Altai	101.9	0.0	8.3	9.0	202.80	3.5	17.22	7.7	102.20	2.3	16.74	1.3	87.05	3.8	0.00	0.5
3	Altai Territory	46.86	3.7	11.3	3.9	451.74	0.9	116.20	3.5	130.90	9.5	19.14	2.1	26.98	0.0	6.24	8.3
4	Amur Region	63.04	1.3	9.7	7.0	241.02	2.6	11.51	8.8	90.40	1.3	21.56	3.3	79.71	1.6	1.57	4.5
5	Arkhangelsk Region	54.55	2.1	10.2	5.9	21.36	9.7	9.18	9.2	112.06	4.8	41.91	8.3	88.92	4.4	2.27	5.9
6	Astrakhan Region	39.78	4.8	10.8	4.8	338.23	1.5	15.48	8.0	109.61	4.1	26.91	4.7	94.95	7.3	2.74	6.3
7	Bashkortostan	65.99	1.1	13.3	1.2	129.12	6.0	92.49	3.9	122.99	8.4	30.00	5.8	81.27	2.0	0.96	2.8
8	Belgorod Region	29.02	7.1	9.9	6.6	39.31	9.5	18.17	7.6	126.01	8.9	34.87	7.1	98.78	9.1	0.72	2.4
9	Bryansk Region	28.43	7.2	10	6.4	52.32	9.3	48.42	5.6	127.28	9.0	31.43	6.1	88.94	4.4	0.61	2.2
10	Buryatia	87.62	0.1	14.3	1.0	91.36	7.8	202.81	2.0	97.00	1.8	19.37	2.2	97.03	8.5	1.88	5.3
11	Chelyabinsk Region	50.26	2.9	10.9	4.6	267.70	2.2	359.40	1.3	119.18	7.2	28.78	5.4	84.85	3.1	4.63	7.6
12	Chita Region	77.25	0.5	11.3	3.9	114.26	6.6	116.79	3.5	96.15	1.8	15.12	1.0	74.38	0.8	1.21	3.5
13	Chukotka Autonomous Area	76.48	0.5	13.3	1.2	72.39	8.9	9.73	9.1	55.28	0.0	39.51	7.8	93.63	6.6	4.32	7.3
14	Chuvash	52.81	2.3	8.6	8.3	87.10	8.4	33.72	6.4	120.03	7.4	19.37	2.2	85.56	3.4	2.81	6.4
15	Daghestan	3.20	9.8	10.4	5.5	129.37	6.0	13.03	8.4	105.50	3.2	7.29	0.1	67.01	0.1	1.00	2.9

16	Irkutsk Region	57.95	1.8	14.9	0.8	535.00	0.6	639.31	0.1	100.81	2.1	25.06	4.3	79.05	1.5	0.59	2.2
17	Ivanovo Region	45.64	3.9	8.4	8.9	100.61	7.2	210.48	1.9	112.61	5.0	13.21	0.6	87.68	4.0	4.77	7.7
18	Jewish Autonomous Region	65.72	1.1	10.1	6.1	112.31	6.6	8.43	9.4	89.67	1.2	17.79	1.5	76.11	1.1	0.00	0.5
19	Kabardino-Balkarian	14.33	9.2	8.6	8.3	131.54	5.8	17.02	7.7	122.27	8.0	14.85	0.9	87.98	4.1	1.80	5.0
20	Kaliningrad Region	34.46	6.2	8.4	8.9	207.95	3.3	457.35	0.8	102.74	2.4	20.81	3.0	94.54	7.1	0.14	0.8
21	Kalmykia	50.48	2.8	11.8	3.1	96.28	7.5	50.58	5.4	114.72	5.6	10.91	0.4	100.00	9.4	1.25	3.7
22	Kaluga Region	30.18	6.8	8.6	8.3	111.07	6.7	80.19	4.4	112.22	4.8	23.70	4.0	91.83	5.6	0.42	1.6
23	Kamchatka Region	39.85	4.8	11.1	4.2	133.49	5.7	14.38	8.2	91.20	1.3	41.32	8.2	92.01	5.8	3.42	6.8
24	Karachayevo-Circassian	11.19	9.4	10.2	5.9	216.80	3.0	7.56	9.5	122.11	7.9	16.22	1.2	72.53	0.6	0.56	2.1
25	Karelia	51.13	2.7	9.4	7.6	88.67	8.2	29.21	6.7	111.00	4.4	53.81	9.2	88.94	4.4	1.95	5.5
26	Kemerovo Region	50.61	2.8	12.2	2.4	529.01	0.6	206.36	2.0	117.60	6.8	42.32	8.4	97.72	8.7	3.09	6.6
27	Khabarovsk Territory	49.49	3.1	12.2	2.4	310.22	1.8	45.97	5.7	85.11	0.9	32.60	6.5	98.85	9.1	0.59	2.2
28	Khakassia	60.11	1.6	10.5	5.3	143.76	5.0	17.87	7.7	108.41	3.9	23.69	4.0	74.94	0.9	1.07	3.1
29	Kirov Region	61.38	1.5	8.2	9.1	27.47	9.6	10.61	9.0	118.66	7.1	22.33	3.7	83.11	2.6	43.61	9.6
30	Komi	55.30	2.0	17.4	0.5	104.61	7.0	47.73	5.6	103.52	2.6	53.07	9.2	90.54	4.9	8.67	8.7
31	Kostroma Region	48.77	3.3	9.7	7.0	101.69	7.2	93.44	3.9	116.16	6.1	22.75	3.9	94.54	7.0	4.84	7.7
32	Krasnodar Territory	31.39	6.6	12.8	1.8	405.87	1.1	84.97	4.2	113.70	5.3	22.39	3.7	91.05	5.0	1.67	4.7
33	Krasnoyarsk Territory	37.37	5.6	14.3	1.0	333.56	1.6	180.18	2.3	103.02	2.5	33.07	6.6	72.48	0.6	6.12	8.3
34	Kurgan Region	60.49	1.6	13	1.6	246.59	2.5	132.87	3.3	107.68	3.7	16.09	1.1	91.96	5.7	0.14	0.8
35	Kursk Region	27.72	7.4	9.9	6.6	57.15	9.2	14.66	8.1	93.98	1.6	27.04	4.7	78.31	1.4	1.85	5.2
36	Leningrad Region	38.23	5.1	7.7	9.5	225.50	2.8	414.78	1.1	111.51	4.5	19.70	2.4	93.54	6.6	2.03	5.6
37	Lipetsk Region	25.70	7.8	11.6	3.5	97.05	7.4	7.08	9.5	126.71	9.0	35.62	7.2	93.17	6.4	0.00	0.5
38	Magadan Region	24.43	8.0	12.5	2.1	137.82	5.4	21.87	7.3	99.50	2.0	61.10	9.5	100.00	9.4	0.31	1.1
39	Mari El	71.66	0.7	10.9	4.6	91.24	7.8	51.53	5.3	122.44	8.2	12.48	0.5	85.58	3.4	1.80	5.0
40	Mordovia	29.05	7.1	9.2	7.8	77.97	8.7	36.41	6.3	115.69	6.0	18.50	1.7	100.00	9.4	5.64	8.1
41	Moscow	10.97	9.4	50	0.0	224.95	2.8	176.33	2.4	83.37	0.8	104.46	10.0	93.18	6.4	4.09	7.1
42	Moscow Region	25.14	7.9	10.7	5.0	143.99	5.0	348.26	1.3	109.54	4.1	39.04	7.8	96.37	8.3	6.16	8.3
43	Murmansk Region	35.03	6.1	11.7	3.3	166.20	4.3	140.34	3.0	89.27	1.1	46.05	8.9	81.87	2.1	5.46	8.0
44	Nizhny Novgorod Region	35.37	6.0	9.6	7.2	150.44	4.8	81.08	4.4	114.44	5.5	33.67	6.8	96.11	8.2	10.53	8.9
45	North Ossetia – Alania	4.80	9.7	11.1	4.2	141.35	5.2	45.69	5.7	138.17	9.9	28.05	5.0	85.13	3.2	1.33	3.9

(continued)

(continued)

		The incidence of death from suicide		Ratio of funds		The prevalence of substance abuse		Prevalence of HIV		The purchasing power of pensions		The total amount of social payments to poor people		Coverage of the elderly and disabled social care at home		Coverage of the elderly and disabled social service centres, social security	
	Formula for unification	(2.3)		(2.4)		(2.3)		(2.3)		(2.2)		(2.2)		(2.2)		(2.2)	
1	2	3		4		5		6		7		8		9		10	
46	Novgorod Region	42.19	4.4	10.7	5.0	197.63	3.9	79.99	4.4	113.13	5.1	29.00	5.6	91.50	5.3	1.10	3.2
47	Novosibirsk Region	37.38	5.6	11.3	3.9	453.96	0.9	24.62	7.0	104.64	3.0	20.19	2.7	92.25	6.1	0.40	1.5
48	Omsk Region	38.17	5.2	12.8	1.8	332.76	1.6	11.95	8.7	119.27	7.3	31.30	6.1	94.97	7.3	2.46	6.1
49	Orel Region	22.83	8.4	11.1	4.2	90.53	7.9	77.41	4.5	123.38	8.5	28.56	5.3	91.27	5.2	1.36	4.0
50	Orenburg Region	51.21	2.6	9	8.0	192.90	3.9	537.76	0.4	118.61	7.1	22.06	3.5	95.32	7.7	1.44	4.2
51	Penza Region	38.47	5.0	8.4	8.9	147.36	4.9	49.37	5.5	117.56	6.8	21.27	3.2	87.20	3.9	1.23	3.6
52	Perm Region	56.49	1.9	15.5	0.7	255.48	2.4	178.04	2.4	107.55	3.7	33.73	6.8	78.73	1.5	12.02	8.9
53	Primorye Territory	39.19	4.9	9.4	7.6	556.90	0.5	231.12	1.7	80.52	0.6	20.02	2.6	82.28	2.3	0.06	0.6
54	Pskov Region	47.02	3.6	9.1	7.9	72.33	8.9	25.41	7.0	130.92	9.5	28.67	5.4	94.14	6.8	3.71	6.9
55	Rostov Region	14.61	9.1	11.7	3.3	300.15	1.9	63.82	4.8	116.72	6.4	32.46	6.4	90.27	4.8	0.68	2.3
56	Ryazan Region	16.04	9.0	9.6	7.2	64.13	9.0	129.45	3.3	114.96	5.7	29.00	5.6	86.05	3.5	0.38	1.4
57	Sakha (Yakutia)	48.80	3.3	12.8	1.8	87.94	8.3	38.36	6.2	78.83	0.5	54.88	9.3	82.91	2.5	1.80	5.0
58	Sakhalin Region	23.45	8.2	12	2.7	203.24	3.4	14.68	8.1	73.74	0.4	40.26	8.0	96.27	8.3	4.50	7.4
59	Samara Region	27.99	7.3	17.6	0.5	639.08	0.3	609.65	0.2	105.62	3.2	39.09	7.8	94.69	7.1	31.89	9.4
60	Saratov Region	34.52	6.2	9.4	7.6	228.56	2.8	221.21	1.8	113.59	5.3	23.46	4.0	91.83	5.6	9.74	8.8
61	Smolensk Region	41.19	4.5	9.8	6.8	152.62	4.7	34.39	6.4	117.06	6.6	32.39	6.4	91.89	5.6	1.49	4.3

62	St. Petersburg	17.78	8.9	13	1.6	188.14	4.0	484.07	0.6	105.53	3.2	65.96	9.6	98.65	9.0	1.35	3.9
63	Stavropol Territory	21.69	8.5	10.2	5.9	137.32	5.4	11.19	8.9	110.03	4.2	19.05	2.0	82.99	2.5	36.29	9.5
64	Sverdlovsk Region	45.46	3.9	12.3	2.3	273.30	2.2	508.15	0.5	107.12	3.6	43.17	8.5	83.49	2.7	0.56	2.1
65	Tambov Region	30.52	6.7	11.9	2.9	111.64	6.7	28.39	6.7	131.50	9.5	34.79	7.1	85.18	3.2	0.97	2.8
66	Tatarstan	37.64	5.4	13.1	1.5	199.10	3.7	151.00	2.8	129.51	9.3	36.73	7.4	84.63	3.1	5.03	7.8
67	Tomsk Region	41.99	4.4	11.8	3.1	604.34	0.4	67.55	4.7	103.86	2.7	36.18	7.3	80.18	1.7	2.20	5.8
68	Tula Region	30.89	6.6	8	9.3	152.89	4.7	184.82	2.3	124.55	8.6	47.16	8.9	88.59	4.3	2.61	6.2
69	Tuva	66.67	1.1	10.1	6.1	286.40	2.0	4.57	9.7	85.97	0.9	13.41	0.7	95.45	7.8	10.68	8.9
70	Tver Region	47.64	3.5	7.7	9.5	118.73	6.4	272.15	1.5	111.73	4.6	21.68	3.4	94.93	7.3	1.15	3.4
71	Tyumen Region	34.94	6.1	20.6	0.2	530.93	0.6	461.93	0.8	106.07	3.4	64.25	9.6	94.93	7.3	14.63	9.0
72	Udmurtian	66.62	1.1	7.8	9.5	88.68	8.2	133.06	3.3	115.18	5.8	21.34	3.2	76.26	1.1	31.62	9.4
73	Ulyanovsk Region	18.81	8.8	11.2	4.1	322.27	1.6	428.00	1.0	116.33	6.2	19.70	2.4	98.23	8.9	3.66	6.9
74	Vladimir Region	42.46	4.3	8	9.3	60.01	9.1	84.59	4.2	130.62	9.4	26.50	4.6	97.65	8.7	1.69	4.8
75	Volgograd Region	26.98	7.5	10.3	5.7	126.96	6.1	142.64	3.0	122.68	8.3	31.29	6.1	95.84	8.1	0.38	1.4
76	Vologda Region	52.17	2.4	10.8	4.8	89.13	8.2	51.45	5.3	120.84	7.6	44.94	8.7	100.00	9.4	0.47	1.8
77	Voronezh Region	21.53	8.5	12	2.7	166.63	4.3	11.43	8.8	116.61	6.4	26.92	4.7	91.80	5.5	2.46	6.1
78	Yaroslavl Region	36.31	5.8	10.7	5.0	106.01	7.0	55.30	5.0	121.91	7.8	43.86	8.6	100.00	9.4	48.56	9.7

A.2.7d The quality of the ecological niche, 2003

		The volume of waste-water (per 1 sq. km area)		The ratio of volume of polluted wastewater to the volume of water with-drawn		The total emissions from sta-tionary sources of pollution		Storage of toxic waste production (per 1 sq. km area)		Neutralization of toxic waste production (per 1 sq. km area)		Forests planted to replace the lost (1 sq. km of forest area)		The area of young (1 sq. km of forest area)		Forest fires		The area of pro-tected areas (per 1 sq. km area)		Replacement of the damaged earth	
	Formula for unifi-cation	(2.3)		(2.3)		(2.3)		(2.3)		(2.3)		(2.2)		(2.2)		(2.3)		(2.2)		(2.2)	
1	2	3		4		5		6		7		8		9		10		11		12	
		x	$\tilde{x}$	x	$\tilde{x}$	x	$\tilde{x}$	x	$\tilde{x}$	x	$\tilde{x}$	x	$\tilde{x}$	x	$\tilde{x}$	x	$\tilde{x}$	x	$\tilde{x}$	x	$\tilde{x}$
1	Adygeya	3.29	2.8	0.34	4.9	0.26	8.3	−0.26	9.1	−0.80	8.5	0.29	4.4	2.40	5.6	0.17	8.5	2.19	3.9	1.09	9.5
2	Altai	0.01	9.4	0.08	8.0	0.12	8.9	0.15	8.6	−3.00	9.5	-0.11	2.8	1.24	2.1	2.44	1.3	2.18	1.9	1.05	9.5
3	Altai Territory	0.21	8.3	0.06	8.4	1.47	4.0	2.55	3.3	2.16	2.8	2.00	8.9	2.59	6.1	1.82	2.6	2.19	2.6	−0.62	2.2
4	Amur Region	0.25	8.1	0.81	0.5	0.25	8.3	−0.22	9.1	−3.00	9.5	-1.31	1.0	1.44	2.8	2.91	0.0	2.50	9.4	0.05	5.0
5	Arkhangelsk Region	0.81	6.7	0.64	1.7	0.50	7.5	2.30	4.6	1.06	5.4	0.28	4.3	2.85	6.8	1.18	4.4	1.79	1.0	−0.23	3.9
6	Astrakhan Region	1.61	5.0	0.05	8.9	2.70	2.0	1.01	7.2	−0.50	7.9	0.97	6.6	3.01	7.2	0.46	8.0	2.19	4.4	0.29	6.1
7	Bashkortostan	2.84	3.6	0.47	3.6	3.22	1.6	3.72	0.6	2.92	0.8	1.68	8.4	4.11	8.9	0.73	6.5	2.19	3.4	0.43	7.0
8	Belgorod Region	0.37	7.6	0.02	9.4	3.62	1.5	2.51	3.7	2.12	3.0	0.96	6.6	1.30	2.3	0.98	4.9	2.20	6.2	−0.97	1.7
9	Bryansk Region	2.69	3.8	0.64	1.7	1.58	3.7	−0.46	9.4	−1.24	8.9	1.28	7.3	2.76	6.6	0.93	5.2	2.19	3.3	−0.46	2.9
10	Buryatia	0.19	8.4	0.16	6.6	0.24	8.3	0.11	8.7	−2.15	9.3	−1.76	0.7	1.35	2.4	3.00	0.0	2.18	1.7	0.80	9.1
11	Chelyabinsk Region	8.20	1.0	0.75	0.7	10.35	0.0	4.22	0.0	3.47	0.3	0.80	6.0	2.19	5.0	2.13	1.9	2.20	6.6	0.37	6.6
12	Chita Region	0.17	8.5	0.15	6.7	0.36	8.0	2.90	2.3	2.23	2.5	−4.96	0.0	2.25	5.1	3.52	0.0	2.22	7.6	−0.04	4.6
13	Chukotka Autono-mous Area	0.01	9.4	0.15	6.7	0.05	9.4	0.79	7.7	−1.79	9.1	−3.77	0.0	0.08	0.5	2.58	0.5	2.19	3.9	−2.00	0.0
14	Chuvash	6.89	1.4	0.73	0.9	2.90	1.9	2.91	2.2	2.02	3.4	2.02	8.9	5.15	9.5	0.60	7.3	2.32	9.0	0.27	6.0
15	Daghestan	1.59	5.0	0.02	9.5	0.52	7.5	−0.40	9.3	−1.92	9.2	2.88	9.5	5.07	9.5	0.48	7.9	2.19	2.8	−0.42	3.1
16	Irkutsk Region	1.07	6.1	0.59	2.4	0.68	6.8	3.36	1.5	1.00	5.5	−0.85	1.5	2.75	6.5	2.50	1.1	2.20	5.6	0.34	6.4
17	Ivanovo Region	7.25	1.3	0.60	2.3	1.97	3.0	0.77	7.7	1.30	4.9	1.30	7.3	3.18	7.8	0.89	5.6	2.19	2.2	−2.00	0.0
18	Jewish Autono-mous Region	0.33	7.7	0.57	2.8	0.83	6.1	0.76	7.7	1.27	5.0	−0.49	2.3	4.96	9.4	2.54	0.7	2.51	9.4	0.41	6.8

19	Kabardino-Balkarian	4.16	2.3	0.05	8.7	0.16	8.7	2.76	2.8	4.01	0.0	-1.53	0.9	1.51	3.3	−4.0	9.4	2.19	2.9	1.17	9.6
20	Kaliningrad Region	9.93	0.6	0.69	1.1	2.32	2.5	0.71	7.8	−1.03	8.8	1.28	7.3	3.01	7.2	0.86	5.8	2.19	2.6	−0.36	3.3
21	Kalmykia	0.57	7.1	0.88	0.2	0.04	9.5	-4.00	10.0	−3.00	9.5	0.31	4.4	31.2	10.0	−4.0	9.4	2.19	2.1	0.20	5.5
22	Kaluga Region	3.61	2.6	0.62	2.0	0.43	7.7	1.32	6.6	−0.08	7.3	0.64	5.5	1.56	3.4	0.49	7.9	2.20	6.1	0.59	8.2
23	Kamchatka Region	0.07	8.9	0.16	6.7	0.10	9.0	1.51	6.1	−0.65	8.3	-1.18	1.1	0.25	0.7	2.16	1.9	1.80	1.0	0.49	7.5
24	Karachayevo-Circassian	4.04	2.3	0.02	9.4	0.99	5.6	2.45	4.0	2.97	0.7	0.10	3.7	0.22	0.6	−4.0	9.4	2.19	3.6	0.87	9.2
25	Karelia	1.13	6.0	0.84	0.4	0.77	6.4	2.53	3.3	0.46	6.7	0.75	5.9	4.58	9.2	1.55	3.2	2.23	8.0	0.46	7.3
26	Kemerovo Region	7.52	1.2	0.33	5.0	12.65	0.0	3.68	0.7	2.69	1.1	0.33	4.5	2.47	5.7	1.27	4.1	2.27	8.9	0.94	9.4
27	Khabarovsk Territory	0.28	8.0	0.44	4.0	0.20	8.5	3.49	1.1	0.59	6.5	-0.02	3.2	1.91	4.5	2.49	1.1	2.00	1.1	0.33	6.4
28	Khakassia	0.92	6.4	0.30	5.2	1.50	3.9	1.78	5.6	−0.03	7.3	0.51	5.2	1.45	2.8	1.38	3.8	2.14	1.6	0.64	8.5
29	Kirov Region	1.42	5.4	0.65	1.5	0.61	7.2	1.50	6.1	1.29	5.0	0.68	5.7	3.13	7.6	0.05	8.8	2.22	7.5	−1.31	1.5
30	Komi	0.34	7.7	0.23	5.8	1.52	3.9	2.47	4.0	1.15	5.2	0.05	3.5	1.45	2.8	0.57	7.6	2.61	9.6	0.80	9.1
31	Kostroma Region	1.06	6.1	0.04	9.1	0.82	6.2	-1.00	9.7	−1.13	8.9	0.50	5.1	3.32	8.2	-0.41	8.9	2.24	8.4	−2.00	0.0
32	Krasnodar Territory	16.4	0.0	0.29	5.3	1.51	3.9	2.52	3.4	2.06	3.4	0.39	4.7	1.60	3.5	0.91	5.5	2.22	7.5	0.68	8.8
33	Krasnoyarsk Territory	0.25	8.1	0.19	6.1	1.06	5.1	4.26	0.0	2.30	2.3	0.02	3.4	0.69	1.2	2.12	1.9	2.22	7.3	0.61	8.4
34	Kurgan Region	0.80	6.7	0.75	0.8	0.96	5.7	0.73	7.8	−0.92	8.6	-0.08	2.9	2.76	6.6	1.62	3.1	2.19	4.8	−2.00	0.0
35	Kursk Region	1.17	5.9	0.11	7.1	0.67	6.9	2.09	5.0	2.15	2.8	1.54	8.0	1.25	2.1	0.56	7.7	2.19	5.3	−0.82	1.9
36	Leningrad Region	4.54	2.1	0.06	8.3	2.16	2.8	3.07	2.0	2.64	1.5	1.03	6.8	3.08	7.4	0.62	7.2	2.66	9.6	−0.32	3.5
37	Lipetsk Region	5.23	1.9	0.43	4.0	16.43	0.0	2.90	2.3	2.60	1.7	1.48	7.8	2.73	6.5	0.94	5.0	2.19	2.4	−0.59	2.3
38	Magadan Region	0.09	8.8	0.45	3.7	0.07	9.2	2.94	2.2	0.78	6.1	-1.30	1.0	0.36	0.8	2.54	0.7	2.70	9.7	0.28	6.1
39	Mari El	2.84	3.6	0.57	2.8	1.08	5.0	-0.60	9.5	0.05	7.2	1.34	7.4	3.17	7.8	0.83	6.0	2.24	8.2	0.41	6.8
40	Mordovia	2.14	4.3	0.67	1.4	1.60	3.7	0.75	7.8	−0.65	8.3	1.13	7.0	3.13	7.7	0.90	5.6	2.19	4.8	0.42	6.9
41	Moscow																				
42	Moscow Region	13.64	0.1	0.20	6.1	3.73	1.4	3.43	1.3	3.24	0.5	-0.63	1.9	1.92	4.5	0.69	6.7	2.18	1.7	−0.54	2.6
43	Murmansk Region	2.34	4.1	0.20	6.1	2.19	2.7	1.30	6.7	−0.58	8.1	-0.19	2.7	1.01	1.6	1.81	2.6	1.74	0.9	0.38	6.6
44	Nizhny Novgorod Region	5.99	1.7	0.35	4.8	2.18	2.8	2.20	4.8	2.19	2.7	1.02	6.7	2.21	5.0	1.09	4.6	2.19	3.6	−0.21	3.9
45	North Ossetia – Alania	12.1	0.3	0.06	8.6	1.00	5.5	2.50	3.8	2.67	1.3	0.14	3.8	0.68	1.2	−4.0	9.4	2.19	5.3	1.02	9.5
46	Novgorod Region	1.46	5.3	0.62	1.9	0.90	5.9	2.52	3.5	2.13	3.0	0.95	6.5	1.45	2.8	0.56	7.7	2.30	8.9	0.55	7.9
47	Novosibirsk Region	0.44	7.4	0.08	7.8	1.16	4.4	2.51	3.7	0.95	5.7	0.36	4.6	1.25	2.1	2.04	2.1	2.19	3.6	−2.00	0.0
48	Omsk Region	1.60	5.0	0.64	1.7	1.70	3.4	3.28	1.7	2.55	1.8	0.31	4.4	1.56	3.4	1.78	2.7	2.21	6.8	−2.00	0.0
49	Orel Region	3.20	3.0	0.58	2.7	0.61	7.2	2.77	2.8	1.97	3.5	-6.00	0.0	1.30	2.3	-4.00	9.4	2.19	5.0	0.53	7.8
50	Orenburg Region	1.23	5.8	0.08	7.8	5.16	0.9	3.86	0.4	3.43	0.4	1.68	8.4	1.79	3.9	1.58	3.2	2.20	5.8	−0.76	2.0
51	Penza Region	3.31	2.8	0.44	3.9	0.49	7.6	2.00	5.2	1.27	5.0	0.83	6.1	4.75	9.3	1.13	4.5	2.20	5.7	−0.33	3.4
52	Perm Region	2.37	4.1	0.14	6.8	4.50	1.1	1.69	5.8	1.60	4.2	0.46	5.0	2.91	7.0	0.07	8.7	2.27	8.8	0.24	5.7
53	Primorye Territory	2.34	4.1	0.60	2.3	1.83	3.2	3.30	1.6	2.73	1.1	-2.17	0.5	1.90	4.4	2.44	1.3	2.20	6.1	0.62	8.4
54	Pskov Region	1.32	5.6	0.21	6.0	0.33	8.1	1.04	7.2	0.85	6.0	0.65	5.6	1.19	2.0	0.92	5.3	2.19	4.5	0.20	5.5
55	Rostov Region	3.34	2.8	0.09	7.4	1.48	3.9	3.28	1.7	1.57	4.3	2.10	9.0	2.94	7.0	1.49	3.4	2.19	5.3	−1.02	1.7

(continued)

(continued)

		The volume of waste-water (per 1 sq. km area)		The ratio of volume of polluted wastewater to the volume of water with-drawn		The total emissions from sta-tionary sources of pollution		Storage of toxic waste production (per 1 sq. km area)		Neutralization of toxic waste production (per 1 sq. km area)		Forests planted to replace the lost (1 sq. km of forest area)		The area of young (1 sq. km of forest area)		Forest fires		The area of pro-tected areas (per 1 sq. km area)		Replacement of the damaged earth	
	Formula for unifi-cation	(2.3)		(2.3)		(2.3)		(2.3)		(2.3)		(2.2)		(2.2)		(2.3)		(2.2)		(2.2)	
1	2	3		4		5		6		7		8		9		10		11		12	
56	Ryazan Region	0.48	7.3	0.08	8.0	3.03	1.8	0.66	7.9	0.79	6.1	1.50	7.8	2.21	5.0	0.75	6.4	2.21	6.9	0.60	8.3
57	Sakha (Yakutia)	0.03	9.3	0.47	3.6	0.04	9.4	0.61	8.0	−3.00	9.5	−0.76	1.6	0.61	1.1	1.88	2.4	2.23	8.0	−0.13	4.2
58	Sakhalin Region	0.60	7.0	0.16	6.5	1.00	5.5	2.07	5.0	1.61	4.2	−0.73	1.7	3.38	8.3	2.51	1.0	2.28	8.9	−0.19	4.0
59	Samara Region	9.12	0.8	0.40	4.3	6.01	0.6	2.47	4.0	2.19	2.6	1.53	7.9	2.54	5.9	1.70	2.9	2.21	7.0	0.47	7.3
60	Saratov Region	2.69	3.8	0.38	4.5	1.09	4.9	3.28	1.7	2.38	2.1	3.24	9.6	4.32	9.0	0.69	6.7	2.19	2.2	−0.59	2.3
61	Smolensk Region	1.83	4.7	0.40	4.3	0.74	6.5	1.33	6.6	0.50	6.6	−0.16	2.7	1.81	4.0	0.63	7.1	2.12	1.5	0.47	7.3
62	St. Petersburg																				
63	Stavropol Territory	3.34	2.8	0.53	3.0	1.10	4.8	1.31	6.7	−0.42	7.8	4.99	10.0	3.77	8.6	1.82	2.5	2.21	6.7	−2.00	0.0
64	Sverdlovsk Region	4.46	2.1	0.67	1.4	6.49	0.5	2.66	3.0	2.70	1.1	0.31	4.4	3.54	8.4	0.77	6.3	2.25	8.6	−0.11	4.3
65	Tambov Region	1.60	5.0	0.29	5.3	0.73	6.6	2.14	4.9	2.68	1.2	0.91	6.4	2.39	5.5	1.42	3.7	2.19	5.4	−0.52	2.7
66	Tatarstan	7.85	1.1	0.55	2.9	3.97	1.3	1.40	6.4	1.13	5.3	3.48	9.7	6.30	9.9	0.52	7.8	2.23	7.9	−0.41	3.1
67	Tomsk Region	0.06	9.0	0.03	9.3	1.06	5.1	2.41	4.2	1.74	3.9	−0.61	2.0	1.43	2.7	2.51	1.0	1.89	1.0	−2.00	0.0
68	Tula Region	10.0	0.5	0.57	2.8	5.95	0.7	3.61	0.8	2.43	2.0	−1.09	1.2	1.51	3.3	−0.60	9.0	2.19	4.2	−2.00	0.0
69	Tuva	0.05	9.1	0.10	7.3	0.18	8.6	−4.00	10.0	−0.70	8.4	−0.59	2.0	0.53	1.0	2.42	1.4	2.18	1.8	0.56	8.0
70	Tver Region	1.02	6.2	0.07	8.3	0.61	7.2	2.14	4.9	0.98	5.6	−0.71	1.7	2.25	5.1	1.23	4.2	2.20	6.2	−0.54	2.6
71	Tyumen Region	0.10	8.8	0.09	7.4	2.38	2.4	3.73	0.6	0.63	6.4	−0.32	2.5	0.84	1.4	0.58	7.4	0.70	0.4	0.22	5.5
72	Udmurtian	0.62	7.0	0.08	7.8	4.58	1.1	2.02	5.1	2.11	3.1	1.92	8.8	4.13	8.9	0.80	6.1	2.23	7.8	−0.31	3.5
73	Ulyanovsk Region	2.90	3.5	0.47	3.6	1.10	4.8	1.19	6.9	0.80	6.1	2.06	8.9	3.38	8.3	1.37	3.8	2.19	4.6	−1.02	1.7
74	Vladimir Region	6.00	1.7	0.73	0.9	1.14	4.5	1.61	5.9	1.86	3.7	1.63	8.2	3.19	7.8	1.16	4.4	2.20	6.5	0.61	8.4
75	Volgograd Region	1.77	4.7	0.16	6.6	1.97	3.0	1.84	5.5	0.19	7.0	2.57	9.3	2.62	6.1	1.11	4.5	2.19	3.2	−2.00	0.0
76	Vologda Region	1.41	5.5	0.34	4.9	3.05	1.8	3.50	1.1	2.63	1.6	0.50	5.1	1.81	4.0	0.70	6.7	2.25	8.5	0.21	5.5
77	Voronezh Region	3.30	2.8	0.28	5.4	0.90	5.9	2.90	2.3	1.56	4.3	1.84	8.6	2.22	5.0	1.40	3.8	2.19	2.8	−0.32	3.4
78	Yaroslavl Region	8.41	1.0	0.81	0.5	2.36	2.4	1.29	6.7	1.44	4.6	1.05	6.8	1.82	4.0	0.94	5.0	2.19	3.9	0.28	6.0

A2.8 The values of the block and consolidated II for the regions of the Russian Federation for each synthetic category

A2.8a The quality of population

i	Region	Block II (weight)			The distance to the reference	$y_i^{(I)}$ Consolidated II of the population-quality	Rank region on population quality
		$y_i^{(1)}(1)$ (0.409)	$y_i^{(1)}(2)$ (0.407)	$y_i^{(1)}(3)$ (0.184)			
1	2	3	4	5	6	7	8
1	Adygeya	7.51	5.03	6.58	3.841	6.16	18
2	Altai	1.93	6.74	5.64	5.873	4.13	48
3	Altai Territory	6.02	4.80	3.79	4.958	5.04	32
4	Amur Region	2.42	6.74	4.29	5.820	4.18	47
5	Arkhangelsk Region	4.04	4.70	2.12	6.112	3.89	51
6	Astrakhan Region	6.35	6.55	5.09	3.837	6.16	17
7	Bashkortostan	7.51	7.61	3.74	3.473	6.53	8
8	Belgorod Region	8.97	3.77	5.22	4.516	5.48	25
9	Bryansk Region	5.35	2.48	4.18	6.168	3.83	53
10	Buryatia	2.42	7.34	3.15	5.916	4.08	49
11	Chelyabinsk Region	4.72	5.02	6.71	4.848	5.15	28
12	Chita region	1.43	6.72	2.06	6.788	3.21	64
13	Chukotka Autonomous Area	3.15	9.39	3.30	5.259	4.74	37
14	Chuvash	5.40	7.14	6.56	3.766	6.23	15
15	Daghestan	8.30	8.86	1.70	3.793	6.21	16
16	Irkutsk Region	1.40	6.44	6.90	6.102	3.90	50
17	Ivanovo Region	3.28	1.76	5.66	7.040	2.96	69
18	Jewish Autonomous Region	2.75	6.25	1.31	6.412	3.59	58
19	Kabardino-Balkarian	7.55	8.81	4.98	2.769	7.23	3
20	Kaliningrad Region	2.30	5.31	4.66	6.202	3.80	54
21	Kalmykia	6.58	8.57	6.99	2.702	7.30	2
22	Kaluga Region	5.36	2.25	4.45	6.235	3.77	56

(continued)

(continued)

i	Region	Block II (weight)			The distance to the reference	$y_i^{(l)}$ Consolidated II of the population-quality	Rank region on population quality
		$y_i^{(1)}(1)$ (0.409)	$y_i^{(1)}(2)$ (0.407)	$y_i^{(1)}(3)$ (0.184)			
1	2	3	4	5	6	7	8
23	Kamchatka region	5.27	8.53	5.95	3.613	6.39	11
24	Karachayevo-Circassian	7.91	6.98	5.02	3.170	6.83	6
25	Karelia	3.41	2.66	2.56	7.063	2.94	70
26	Kemerovo Region	2.32	4.77	3.22	6.611	3.39	62
27	Khabarovsk Territory	2.52	5.99	9.08	5.440	4.56	39
28	Khakassia	1.34	5.50	5.32	6.556	3.44	60
29	Kirov Region	4.06	3.50	4.28	6.138	3.86	52
30	Komi	3.65	7.33	1.48	5.725	4.28	46
31	Kostroma Region	3.71	1.76	1.46	7.566	2.43	73
32	Krasnodar Territory	7.57	5.07	3.57	4.461	5.54	24
33	Krasnoyarsk Territory	3.26	6.59	6.66	5.040	4.96	34
34	Kurgan Region	3.52	5.28	1.79	6.215	3.88	55
35	Kursk Region	6.15	2.51	7.38	5.491	4.51	42
36	Leningrad Region	3.12	1.27	1.86	7.912	2.09	75
37	Lipetsk Region	7.30	2.94	3.36	5.600	4.40	44
38	Magadan Region	5.23	8.15	5.96	3.700	6.30	12
39	Mari El	4.48	5.77	6.84	4.645	5.35	27
40	Mordovia	7.62	3.23	8.18	4.642	5.36	26
41	Moscow	8.78	2.49	10.00	4.854	5.15	29
42	Moscow Region	5.90	2.51	8.46	5.486	4.51	41
43	Murmansk Region	6.67	7.96	2.68	4.013	5.99	19
44	Nizhny Novgorod Region	4.67	1.50	6.23	6.607	3.39	61
45	North Ossetia – Alania	7.69	7.30	6.58	2.701	7.30	1
46	Novgorod Region	2.06	0.61	3.61	8.314	1.69	77
47	Novosibirsk Region	6.20	5.65	8.81	3.722	6.28	14

(continued)

i	Region	Block II (weight)			The distance to the reference	$y_i^{(I)}$ Consolidated II of the population-quality	Rank region on population quality
		$y_i^{(1)}(1)$ (0.409)	$y_i^{(1)}(2)$ (0.407)	$y_i^{(1)}(3)$ (0.184)			
1	2	3	4	5	6	7	8
48	Omsk Region	6.32	5.44	5.76	4.159	5.84	21
49	Orel Region	5.71	2.46	8.09	5.596	4.40	43
50	Orenburg Region	5.63	4.82	3.67	5.113	4.89	36
51	Penza Region	5.90	4.03	6.01	4.929	5.07	31
52	Perm Region	2.44	4.54	2.02	6.876	3.12	68
53	Primorye Territory	3.52	6.32	6.10	5.050	4.95	35
54	Pskov Region	2.03	0.89	2.22	8.420	1.58	78
55	Rostov Region	7.22	3.85	7.78	4.412	5.59	23
56	Ryazan Region	4.47	1.19	6.57	6.801	3.20	65
57	Sakha (Yakutia)	6.37	9.35	4.76	3.258	6.74	7
58	Sakhalin Region	2.94	6.15	1.52	6.298	3.70	57
59	Samara Region	6.97	5.57	8.44	3.492	6.51	10
60	Saratov Region	6.60	4.87	7.77	4.044	5.96	20
61	Smolensk Region	3.48	1.60	3.67	7.312	2.69	72
62	St. Petersburg	7.83	1.75	9.71	5.443	4.56	40
63	Stavropol Territory	7.94	5.28	6.03	3.700	7.30	13
64	Sverdlovsk Region	4.13	4.36	4.98	5.627	4.37	45
65	Tambov Region	5.52	1.35	3.73	6.772	3.23	63
66	Tatarstan	8.11	6.65	6.70	2.834	7.17	5
67	Tomsk Region	5.46	7.01	9.44	3.483	6.52	9
68	Tula Region	3.42	1.06	2.11	7.854	2.15	74
69	Tuva	0.55	8.23	3.09	6.825	3.17	67
70	Tver Region	2.14	1.45	2.94	8.009	1.99	76
71	Tyumen Region	7.83	9.17	4.39	2.830	7.17	4
72	Udmurtian	3.69	7.19	5.29	4.854	5.15	30
73	Ulyanovsk Region	6.08	3.42	5.06	5.328	4.67	38
74	Vladimir Region	4.96	1.53	6.22	6.497	3.50	59
75	Volgograd Region	6.21	3.74	5.65	5.029	4.97	33

(continued)

(continued)

i	Region	Block II (weight)			The distance to the reference	$y_i^{(I)}$ Consolidated II of the population-quality	Rank region on population quality
		$y_i^{(1)}(1)$ (0.409)	$y_i^{(1)}(2)$ (0.407)	$y_i^{(1)}(3)$ (0.184)			
1	2	3	4	5	6	7	8
76	Vologda Region	3.41	2.31	2.07	7.304	2.70	71
77	Voronezh Region	8.00	3.81	8.75	4.187	5.81	22
78	Yaroslavl Region	3.89	2.01	4.71	6.813	3.19	66

A2.8b The level of well-being

i	Region	Block II (weight)		The distance to the reference	$y_i^{(II)}$ Consolidated II welfare	Rank region on Wellbeing of Population
		$y_i^{(1)}(1)$ (0.595)	$y_i^{(1)}(2)$ (0.405)			
1	2	3	4	5	6	7
1	Adygeya	2.71	7.11	5.917	4.08	55
2	Altai	1.85	2.12	8.040	2.96	77
3	Altai Territory	3.95	4.47	5.843	4.16	52
4	Amur Region	3.13	2.02	7.342	2.66	72
5	Arkhangelsk Region	6.35	1.57	6.062	3.94	58
6	Astrakhan Region	6.73	4.02	4.566	5.43	26
7	Bashkortostan	8.48	7.44	2.008	8.99	5
8	Belgorod Region	6.66	8.73	2.699	7.30	8
9	Bryansk Region	4.47	6.98	4.676	5.33	31
10	Buryatia	2.82	2.38	7.363	2.64	73
11	Chelyabinsk Region	5.10	5.94	4.575	5.42	27
12	Chita Region	3.13	2.43	7.164	2.84	70
13	Chukotka Autonomous Area	3.65	5.21	5.768	4.23	51
14	Chuvash	3.35	8.35	5.231	4.77	43
15	Daghestan	1.07	3.94	7.892	2.11	76
16	Irkutsk Region	5.49	1.82	6.258	3.74	62
17	Ivanovo Region	1.08	5.49	7.451	2.55	74

(continued)

i	Region	Block II (weight)		The distance to the reference	$y_i^{(II)}$ Consolidated II welfare	Rank region on Wellbeing of Population
		$y_i^{(1)}(1)$ (0.595)	$y_i^{(1)}(2)$ (0.405)			
1	2	3	4	5	6	7
18	Jewish Autonomous Region	4.03	1.29	7.207	2.79	71
19	Kabardino-Balkarian	3.51	6.54	5.466	4.53	49
20	Kaliningrad Region	4.58	6.90	4.622	5.38	28
21	Kalmykia	1.91	4.90	7.037	2.96	67
22	Kaluga Region	4.24	5.49	5.288	4.71	45
23	Kamchatka Region	4.83	1.50	6.725	3.23	65
24	Karachayevo-Circassian	2.14	5.96	6.586	3.41	64
25	Karelia	7.38	3.09	4.840	5.16	36
26	Kemerovo Region	7.22	3.29	4.782	5.22	32
27	Khabarovsk Territory	6.41	2.74	5.391	4.61	47
28	Khakassia	4.42	3.23	6.093	3.91	60
29	Kirov Region	3.07	3.22	6.865	3.13	66
30	Komi	8.67	2.23	5.053	4.95	39
31	Kostroma Region	3.30	4.19	6.354	3.65	63
32	Krasnodar Territory	6.35	6.91	3.433	6.57	11
33	Krasnoyarsk Territory	7.26	3.22	4.808	5.19	33
34	Kurgan Region	1.83	2.77	7.800	2.20	75
35	Kursk Region	3.85	8.36	4.859	5.14	37
36	Leningrad Region	3.09	6.15	5.864	4.14	53
37	Lipetsk Region	7.83	8.70	1.871	8.13	4
38	Magadan Region	7.45	1.22	5.926	4.07	56
39	Mari El	1.45	6.12	7.040	2.96	68
40	Mordovia	2.52	6.50	6.184	3.82	61
41	Moscow	9.09	9.27	0.842	9.16	1
42	Moscow Region	6.68	9.45	2.585	7.41	7
43	Murmansk Region	7.51	3.32	4.665	5.33	30
44	Nizhny Novgorod Region	6.98	6.12	3.392	6.61	10
45	North Ossetia – Alania	4.74	7.60	4.337	5.66	23
46	Novgorod Region	6.00	5.27	4.310	5.69	21
47	Novosibirsk Region	4.94	4.47	5.256	4.74	44
48	Omsk Region	6.94	4.85	4.040	5.96	18
49	Orel Region	5.85	6.86	3.775	6.22	15
50	Orenburg Region	4.85	4.90	5.130	4.87	40
51	Penza Region	3.03	5.97	5.956	4.04	57
52	Perm Region	7.83	3.23	4.622	5.38	29
53	Primorye Territory	2.81	3.09	7.076	2.93	69

(continued)

(continued)

i	Region	Block II (weight)		The distance to the reference	$y_i^{(II)}$ Consolidated II welfare	Rank region on Wellbeing of Population
		$y_i^{(1)}(1)$ (0.595)	$y_i^{(1)}(2)$ (0.405)			
1	2	3	4	5	6	7
54	Pskov Region	4.84	5.76	4.812	5.19	35
55	Rostov Region	6.68	5.66	3.768	6.23	14
56	Ryazan Region	5.13	6.61	4.330	5.67	22
57	Sakha (Yakutia)	7.85	2.53	5.038	4.96	38
58	Sakhalin Region	6.68	1.35	6.076	3.92	59
59	Samara Region	8.71	6.84	2.243	7.76	6
60	Saratov Region	4.97	4.60	5.186	4.83	42
61	Smolensk Region	6.22	7.48	3.324	6.68	9
62	St. Petersburg	8.05	9.29	1.572	8.43	2
63	Stavropol Territory	4.44	6.57	4.810	5.19	34
64	Sverdlovsk Region	8.47	4.88	3.467	6.53	12
65	Tambov Region	6.55	6.02	3.677	6.32	13
66	Tatarstan	8.49	7.85	1.800	8.20	3
67	Tomsk Region	6.31	2.91	5.335	4.66	46
68	Tula Region	5.98	4.88	4.500	5.50	25
69	Tuva	1.07	0.94	8.984	1.02	78
70	Tver Region	3.07	6.10	5.891	4.11	54
71	Tyumen Region	9.35	3.00	4.484	5.52	24
72	Udmurtian	4.41	5.58	5.152	4.85	41
73	Ulyanovsk Region	3.91	5.72	5.431	4.57	48
74	Vladimir Region	3.39	6.61	5.532	4.47	50
75	Volgograd Region	7.11	4.63	4.083	5.92	19
76	Vologda Region	7.15	4.30	4.239	5.76	20
77	Voronezh Region	5.03	8.33	3.980	6.02	17
78	Yaroslavl Region	7.23	4.93	3.869	6.13	16

A2.8c The quality of social services

i	Region	Block II (weight)				The distance to the reference	$y_i^{(III)}$ consolidated II-quality of social services	Rank region on QSS
		$y_i^{(1)}(1)$ (0.260)	$y_i^{(1)}(2)$ (0.452)	$y_i^{(1)}(3)$ (0.156)	$y_i^{(1)}(4)$ (0.132)			
1	2	3	4	5	6	7	8	9
1	Adygeya	2.701	7.835	5.812	1.755	5.259	4.74	40
2	Altai	2.600	1.968	6.249	1.899	7.365	2.73	75
3	Altai Territory	2.467	3.590	2.505	2.854	6.988	3.01	69
4	Amur Region	2.582	2.948	5.617	3.035	6.796	3.20	66
5	Arkhangelsk Region	4.766	4.275	8.491	6.574	4.884	5.12	32
6	Astrakhan Region	3.884	5.017	4.307	5.830	5.321	5.68	42
7	Bashkortostan	6.247	6.328	4.041	4.031	4.470	5.43	24
8	Belgorod Region	7.286	8.161	8.147	6.704	2.326	7.67	1
9	Bryansk Region	4.565	6.429	7.406	4.823	4.246	5.75	17
10	Buryatia	2.735	1.712	4.198	4.769	7.325	2.67	74
11	Chelyabinsk Region	6.473	3.140	2.637	5.207	6.000	4.00	56
12	Chita Region	0.929	1.438	4.939	1.479	8.253	1.75	78
13	Chukotka Autonomous Area	6.568	3.190	6.781	7.259	5.161	4.84	39
14	Chuvash	5.279	5.924	7.799	3.508	4.428	5.57	23
15	Daghestan	1.007	9.519	6.536	0.729	5.859	4.14	53
16	Irkutsk Region	2.931	2.092	0.498	2.998	7.863	2.14	76
17	Ivanovo Region	6.116	4.101	6.177	3.200	5.293	4.71	41
18	Jewish Autonomous Region	4.817	1.448	7.272	1.167	7.174	2.83	73
19	Kabardino-Balkarian	1.200	9.319	7.082	2.831	5.335	4.66	43
20	Kaliningrad Region	6.308	4.989	4.150	3.741	5.039	4.96	37
21	Kalmykia	0.764	5.180	5.661	3.880	6.371	3.63	62
22	Kaluga Region	7.404	6.510	6.511	3.996	3.727	6.23	10
23	Kamchatka Region	2.355	4.245	5.992	7.086	5.816	4.18	51
24	Karachayevo-Circassian	1.260	7.505	5.691	1.253	5.971	4.03	55
25	Karelia	6.056	2.938	7.596	6.933	5.360	4.64	44
26	Kemerovo Region	4.932	2.566	1.501	8.074	6.591	3.41	63
27	Khabarovsk Territory	5.293	1.398	3.083	6.293	6.964	3.04	68
28	Khakassia	3.685	2.448	5.858	2.904	6.742	3.26	65
29	Kirov Region	5.893	4.806	9.294	4.620	4.524	5.48	26
30	Komi	2.734	1.767	4.774	7.703	7.025	2.97	70
31	Kostroma Region	5.919	4.531	6.186	5.669	4.753	5.25	30
32	Krasnodar Territory	6.235	8.124	2.201	4.320	4.360	5.64	20
33	Krasnoyarsk Territory	3.049	3.710	1.618	5.094	6.679	3.32	64
34	Kurgan Region	3.246	3.499	2.471	2.480	6.875	3.12	67
35	Kursk Region	5.116	6.479	8.152	3.794	4.172	5.83	14
36	Leningrad Region	7.670	4.278	4.244	4.374	5.054	4.95	38

(continued)

(continued)

i	Region	Block II (weight)				The distance to the reference	$y_i^{(III)}$ consolidated II-quality of social services	Rank region on QSS
		$y_i^{(1)}(1)$ (0.260)	$y_i^{(1)}(2)$ (0.452)	$y_i^{(1)}(3)$ (0.156)	$y_i^{(1)}(4)$ (0.132)			
1	2	3	4	5	6	7	8	9
37	Lipetsk Region	7.707	6.689	6.909	5.598	3.219	6.78	2
38	Magadan Region	3.390	3.769	5.000	7.631	5.795	4.20	50
39	Mari El	2.822	4.104	6.181	2.426	6.241	3.76	58
40	Mordovia	4.199	6.362	7.749	5.428	4.276	5.72	18
41	Moscow	9.363	8.527	1.914	8.215	3.422	6.58	4
42	Moscow Region	8.891	6.597	3.948	8.016	3.434	6.57	5
43	Murmansk Region	3.244	5.495	3.683	6.573	5.370	4.63	45
44	Nizhny Novgorod Region	8.253	6.163	5.352	7.638	3.399	6.60	3
45	North Ossetia – Alania	2.533	8.655	5.072	4.281	4.840	5.16	31
46	Novgorod Region	7.410	4.280	4.346	4.988	4.982	5.02	36
47	Novosibirsk Region	3.627	3.950	3.499	3.485	6.269	3.73	60
48	Omsk Region	4.310	4.698	3.670	6.479	5.388	4.61	46
49	Orel Region	5.541	6.795	5.888	5.009	3.967	6.03	11
50	Orenburg Region	4.557	4.647	4.090	4.980	5.425	4.57	47
51	Penza Region	4.440	7.898	6.185	3.518	4.223	5.78	15
52	Perm Region	5.543	1.538	1.925	5.604	7.090	2.91	72
53	Primorye Territory	4.171	2.674	2.830	2.057	7.030	2.97	71
54	Pskov Region	5.537	5.219	8.055	6.188	4.245	5.74	16
55	Rostov Region	5.718	7.143	3.123	5.068	4.364	5.64	21
56	Ryazan Region	6.293	7.488	6.903	4.067	3.542	6.46	7
57	Sakha (Yakutia)	4.219	2.683	5.865	6.221	6.120	3.88	57
58	Sakhalin Region	5.310	2.825	4.582	7.874	5.846	4.15	52
59	Samara Region	7.659	5.866	0.316	7.878	4.938	5.06	35
60	Saratov Region	4.888	6.510	3.839	5.503	4.572	5.43	27
61	Smolensk Region	4.538	5.439	5.762	5.732	4.730	5.27	29
62	St. Petersburg	8.863	7.731	2.370	8.159	3.493	6.51	6
63	Stavropol Territory	5.760	7.765	6.554	3.769	3.727	6.27	9
64	Sverdlovsk Region	7.139	2.612	1.734	5.315	6.353	3.65	61
65	Tambov Region	2.908	8.106	5.621	5.028	4.579	5.42	28
66	Tatarstan	7.338	6.949	2.819	6.210	3.999	6.00	12
67	Tomsk Region	3.577	3.954	2.405	5.264	6.263	3.74	59
68	Tula Region	7.569	6.101	5.302	6.962	3.615	6.38	8
69	Tuva	0.812	1.421	5.390	4.581	7.902	2.10	77
70	Tver Region	6.589	4.294	5.908	4.561	4.924	5.08	34
71	Tyumen Region	6.848	3.622	0.528	8.692	5.934	4.07	54
72	Udmurtian	5.949	3.188	7.173	3.923	5.599	4.40	49
73	Ulyanovsk Region	4.119	5.457	2.150	5.364	5.548	4.45	48
74	Vladimir Region	5.007	5.418	7.785	5.934	4.350	5.65	19

(continued)

i	Region	Block II (weight)				The distance to the reference	$y_i^{(III)}$ consolidated II-quality of social services	Rank region on QSS
		$y_i^{(1)}(1)$ (0.260)	$y_i^{(1)}(2)$ (0.452)	$y_i^{(1)}(3)$ (0.156)	$y_i^{(1)}(4)$ (0.132)			
1	2	3	4	5	6	7	8	9
75	Volgograd Region	5.934	5.351	5.098	5.719	4.498	5.50	25
76	Vologda Region	7.780	3.370	6.403	7.452	4.902	5.10	33
77	Voronezh Region	4.620	7.729	5.162	5.277	4.057	5.94	13
78	Yaroslavl Region	8.260	4.131	5.843	9.056	4.379	5.62	22

A2.8d The quality of the ecological niche

i	Region	Block II (weight)				The distance to the reference	$y_i^{(III)}$ Consolidated II of the environmental quality	Rank region on QEN
		$y_i^{(1)}(1)$ (0.392)	$y_i^{(1)}(2)$ (0.276)	$y_i^{(1)}(3)$ (0.185)	$y_i^{(1)}(4)$ (0.147)			
1	2	3	4	5	6	7	8	9
1	Adygeya	7.687	5.546	9.542	3.889	3.620	6.38	4
2	Altai	9.109	2.259	9.502	1.925	5.146	4.85	27
3	Altai Territory	4.082	6.715	2.196	2.641	5.999	4.00	52
4	Amur Region	8.893	1.486	5.000	9.435	5.014	4.99	22
5	Arkhangelsk Region	5.869	5.240	3.881	0.975	5.647	4.35	40
6	Astrakhan Region	5.913	7.071	6.114	4.422	4.037	5.96	9
7	Bashkortostan	1.394	8.221	7.011	3.371	6.163	3.84	60
8	Belgorod Region	3.606	4.690	1.719	6.201	6.216	3.78	62
9	Bryansk Region	7.019	6.658	2.949	3.283	4.738	5.26	17
10	Buryatia	8.741	1.223	9.098	1.664	5.678	4.32	41
11	Chelyabinsk Region	0.281	4.897	6.589	6.588	6.928	3.07	69
12	Chita Region	4.673	1.875	4.585	7.554	5.968	4.03	51
13	Chukotka Autonomous Area	8.848	0.276	0.000	3.889	7.118	2.88	72
14	Chuvash	2.422	8.820	5.989	8.998	5.097	4.92	24
15	Daghestan	8.145	9.200	3.070	2.771	4.262	5.74	11
16	Irkutsk Region	4.831	3.291	6.383	5.645	5.299	4.70	33
17	Ivanovo Region	4.653	7.209	0.000	2.232	6.389	3.61	65
18	Jewish Autonomous Region	6.444	4.607	6.838	9.448	3.855	6.14	7
19	Kabardino-Balkarian	3.111	3.310	9.614	2.920	6.190	3.81	61

(continued)

(continued)

i	Region	Block II (weight)				The distance to the reference	$y_i^{(III)}$ Consolidated II of the environmental quality	Rank region on QEN
		$y_i^{(1)}(1)$ (0.392)	$y_i^{(1)}(2)$ (0.276)	$y_i^{(1)}(3)$ (0.185)	$y_i^{(1)}(4)$ (0.147)			
1	2	3	4	5	6	7	8	9
20	Kaliningrad Region	5.737	7.014	3.274	2.572	5.110	4.89	25
21	Kalmykia	9.229	7.377	5.480	2.129	3.882	6.12	8
22	Kaluga Region	6.457	5.192	8.207	6.115	3.755	6.24	6
23	Kamchatka Region	7.955	1.103	7.496	0.984	6.049	3.95	56
24	Karachayevo-Circassian	3.000	3.606	9.219	3.571	6.053	3.95	57
25	Karelia	5.615	6.607	7.264	8.025	3.558	6.44	3
26	Kemerovo Region	0.765	4.901	9.356	8.874	6.386	3.61	64
27	Khabarovsk Territory	5.746	3.296	6.371	1.148	5.785	4.21	43
28	Khakassia	5.918	4.056	8.517	1.559	5.213	4.79	29
29	Kirov Region	5.851	6.946	1.496	7.479	4.867	5.13	20
30	Komi	4.996	3.985	9.100	9.575	4.465	5.54	12
31	Kostroma Region	8.042	6.923	0.000	8.361	4.806	5.19	18
32	Krasnodar Territory	2.930	4.425	8.795	7.509	5.412	4.59	34
33	Krasnoyarsk Territory	3.272	2.348	8.354	7.270	5.953	4.05	50
34	Kurgan Region	7.407	4.271	0.000	4.775	5.856	4.14	47
35	Kursk Region	4.846	5.767	1.919	5.323	5.540	4.46	37
36	Leningrad Region	2.021	7.089	3.452	9.628	5.935	4.06	48
37	Lipetsk Region	1.515	6.793	2.287	2.420	7.108	2.89	71
38	Magadan Region	6.211	0.906	6.075	9.681	5.593	4.41	39
39	Mari El	6.704	7.293	6.837	8.177	2.936	7.06	1
40	Mordovia	6.430	6.977	6.943	4.821	3.633	6.37	5
41	Moscow							
42	Moscow Region	0.847	3.714	2.578	1.657	8.010	1.99	75
43	Murmansk Region	5.846	2.259	6.615	0.942	6.123	3.88	59
44	Nizhny Novgorod Region	3.091	5.721	3.932	3.633	6.045	4.95	55
45	North Ossetia – Alania	2.777	3.873	9.480	5.323	5.833	4.17	46
46	Novgorod Region	4.153	5.373	7.859	8.949	4.504	5.50	13
47	Novosibirsk Region	5.143	3.244	0.000	3.645	6.808	3.19	68
48	Omsk Region	2.660	3.765	0.000	6.777	7.204	2.80	73
49	Orel Region	4.069	2.528	7.759	5.000	5.810	4.19	44
50	Orenburg Region	1.397	5.788	2.005	5.788	6.952	3.05	70
51	Penza Region	5.277	6.988	3.376	5.741	4.697	5.30	16
52	Perm Region	3.902	6.391	5.663	8.848	4.672	5.33	15
53	Primorye Territory	2.210	2.077	8.375	6.081	6.617	3.38	67
54	Pskov Region	6.718	4.199	5.471	4.470	4.670	5.33	14

(continued)

i	Region	Block II (weight)				The distance to the reference	$y_i^{(III)}$ Consolidated II of the environmental quality	Rank region on QEN
		$y_i^{(1)}(1)$ (0.392)	$y_i^{(1)}(2)$ (0.276)	$y_i^{(1)}(3)$ (0.185)	$y_i^{(1)}(4)$ (0.147)			
1	2	3	4	5	6	7	8	9
55	Rostov Region	3.250	7.257	1.655	5.265	6.012	3.99	54
56	Ryazan Region	5.791	6.537	8.285	6.860	3.497	6.50	2
57	Sakha (Yakutia)	9.029	1.568	4.214	7.964	5.176	4.82	28
58	Sakhalin Region	5.179	3.984	4.030	8.905	5.085	4.91	23
59	Samara Region	2.247	6.298	7.334	6.963	5.474	4.53	35
60	Saratov Region	2.915	8.875	2.255	2.215	6.331	3.67	63
61	Smolensk Region	6.281	3.955	7.339	1.501	5.239	4.76	31
62	St. Petersburg							
63	Stavropol Territory	5.976	8.158	0.000	6.740	5.237	4.76	30
64	Sverdlovsk Region	1.656	6.234	4.314	8.568	6.120	3.88	58
65	Tambov Region	4.039	5.604	2.689	5.390	5.683	4.32	42
66	Tatarstan	4.004	9.432	3.112	7.908	4.860	5.14	19
67	Tomsk Region	5.089	2.080	0.000	1.044	7.559	2.44	74
68	Tula Region	1.145	3.393	0.000	4.238	8.135	1.86	76
69	Tuva	8.973	1.541	7.976	1.849	5.540	4.46	36
70	Tver Region	5.883	3.415	2.599	6.164	5.561	4.44	38
71	Tyumen Region	4.301	2.971	5.528	0.401	6.605	3.39	66
72	Udmurtian	3.841	8.331	3.504	7.750	4.919	5.08	21
73	Ulyanovsk Region	5.590	7.800	1.656	4.552	5.125	4.87	26
74	Vladimir Region	4.137	7.400	8.357	6.451	4.203	5.80	10
75	Volgograd Region	5.326	7.285	0.000	3.179	6.003	3.00	53
76	Vologda Region	2.128	4.973	5.483	8.488	5.944	4.06	49
77	Voronezh Region	3.889	6.423	3.434	2.803	5.812	4.19	45
78	Yaroslavl Region	4.059	5.460	6.022	3.889	5.285	4.71	32

A2.9 The values of consolidated QOL index and corresponding ranks for the regions of the Russian Federation

i	Region	$\hat{y}_i^{cons}$	Rank
1	2	3	4
1	Adygeya	5.525	9
2	Altai	3.921	63
3	Altai Territory	3.991	62
4	Amur Region	3.918	64
5	Arkhangelsk Region	4.395	52
6	Astrakhan Region	5.690	5
7	Bashkortostan	6.005	3
8	Belgorod Region	5.887	4
9	Bryansk Region	5.042	26
10	Buryatia	4.007	61
11	Chelyabinsk Region	4.339	54
12	Chita Region	3.035	76
13	Chukotka Autonomous Area	4.367	53
14	Chuvash	5.175	22
15	Daghestan	4.510	46
16	Irkutsk Region	3.731	70
17	Ivanovo Region	3.264	79
18	Jewish Autonomous Region	3.858	66
19	Kabardino-Balkarian	5.150	23
20	Kaliningrad Region	4.843	31
21	Kalmykia	5.224	19
22	Kaluga Region	5.217	21
23	Kamchatka Region	4.952	29
24	Karachayevo-Circassian	4.593	43
25	Karelia	4.727	37
26	Kemerovo Region	3.903	65
27	Khabarovsk Territory	4.413	50
28	Khakassia	4.171	57
29	Kirov Region	4.073	60
30	Komi	4.600	42
31	Kostroma Region	3.844	67
32	Krasnodar Territory	5.639	6
33	Krasnoyarsk Territory	4.452	49
34	Kurgan Region	3.192	74
35	Kursk Region	4.811	34
36	Leningrad Region	3.587	71
37	Lipetsk Region	5.536	8
38	Magadan Region	4.920	30
39	Mari El	4.689	39
40	Mordovia	5.265	17

(continued)

i	Region	$\hat{y}_i^{cons}$	Rank
1	2	3	4
41	Moscow*	6.004	
42	Moscow Region	5.146	24
43	Murmansk Region	5.392	13
44	Nizhny Novgorod Region	5.103	25
45	North Ossetia – Alania	5.633	7
46	Novgorod Region	4.411	51
47	Novosibirsk Region	4.508	47
48	Omsk Region	4.674	40
49	Orel Region	5.313	16
50	Orenburg Region	4.112	59
51	Penza Region	4.838	32
52	Perm Region	4.265	56
53	Primorye Territory	3.540	72
54	Pskov Region	4.543	45
55	Rostov Region	5.222	20
56	Ryazan Region	5.413	11
57	Sakha (Yakutia)	5.405	12
58	Sakhalin Region	4.132	58
59	Samara Region	6.054	2
60	Saratov Region	4.816	33
61	Smolensk Region	5.230	18
62	St. Petersburg*	5.316	
63	Stavropol Territory	5.365	14
64	Sverdlovsk Region	4.619	41
65	Tambov Region	4.737	35
66	Tatarstan	6.662	1
67	Tomsk Region	4.485	48
68	Tula Region	3.787	69
69	Tuva	3.176	75
70	Tver Region	3.838	68
71	Tyumen Region	5.346	15
72	Udmurtian	4.691	38
73	Ulyanovsk Region	4.546	44
74	Vladimir Region	4.728	36
75	Volgograd Region	5.019	27
76	Vologda Region	4.278	55
77	Voronezh Region	5.454	10
78	Yaroslavl Region	4.991	28

* Specificity of metropolises of Moscow and St.Petersburg led to their exclusion from the regions, which were all compared by single (combined) II QOL

A3 Appendices to Chapter 3

A3.1 Resulting values for indicators (criterion) of synthetic quality categories and lifestyle $y^{(1)}$–$y^{(4)}$ and relevant explanatory variables $x^{(1)}$–$x^{(17)}$ for 60 countries and regions (source of data [WCY, 2004])

No	Country (region)	Resulting synthetic categories				Explanatory variables –				
		$y^{(1)}$ 4.4.08	$y^{(2)}$ 1.1.24	$y^{(3)}$ 2.5.05	$y^{(4)}$ 4.4.17	$x^{(1)}$ 4.4.01	$x^{(2)}$ 4.4.09	$x^{(3)}$ 44.15	$x^{(4)}$ 4.4.12	$x^{(5)}$ 4.5.01
1	Argentina*	0.849	2015	4.80	2.75	9.500	6.19	4.75	503.3	4.390
2	Australia*	0.939	11971	7.90	9.35	9.200	6.94	8.08	822.2	5.216
3	Austria*	0.929	17761	7.70	9.44	7.700	8.22	8.74	288.2	5.609
4	Bavaria	0.921	14633	7.31	9.33	10.700	7.11	7.70	251.5	4.189
5	Belgium*	0.937	16016	6.80	8.43	9.000	7.27	6.50	472.7	5.198
6	Brazil*	0.777	1569	6.50	3.67	7.600	6.13	4.83	577.6	1.645
7	Canada*	0.937	15632	8.33	9.24	9.700	7.52	7.18	739.9	8.127
8	Catalonia	0.918	10957	6.22	8.28	7.500	7.37	6.39	451.5	4.384
9	Chile*	0.831	2842	7.62	6.43	7.000	6.88	5.62	808.3	4.220
10	China Mainland*	0.721	456	7.15	3.96	5.500	6.39	2.91	3077.7	3.141
11	Colombia*	0.779	1212	7.17	3.91	5.500	6.67	5.81	655.6	3.819
12	Czech Republic*	0.861	4449	6.80	6.25	7.300	6.75	5.95	2012.0	4.607
13	Denmark*	0.930	18629	8.21	8.72	8.600	8.58	8.53	308.0	8.804
14	Estonia	0.833	3485	4.25	4.76	5.500	4.98	6.40	2827.1	6.467
15	Finland*	0.930	16235	8.49	8.68	7.000	7.78	8.65	452.1	6.346
16	France*	0.925	16281	6.89	8.03	9.500	7.46	6.63	250.3	6.287
17	Germany*	0.921	17068	7.42	8.68	10.700	7.32	8.09	389.8	4.189
18	Greece*	0.892	11969	6.74	6.40	9.400	8.18	5.73	678.8	4.334
19	Hong Kong*	0.889	13327	5.53	6.43	KONG	8.16	4.98	137.0	4.349
20	Hungary*	0.837	5275	5.45	4.94	6.800	6.31	5.42	1205.9	6.207
21	Iceland*	0.942	20144	6.92	9.52	9.200	8.08	9.08	249.6	6.885
22	Ile-De-France	0.925	24692	6.48	8.07	9.500	7.16	5.82	121.6	6.287
23	India*	0.590	317	6.42	4.93	5.100	7.06	4.46	2208.5	3.996
24	Indonesia*	0.682	679	5.32	3.35	2.400	6.04	3.17	1722.9	0.915
25	Ireland*	0.930	18087	7.27	7.59	7.858	5.30	6.37	413.3	5.730
26	Israel*	0.905	9775	4.74	5.68	8.700	8.11	5.79	546.9	8.600
27	Italy*	0.916	15169	3.61	6.39	8.400	7.48	4.66	356.2	4.881
28	Japan*	0.932	19184	6.38	6.70	8.000	8.35	7.16	260.2	3.604
29	Jordan	0.743	1396	7.03	5.00	7.781	8.00	6.49	1646.6	5.759
30	Korea*	0.879	6805	5.21	5.67	6.000	6.75	5.46	921.9	3.571

(continued)

(continued)

No	Country (region)	Resulting synthetic categories				Explanatory variables –				
		$y^{(1)}$ 4.4.08	$y^{(2)}$ 1.1.24	$y^{(3)}$ 2.5.05	$y^{(4)}$ 4.4.17	$x^{(1)}$ 4.4.01	$x^{(2)}$ 4.4.09	$x^{(3)}$ 44.15	$x^{(4)}$ 4.4.12	$x^{(5)}$ 4.5.01
31	Lombardy	0.916	14349	4.30	6.75	8.400	6.56	4.68	356.2	4.881

* The country is on the list of countries for which a comparative analysis of the dynamics was carried out for macroeconomic indicators and synthetic categories of life quality and lifestyle (see point 3.2.6).

– characteristics of institutional development and socio-economic policy											
$x^{(6)}$ 4.3.02	$x^{(7)}$ 4.3.10	$x^{(8)}$ 2.4.17	$x^{(9)}$ 3.3.18	$x^{(10)}$ 2.3.16	$x^{(11)}$ 2.3.17	$x^{(12)}$ $\frac{2.5.08}{2.5.07}$	$x^{(13)}$ 2.4.02	$x^{(14)}$ 2.5.01	$x^{(15)}$ 3.4.09	$x^{(16)}$ 2.5.02	$x^{(17)}$ 2.5.13
10.5	3.40	3.39	3.14	1.14	0.65	11.52	4.51	1.35	3.40	0.84	4.78
313.8	7.01	8.00	8.21	5.11	8.44	7	7.95	9.32	6.66	9.14	8.55
492.1	5.85	8.30	7.32	4.30	8.22	5.41	7.89	9.04	7.38	9.52	9.26
789.6	7.56	7.63	7.26	2.52	6.22	7.84	6.59	8.37	6.07	9.63	8.59
478.0	6.64	6.88	6.85	2.41	5.29	4.49	7.56	5.40	6.53	6.03	7.93
37.3	3.92	6.38	5.67	1.70	2.51	29.14	5.87	4.08	5.71	3.29	5.88
422.6	7.44	8.25	7.56	4.89	7.47	5.38	7.33	8.75	6.90	9.08	8.75
236.5	3.50	7.28	6.57	3.83	5.67	5.37	7.00	6.17	5.61	7.28	7.89
22.3	4.41	8.48	7.59	4.58	5.88	19.16	8.29	6.38	6.23	7.10	8.35
12.1	7.19	6.07	3.98	1.57	1.17	7.9	6.57	5.26	5.59	5.39	5.89
7.9	3.56	7.47	7.26	3.16	2.68	20.3	6.62	3.52	6.26	3.87	5.26
88.5	3.90	6.25	4.97	2.25	2.50	3.49	6.15	4.30	4.30	5.85	8.85
711.2	6.36	8.52	7.79	6.41	9.12	4.31	8.31	8.98	7.69	9.00	9.12
38.6	4.98	7.35	6.76	4.84	4.96	6.44	7.71	6.98	4.76	5.78	8.15
798.2	7.38	8.62	8.25	6.09	9.38	3.47	8.31	8.77	6.78	9.17	9.23
527.0	6.09	7.10	6.40	2.77	6.17	5.58	7.00	6.77	6.29	7.40	7.66
603.9	6.49	7.26	6.59	2.10	6.38	7.84	7.25	8.17	5.78	8.68	8.55
75.6	4.07	5.53	5.70	1.65	2.74	6.14	6.04	6.30	4.96	7.48	8.02
135.5	5.41	7.67	7.41	5.45	6.88	14.16	8.04	7.73	5.65	8.84	8.40
65.6	5.48	6.55	5.39	2.42	3.15	3.44	6.24	5.39	3.76	6.79	8.67
907.2	6.32	6.48	6.24	6.40	8.72	4.72	6.83	8.96	6.16	9.20	9.76
1059.8	5.89	6.22	5.68	2.31	6.57	5.58	6.41	7.00	6.21	7.38	7.45
3.7	5.69	6.03	6.53	2.86	1.75	5.69	5.56	5.75	5.15	6.81	6.44
0.3	3.25	3.73	4.54	1.50	0.73	5.15	4.50	1.96	4.26	3.96	4.79
311.0	6.80	6.77	6.23	3.93	5.46	6.4	7.39	7.40	5.80	7.37	7.47
746.0	6.89	7.00	6.26	2.42	4.58	6.42	5.51	7.53	4.74	6.68	7.79
199.9	3.01	3.79	3.34	2.00	2.85	7.1	6.35	4.57	4.05	6.31	7.06
1006.2	7.23	5.06	4.44	2.87	5.44	3.37	5.18	7.33	5.92	7.92	7.03
5.7	4.38	6.96	6.11	3.41	4.91	5.84	6.19	6.43	4.53	8.51	7.92
290.7	6.96	5.83	5.21	3.11	2.94	5.39	4.88	5.38	5.74	6.96	7.13
282.9	4.09	3.48	3.30	2.24	3.06	7.1	6.62	3.94	4.75	5.76	6.72

No	Country (region)	Resulting synthetic categories				Explanatory variables –				
		$y^{(1)}$ 4.4.08	$y^{(2)}$ 1.1.24	$y^{(3)}$ 2.5.05	$y^{(4)}$ 4.4.17	$x^{(1)}$ 4.4.01	$x^{(2)}$ 4.4.09	$x^{(3)}$ 4.4.15	$x^{(4)}$ 4.4.12	$x^{(5)}$ 4.5.01
32	Luxembourg*	0.930	24519	8.16	8.84	6.000	6.63	7.53	422.8	5.090
33	Maharashtra	0.590	425	6.62	4.96	5.100	6.31	4.31	2208.5	3.996
34	Malaysia*	0.790	1784	8.45	7.41	3.800	6.41	6.29	1279.9	8.026
35	Mexico*	0.800	4090	5.18	4.19	6.000	5.10	3.85	740.2	4.427
36	Netherlands*	0.938	15559	7.37	8.38	8.900	6.99	7.16	418.0	4.962
37	New Zealand*	0.917	11399	7.12	8.68	8.100	5.36	7.92	567.9	7.041
38	Norway*	0.944	22331	8.15	9.31	8.000	8.26	8.38	241.6	6.565
39	Philippines*	0.751	679	4.98	3.47	3.300	4.51	2.74	870.6	2.660
40	Poland*	0.841	3197	2.88	3.12	6.300	4.34	3.79	1918.9	6.022
41	Portugal*	0.896	7170	6.52	6.12	9.200	6.64	6.55	530.9	6.911
42	Rhone-Alps	0.925	13000	6.42	8.25	9.500	5.75	6.50	250.3	6.287
43	Romania	0.773	1807	4.67	2.83	6.500	4.37	3.49	2433.1	3.178
44	Russia*	0.779	1199	4.00	3.03	5.400	3.90	4.81	7676.5	3.756
45	Sao Paulo	0.777	2726	6.55	4.29	7.600	6.93	5.50	577.6	1.645
46	Scotland	0.930	14042	6.61	7.25	7.600	4.17	5.67	592.4	4.694
47	Singapore*	0.884	9415	8.95	8.14	3.900	8.93	8.43	645.8	4.214
48	Slovak Republic	0.836	3430	4.05	4.05	5.700	6.11	5.46	1950.1	3.720
49	Slovenia	0.881	7040	5.06	6.25	8.400	8.11	5.59	749.6	6.313
50	South Africa*	0.684	2136	6.54	4.89	8.600	5.11	5.86	2643.7	5.935
51	Spain*	0.918	11944	6.74	7.77	7.500	7.05	6.48	451.5	4.384
52	Sweden*	0.932	22268	7.65	9.35	11.100	7.84	8.00	154.4	6.734
53	Switzerland*	0.941	16585	7.31	8.27	7.705	7.19	8.00	191.8	7.699
54	Taiwan*	0.781	7952	5.03	5.93	5.912	7.57	6.05	711.0	4.494
55	Thailand*	0.768	1258	6.39	5.73	3.700	5.73	5.12	1271.3	4.215
56	Turkey*	0.734	1748	5.78	4.22	5.000	8.22	5.42	992.0	3.927
57	United Kingdom*	0.930	19698	6.49	7.42	7.600	4.95	6.12	366.6	4.694
58	USA*	0.937	25548	5.74	8.11	13.900	5.95	6.25	602.2	7.880
59	Venezuela*	0.775	2486	0.97	2.00	6.000	5.33	5.22	1161.5	5.164
60	Zhejiang	0.721	680	7.26	6.93	5.500	7.78	3.96	3077.7	3.141

* The country is on the list of countries for which a comparative analysis of the dynamics is carried out for macroeconomic indicators and synthetic categories of life quality and lifestyle (see point 3.2.6).

– characteristics of institutional development and socio-economic policy											
$x^{(6)}$ 4.3.02	$x^{(7)}$ 4.3.10	$x^{(8)}$ 2.4.17	$x^{(9)}$ 3.3.18	$x^{(10)}$ 2.3.16	$x^{(11)}$ 2.3.17	$x^{(12)}$ $\frac{2.5.08}{2.5.07}$	$x^{(13)}$ 2.4.02	$x^{(14)}$ 2.5.01	$x^{(15)}$ 3.4.09	$x^{(16)}$ 2.5.02	$x^{(17)}$ 2.5.13
758.4	5.16	8.16	7.19	4.35	7.06	4.96	7.38	7.44	6.66	8.28	8.16
5.0	5.38	6.62	6.23	2.08	1.38	5.69	5.38	6.15	4.85	6.92	5.77

(continued)

(continued)

– characteristics of institutional development and socio-economic policy											
$x^{(6)}$	$x^{(7)}$	$x^{(8)}$	$x^{(9)}$	$x^{(10)}$	$x^{(11)}$	$x^{(12)}$	$x^{(13)}$	$x^{(14)}$	$x^{(15)}$	$x^{(16)}$	$x^{(17)}$
4.3.02	**4.3.10**	**2.4.17**	**3.3.18**	**2.3.16**	**2.3.17**	$\frac{2.5.08}{2.5.07}$	**2.4.02**	**2.5.01**	**3.4.09**	**2.5.02**	**2.5.13**
27.5	7.22	7.64	6.94	4.82	4.06	12.34	6.35	7.00	6.55	7.49	6.49
25.4	3.25	4.99	4.40	1.76	1.68	16.94	5.76	2.47	4.26	2.74	5.31
453.6	6.23	7.82	6.99	3.80	6.76	5.49	7.47	8.20	6.44	7.29	7.82
156.7	5.92	8.24	7.48	3.43	8.56	6.84	8.36	8.16	6.57	7.80	7.88
602.4	4.94	7.89	7.74	4.56	7.48	3.69	6.44	8.67	7.15	8.78	9.06
0.7	3.09	5.58	4.56	1.86	1.23	9.69	4.75	3.69	5.54	4.31	5.72
30.7	4.22	3.94	4.52	1.42	1.39	5.09	3.61	2.55	3.42	2.70	5.38
110.1	3.39	7.52	6.48	1.94	4.00	7.91	7.05	4.06	4.33	7.52	8.46
522.6	7.33	5.83	6.58	2.17	6.17	5.58	6.42	6.70	5.39	6.70	6.67
8.0	5.76	5.39	3.86	1.14	1.01	4.68	5.23	2.75	3.92	4.29	4.84
29.9	6.26	4.40	3.90	1.85	1.45	10.47	4.65	3.84	3.48	3.02	5.31
57.7	4.00	6.71	6.34	2.14	3.52	29.14	5.72	3.72	6.34	3.03	6.28
355.2	7.00	6.00	6.08	1.74	6.87	7.08	6.43	7.13	5.22	6.78	6.17
455.7	7.90	8.49	7.67	5.95	8.54	9.8	7.37	8.24	6.67	9.19	8.48
26.0	3.65	6.76	6.11	2.70	2.11	3.95	4.97	3.61	4.11	5.19	7.37
127.1	4.32	5.68	4.76	1.93	3.68	4.14	4.86	4.30	5.16	5.95	7.90
19.6	5.07	7.53	7.06	2.98	3.42	33.25	5.91	6.37	6.86	3.07	5.95
139.0	4.48	6.98	5.83	3.78	5.80	5.37	6.98	5.50	5.03	6.72	7.51
882.2	7.77	7.68	6.37	4.52	7.50	5.84	5.21	8.23	5.84	9.23	8.58
1052.2	6.34	7.86	6.90	4.85	7.47	3.79	7.76	8.47	7.10	7.49	8.51
288.2	6.05	5.91	5.75	4.70	4.89	6.39	6.61	5.93	6.34	7.42	7.08
5.2	4.63	6.34	5.93	3.93	2.99	8.2	6.04	6.02	5.84	6.49	7.11
19.0	3.35	4.77	4.09	2.00	2.41	7.66	6.03	3.42	5.02	6.43	7.88
453.0	6.45	6.44	6.46	2.51	6.83	7.08	7.03	6.53	5.33	5.79	6.30
964.7	8.14	7.13	7.23	4.51	6.63	8.92	6.19	7.23	5.83	7.79	7.58
23.2	2.93	3.88	4.24	0.58	0.66	17.73	3.07	0.81	4.24	0.91	3.28
14.5	7.63	5.15	4.15	2.26	3.78	7.9	6.96	7.26	7.81	7.93	7.43

A3.2 Dynamics of the determinant variables measured in physical units* (data sources – yearbooks [WCY, 1998–2004])

A3.2a The dynamics of households consumption in Russia per year (in US $ in terms of purchasing power parity rate)

	Years, *t*							
	1996	**1997**	**1998**	**1999**	**2000**	**2001**	**2002**	**2003**
$y_t^{(2)}$	1.114	1.388	1.533	657	770	1.021	1.067	1.199
$y_{\max}^{(2)}(t)$	213	207	212	289	241	297	311	317

(continued)

	Years, t							
	1996	1997	1998	1999	2000	2001	2002	2003
$y^{(2)}_{max}(t)$	24.498	21.683	21.995	24.513	21.953	24.794	25.548	25.548
$y^{(2)}_{opt}(t)$	24.498	21.683	21.995	24.513	21.953	24.794	25.548	25.548

* The meaning and method of determining (in each year) minimum, maximum and optimal values of the analyzed variables are described earlier (see point 2.3.3). However, when analyzing the dynamics are not considered for 60 countries and regions represented in WCY [WCY, 2004], but only 46 of them are marked with an asterisk in Appendix 3.1.

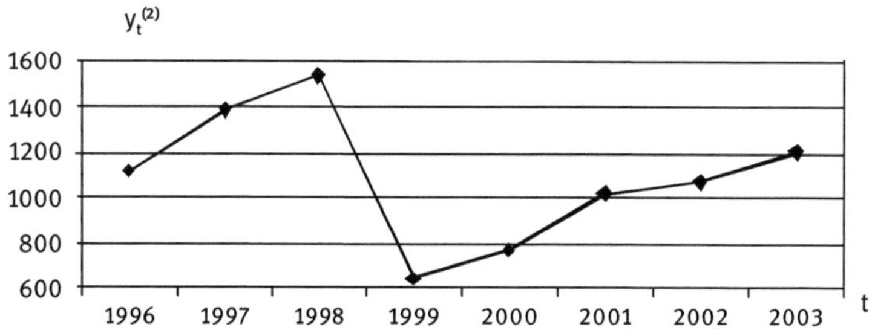

A3.2b Dynamics of total expenditures on health in Russia (in % of GDP)

	Years, t				
	1997	1998	1999	2000	2001
$x^{(1)}_t$	3.2	3.2	4.4	5.1	5.4
$x^{(1)}_{min}(t)$	0.16	0.15	0.10	2.5	2.4
$x^{(1)}_{max}(t)$	13.5	12.9	12.9	13.0	13.9
$x^{(1)}_{opt}(t)$	8.2	8.1	8.4	9.0	9.0

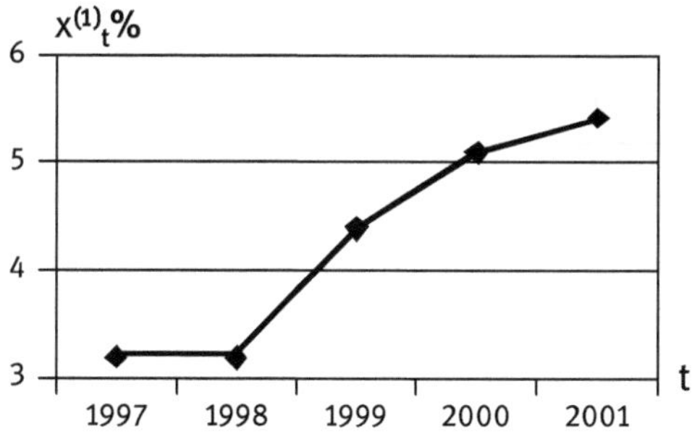

A3.2c Dynamics of the total expenditure on R&D and development activities in Russia per capita (in US $ in terms of purchasing power parity rate)

	Years, *t*							
	1995	**1996**	**1997**	**1998**	**1999**	**2000**	**2001**	**2002**
$x_t^{(6)}$	26.9	25.4	28.7	17.6	13.4	18.7	30.6	29.9
$x_{min}^{(6)}(t)$	1.4	0.9	0.2	0.2	0.7	0.7	0.3	0.3
$x_{max}^{(6)}(t)$	1065.9	1231.4	1143.2	1143.2	1143.2	1170.5	1119.4	1052.2
$x_{opt}^{(6)}(t)$	493	488	554	531	565	607	576	639

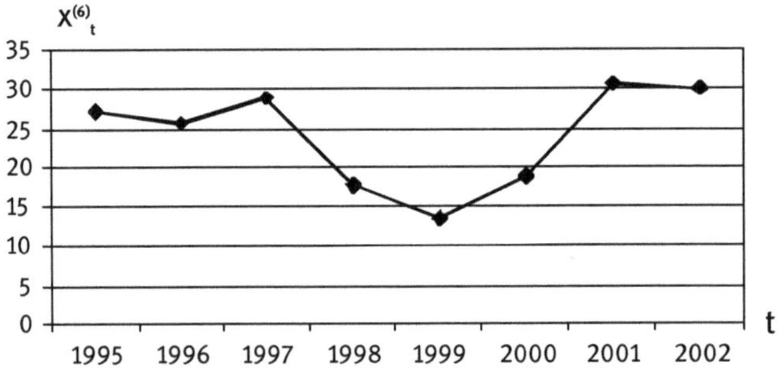

A3.2d Dynamics of a 20% coefficient of funds (the ratio of incomes of the 20% richest people from the population to 20% of the poorest from the population) in Russia (in times)

	Years, *t*						
	1995	**1996**	**1997**	**1998**	**1999**	**2000**	**2001**
$x_t^{(12)}$	9.8	12.3	12.6	12.2	12.2	12.2	10.5
$x_{min}^{(12)}(t)$	3.9	3.5	3.5	3.6	3.6	3.6	3.4
$x_{max}^{(12)}(t)$	32.1	25.7	25.7	25.2	24.2	29.1	29.1
$x_{opt}^{(12)}(t)$	6.9	4.8	4.9	4.8	5.0	6.2	5.5

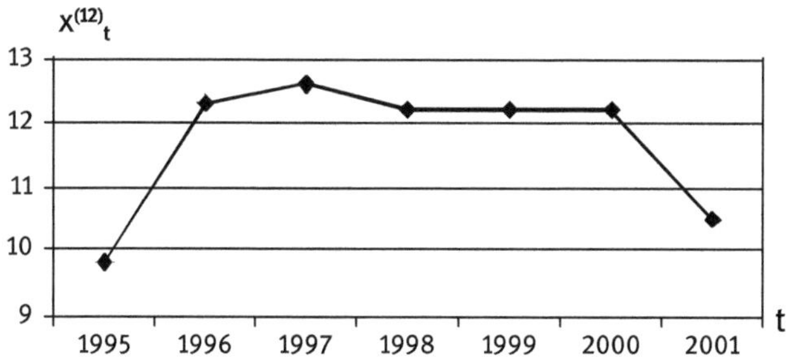

A3.3 A priori set of indicators of social and economic policy and institutional development in the region of the Russian Federation

Variable name	Source	Years	Index	Units
1	2	3	4	5
The conditions and characteristics of entrepreneurial activity				
ASB	**See the method of calculating**	**2002–2006**	Activity of small business	without units
The effectiveness of bureaucratic system (including corruption)				
IAORE	**Support of Russia**	**2005**	Illegal actions by officials in relation to the entrepreneurs. Index (higher value corresponds to better situation in the region)	without units
CORRUP.	**INDEM**	**2002**	Integrated index of corruption (lowest levels of index correspond to the least corrupt [by estimates] region)	without units
CORR. VOL.	**INDEM**	**2002**	Integrated index of the corruption volume (the lowest value of the index corresponds to the least corrupt [by estimates] region)	without units

(continued)

(continued)

Variable name	Source	Years	Index	Units
1	2	3	4	5
The level of protection of contracts and property rights				
DLIC	**Support of Russia**	**2005**	Chances of the entrepreneur to defend legitimate interests in court if his opponent will be the regional authorities. Index (higher value corresponds to better situation in the region)	without units
The availability of financial resources, particularly credit resources (number of the opened bank branches)				
CREDIT	**Rosstat, our calculations**	**2002–2006**	Number of credit institutions per capita	pcs
Information and communication technologies				
PC	**Rosstat**	**2004–2006**	The number of personal computers per 100 employees in public institutions	pcs
PCI	**Rosstat**	**2004–2006**	The number of personal computers with Internet access per 100 employees in public institutions	pcs
ICT	**Rosstat**	**2004–2006**	The share of the cost on information and communication technologies in the costs of the regional budget	%
SOFT	**Rosstat, our calculations**	**2004–2006**	The share of the cost for acquisition of software in the costs of the regional budget	%

(continued)

Variable name	Source	Years	Index	Units
1	2	3	4	5
Criminality				
SECUR.	**See the method**	**2002–2007**	Physical security	Without units
Foreign trade				
EXP. DIST.	**Rosstat**	**2002–2006**	Far abroad countries export	$ mln
Imp. DIST.	**Rosstat**	**2002–2006**	Import	$ mln
EXP. CIS	**Rosstat**	**2002–2006**	CIS countries export	$ mln
IMP. CIS	**Rosstat**	**2002–2006**	Import	$ mln
AFER	**INDEM**		The indicator "active foreign economic relations"	Without units
Budget indicators				
a) *Receipt in the regional budget from the federal*				
FFSR	**Min. of Fin.**	**2002–2006**	Fund for Financial Support of the Regions (FFSR)	Thousands of roubles
Inc. from FFSR	**Min. of Fin., our calculations**	**2002–2006**	The share of revenue from FFSR in SSA region	%
b) *Consolidated budgets of the regions in the Russian Federation*				
EXPEM.	**Rosstat**	**2000–2006**	Expenditures of the consolidated budget in the Russian Federation	millions of rubles
REVEN	**Rosstat**	**2000–2006**	Revenues of consolidated budget in the Russian Federation	millions of rubles
EXP. EDU.	**Rosstat**	**2000–2006**	Expenditures on education (regional budget)	millions of rubles
EXP. HEALTH	**Rosstat**	**2000–2006**	Expenditures on health and sports (regional budget)	millions of rubles
EXP. NE	**Rosstat**	**2000–2006**	Expenditures on the national economy (regional budget)	millions of rubles
EXP. NCS.	**Rosstat**	**2000–2006**	Expenditures on utilities (regional budget)	millions of rubles

(continued)

(continued)

Variable name	Source	Years	Index	Units
1	2	3	4	5
NE	**Rosstat, our calculations**	**2001–2006**	Growth in the share of budget expenditures on the national economy in relation to the previous year (index)	without limits
EDU.	**Rosstat, our calculations ы**	**2001–2006**	Growth in the share of budget expenditures on education with respect to the previous year (index)	without units
HEALTH	**Rosstat, our calculations**	**2001–2006**	Growth in the share of budget expenditures on sound-protection and sport with respect to the preceding year (index)	without units
HCS	**Rosstat, our calculations**	**2001–2006**	The growing share of the cost of housing and communal services in the total expenditures of the budget compared to the previous year (index)	without units
HCS 2001	**Rosstat, our calculations**	**2002–2006**	The growing share of the cost of housing and communal services in the total expenditures of the budget relative to the base, 2001 (index)	without units
SSF	**Rosstat, our calculations**	**2001–2006**	Growth in spendings from Social Insurance Fund in relation to the previous year (index)	without units
GRTB	**Rosstat, our calculations**	**2001–2006**	The growing share of gratuitous revenues in total budget revenues compared to the previous year (index)	without units

(continued)

Variable name	Source	Years	Index	Units
1	2	3	4	5
TRTB	**Rosstat, our calculations**	**2001–2006**	The growing share of tax revenues in total budget revenues compared to the previous year (index)	without units
IRBF	**INDEM**	**2002**	The indicator "Independence of regional budgets from the federal" (lower index value corresponds to a greater dependence on federal budget)	without units
Democracy				
The electoral indicators				
Index Rae	**Our calculations**	**1999. 2003**	Index Rae factionalization votes in the elections to the State Duma in 1999	without units
Index of democr.	**Atlas of Russian regions**	**1999–2003**	Index of the democratic elections in 1999–2003. As a result of expert estimates (higher value corresponds to greater democracy)	without units
IEM	**Our calculations**	**1995–2004**	Index of the electoral manageability (10.000 – the worst value, 0 – Better)	without units
Demography, human potential				
RESEAR.	**Rosstat, our calculations**	**2002–2006**	Number of researchers with advanced degrees per 1.000 people of the population	without units
UNEMP.	**Rosstat**	**2002–2006**	Unemployment rate	%
ETNHO	**Rosstat**	**2002**	Ethno-linguistic diversity	without units
DENS.	**Rosstat, our calculations**	**2002–2006**	The relative density of the population	without units

(continued)

(continued)

Variable name	Source	Years	Index	Units
1	2	3	4	5
The differentiation of the population by income				
CF	**Rosstat**	**2002–2006**	Ratio of funds	times
GINI	**See methodology**	**2002–2006**	Unified Gini coefficient	without units

A3.4 Initial data for econometric analysis of dependences of the integral life quality indicators on the characteristics of social and economic policy and institutional development in the region

A3.4a Values of the explanatory variables*

	ASB					IADRE	DLIC	CORRUP.	CORR. VOL.	SECUR.
	2002	2003	2004	2005	2006	2005	2005	2002	2002	2002
Adygeya	1.88	2.04	1.64	1.78	1.77	12	4			8.803
Altai	1.93	1.71	1.5	2.04	1.97	72	32			7.68
Altai Territory	2.14	2.47	2.17	2.59	3.1	74	32	0.551	0.721	8.215
Amur Region	1.84	1.34	1.1	1.29	1.68	52	32	0.633	0.299	8.585
Arkhangelsk Region	1.41	1.11	1.23	1.2	1.75	76	46.2	0.128	0.068	8.684
Astrakhan Region	1.92	1.83	1.77	2.01	2.27	44.2	50			8.603
Bashkortostan	1.37	1.82	2.01	2.58	3.11	77.6	12	0	0.114	9.181
Belgorod Region	1.83	2.13	1.87	1.96	2.06	70	44	0.435	0.403	9.628
Bryansk Region	1.2	1.03	0.92	1.51	1.64	54	40			9.247
Buryatia	0.96	0.88	1.05	1.28	1.41	66	17.6			7.163
Chechen				0			29.2			
Chelyabinsk Region	2.17	2.12	1.87	2.5	2.6	34		0.556	0.853	7.716
Chita Region	0.88	0.39	0.63	0.87	1.05	22	26			6.478
Chuvash	1.74	1.78	1.63	2.48	2.61	16	54.9			9.027
Dagestan	0.43	0.27	0.35	1.13	0.95	83.7				9.857
Ingushetia	0.35	0.2	0.06	0.33	0.3	176	47.1			
Irkutsk Region	1.31	1.34	1.28	1.59	1.84	51	32			6.608
Ivanovo Region	1.84	2.15	1.24	2.05	2.12	29.4	36			8.504

(continued)

	ASB					IADRE	DLIC	CORRUP.	CORR. VOL.	SECUR.
	2002	2003	2004	2005	2006	2005	2005	2002	2002	2002
Jewish Autonomous Region	0.39	0.71	1.26	1.7	2.14		49			6.689
Kabardino-Balkaria	1.1	0.79	0.78	0.95	1	38	32			9.501
Kaliningrad Region	3.43	4.76	3.53	6.51	7.04	28	38			8.63
Kalmykia	0.42	0.3	0.36	0.39	0.57	46	34.6			8.549
Kaluga Region	3.55	3.47	2.75	3.39	3.16	58.8	64			9.022
Kamchatka Region	2.53	2.89	2.01	2.23	2.39	74	48			8.654
Karachayevo–Circassia	1.3	0.87	0.97	1.41	1.42	64	37.7			9.097
Karelia	2.66	2.52	2.36	2.67	2.66	26.4	27.8	0.864	0	8.855
Kemerovo Region	2.06	2.09	1.84	2.15	2.33	29.6	44	0.269	0.664	7.607
Khabarovsk Territory	3.06	2.45	2.53	2.77	2.83	20	36	0.644	0.782	7.4
Khakassia	2.02	1.59	1.42	2.03	2.31	82	30			7.88
Kirov Region	1.65	1.63	1.61	2.22	2.18	24	26			9.214
Komi	1.66	2.1	1.86	2.09	2.56	28	40.4			8.365
Kostroma Region	1.5	1.41	1.43	1.64	1.89	67.3	22			8.612
Krasnodar Territory	2.57	3.14	2.77	3.04	3.08	52	22	1	0.681	9.331
Krasnoyarsk Territory	1.05	1.28	1.45	1.39	1.57	22	42	0.331	0.117	8.219
Kurgan Region	1.2	1	1.05	1.39	1.54	38	22	0.658	0.253	7.567
Kursk Region	1.02	1.01	1.1	1.58	1.45	60	34			8.96
Leningrad Region	3.45	3.51	3.64	3.29	3.2	72	22	0.53	0.34	8.109
Lipetsk Region	1.63	1.37	1.68	1.84	1.66	18	34.9			9.342
Magadan Region	5.91	4.28	4.8	4.4	3.86	90.5	44			8.109
Mari El	2.33	2.05	2.31	2.53	2.63	98	10.2			8.596
Mordovia	1.18	0.8	1.03	1.38	1.44	40	22.7			8.74
Moscow	7.96	7.94	8.9	7.73	7.72	24.7	78	0.634	0.864	9.359
Moscow Region	3.35	3.5	2.51	3.48	3.41	30	42	0.754	1	8.728

(continued)

(continued)

	ASB					IADRE	DLIC	CORRUP.	CORR. VOL.	SECUR.
	2002	2003	2004	2005	2006	2005	2005	2002	2002	2002
Murmansk Region	1.52	1.43	1.4	1.72	1.6	42	41.2			9.211
Nizhny Novgorod Region	4.79	3.62	3.07	4.49	4.06	65.4	38	0.712	0.929	8.985
North Ossetia–Alania	1.96	2.23	1.22	1.09	1.03		42			9.504
Novgorod Region	2.03	2.57	1.84	2.2	1.87	76	40	0.658	0.181	8.465
Novosibirsk Region	3.21	4.17	3.3	3.34	3.38	76	68.6	0.643	0.275	8.695
Omsk Region	2.44	2.4	2.32	2.66	2.46	39.2	24	0.542	0.074	8.761
Orel Region	1.57	1.22	1.27	1.61	1.66	57.1	28			8.882
Orenburg Region	2.17	1.95	2.1	1.85	1.83	46	27.1			8.472
Penza Region	2.66	2.53	2.21	2.85	2.7	16	42			9.529
Perm Region	1.36	1.04	1.85	1.74	1.82	70	54	0.47	0.115	7.24
Primorye Territory	2.66	2.95	3.17	3.27	3.08	132	29.4	0.868	0.201	7.993
Pskov Region	2.98	3.11	2.14	2.64	2.51	60.8	52	0.595	0.542	8.98
Rostov Region	2.92	3.02	2.15	3.33	3.29	26	34.6	0.747	0.753	9.239
Ryazan Region	3.68	3.36	2.67	2.65	2.98	45.1	15.7	0.558	0.395	9.235
Sakha (Yakutia)	0.72	0.67	0.82	1.03	1.05	28	51			7.801
Sakhalin Region	3.59	3.61	3.03	2.36	3.09	37.3	44			8.034
Samara Region	4.25	4.6	4.19	4.12	3.94	62.7	29.1	0.731	0.2	8.783
Saratov Region	2.15	1.75	1.77	1.95	1.75	32	56	0.913	0.867	8.986
Smolensk Region	1.07	0.99	0.85	1.48	1.72	62	57.1			8.876
St. Petersburg	8.38	8.63	8.96	7.52	7.45	41.3	20.4	0.412	0.843	9.226
Stavropol Territory	1.73	2.54	1.98	2.25	2.36	86	48	0.707	0.501	8.904
Sverdlovsk Region	2.39	2.79	2.61	3.26	3.39	56	10	0.582	0.683	8.054
Tambov Region	1	0.88	0.91	1.37	1.37	144	16	0.489	0.58	9.392
Tatarstan	1.58	1.3	1.69	1.94	2.3	62	33.3	0.658	0.245	9.132
Tomsk Region	2.72	3.44	3.22	2.83	3.44	42	54.2	0.645	0.352	8.526
Tula Region	1.66	1.64	1.55	1.85	1.93	32		0.486	0.554	8.657
Tuva	0.34	0.31	0.29	0.79	0.8		38.6			0
Tver Region	2.23	1.91	1.53	1.89	1.79	27.5	34	0.629	0.16	8.452
Tyumen Region	1.35	1.14	1.69	1.65	1.76	32.7	40	0.283	0.033	8.188
Udmurtian	1.94	2.15	1.64	1.45	1.84	32	36	0.872	0.333	8.57

(continued)

	ASB					IADRE	DLIC	CORRUP.	CORR. VOL.	SECUR.
	2002	2003	2004	2005	2006	2005	2005	2002	2002	2002
Ulyanovsk Region	1.52	1.48	1.38	2.11	2.55	92	48	0.58	0.552	8.177
Vladimir Region	2.3	1.98	1.91	2.21	2.18	72.5	34.6			8.603
Volgograd Region	2.2	2.58	2.11	2.33	2.25	28.3	32	0.803	0.801	8.381
Vologda Region	2.21	2.47	2.22	2.03	1.97	76	16			8.467
Voronezh Region	3.32	3.35	2.6	2.91	2.66	42	44	0.626	0.39	9.139
Yaroslavl Region	2.55	2.41	2.42	2.5	2.91	37.3				8.218
Russian Federation	**2.74**	**2.76**	**2.73**	**2.95**	**3**			**0.295**	**0.01**	

*Explanation of the variables are presented in Appendix 3.3.

	SECUR.				EXPEN.				
	2003	2004	2005	2006	2002	2003	2004	2005	2006
Adygeya	9.253	9.391	9.528	9.3	3747.9	4001.2	3866.2	4626.7	5545.5
Altai	6.957	7.129	7.056	6.634	3606.8	4279.5	5086.8	6001.1	7255.4
Altai Territory	8.55	8.728	8.61	8.731	17054.9	20953.5	24680.3	28301.3	34964.8
Amur Region	8.513	8.261	8.099	7.624	10562.7	13554.7	15773.5	18000.6	21268
Arkhangelsk Region	8.809	8.536	8.462	8.494	15128.5	18777.5	21697.9	22230.8	27129
Astrakhan Region	8.826	8.843	8.66	8.56	7059	8693.8	9327.1	12529.3	16868.7
Bashkortostan	9.208	9.164	9.14	8.999	38057.1	48106.9	48252.3	55848.1	74930.1
Belgorod Region	9.703	9.678	9.563	9.591	10560.2	11759.5	15820.8	25783.9	31098.5
Bryansk Region	9.344	9.098	9.102	9.146	8419.9	9740.7	11670.5	14105.8	17786.2
Buryatia	7.147	7.153	7.01	6.909	11714.7	12557.9	14165.1	15836.2	21141.4
Chechen					6683.3	10187.4	24668.5	15357.2	27376.1
Chelyabinsk Region	7.941	7.995	8.409	8.353	24393.4	28727.6	37012.3	46836.1	59100
Chita Region	6.404	6.194	6.185	6.453	13241	17652.9	18462	16814	21293.7
Chuvash	9.135	9.113	9.031	9.199	9205.6	11482.6	13165.8	15679.4	19670.7
Dagestan	9.734	9.864	9.9	9.735	15972.3	17170.5	19564.3	23053.5	26554.6
Ingushetia					4741.4	4097.7	4188.8	5633.4	6425

(continued)

(continued)

	SECUR.				EXPEN.				
	2003	2004	2005	2006	2002	2003	2004	2005	2006
Irkutsk Region	6.965	7.262	6.997	6.868	25526.7	30304.7	34820.2	43256	51979.6
Ivanovo Region	8.813	8.549	8.687	8.72	8067.2	9079.4	10476.7	13140.4	17474.3
Jewish Autonomous Region	6.988	7.56	7.826	7.784	2312.7	2745.4	3001.1	3342	4463.4
Kabardino-Balkaria	9.728	9.722	9.666	9.78	6570	5743.6	6466.3	8959.3	10561.6
Kaliningrad Region	8.941	8.92	9.35	9.147	7936.5	9712.3	11442.7	15166.8	20906.5
Kalmykia	7.926	8.139	7.738	8.14	2778.2	3869.4	4767.6	4338.9	4645.6
Kaluga Region	9.113	8.684	8.452	8.656	8769.5	10574.5	12635.8	14953.7	18800
Kamchatka Region	8.513	8.944	8.314	8.318	9250	10849.2	12530	12702.6	16763.5
Karachayevo-Circassia	8.861	9.025	9.157	9.27	3425.2	3584.7	4630.1	5260.8	6325.8
Karelia	8.548	8.582	8.72	8.836	8446.7	9857.6	10917	13536.1	16646.3
Kemerovo Region	7.941	8.09	8.074	7.936	26091.6	29214.2	43641.9	57655.4	66026.8
Khabarovsk Territory	7.025	7.571	7.605	7.021	21929.8	28184.2	33813.7	35083.8	37638.8
Khakassia	8.052	8.188	8.281	7.945	4768.5	5440.6	6200.3	7151.8	8862.9
Kirov Region	8.99	8.848	9.011	9.313	10498.4	12796.8	33506.5	18255.6	21980.2
Komi	7.742	7.711	7.737	7.368	17211.6	17282.6	20448.8	23506.5	29394.8
Kostroma Region	8.808	8.538	8.76	8.713	5783	7397.5	7583.1	8959.6	10910.8
Krasnodar Territory	9.477	9.525	9.548	9.378	40181.7	44914.6	53764.1	60032.7	75884.5
Krasnoyarsk Territory	8.294	8.269	8.137	8.161	44325	53786.1	64530.1	72292.4	89483.9
Kurgan Region	7.7	7.462	7.685	7.922	7050.8	8408.8	9336.4	11090.9	13699
Kursk Region	9.221	9.19	9.209	9.342	7486.6	8600.3	11483.1	14744.7	16398.8
Leningrad Region	8.13	8.303	8.384	8.268	15963.1	18970.5	23462.4	29285.1	34603.1
Lipetsk Region	9.492	9.343	9.358	9.362	9812.6	14051.2	18642.9	24550.6	27287.4
Magadan Region	8.692	8.079	7.578	7.553	6998.6	7546.3	7805.1	9121.6	11316.7
Mari El	8.401	8.397	8.568	8.469	4897.7	5506.2	7356.9	8197	10607
Mordovia	8.895	8.876	8.975	8.92	9319.9	12127.9	19447.9	19855.3	16236.2
Moscow	9.305	9.555	9.527	9.528	303929.2	356061.5	407855.1	503760.2	691329.8
Moscow Region	8.69	8.739	8.64	8.677	64188.8	82804.3	103127.2	140922.8	181416.7

(continued)

	SECUR.				EXPEN.				
	2003	2004	2005	2006	2002	2003	2004	2005	2006
Murmansk Region	9.272	9.347	9.272	9.281	12264.5	13679.6	15976.9	17237.8	28997
Nizhny Novgorod Region	9.24	9.066	9.01	9.044	24353.8	28128.5	22362.8	41448.5	54277.1
North Ossetia–Alania	9.474	9.686	9.578	9.382	5854.5	6354.1	8092	10238.3	11316.8
Novgorod Region	8.474	8.604	8.823	8.796	4943.1	6357.3	7476.1	10581	12343.3
Novosibirsk Region	8.687	8.646	8.679	8.809	22381.2	29416.3	33745.3	41698.6	50061.9
Omsk Region	8.703	8.776	8.789	8.844	14779.5	20240.9	30107.6	41548.8	44951.8
Orel Region	9.117	8.995	9.088	8.989	6109.3	7405.4	8675.9	9783.1	10930.4
Orenburg Region	8.564	8.468	8.113	8.186	15819.8	18279.6	12405.3	29407.3	35582.3
Penza Region	9.499	9.35	9.43	9.443	8447.2	10397	38851.1	14625.3	19706.8
Perm Region	7.298	7.22	7.159	7.338	27604.4	30900.3	15091.7	43856.1	54212.2
Primorye Territory	7.956	7.991	7.802	7.357	20510.5	24231.8	28741.6	33116.7	41028.2
Pskov Region	9.019	8.768	8.82	8.985	6120.3	7571.7	8307.7	8635.4	10069.2
Rostov Region	9.213	9.342	9.406	9.353	26567.6	30218.8	35346.6	46328.1	58120.9
Ryazan Region	9.184	9.254	8.95	9.224	9481.8	11402.6	11896.1	14825.1	18051.4
Sakha (Yakutia)	7.688	7.711	7.772	7.197	37042	41557.4	46342.3	64514.4	60737.1
Sakhalin Region	7.742	7.834	7.668	7.142	9808.3	12680.5	15414.7	17241.7	23845.5
Samara Region	8.591	8.86	8.877	8.79	30171.2	34580.5	42029.9	52914.4	70833.9
Saratov Region	9.055	9.32	9.268	9.284	16178.9	20451.8	23286.6	27906.6	33728
Smolensk Region	9.021	8.417	8.531	8.552	7296.5	7853.5	9393.7	10892.1	13292.7
St. Petersburg	9.308	9.249	9.208	9.124	65552.4	78329.7	94043.5	136264.4	186136.8
Stavropol Territory	9.113	9.19	9.428	9.289	18320.7	18953.4	22896.6	29289.2	34486.2
Sverdlovsk Region	7.974	8.085	8.105	7.918	36091.3	41885.5	52584.3	64736.4	91074.3
Tambov Region	9.534	9.241	9.204	9.211	7281.7	8516.6	9844.3	12612.2	14907.4
Tatarstan	9.151	9.206	9.335	9.329	54881.3	58317.1	79373.9	87822.9	86262
Tomsk Region	8.71	8.632	8.805	8.824	10762.9	14170.2	16360.6	20247.5	25523.1
Tula Region	8.834	8.798	8.868	8.87	12723.1	15020.1	16379.9	20495.7	25183
Tuva	0	0	0	0	4720.2	5404.6	5957.3	6751	7585.2
Tver Region	8.535	8.63	8.291	7.922	10776.2	11739.6	14501.8	18903.5	24110.8

(continued)

(continued)

	SECUR.				EXPEN.				
	2003	2004	2005	2006	2002	2003	2004	2005	2006
Tyumen Region	8.379	8.241	8.431	8.336	152038.3	171687.4	203103.2	98704.2	145808.7
Udmurtian	8.508	8.395	8.322	8.47	14757.3	15883.2	18281.8	21736.6	26433.3
Ulyanovsk Region	8.225	8.197	8.572	8.436	9192.3	9728	11267.8	13349.6	19205.6
Vladimir Region	8.838	8.735	8.588	8.766	10078.9	11966.9	13327.2	15413.1	21023.1
Volgograd Region	8.596	8.783	8.939	8.727	16364.5	20215.3	24041.4	30044.8	40542.8
Vologda Region	8.506	8.446	8.528	8.659	12998.4	16748.6	21287.8	28099.3	32161.5
Voronezh Region	9.303	9.252	9.42	9.463	15624.7	17800.8	20627.8	24463.6	30041.8
Yaroslavl Region	8.611	8.489	8.474	8.709	13751.7	16449.6	19320.3	23505.7	28337.6
Russian Federation					**1687.2**	**1984.3**	**2373**	**2941.2**	**3657.7**

	EXP. EDU.					EXP. HEALTH		
	2002	2003	2004	2005	2006	2002	2003	2004
Adygeya	666.197	705.4	900.1	1129.7	1409.9	562.3	608	786.8
Altai	819.4	955	1153.6	1976.3	2295	416.6	513	588.2
Altai Territory	4737.508	5291.5	6369.2	7705	10061.5	3237.6	3694	4466.2
Amur Region	2365.837	3053.6	4034.9	5108.7	5863.2	1672.1	2000	2645.4
Arkhangelsk Region	3278.42	4647.7	5901.9	6722.5	8052.7	2015.5	2985	3666.2
Astrakhan Region	1824.212	2018.5	2456.5	2987.3	3925.2	1160.6	1243	1585.5
Bashkortostan	8720.904	10160	11887.5	16230.8	20842.1	5302.4	6227	7170.8
Belgorod Region	2487.48	2755.5	3656	5422.6	7415.3	1634.3	1852	2320.7
Bryansk Region	2081.051	2302	2869.9	3931.3	5119.5	1410.6	1690	1983.8
Buryatia	2694.732	3143.2	3881.7	4851.3	6342.5	1758.8	1830	2356.7
Chechen		1489.2	1745.7	2380.4	5003.4		978	1355.7
Chelyabinsk Region	5693.495	6628.3	8485.3	11350.7	15939.2	3834.4	4613	6071.4
Chita Region	3268.962	4259.4	4970.6	5220.1	6958.2	2165.6	2624	3025.3
Chuvash	2151.855	2430.6	2971.6	3719	4725.5	1343.2	1684	2073.4
Dagestan	4082.415	4296.7	5391.7	6596.2	8634.6	2361.7	2537	2842.9
Ingushetia	546.428	632.4	671.9	870.5	1198.6	511.1	716	668.1
Irkutsk Region	6341.83	7768.1	9281.6	12718.9	15331.9	3668.9	4660	5810.8
Ivanovo Region	1650.321	1729.7	2199.5	2971.8	3840.4	1192.2	1353	1719.2
Jewish Autonomous Region	545.659	610.1	700	986	1226.6	352.0	403	487.9
Kabardino-Balkaria	1573.255	1552.4	1902.2	2612.9	2931.9	910.8	942	1140.6

(continued)

	EXP. EDU.					EXP. HEALTH		
	2002	2003	2004	2005	2006	2002	2003	2004
Kaliningrad Region	1795.01	2119.4	2595.7	3443.4	5565.1	1000.7	1233	1486
Kalmykia	720.262	829.1	1135.4	1081.9	1312.1	469.5	527	708.4
Kaluga Region	2004.644	2337.3	3098.7	4164.1	5130	1145.0	1380	1822.4
Kamchatka Region	1646.374	2205.4	2776.4	3168.6	4463	714.5	1064	1358
Karachayevo–Circassia	724.676	780.4	966.1	1154.5	1591.7	505.4	548	664.8
Karelia	2251.286	2469.6	3208.2	3865.5	4828.9	1392.6	1530	1946.4
Kemerovo Region	7617.018	8163.5	11588.9	13942.8	17327.8	3934.3	4681	6476.9
Khabarovsk Territory	4502.983	5144.8	6560.5	8495	10177.7	2859.3	3742	4452.4
Khakassia	1343.181	1466.2	1836.9	2388.1	2954.3	654.7	787	1039.1
Kirov Region	2729.116	3322.7	4129.3	5274.5	6209.5	1988.2	2419	3123.4
Komi	4158.887	4544	5751	6551.3	8058.7	2507.8	2850	3651.3
Kostroma Region	1348.437	1522.9	1817.3	2368.1	2930.6	814.8	860	1042.1
Krasnodar Territory	7586.141	8905.8	11365.1	14493.3	19215	6040.4	7414	9704.2
Krasnoyarsk Territory	9643.476	11429.1	14447.5	17584.7	23238.3	5089.5	7407	8858.5
Kurgan Region	2116.888	2350.1	2861.7	3549.3	4388.1	1183.3	1324	1524.3
Kursk Region	1925.306	1910.6	2971.1	3419.5	4278.7	947.8	1090	1781.8
Leningrad Region	3196.532	3995.4	5523.7	8201.6	9699.5	1870.4	2513	3588.1
Lipetsk Region	2050.303	2811.8	3676	5430.5	6376.7	1822.9	2333	3226.5
Magadan Region	1009.114	1084.1	1267.3	1698.3	2195.2	917.0	1048	1174.3
Mari El	1373.109	1444.4	1882	2112	2780.6	758.1	861	1151.6
Mordovia	1446.947	1563.5	1978.3	2488.5	2820.8	1140.2	1335	1717.5
Moscow	26247.136	31831	39157.1	58683.7	87999.4	24145.3	26122	32668.9
Moscow Region	13302.659	16385.8	22729.7	32024.7	44145.9	10364.6	14037	17072.5
Murmansk Region	2988.484	3340.4	4043.6	4828.7	8074.4	1711.4	2218	2686
Nizhny Novgorod Region	5596.741	6170.7	8102.7	10220.2	13537.4	3788.5	4095	5387.6
North Ossetia–Alania	1178.457	1279.2	1619.4	2523.2	2431	968.2	1003	1244.1
Novgorod Region	1315.694	1530.5	1911.3	2517.7	3231	667.3	797	1010.8
Novosibirsk Region	5100.477	6443.9	8183.4	10642.8	13683	3782.9	4601	5374.4
Omsk Region	3675.573	4160.7	5791.4	7982.5	10176.6	2815.9	3507	5411.4
Orel Region	1629.016	1962.9	2304	2899.1	3527.2	898.2	1134	1385.4
Orenburg Region	4087.278	4700.1	5841.1	7455.2	9287.3	3659.5	3811	4739.8
Penza Region	2065.46	2307.6	2749.5	3475.6	5245.4	1306.9	1545	1982.5
Perm Region	6638.8	7869.8	9717.4	12707.2	15847.7	4102.2	5144	6193.6
Primorye Territory	4898.265	5352.8	6246	7483	9297.4	2984.8	3321	3740.1
Pskov Region	1316.976	1474.6	1896.4	2268.7	2860.7	946.7	1032	1258.6
Rostov Region	5769.997	6851.9	9024.6	12375	15836.2	3911.2	4845	6062.1
Ryazan Region	1871.354	2145.9	2510.3	3381.2	4438	1258.5	1423	1734.3

(continued)

(continued)

	EXP. EDU.					EXP. HEALTH		
	2002	2003	2004	2005	2006	2002	2003	2004
Sakha (Yakutia)	9691.472	10435	12477.4	15806.5	18310.4	4524.4	4844	5325.9
Sakhalin Region	1832.589	2173.1	2652.8	3586.7	5098.5	1578.3	2088	2507.5
Samara Region	5480.313	6124	7831.8	10285.6	13779.7	3247.7	3540	4588.6
Saratov Region	3435.926	3947.9	4863.7	6164.5	7867.4	2720.4	3123	3610.6
Smolensk Region	1752.878	1982.2	2462.2	3014	3779.7	1150.4	1198	1499.1
St. Petersburg	10636.601	13660.4	18472.1	25008.6	31177	6614.2	7874	10709.3
Stavropol Territory	4148.179	4564.9	5810	7076	9069	2388.0	2645	3262
Sverdlovsk Region	8652.509	9982.3	12633.8	16080.5	24481.8	5304.8	6838	8958.6
Tambov Region	1813.004	1999.2	2562.6	3143.1	4040	1222.9	1263	1542.9
Tatarstan	10411.482	11130.1	13020.2	16170.3	19406.4	7970.5	7852	8950.4
Tomsk Region	2547.084	2996.5	3782.3	4888.7	6869	1277.7	1596	1955.8
Tula Region	2625.061	3033.2	3700.4	4844	6223.1	2131.8	2453	3032.7
Tuva	1522.664	1619	1964.3	2292.9	2888.3	898.0	994	1188.6
Tver Region	2296.29	2591.8	3497.3	4397.8	5984.6	1663.5	1805	2440.6
Tyumen Region	4163.14	25224.5	30333.4	10480.8	15401.7	3365.7	19018	23136.4
Udmurtian	3732.568	3954.6	4781.9	6521.9	7854.5	2492.5	2704	3190
Ulyanovsk Region	1991.135	2189.8	2775.8	3369.6	4425.3	1986.4	2179	2661.1
Vladimir Region	2502.432	2983.3	3693.8	4530	6084.3	1539.7	1864	2227.3
Volgograd Region	3792.241	4509.3	5292.6	6724.9	9052.8	2389.3	3039	3827.6
Vologda Region	3526.447	4227.6	5304.7	7012	8196.9	1856.8	2330	3201.3
Voronezh Region	3235.959	3692.3	4768.9	5828.4	7693.4	2869.3	3351	4487.5
Yaroslavl Region	2617.904	3402.1	4164.6	5082.9	6592.6	2031.0	2424	2865.9
Russian Federation	**296.81**	**375.8**	**471.8**	**628.6**	**810.1**	**201.8**	**258**	**321.2**

	EXP. HEALTH		EXP. NE					NE	
	2005	2006	2002	2003	2004	2005	2006	1//0	2//1
Adygeya	973.8	1195.2	752.3	636	357.2	447.3	470.6	2.97	2.53
Altai	932.3	1320.8	197.4	412.9	943.3	912.2	846.5	14.35	0.88
Altai Territory	4981.5	5974.3	1377.4	1253.6	1185.7	3478.4	4901.7	4.55	1.12
Amur Region	3262	3646.8	822.8	1214.3	1246	2120.4	2305.3	2.97	1.29
Arkhangelsk Region	4364.9	4720.1	833.3	1202.4	1209.3	1978.7	2300.5	1.67	0.84
Astrakhan Region	2284.6	3461.9	307.3	659.6	660.2	1400.6	1897.6	1.75	0.84
Bashkortostan	10791.6	14892.9	8365.3	9900.9	7683.6	9101.4	13700.6	13.34	0.76
Belgorod Region	4195.6	5383.7	1047.8	1234.5	1789.9	7047.2	6916.3	16.91	1.04
Bryansk Region	2664.8	3623	705	907.8	1042.5	1792.6	1944.4	2.01	0.71
Buryatia	2762.4	3642.8	638.1	672.9	731.1	1102	2742.1	3.03	1

(continued)

	EXP. HEALTH		EXP. NE					NE	
	2005	2006	2002	2003	2004	2005	2006	1//0	2//1
Chechen	1464.2	3030.2	1422.4	1777.9	1340.6	1533.8	2806.6		
Chelyabinsk Region	8601.7	11633.1	1457.1	2404.8	4341.8	8159.7	6969.5	13.75	0.82
Chita Region	3643.7	4344.7	881.2	1323.2	1454.2	1581.2	1613	3.25	0.91
Chuvash	2632.3	3894.8	1463.1	1932.9	2411.2	2636.1	3705	7.45	1.2
Dagestan	3610.2	3841.6	2598.5	1951.7	2398.1	3488	2896.4	3.63	1.22
Ingushetia	715	737	1297.2	584.1	735	874.5	1029.9	0.74	0.97
Irkutsk Region	7433.9	9146	978.2	1366.9	1578.1	3541	4106.1	7.66	0.86
Ivanovo Region	2102.2	3777	273.4	360.8	511.7	1533.4	1878.6	4.83	1.14
Jewish Autonomous Region	611.5	1015	73	208.8	279.1	339.3	442.8	3.21	0.62
Kabardino-Balkaria	1503	1752.4	887.1	454.6	523.7	1065.4	1956.3	1.92	1.31
Kaliningrad Region	2152.7	2837.2	254.2	1082.5	1178.6	1744.4	2035	5.7	0.69
Kalmykia	633.3	709.5	216	334.3	548.2	794.5	699	19.22	0.68
Kaluga Region	2297	3084.2	425.5	787.4	1137.3	2022.3	3284.8	10.09	0.94
Kamchatka Region	1612	2106.7	975.3	1182.8	1251.7	1149.8	1392.4	1.86	0.79
Karachayevo–Circassia	887	1290.7	335.1	293.1	712.1	1012.6	549.6	3.32	1.38
Karelia	2596.4	3613.1	445.9	514.2	550.2	1608.1	1740	4.68	0.88
Kemerovo Region	8609.3	9957.1	1005.4	1231.8	2017.4	9170.3	9996	4.06	1.16
Khabarovsk Territory	5861.1	7135.5	2337.8	3717	4266.5	3122.9	3013.5	3.16	0.87
Khakassia	1342.6	1572.6	152.4	257.7	265.4	741.6	890.4	13.04	0.79
Kirov Region	3398.3	3865	458.3	550.9	1091.7	2376.8	2885.5	4.1	0.9
Komi	4160.1	4991.6	1091.7	550.5	1096.2	3260.5	4581.2	35.77	0.49
Kostroma Region	1276	1568.6	206.7	468.5	275.8	1162.1	1255.6	4.92	0.74
Krasnodar Territory	11984.4	15314.3	4875.4	4272	4497.4	6788.5	8572.8	6.14	1.91
Krasnoyarsk Territory	10381.2	13157.7	2156.6	2415	3206.4	7860.2	8727	9.69	0.58
Kurgan Region	1757.5	2288	247.5	509.8	406.4	1191.9	1442.5	2.29	0.62
Kursk Region	2098.9	2276.8	399.4	599.6	873.1	3610	3596.8	3.53	0.91
Leningrad Region	4831.1	5847.9	1125	1482.6	2286.2	3199.6	3583.2	3.65	1.01
Lipetsk Region	5016.7	5812	659.3	1329.2	2474.1	3345.7	3626.6	5.56	1
	1492.6	1940.6	259.8	400.7	211	1181.9	1237.5	2.46	1.4

(continued)

(continued)

	EXP. HEALTH		EXP. NE					NE	
	2005	2006	2002	2003	2004	2005	2006	1//0	2//1
Magadan Region									
Mari El	1460.4	2114.5	272.7	547.8	524	1120.3	1412.2	7.13	0.93
Mordovia	2574.4	2776.7	524.6	857.2	2111.3	8139.6	4228.7	11.75	0.89
Moscow	56795.2	86146.8	73020.2	86473.4	102698.2	141122.7	171298.4	11.58	1.15
Moscow Region	25818.9	35870.5	3682.9	5805.4	7728.1	19638.8	23290.1	8.95	0.81
Murmansk Region	3137.7	5531.5	425.1	429.5	394.9	1734.3	1846.8	5.42	0.77
Nizhny Novgorod Region	7168.2	8812.2	751.5	921.3	1333.2	6059.1	7469.4	7.29	0.71
North Ossetia–Alania	1786.9	1924.5	684.3	504.3	846	955	1071.4	21.49	0.98
Novgorod Region	1540	1941.2	113.7	369.4	230.2	1752	1258.8	4.87	0.58
Novosibirsk Region	6381.4	8055.1	1110.1	1625.4	2018.6	4539.6	6225.9	1.88	0.98
Omsk Region	7789.4	10000	762.6	984.3	2834	8724.2	5459.1	8.78	1.05
Orel Region	1619.9	1810.3	317.4	380.2	461	1487.8	1612.9	28.22	0.45
Orenburg Region	5592.2	7922.3	850.5	1092	1560.9	3859.7	5789.2	1.08	0.54
Penza Region	2816.5	4184.7	473.7	772.4	2839.4	2054.1	2916.5	2.95	1.23
Perm Region	7335.4	10094.1	1285.6	1865.2	752.8	8630.6	8417.7	11.5	0.65
Primorye Territory	4906.4	6396.8	1524.6	2315.3	2495.8	3898.6	4447.8	1.88	1
Pskov Region	1436	1724.7	510.8	611.9	344.5	1206	1416.2	3.41	0.69
Rostov Region	8260.1	10876.7	1927.5	2024.6	2599.9	6109.3	6265.8	5.3	1.02
Ryazan Region	2801	3186.7	679.9	962.6	1181.2	1839.1	2445.5	6.77	1.03
Sakha (Yakutia)	7077.1	7706.6	3910.6	5729	6300.8	14177	9089.2	6.72	0.83
Sakhalin Region	3193.9	4595.7	918.2	1434.4	1195	3691.7	4937.6	2.16	0.59
Samara Region	8902.2	14571.3	3000.4	3369.6	4820.1	6923.6	9165	6.53	0.83
Saratov Region	4574	6117.9	867.6	1730.3	1648.8	5105.5	5196	13.21	0.81
Smolensk Region	1803.7	2111.6	222.9	315.2	386.6	1427.7	1525.1	7.59	0.84
St. Petersburg	24340.6	34521.7	6205.9	7930.2	9104	16206.2	20821.8	37.87	0.62
Stavropol Territory	4503.6	5874.3	1869.1	1456	1339.9	5247.5	5021.6	2.89	1.45
Sverdlovsk Region	13586.4	19478.9	1368	2098.7	2877.2	8675	11970.7	4.52	0.85
Tambov Region	1949.9	2524.7	600.4	708.6	759.7	1841.7	1906.3	5.54	1

(continued)

	EXP. HEALTH		EXP. NE					NE	
	2005	2006	2002	2003	2004	2005	2006	1//0	2//1
Tatarstan	16829.6	10171.1	11094.6	14896.7	22940.5	22460	21861.3	37.59	0.64
Tomsk Region	2892.3	3867.4	550	834.6	1197.1	2637.3	2760.1	40.16	0.44
Tula Region	3866.2	5522.1	487.7	828	634.4	2423.5	3023.2	10.98	0.64
Tuva	1254	1480.9	329.2	588.9	540.4	617.9	687.9	9.41	0.77
Tver Region	3140.8	4144.8	493	742.7	527.7	2461.4	2490	2.62	0.9
Tyumen Region	8348.9	13167.1	43787.2	49893.4	54071.3	35019.8	57136.2	5.95	0.95
Udmurtian	4062.5	5606.2	1381	1544.5	1824.4	3020.7	3367.3	12.33	0.61
Ulyanovsk Region	3095.6	3987.6	124	210.7	283.4	1661.3	3528.8	12.58	0.56
Vladimir Region	2356.9	3392.2	394.4	657.8	578.1	1555.5	2341.8	20.3	0.41
Volgograd Region	5325.9	7054.6	1127.8	1207.7	1452.7	4766.6	6265.5	8.43	0.78
Vologda Region	5192	5797.3	376.6	772.3	1302.3	3270.9	3829.6	12.73	0.64
Voronezh Region	5323.2	6566.2	942.5	1112.5	1033.7	3185.1	3441	6.59	1.3
Yaroslavl Region	3902.3	5161.8	1718.2	1715.7	1060.3	4803.7	4634.5	9.46	1.48
Russian Federation	463.8	601.8	219	266	317	515.5	603.9	6.76	0.92

	NE				EDU.						HEALTH	
	3//2	4//3	5//4	6//5	1//0	2//1	3//2	4//3	5//4	6//5	1//0	2//1
Adygeya	0.79	0.58	1.05	0.88	0.83	0.95	0.99	1.32	1.05	1.04	0.82	0.92
Altai	1.76	1.92	0.82	0.77	0.52	1.19	0.98	1.02	1.45	0.96	0.65	1.2
Altai Territory	0.74	0.8	2.56	1.14	0.85	1.23	0.91	1.02	1.05	1.06	0.88	1.23
Amur Region	1.15	0.88	1.49	0.92	0.81	1.18	1.01	1.14	1.11	0.97	0.76	1.36
Arkhangelsk Region	1.16	0.87	1.6	0.95	0.93	1.19	1.14	1.1	1.11	0.98	0.79	1.2
Astrakhan Region	1.74	0.93	1.58	1.01	0.93	1.07	0.9	1.13	0.91	0.98	0.98	1.02
Bashkortostan	0.94	0.77	1.02	1.12	1.08	1.34	0.92	1.17	1.18	0.96	0.97	1.18
Belgorod Region	1.06	1.08	2.42	0.81	0.79	1.11	0.99	0.99	0.91	1.13	0.73	1.04
Bryansk Region	1.11	0.96	1.42	0.86	0.95	1.1	0.96	1.04	1.13	1.03	1.14	1.02
Buryatia	0.98	0.96	1.35	1.86	0.96	1.13	1.09	1.09	1.12	0.98	1.11	1.26
Chechen	0.82	0.31	1.84	1.03				0.48	2.19	1.18		
Chelyabinsk Region	1.4	1.4	1.49	0.68	0.9	1.17	0.99	0.99	1.06	1.11	0.9	1.13

(continued)

(continued)

	NE				EDU.						HEALTH	
	3//2	4//3	5//4	6//5	1//0	2//1	3//2	4//3	5//4	6//5	1//0	2//1
Chita Region	1.13	1.05	1.19	0.81	0.81	1.1	0.98	1.12	1.15	1.05	0.78	1.15
Chuvash	1.06	1.09	0.92	1.12	0.84	1.16	0.91	1.07	1.05	1.01	0.75	1.41
Dagestan	0.7	1.08	1.23	0.72	0.92	1.11	0.98	1.1	1.04	1.14	0.81	1.08
Ingushetia	0.52	1.23	0.88	1.03	1.02	0.95	1.34	1.04	0.96	1.21	0.87	1.09
Irkutsk Region	1.18	1	1.81	0.96	0.99	1.26	1.03	1.04	1.1	1	0.94	1.21
Ivanovo Region	1.17	1.23	2.39	0.92	0.79	1.14	0.93	1.1	1.08	0.97	0.83	0.97
Jewish Autonomous Region	2.41	1.22	1.09	0.98	0.75	1.25	0.94	1.05	1.26	0.93	0.83	1.13
Kabardino-Balkaria	0.59	1.02	1.47	1.56	0.86	1.1	1.13	1.09	0.99	0.95	0.96	0.96
Kaliningrad Region	3.48	0.92	1.12	0.85	1.21	1.06	0.96	1.04	1	1.17	1.04	0.88
Kalmykia	1.11	1.33	1.59	0.82	0.91	1.05	0.83	1.11	1.05	1.13	0.81	0.94
Kaluga Region	1.53	1.21	1.5	1.29	1.07	1.22	0.97	1.11	1.14	0.98	0.95	1.17
Kamchatka Region	1.03	0.92	0.91	0.92	0.93	1.01	1.14	1.09	1.13	1.07	1.04	1
Karachayevo–Circassia	0.84	1.88	1.25	0.45	0.92	0.89	1.03	0.96	1.05	1.15	0.88	0.94
Karelia	0.99	0.97	2.36	0.88	0.96	1.21	0.94	1.17	0.97	1.02	1.08	1.18
Kemerovo Region	1.09	1.1	3.44	0.95	0.95	1.18	0.96	0.95	0.91	1.09	0.92	1.22
Khabarovsk Territory	1.24	0.96	0.71	0.9	0.77	1.39	0.89	1.06	1.25	1.12	0.92	1.19
Khakassia	1.48	0.9	2.42	0.97	0.96	1.09	0.96	1.1	1.13	1	0.8	1.17
Kirov Region	0.99	0.76	4	1.01	0.97	1.03	1	1.05	1.06	0.98	0.9	1.13
Komi	0.5	1.68	2.59	1.12	1.03	1.12	1.09	1.07	0.99	0.98	0.8	1
Kostroma Region	1.77	0.57	3.57	0.89	0.84	1.06	0.88	1.16	1.1	1.02	0.77	1.08
Krasnodar Territory	0.78	0.88	1.35	1	0.89	0.91	1.05	1.07	1.14	1.05	0.88	0.88
Krasnoyarsk Territory	0.92	1.11	2.19	0.9	0.9	1.26	0.98	1.05	1.09	1.07	0.64	1.14
Kurgan Region	1.73	0.72	2.47	0.98	1.04	1.13	0.93	1.1	1.04	1	0.98	1.15
Kursk Region	1.31	1.09	3.22	0.9	1.03	1.24	0.86	1.16	0.9	1.13	0.77	1.17
Leningrad Region	1.11	1.25	1.12	0.95	1.07	1.08	1.05	1.12	1.19	1	0.88	1.09
Lipetsk Region	1.41	1.4	1.03	0.98	1.14	1.11	0.96	0.99	1.12	1.06	1.07	1.05
Magadan Region	1.43	0.51	4.79	0.84	0.84	1.1	1	1.13	1.15	1.04	0.88	1.14
Mari El	1.79	0.72	1.92	0.97	0.83	1.15	0.94	0.98	1.01	1.02	0.94	1.15
Mordovia	1.26	1.54	3.78	0.64	0.77	1.26	0.83	0.79	1.23	1.39	0.71	1.2
Moscow	1.01	1.04	1.11	0.88	0.92	1.14	1.04	1.07	1.21	1.09	0.89	1.04
Moscow Region	1.22	1.07	1.86	0.92	1.06	1.12	0.95	1.11	1.03	1.07	0.98	1.08
Murmansk Region	0.91	0.79	4.07	0.63	0.98	1.04	1	1.04	1.11	0.99	0.83	0.98
Nizhny Novgorod Region	1.06	1.82	2.45	0.94	0.96	1.05	0.95	1.1	1.02	1.01	0.89	1.13

(continued)

	NE				EDU.						HEALTH	
	3//2	**4//3**	**5//4**	**6//5**	**1//0**	**2//1**	**3//2**	**4//3**	**5//4**	**6//5**	**1//0**	**2//1**
North Ossetia–Alania	0.68	1.32	0.89	1.01	0.79	1.29	1	0.99	1.23	0.87	0.78	1.22
Novgorod Region	2.53	0.53	5.38	0.62	0.88	1.17	0.9	1.06	0.93	1.1	0.83	1.3
Novosibirsk Region	1.11	1.08	1.82	1.14	0.77	1.21	0.96	1.11	1.05	1.07	0.8	1.23
Omsk Region	0.94	1.94	2.23	0.58	1.03	1.09	0.83	0.94	1	1.18	0.87	1.25
Orel Region	0.99	1.03	2.86	0.97	1.06	1.31	0.99	1	1.12	1.09	0.97	1.27
Orenburg Region	1.11	2.11	1.04	1.24	1.27	1.14	1	1.02	0.97	1.03	1.17	1.15
Penza Region	1.32	0.98	1.92	1.05	0.83	1.1	0.91	1	1.07	1.12	0.78	1.06
Perm Region	1.3	0.83	3.95	0.79	1.13	1.09	1.06	0.98	1.16	1.01	0.87	1.03
Primorye Territory	1.29	0.91	1.36	0.92	0.99	1.14	0.92	0.98	1.04	1	1.04	1.14
Pskov Region	0.97	0.51	3.37	1.01	0.9	1.21	0.91	1.17	1.15	1.08	0.89	1.19
Rostov Region	0.92	1.1	1.79	0.82	0.86	1.15	1.04	1.13	1.05	1.02	0.78	1.02
Ryazan Region	1.18	1.18	1.25	1.09	0.9	0.99	0.95	1.12	1.08	1.08	0.86	0.99
Sakha (Yakutia)	1.31	0.99	1.62	0.68	0.9	1.3	0.96	1.07	0.91	1.23	0.9	1.17
Sakhalin Region	1.21	0.69	2.76	0.97	0.87	1.19	0.92	1	1.21	1.03	0.84	1.28
Samara Region	0.98	1.18	1.14	0.99	0.77	1.28	0.97	1.05	1.04	1	0.72	1.05
Saratov Region	1.58	0.84	2.58	0.84	0.89	1.11	0.91	1.08	1.06	1.06	0.87	1.1
Smolensk Region	1.31	1.03	3.18	0.88	0.89	1.14	1.05	1.04	1.06	1.03	0.86	1.22
St. Petersburg	1.07	0.96	1.23	0.94	0.94	1.07	1.07	1.13	0.93	0.91	0.76	1.06
Stavropol Territory	0.75	0.76	3.06	0.81	0.9	0.91	1.06	1.05	0.95	1.09	0.9	0.87
Sverdlovsk Region	1.32	1.09	2.45	0.98	0.93	1.13	0.99	1.01	1.03	1.08	0.93	1.03
Tambov Region	1.01	0.93	1.89	0.88	0.91	1.08	0.94	1.11	0.96	1.09	0.91	0.98
Tatarstan	1.26	1.13	0.88	0.99	1.1	1.55	1.01	0.86	1.12	1.22	1.17	1.16
Tomsk Region	1.15	1.24	1.78	0.83	0.88	1.38	0.89	1.09	1.04	1.11	0.98	1.2
Tula Region	1.44	0.7	3.05	1.02	0.96	1.04	0.98	1.12	1.05	1.05	0.91	1.06
Tuva	1.56	0.83	1.01	0.99	0.77	1.22	0.93	1.1	1.03	1.12	0.86	1.09
Tver Region	1.38	0.58	3.58	0.79	0.95	0.98	1.04	1.09	0.96	1.07	1.04	0.96
Tyumen Region	1.01	0.92	1.33	1.1	1.17	1.33	5.37	1.02	0.71	0.99	0.69	1.44
Udmurtian	1.04	1.03	1.39	0.92	0.91	1.17	0.98	1.05	1.15	0.99	0.87	1.2
Ulyanovsk Region	1.61	1.16	4.95	1.48	0.96	1.01	1.04	1.09	1.02	0.91	0.92	0.82
Vladimir Region	1.4	0.79	2.33	1.1	0.89	1.09	1	1.11	1.06	0.98	0.91	1.04
Volgograd Region	0.87	1.01	2.63	0.97	0.97	1.04	0.96	0.99	1.02	1	0.96	1
Vologda Region	1.59	1.33	1.9	1.02	0.9	1.22	0.93	0.99	1	1.02	0.79	1.11
Voronezh Region	1.04	0.8	2.6	0.88	0.92	0.94	1	1.11	1.03	1.07	0.89	1
Yaroslavl Region	0.83	0.53	3.72	0.8	0.95	1.03	1.09	1.04	1	1.08	0.93	1.08
Russian Federation	**1.03**	**1**	**1.31**	**0.94**	**0.98**			**1.05**	**1.07**	**1.04**	**0.92**	

	HEALTH				RESEAR.					UNEMP.		
	3//2	4//3	5//4	6//5	2002	2003	2004	2005	2006	2002	2003	2004
Adygeya	1.01	1.34	1.03	1.02	0.08	0.09	0.09	0.09	0.14	13.5	15.9	17.6
Altai	1.04	0.96	1.34	1.17	0.14	0.14	0.14	0.16	0.18	11.4	10.2	9.6
Altai Territory	0.93	1.03	0.97	0.97	0.1	0.11	0.11	0.12	0.15	8.3	12.1	9
Amur Region	0.93	1.14	1.08	0.95	0.18	0.2	0.19	0.19	0.21	10.8	10.2	11.2
Arkhangelsk Region	1.19	1.06	1.16	0.89	0.08	0.09	0.08	0.1	0.1	8.1	9.9	7.2
Astrakhan Region	0.87	1.19	1.07	1.13	0.2	0.21	0.22	0.23	0.25	10.9	10.1	11
Bashkortostan	0.93	1.15	1.3	1.03	0.28	0.27	0.26	0.25	0.26	8.3	8.1	7.1
Belgorod Region	1.02	0.93	1.11	1.06	0.09	0.08	0.07	0.07	0.07	8.1	8.2	5.9
Bryansk Region	1.04	0.98	1.11	1.08	0.03	0.04	0.04	0.04	0.04	8.6	7.4	8.8
Buryatia	0.97	1.14	1.05	0.99	0.47	0.51	0.54	0.51	0.43	15.4	16.9	15.3
Chechen		0.57	1.73	1.16			0.09	0.1	0.1			
Chelyabinsk Region	1.02	1.02	1.12	1.07	0.2	0.2	0.18	0.22	0.19	6.5	6	5.2
Chita Region	0.91	1.1	1.32	0.94	0.07	0.06	0.07	0.05	0.06	11.1	15.1	12.4
Chuvash	1	1.07	1.07	1.18	0.05	0.05	0.05	0.05	0.04	9.4	8.5	9.9
Dagestan	1	0.98	1.08	0.92	0.19	0.19	0.19	0.2	0.2	24.1	20.6	27.2
Ingushetia	1.62	0.91	0.8	0.9	0.02	0.01	0.01	0.01	0.01	43.7	55.6	46.3
Irkutsk Region	1.07	1.09	1.03	1.02	0.49	0.5	0.51	0.52	0.52	11.1	11.7	10.5
Ivanovo Region	1.01	1.1	0.97	1.35	0.25	0.24	0.23	0.22	0.22	6.9	6.6	4.6
Jewish Autonomous Region	0.97	1.11	1.13	1.24	0.09	0.09	0.49	0.09	0.09	9.1	6.5	8.3
Kabardino-Balkaria	1.18	1.08	0.95	0.99	0.22	0.25	0.26	0.26	0.28	19.1	22.6	25.7
Kaliningrad Region	1.01	1.02	1.09	0.96	0.12	0.12	0.05	0.13	0.12	7.2	7.6	6.5
Kalmykia	0.81	1.09	0.98	1.05	0.25	0.26	0.25	0.27	0.27	18	17.5	21.7
Kaluga Region	1	1.1	1.07	1.07	1.05	1.05	0.98	0.98	0.98	6.7	6.2	6.3
Kamchatka Region	1.27	1.11	1.17	0.99	0.42	0.72	0.55	0.55	0.6	11.6	12.1	11
Karachayevo-Circassia	1.04	0.94	1.17	1.21	0.26	0.25	0.23	0.24	0.25	12.1	19.2	16.9
Karelia	0.94	1.15	1.08	1.13	0.45	0.46	0.49	0.5	0.51	7.8	8.4	7.3
Kemerovo Region	1.06	0.93	1.01	1.01	0.09	0.08	0.08	0.08	0.11	9.3	9.8	9.8
Khabarovsk Territory	1.02	0.99	1.27	1.13	0.25	0.26	0.26	0.24	0.23	7	6.2	6.6
Khakassia	1.05	1.16	1.12	0.95	0.05	0.05	0.06	0.21	0.13	8	10.5	10.3
Kirov Region	1	0.49	2	0.94	0.1	0.1	0.11	0.12	0.12	6.8	7.3	8.7
Komi	1.13	1.08	0.99	0.96	0.38	0.4	0.41	0.42	0.46	9.2	11.9	12.4
Kostroma Region	0.83	1.18	1.04	1.01	0.04	0.03	0.03	0.03	0.03	5	6.2	5.9
Krasnodar Territory	1.1	1.09	1.11	1.01	0.17	0.18	0.18	0.17	0.2	7.6	10.2	8.8
Krasnoyarsk Territory	1.2	1	1.05	1.02	0.26	0.26	0.26	0.26	0.27	7.8	11.1	9.2
Kurgan Region	0.94	1.04	0.97	1.05	0.09	0.1	0.09	0.1	0.11	10.7	9.4	12.5
Kursk Region	1	1.22	0.92	0.98	0.09	0.1	0.08	0.08	0.11	7.1	8.5	7.5
Leningrad Region	1.13	1.15	1.08	1.02	0.38	0.38	0.07	0.37	0.37	7	8.7	6.8
Lipetsk Region	0.89	1.04	1.18	1.04	0.02	0.03	0.04	0.06	0.09	4.9	4.4	4.2
Magadan Region	1.06	1.08	1.09	1.05	0.65	0.69	.	0.65	0.64	8.2	10.1	7.9
Mari El	1.01	1	1.14	1.12	0.07	0.07	0.07	0.05	0.05	13.6	12	9.1
Mordovia	0.9	0.8	1.47	1.32	0.09	0.08	0.1	0.08	0.12	9.4	7.4	6
Moscow	0.92	1.09	1.41	1.11	4.5	4.42	4.26	4.21	4.18	1.4	1.3	1.6
Moscow Region	1.05	0.98	1.11	1.08	1.21	1.25	1.22	1.22	1.24	4.4	4.4	3.7

(continued)

	HEALTH				RESEAR.					UNEMP.		
	3//2	4//3	5//4	6//5	2002	2003	2004	2005	2006	2002	2003	2004
Murmansk Region	1.16	1.04	1.08	1.05	0.52	0.54	0.72	0.57	0.57	10.3	10.2	11
Nizhny Novgorod Region	0.94	1.65	0.72	0.94	0.58	0.59	0.61	0.65	0.66	7.9	6.2	7.4
North Ossetia–Alania	0.95	0.97	1.14	0.97	0.15	0.16	0.16	0.16	0.17	12.6	10.2	11.7
Novgorod Region	0.93	1.08	1.08	1.08	0.04	0.04	0.72	0.03	0.03	6.3	5	6.2
Novosibirsk Region	0.93	1.02	0.96	1.05	1.88	1.91	1.9	1.92	1.9	11.4	11.1	9
Omsk Region	0.91	1.04	1.04	1.19	0.16	0.16	0.16	0.16	0.15	9.5	9.5	9.1
Orel Region	1.04	1.04	1.04	1	0.17	0.16	0.12	0.13	0.16	6.7	7.8	5.9
Orenburg Region	0.9	1.83	0.5	1.17	0.08	0.09	0.09	0.08	0.09	10.3	11.4	10.8
Penza Region	0.96	0.34	3.77	1.1	0.06	0.08	0.07	0.08	0.11	7.3	9	6.9
Perm Region	1.12	2.47	0.41	1.11	0.21	0.22	0.21	0.22	0.21	9.1	7.1	7
Primorye Territory	0.94	0.95	1.14	1.05	0.65	0.67	0.69	0.7	0.72	9.1	8.1	9.6
Pskov Region	0.88	1.11	1.1	1.03	0.1	0.1	0.04	0.05	0.08	8.1	8.1	5.6
Rostov Region	1.09	1.07	1.04	1.05	0.26	0.25	0.26	0.27	0.27	11.4	12.6	8.6
Ryazan Region	0.94	1.17	1.3	0.93	0.12	0.12	0.12	0.13	0.13	8.1	8.3	5.8
Sakha (Yakutia)	0.95	0.99	0.95	1.16	0.62	0.61	0.61	0.6	0.6	7.1	9.3	8.8
Sakhalin Region	1.02	0.99	1.14	1.04	0.16	0.17	0.21	0.2	0.19	9.4	9.3	7.5
Samara Region	0.95	1.07	1.54	1.22	0.21	0.18	0.18	0.18	0.17	5.4	4.4	5.3
Saratov Region	0.91	1.02	1.06	1.11	0.29	0.29	0.28	0.28	0.32	7.6	10.6	9.9
Smolensk Region	0.97	1.05	1.04	0.96	0.05	0.05	0.06	0.06	0.08	11.1	10.8	9.1
St. Petersburg	1	1.13	1.57	1.04	2.74	2.74	2.7	2.66	2.6	3.4	4.2	2.7
Stavropol Territory	1.07	1.02	1.08	1.11	0.11	0.1	0.09	0.1	0.1	9.7	10.5	9.6
Sverdlovsk Region	1.11	1.04	1.23	1.02	0.53	0.53	0.53	0.52	0.51	8.5	7.6	7.3
Tambov Region	0.88	1.06	0.99	1.1	0.14	0.14	0.14	0.15	0.15	9.5	9	9.7
Tatarstan	0.93	0.84	1.7	0.62	0.38	0.38	0.37	0.37	0.36	5.3	6.7	7.4
Tomsk Region	0.95	1.06	1.19	1.06	1.43	1.41	1.46	1.45	1.48	11.2	13.9	10.1
Tula Region	0.97	1.13	1.02	1.16	0.12	0.13	0.12	0.12	0.12	6.2	5.3	4.6
Tuva	0.97	1.09	0.93	1.05	0.16	0.18	0.19	0.23	0.21	20.3	20.7	19.7
Tver Region	1	1.09	0.99	1.03	0.34	0.33	0.33	0.34	0.34	4.7	6.6	5.3
Tyumen Region	5	1.03	0.74	1.07	0.15	0.16	0.15	0.15	0.18	8.7	8.3	8.7
Udmurtian	1.01	1.02	1.07	1.13	0.14	0.14	0.14	0.15	0.15	7.9	6.7	8
Ulyanovsk Region	1.04	1.05	0.98	0.9	0.14	0.14	0.17	0.18	0.19	6.1	7.3	9.5
Vladimir Region	1.02	1.07	0.91	1.06	0.27	0.27	0.24	0.24	0.23	10.4	10.1	9.1
Volgograd Region	1.03	1.06	1.11	0.98	0.14	0.15	0.16	0.16	0.16	8.4	11	9.4
Vologda Region	0.97	1.08	1.23	0.98	0.04	0.04	0	0.04	0.05	6	4.7	6.3
Voronezh Region	1.03	1.16	1	1	0.28	0.29	0.32	0.39	0.4	8.9	8.1	8.6
Yaroslavl Region	1	1.01	1.12	1.1	0.27	0.27	0.27	0.27	0.27	3.8	5.7	4.7
Russian Federation					**0.71**	**0.71**	**0.7**	**0.7**	**0.7**	**7.9**	**8.2**	**7.8**

	UNEMP.		ETHNO	CF					GINI				
	2005	2006	2002	2002	2003	2004	2005	2006	2002	2003	2004	2005	2006
Adygeya	12.9	13.7	0.52	8.9	9.3	9.4	9.3	9.2	0.85	0.85	0.73	0.73	0.7
Altai	10	11.6	0.57	8.1	8.3	8.4	8.5	8.1	0.8	0.78	0.65	0.66	0.59
Altai Territory	9	8.8	0.15	10.8	11.3	11.5	11.4	11.9	0.97	0.96	0.89	0.9	0.9
Amur Region	10.3	8.2	0.15	9.5	9.7	10.1	9.3	8.9	0.89	0.87	0.79	0.73	0.67
Arkhangelsk Region	5.5	5.9	0.11	9	10.2	10.7	11	11.2	0.86	0.9	0.84	0.87	0.86
Astrakhan Region	12	7.9	0.49	9.2	10.8	11.5	11.8	12.7	0.88	0.94	0.89	0.93	0.96
Bashkortostan	7.1	6.5	0.72	11.3	13.3	14.1	14.7	15.6	1	0.94	0.95	0.9	0.87
Belgorod Region	6	5.6	0.13	9.4	9.9	10.2	10.5	11.2	0.88	0.88	0.8	0.83	0.86
Bryansk Region	6.7	6.8	0.07	8.7	10	10.3	10.3	10.8	0.84	0.89	0.81	0.82	0.83
Buryatia	14.2	13.4	0.46	14.1	14.3	13.4	13.6	13.9	0.87	0.9	0.99	0.96	0.97
Chechen		66.9	0.19										
Chelyabinsk Region	5.4	5.1	0.31	10	10.9	11.1	11.6	12.8	0.93	0.95	0.87	0.91	0.97
Chita Region	11.1	8.8	0.47	8.1	11.3	11.5	11.9	12.5	0.8	0.97	0.89	0.93	0.94
Chuvash	11.4	8.6	0.08	7.5	8.6	9.1	9	9.9	0.75	0.8	0.71	0.71	0.76
Dagestan	22.3	22.3	0.84	10.5	10.4	11	12.2	12.6	0.95	0.91	0.86	0.95	0.95
Ingushetia	64.9	58.5	0.36	8	8	8.3	8.3	9.4	0.79	0.75	0.64	0.64	0.71
Irkutsk Region	10	8.9	0.18	14.2	14.9	15.1	15	15.4	0.87	0.87	0.9	0.89	0.89
Ivanovo Region	6.8	4.2	0.08	7.4	8.4	8.4	7.6	8.4	0.75	0.79	0.65	0.58	0.62
Jewish Autonomous Region	7.9	9.8	0.19	9.9	10.1	12.1	12.8	11.2	0.92	0.9	0.93	0.99	0.86
Kabardino-Balkaria	23.4	20.7	0.62	8.7	8.6	9.6	9.6	10.1	0.84	0.8	0.76	0.76	0.77
Kaliningrad Region	6.6	4.5	0.3	8.2	8.4	11.2	11.2	10.3	0.8	0.79	0.87	0.88	0.79
Kalmykia	18	16.7	0.6	11.7	11.8	11.1	11	11.2	0.98	0.99	0.87	0.87	0.86
Kaluga Region	5.7	5.6	0.12	7.8	8.6	8.8	9.2	10.1	0.77	0.79	0.69	0.73	0.77
Kamchatka Region	9.5	9.1	0.3	10	11.1	11.1	9.8	10.3	0.93	0.95	0.87	0.78	0.79
Karachayevo–Circassia	13.6	19.4	0.72	10.1	10.2	10.3	10.3	10.7	0.93	0.9	0.81	0.82	0.82
Karelia	8.8	3.6	0.39	8.8	9.4	9.7	9.2	10.1	0.85	0.85	0.76	0.73	0.78
Kemerovo Region	8.6	7.3	0.15	11.3	12.2	13.6	13.7	14.5	1	0.99	0.98	0.96	0.94
Khabarovsk Territory	5.7	6	0.19	10.6	12.2	12.4	12.3	13.3	0.96	0.99	0.95	0.96	1
Khakassia	8.9	9.1	0.34	9.7	10.5	10.6	9.6	10.3	0.91	0.92	0.83	0.76	0.79
Kirov Region	7.1	7.9	0.17	7.4	8.2	8.4	8.3	9.1	0.74	0.77	0.65	0.64	0.69
Komi	11.5	12.4	0.57	15.4	17.4	17.4	17.3	17.3	0.82	0.79	0.8	0.78	0.79
Kostroma Region	4.8	5	0.07	9.2	9.7	9.9	9.6	10.4	0.88	0.87	0.78	0.76	0.8

(continued)

	UNEMP.		ETHNO	CF					GINI				
	2005	2006	2002	2002	2003	2004	2005	2006	2002	2003	2004	2005	2006
Krasnodar Territory	7.5	7.4	0.24	12.1	12.8	12.7	12.9	13.6	0.96	0.96	0.97	0.99	0.99
Krasnoyarsk Territory	9	9.9	0.2	13.9	14.3	14.5	14.3	15.2	0.88	0.9	0.94	0.93	0.9
Kurgan Region	11.3	12.4	0.16	11.9	13	12.5	12.4	13.3	0.97	0.95	0.96	0.97	1
Kursk Region	7.1	7.3	0.08	9	9.9	9.9	9.7	10.4	0.86	0.89	0.78	0.77	0.79
Leningrad Region	7.4	5.9	0.16	7.3	7.7	8.7	9.1	10.7	0.73	0.74	0.68	0.72	0.82
Lipetsk Region	8	4.9	0.07	10.9	11.6	11.8	10.9	11.7	0.98	0.98	0.91	0.87	0.89
Magadan Region	7	5.4	0.34	8.5	12.5	9.7	9.1	12.8	0.83	0.97	0.77	0.72	0.97
Mari El	9.9	10.2	0.58	9.7	10.9	10.6	10.4	11.1	0.91	0.95	0.83	0.83	0.86
Mordovia	7	4.7	0.51	8.8	9.2	9.1	9.1	9.4	0.85	0.84	0.71	0.72	0.71
Moscow	0.8	1.6	0.22	51	51.8	44	38.6	41.4	0	0	0	0	0
Moscow Region	3.3	3	0.13	9.6	10.7	11.4	10.9	11.6	0.9	0.93	0.88	0.87	0.89
Murmansk Region	8.8	6.7	0.25	11	11.7	8.2	10	12.1	0.98	0.99	0.63	0.79	0.92
Nizhny Novgorod Region	6	5.3	0.1	9.5	9.6	10	10	11	0.9	0.87	0.79	0.79	0.84
North Ossetia–Alania	8.8	8.5	0.55	8.6	11.1	10.8	10.7	10.2	0.84	0.95	0.85	0.85	0.78
Novgorod Region	5.8	5.5	0.1	10.1	10.7	11.6	11.6	12.4	0.93	0.93	0.9	0.91	0.94
Novosibirsk Region	7.8	7.4	0.13	10	11.3	12	12.7	13.1	0.92	0.96	0.93	0.98	0.99
Omsk Region	8.6	9.3	0.3	10.9	12.8	13.8	13.6	13.7	0.98	0.96	0.97	0.97	0.98
Orel Region	6.1	6	0.08	10.6	11.1	11.6	11.6	12.3	0.96	0.95	0.9	0.91	0.94
Orenburg Region	9.4	6.5	0.44	8.1	9	10	10.5	11.5	0.8	0.83	0.78	0.83	0.88
Penza Region	6.5	6.5	0.25	7.5	8.4	8.5	8.7	9.5	0.76	0.79	0.66	0.68	0.72
Perm Region	7	6.9	0.26	14	15.5	15.9	16.5	17.6	0.88	0.85	0.87	0.82	0.78
Primorye Territory	8	8	0.17	9.2	9.4	10.1	10.8	11	0.88	0.85	0.79	0.85	0.84
Pskov Region	6.6	7.4	0.11	7.9	9.1	10.3	10.5	10.1	0.78	0.83	0.81	0.83	0.77
Rostov Region	8.6	8	0.2	11.1	11.7	12.3	12.8	13	0.98	0.99	0.94	0.99	0.98
Ryazan Region	5.3	5.2	0.09	8.8	9.6	9.3	9.1	10.1	0.85	0.87	0.73	0.72	0.77
Sakha (Yakutia)	8.9	9.5	0.62	10.7	12.8	12.9	13.3	13.4	0.97	0.96	0.98	0.98	0.99
Sakhalin Region	7.6	4.6	0.28	10.4	12	12.7	12.5	13.9	0.95	1	0.97	0.97	0.97
Samara Region	5.3	4.3	0.29	16.9	17.7	17.8	18.8	19	0.77	0.77	0.79	0.72	0.72
Saratov Region	9.1	8.2	0.25	8.2	9.4	9.6	10.2	11	0.8	0.85	0.75	0.81	0.84

(continued)

(continued)

	UNEMP.		ETHNO	CF					GINI				
	2005	2006	2002	2002	2003	2004	2005	2006	2002	2003	2004	2005	2006
Smolensk Region	7.8	8	0.11	9.1	9.8	10	9.8	9.8	0.87	0.88	0.79	0.78	0.75
St. Petersburg	2.2	2.4	0.15	9.8	13	9.5	9.6	17.8	0.91	0.95	0.74	0.76	0.77
Stavropol Territory	7	8.9	0.33	9.4	10.2	10.8	11.5	12.3	0.89	0.9	0.85	0.91	0.94
Sverdlovsk Region	6.7	7	0.19	10.4	12.3	14.1	14.7	15.8	0.95	0.98	0.95	0.9	0.86
Tambov Region	8.5	8.7	0.06	10.8	11.9	12	11.5	12.2	0.97	1	0.93	0.91	0.93
Tatarstan	6.7	5.6	0.56	11.8	13.1	13.3	13.3	14.2	0.98	0.95	0.99	0.98	0.95
Tomsk Region	10.5	9	0.17	11.3	11.8	12.1	12.6	13.6	1	0.99	0.94	0.98	0.98
Tula Region	5	2.7	0.09	7.7	8	8.5	8	8.9	0.76	0.76	0.66	0.61	0.66
Tuva	21.8	20.5	0.37	9.8	10.1	10.3	10.6	10.8	0.91	0.9	0.81	0.84	0.83
Tver Region	5.9	4.5	0.13	7.3	7.7	7.9	8.7	9.5	0.74	0.73	0.6	0.68	0.72
Tyumen Region	6.7	6.8	0.47	20.3	20.6	20.6	20.8	21.4	0.67	0.69	0.68	0.64	0.63
Udmurtian	7.8	8.4	0.55	7.5	7.8	8.5	8.4	9.3	0.75	0.74	0.66	0.65	0.71
Ulyanovsk Region	7.7	6.9	0.45	10.7	11.2	11	11.3	12.1	0.97	0.96	0.86	0.89	0.92
Vladimir Region	9	10.9	0.08	7.1	8	7.8	8	8.1	0.72	0.75	0.6	0.61	0.59
Volgograd Region	6.8	8.6	0.2	8.6	10.3	11.3	11.6	12.3	0.83	0.91	0.88	0.91	0.93
Vologda Region	5.2	5.3	0.06	8.8	10.8	14	15.9	11.4	0.85	0.94	0.96	0.84	0.87
Voronezh Region	7.5	5.5	0.1	10.2	12	12.2	13.4	14.4	0.94	1	0.94	0.98	0.94
Yaroslavl Region	4	3	0.35	11.7	13.3	11.4	11.4	12.2	0.98	0.94	0.88	0.9	0.93
Russian Federation	**7.2**	**7.2**		**14**	**14.3**	**15**	**14.8**	**15.3**	**0.88**	**0.9**	**0.91**	**0.9**	**0.89**

	DENS.					ln DENS.				
	2002	2003	2004	2005	2006	2002	2003	2004	2005	2006
Adygeya	6.75	6.7	6.71	6.76	6.77	1.91	1.9	1.9	1.91	1.91
Altai	0.26	0.26	0.26	0.26	0.26	−1.35	−1.35	−1.35	−1.35	−1.35
Altai Territory	1.86	1.84	1.83	1.83	1.82	0.62	0.61	0.6	0.6	0.6
Amur Region	0.32	0.3	0.3	0.29	0.29	−1.14	−1.2	−1.2	−1.24	−1.24
Arkhangelsk Region	0.29	0.27	0.27	0.27	0.26	−1.24	−1.31	−1.31	−1.31	−1.35
Astrakhan Region	2.44	2.41	2.41	2.42	2.42	0.89	0.88	0.88	0.88	0.88
Bashkortostan	3.38	3.37	3.37	3.39	3.4	1.22	1.21	1.21	1.22	1.22

(continued)

	DENS.					ln DENS.				
	2002	2003	2004	2005	2006	2002	2003	2004	2005	2006
Belgorod Region	6.48	6.5	6.53	6.58	6.62	1.87	1.87	1.88	1.88	1.89
Bryansk Region	4.84	4.71	4.68	4.65	4.63	1.58	1.55	1.54	1.54	1.53
Buryatia	0.35	0.33	0.33	0.33	0.33	−1.05	−1.11	−1.11	−1.11	−1.11
Chechen	0	7.47	8.07	8.35	8.52	−2.3	2.01	2.09	2.12	2.14
Chelyabinsk Region	4.87	4.81	4.8	4.8	4.79	1.58	1.57	1.57	1.57	1.57
Chita Region	0.34	0.32	0.32	0.31	0.31	−1.08	−1.14	−1.14	−1.17	−1.17
Chuvash	8.72	8.48	8.47	8.46	8.46	2.17	2.14	2.14	2.14	2.14
Dagestan	5	5.78	5.92	6.05	6.14	1.61	1.75	1.78	1.8	1.81
Ingushetia	11.58	14.48	14.87	15.37	15.68	2.45	2.67	2.7	2.73	2.75
Irkutsk Region	0.42	0.4	0.39	0.39	0.39	−0.87	−0.92	−0.94	−0.94	−0.94
Ivanovo Region	6.69	6.44	6.37	6.31	6.26	1.9	1.86	1.85	1.84	1.83
Jewish Autonomous Region	0.64	0.62	0.62	0.62	0.62	−0.45	−0.48	−0.48	−0.48	−0.48
Kabardino-Balkaria	7.38	8.29	8.4	8.5	8.53	2	2.12	2.13	2.14	2.14
Kaliningrad Region	7.38	7.41	7.43	7.45	7.46	2	2	2.01	2.01	2.01
Kalmykia	0.49	0.48	0.47	0.46	0.46	−0.71	−0.73	−0.76	−0.78	−0.78
Kaluga Region	4.25	4.16	4.13	4.11	4.1	1.45	1.43	1.42	1.41	1.41
Kamchatka Region	0.1	0.09	0.09	0.09	0.09	−2.3	−2.41	−2.41	−2.41	−2.41
Karachayevo–Circassia	3.54	3.6	3.61	3.62	3.62	1.26	1.28	1.28	1.29	1.29
Karelia	0.5	0.47	0.47	0.47	0.47	−0.69	−0.76	−0.76	−0.76	−0.76
Kemerovo Region	3.66	3.59	3.58	3.57	3.56	1.3	1.28	1.28	1.27	1.27
Khabarovsk Territory	0.22	0.22	0.22	0.21	0.21	−1.51	−1.51	−1.51	−1.56	−1.56
Khakassia	1.1	1.05	1.05	1.04	1.05	0.1	0.05	0.05	0.04	0.05
Kirov Region	1.55	1.49	1.48	1.47	1.46	0.44	0.4	0.39	0.39	0.38
Komi	0.32	0.29	0.29	0.29	0.29	−1.14	−1.24	−1.24	−1.24	−1.24
Kostroma Region	1.52	1.46	1.45	1.44	1.43	0.42	0.38	0.37	0.36	0.36
Krasnodar Territory	7.79	7.95	7.98	8	8.02	2.05	2.07	2.08	2.08	2.08
Krasnoyarsk Territory	0.15	0.15	0.15	0.15	0.15	−1.9	−1.9	−1.9	−1.9	−1.9
Kurgan Region	1.8	1.71	1.69	1.68	1.67	0.59	0.54	0.52	0.52	0.51

(continued)

(continued)

	DENS.					ln DENS.				
	2002	2003	2004	2005	2006	2002	2003	2004	2005	2006
Kursk Region	5.14	4.94	4.88	4.84	4.8	1.64	1.6	1.59	1.58	1.57
Leningrad Region	2.33	2.34	2.34	2.34	2.35	0.85	0.85	0.85	0.85	0.85
Lipetsk Region	6.07	5.98	5.97	5.95	5.93	1.8	1.79	1.79	1.78	1.78
Magadan Region	0.06	0.05	0.05	0.05	0.05	−2.81	−3	−3	−3	−3
Mari El	3.81	3.69	3.68	3.66	3.66	1.34	1.31	1.3	1.3	1.3
Mordovia	4.18	4.07	4.03	4	3.98	1.43	1.4	1.39	1.39	1.38
Moscow	911.64	1074.54	1096.03	1113.75	1120.33	6.82	6.98	7	7.02	7.02
Moscow Region	16.58	16.88	16.94	17.04	17.15	2.81	2.83	2.83	2.84	2.84
Murmansk Region	0.81	0.74	0.73	0.72	0.72	−0.21	−0.3	−0.31	−0.33	−0.33
Nizhny Novgorod Region	5.62	5.48	5.45	5.41	5.39	1.73	1.7	1.7	1.69	1.68
North Ossetia–Alania	9.84	10.28	10.4	10.47	10.48	2.29	2.33	2.34	2.35	2.35
Novgorod Region	1.57	1.52	1.51	1.5	1.49	0.45	0.42	0.41	0.41	0.4
Novosibirsk Region	1.81	1.78	1.78	1.78	1.78	0.59	0.58	0.58	0.58	0.58
Omsk Region	1.8	1.75	1.74	1.73	1.73	0.59	0.56	0.55	0.55	0.55
Orel Region	4.27	4.15	4.13	4.1	4.08	1.45	1.42	1.42	1.41	1.41
Orenburg Region	2.11	2.08	2.08	2.07	2.07	0.75	0.73	0.73	0.73	0.73
Penza Region	4.14	4	3.97	3.94	3.92	1.42	1.39	1.38	1.37	1.37
Perm Region	2.17	2.09	2.08	2.07	2.07	0.77	0.74	0.73	0.73	0.73
Primorye Territory	1.55	1.5	1.49	1.48	1.48	0.44	0.41	0.4	0.39	0.39
Pskov Region	1.7	1.65	1.63	1.61	1.6	0.53	0.5	0.49	0.48	0.47
Rostov Region	5.05	5.14	5.13	5.13	5.13	1.62	1.64	1.64	1.64	1.64
Ryazan Region	3.81	3.72	3.68	3.65	3.62	1.34	1.31	1.3	1.29	1.29
Sakha (Yakutia)	0.04	0.04	0.04	0.04	0.04	-3.22	-3.22	-3.22	-3.22	−3.22
Sakhalin Region	0.81	0.75	0.74	0.74	0.73	−0.21	−0.29	−0.3	−0.3	−0.31
Samara Region	7.22	7.14	7.13	7.12	7.12	1.98	1.97	1.96	1.96	1.96
Saratov Region	3.14	3.12	3.11	3.1	3.1	1.14	1.14	1.13	1.13	1.13
Smolensk Region	2.66	2.54	2.51	2.48	2.46	0.98	0.93	0.92	0.91	0.9
St. Petersburg	391.08	393.59	393.1	392.26	391.72	5.97	5.98	5.97	5.97	5.97
Stavropol Territory	4.72	4.84	4.85	4.87	4.88	1.55	1.58	1.58	1.58	1.59

(continued)

	DENS.					ln DENS.				
	2002	2003	2004	2005	2006	2002	2003	2004	2005	2006
Sverdlovsk Region	2.78	2.73	2.73	2.72	2.72	1.02	1	1	1	1
Tambov Region	4.33	4.11	4.06	4.01	3.98	1.47	1.41	1.4	1.39	1.38
Tatarstan	6.55	6.53	6.55	6.57	6.6	1.88	1.88	1.88	1.88	1.89
Tomsk Region	0.4	0.39	0.39	0.39	0.39	−0.92	−0.94	−0.94	−0.94	−0.94
Tula Region	7.95	7.82	7.74	7.66	7.59	2.07	2.06	2.05	2.04	2.03
Tuva	0.22	0.21	0.21	0.21	0.22	−1.51	−1.56	−1.56	−1.56	−1.51
Tver Region	2.23	2.11	2.08	2.05	2.03	0.8	0.75	0.73	0.72	0.71
Tyumen Region	0.26	0.26	0.26	0.26	0.27	−1.35	−1.35	−1.35	−1.35	−1.31
Udmurtian	4.55	4.41	4.4	4.39	4.39	1.52	1.48	1.48	1.48	1.48
Ulyanovsk Region	4.62	4.44	4.41	4.37	4.35	1.53	1.49	1.48	1.47	1.47
Vladimir Region	6.47	6.26	6.21	6.16	6.13	1.87	1.83	1.83	1.82	1.81
Volgograd Region	2.79	2.82	2.82	2.81	2.81	1.03	1.04	1.04	1.03	1.03
Vologda Region	1.07	1.04	1.04	1.03	1.03	0.07	0.04	0.04	0.03	0.03
Voronezh Region	5.53	5.42	5.39	5.36	5.35	1.71	1.69	1.68	1.68	1.68
Yaroslavl Region	4.58	4.5	4.47	4.44	4.43	1.52	1.5	1.5	1.49	1.49

	Inc. from FFSR					FFSR				
	2002	2003	2004	2005	2006	2002	2003	2004	2005	2006
Adygeya	12.65	12.02	10.43	11.1	12.57	154.69	110.93	99.44	144.89	113.77
Altai	32.68	30.83	27.46	31.48	33.81	144.35	117.26	107.13	125.67	132.23
Altai Territory	10.74	9.91	7.95	7.65	6.8	152.31	106.63	103.93	124.07	123.67
Amur Region	7.24	7.81	6.9	5.57	4.89	149.68	110.12	106.17	126.42	114.7
Arkhangelsk Region	3.49	3.13	2.11	2.05	1.26	125.88	107.8	104.75	118.67	119.64
Astrakhan Region	1	0.9	0.71	0	0	113.77	102.49	109.02	122.93	120.83
Bashkortostan	0	0	0	0	0	120.55	99.28	102.07	132.97	117.27
Belgorod Region	0.53	0.77	0.58	0.69	0.46	126.02	106.5	106.03	121.4	123.12
Bryansk Region	7.65	7.32	6.12	5.32	5.49	150.78	109.3	109.79	122.38	110.11
Buryatia	12.92	10.59	8.08	6.38	7.34	141.07	110.71	101.81	119.09	131.86
Chechen	,	,	,	30.98	32.17	98.94	124.84	138.88	133.33	150.03

(continued)

(continued)

	Inc. from FFSR					FFSR				
	2002	2003	2004	2005	2006	2002	2003	2004	2005	2006
Chelyabinsk Region	0.42	0.37	0.27	0.13	0	115.21	108	102	123.1	114.31
Chita Region	10.22	9.6	7.49	4.97	5.48	140.53	108.8	107.9	125.01	120.05
Chuvash	6.81	6.29	4.96	4.91	4.14	147.08	106.66	103.33	121.75	125.14
Dagestan	23	18.53	14.71	14.37	12.52	176.88	112.71	107.9	121.11	142.08
Ingushetia	55.56	47.01	37.66	37.19	46.38	179.2	131.63	111.37	130.97	131.03
Irkutsk Region	0.07	0.6	0.54	0.99	1.17	117.32	102.86	102.02	122.41	120.03
Ivanovo Region	9.67	8.8	7.36	7.77	8.74	125.12	108.42	99.75	118.71	113.95
Jewish Autonomous Region	18.18	16.26	11.7	8.75	8.89	157.38	109.1	100.9	116.98	129.41
Kabardino-Balkaria	10.34	10.15	9.9	9.15	9.62	139.58	108.28	102.98	139.88	111.83
Kaliningrad Region	2.03	1.95	1.25	0.82	0.78	117.13	106.06	109.51	137.38	110.17
Kalmykia	12.78	15.92	11.09	8.47	10.8	144.5	102.75	108.95	132.8	107.71
Kaluga Region	4.41	3.77	2.87	1.64	1.42	128.84	110.26	101.28	135.92	116.04
Kamchatka Region	9.56	11.69	10.22	11.09	12.01	130.76	100.14	101.74	117.38	111.91
Karachayevo–Circassia	14.35	14.33	12.83	12.04	12.63	167.21	130.71	109.27	132.25	118.34
Karelia	1.8	1.79	1.58	1.15	1.29	127.91	108.96	95.09	115.62	130.02
Kemerovo Region	0.86	0.95	0.73	0.63	0.41	133.29	108.35	108.26	128.64	118.04
Khabarovsk Territory	2.93	3.07	2.82	1.97	1.89	124.26	107.52	104.33	135.67	99.45
Khakassia	1.7	2.5	2.39	1.84	1.51	117.19	107.84	103.79	115.65	119.66
Kirov Region	1.15	0.31	0.29	0.33	0.34	128.8	103.08	102.76	122.29	120.61
Komi	0	0	0	0	0	142.67	107.57	99.6	108.93	110.59
Kostroma Region	4.02	3.94	2.89	2.44	2.87	129.01	106.48	101.59	121.9	120.54
Krasnodar Territory	1.44	1.46	1.25	0.99	1.12	129.4	96.66	105.52	147.13	112.82
Krasnoyarsk Territory	0	0	0	0	0	103.62	98.12	102.67	122.9	108.66
Kurgan Region	9.06	8.2	7.47	6.86	5.47	138.88	110.66	105.26	119.79	113.15
Kursk Region	2.15	2.07	1.68	2.02	1.67	129.78	108.89	105.32	138.16	109.43
Leningrad Region	0	0	0	0	0	112.03	109.58	102.94	136.82	107.78
Lipetsk Region	0	0	0	0	0	119.53	108.92	107.47	128.03	122.2
Magadan Region	10.8	11.13	9.91	10.41	7.83	132.61	104.8	102.15	103.27	112.31
Mari El	10.06	8.73	7.01	6.94	6.94	140.19	105.38	107.51	138.25	113.29
Mordovia	6.69	6.2	5.41	4.71	4.96	155.28	110.8	103.57	122.88	128.46
Moscow	0	0	0	0	0	116.4	80.5	104.59	214.73	69.69
Moscow Region	0.31	0.48	0.41	0.27	0.46	122.33	115.12	103	138.88	101.1
Murmansk Region	0.48	0.66	0.43	0.61	0.67	114.02	107.03	100.64	113.59	116.59
Nizhny Novgorod Region	0.49	0	0	0	0	123.96	105.2	104.38	131.59	113.63
North Ossetia–Alania	13.46	12.75	10.65	7.84	7.95	168.45	103.89	109.86	136	123.32
Novgorod Region	1.07	1.43	1.21	0.94	1.16	129.39	112.8	102.09	132.55	125.11

(continued)

	Inc. from FFSR					FFSR				
	2002	2003	2004	2005	2006	2002	2003	2004	2005	2006
Novosibirsk Region	2.23	2.09	1.8	1.34	0.98	137.46	103.25	106.89	133.32	116.3
Omsk Region	2.53	2.28	1.52	1.92	1.08	145.74	110.14	102.93	130.22	119.34
Orel Region	2.51	2.38	2.12	1.79	1.51	131.85	100.42	100.78	141.39	109.22
Orenburg Region	0	6.93	5.16	4.56	5.49	118.46	103.3	100.59	127.38	120.9
Penza Region	1.67	0	0	0	0	135.71	105.8	101.87	131.41	122.08
Perm Region	0	3.8	3.1	3.83	3.84	112.68	99.22	101.37	125.56	115.44
Primorye Territory	5.82	5.29	3.82	3.58	3.13	132.78	107.83	102.91	117.9	118.26
Pskov Region	7.93	7.26	5.44	4.75	4.93	139.83	107.99	99.76	128.6	117.8
Rostov Region	3.26	2.99	2.47	2.82	2.66	147.99	109.2	105.12	124.24	124.43
Ryazan Region	3.13	2.7	2.15	1.57	1.35	137.53	107.37	102.86	125.21	122.98
Sakha (Yakutia)	6.95	7.41	5.92	5.7	7.1	128.62	106.68	102.84	105.94	116.24
Sakhalin Region	3.24	2.71	1.65	0.4	0.84	135.16	109.79	107.22	120.4	103.03
Samara Region	0	0	0	0	0	113.3	106.41	103.06	128.18	113.68
Saratov Region	1.64	1.57	1.36	1.29	1.66	140.55	103.93	105.36	129.13	114.44
Smolensk Region	2.31	2.22	1.78	1.11	1.46	124.58	112.88	103.97	117.37	108.33
St. Petersburg	0	0	0	0	0	121.37	92.85	100.27	187.24	83.84
Stavropol Territory	3.6	3.21	2.92	3.75	3.49	139.81	103.57	105.75	136.03	121.67
Sverdlovsk Region	0	0	0.01	0	0	115	110.33	104.06	124.22	118.57
Tambov Region	5.52	5.13	3.92	4.36	4.12	138.83	111.01	103.29	141.73	106.12
Tatarstan	0	0	0	0	0	110.92	97.87	103.54	139.37	110.25
Tomsk Region	1.16	1.01	0.75	0.32	0	126.93	97.04	104.84	121.48	121.92
Tula Region	1.71	1.73	1.35	1.17	0.94	133.85	109.83	101.68	122.11	111.48
Tuva	45.33	42.81	38.92	32.88	31.77	164.53	111.14	103.86	118.42	138.85
Tver Region	2.73	2.48	1.9	1.72	1.73	132.92	115.23	101.13	127.28	115.73
Tyumen Region	0	0	0	0	0	88.7	93.28	97.66	137.16	95.43
Udmurtian	0.18	0.18	0	0	0.17	125.57	110.44	103.98	115.98	123.46
Ulyanovsk Region	2.32	2.35	2.12	1.86	2.68	115.09	105.05	99.16	136.23	115.8
Vladimir Region	2.98	2.72	2.1	1.68	1.47	137.02	104.02	100.62	120	119.77
Volgograd Region	0.77	0.86	0.81	0.87	0.9	129.59	105.38	104.1	119.38	120.81
Vologda Region	0	0	0	0	0	105.09	106.58	102.35	125.54	121.89
Voronezh Region	2.76	2.62	2.44	2.32	2.58	137.56	108.56	105.06	124.04	118.37
Yaroslavl Region	0	0	0	0	0	131.95	109.48	100.26	129.38	109.09

	GRTB				
	2002	2003	2004	2005	2006
Adygeya	119.6	74.41	105.92	112.43	98.54
Altai	103.32	102.7	105.73	106.42	101.36
Altai Territory	106.56	87.31	87.67	114.5	92.62
Amur Region	91.8	92.1	90.02	110.73	85.88

(continued)

(continued)

	GRTB				
	2002	2003	2004	2005	2006
Arkhangelsk Region	107.95	71.25	71.38	172.93	98.37
Astrakhan Region	122.59	81.11	87.66	165.28	113.35
Bashkortostan	80.65	83.41	101.4	98.61	94.87
Belgorod Region	136.82	34.79	40.83	646.91	106.83
Bryansk Region	83.6	98.02	84.49	116.48	96.27
Buryatia	117.65	98.84	87.37	98.02	100.94
Chechen	83.08	106	106	93.5	102.26
Chelyabinsk Region	137.89	78.08	84.28	106.53	105.25
Chita Region	102.2	75.87	100.17	105.92	110.04
Chuvash	107.33	92.12	87.34	103.35	90.77
Dagestan	96.26	98.8	98.2	102	96.8
Ingushetia	110.85	98.25	99.99	102.52	98.52
Irkutsk Region	100.28	81.45	92.23	147.3	95.63
Ivanovo Region	107.74	86.17	80.74	113.72	90.87
Jewish Autonomous Region	89.18	100.03	84.67	95.14	94.61
Kabardino-Balkaria	122.85	104.89	89.39	100.46	88.18
Kaliningrad Region	111.22	84.42	103.59	116.44	102.04
Kalmykia	105.43	46.79	85.37	230.63	91.29
Kaluga Region	106.56	90.99	70.94	110.91	81.19
Kamchatka Region	104.55	92.3	103.42	102.4	95.76
Karachayevo–Circassia	113.32	92.64	85.71	102.23	105.82
Karelia	136.59	92.8	64.23	102.54	101.69
Kemerovo Region	89.07	43.65	161.63	113.07	84.52
Khabarovsk Territory	96.88	94.79	83.33	118.34	85.21
Khakassia	94.87	92.36	88.3	119.46	87.13
Kirov Region	99.96	99.68	60.5	161.88	99.06
Komi	183.63	11.96	614.05	126	65.53
Kostroma Region	125.28	83.29	67.31	129.22	94.49
Krasnodar Territory	147.12	82.7	52.09	162.84	98.59
Krasnoyarsk Territory	108.15	103.2	43.37	283.97	99.02
Kurgan Region	120.27	89.06	89.97	113.76	86.2
Kursk Region	124.77	67.57	80.17	168.98	91.55
Leningrad Region	97.25	67.23	26.8	439.41	96.13
Lipetsk Region	93.29	97.21	41.78	234.49	105.55
Magadan Region	105.83	97.38	101.88	104.84	92.31
Mari El	112.24	96.35	78.91	105.1	93.89
Mordovia	113.79	79.46	42.77	206.59	132.43
Moscow	225.45	45.76	127.96	−34.44	97.1
Moscow Region	114.07	0.72	1854.13	1009.26	118.64
Murmansk Region	117.6	68.73	115.29	149.28	130.85
Nizhny Novgorod Region	115.83	124.17	16.66	561.89	89.05
North Ossetia–Alania	115.19	110.21	90.29	99.14	95.27
Novgorod Region	111.28	97.56	65.03	145.07	94.81

(continued)

	GRTB				
	2002	2003	2004	2005	2006
Novosibirsk Region	110.79	83.47	57.45	159.23	73.96
Omsk Region	113.8	81.22	53.1	133.62	152.75
Orel Region	118.06	79.92	87.72	124.76	95.19
Orenburg Region	86.38	76.82	108.96	113.15	82.89
Penza Region	110.96	96.26	56.31	154.64	95.85
Perm Region	127.27	36.38	164.33	108.1	97.55
Primorye Territory	87.39	78.06	80.74	135.93	98.35
Pskov Region	112.47	81.74	71.38	125	88.64
Rostov Region	109.21	97.26	78.3	111.91	94.44
Ryazan Region	66.59	89.49	90.48	145.55	85.1
Sakha (Yakutia)	100.38	101.61	79.56	132.28	110.9
Sakhalin Region	94.68	71.65	80.91	110.08	116.69
Samara Region	132.49	104.58	−25.15	−280.81	139.32
Saratov Region	112.01	101.5	81.8	135.58	97.5
Smolensk Region	115.24	90	66.37	149.99	91.06
St. Petersburg	99.88	4.38	−1084.65	−182.26	89.41
Stavropol Territory	130.71	77.9	70.5	162.21	85.05
Sverdlovsk Region	102.43	107.51	−13.78	−650.52	120.87
Tambov Region	103.62	95.38	71.09	123.09	98.61
Tatarstan	98.01	68.88	89.32	169.12	64.93
Tomsk Region	154.15	57.73	100.75	113.4	92.71
Tula Region	104.17	94.27	50.13	204.37	92.85
Tuva	97.35	97.18	97.05	99.08	97.01
Tver Region	83.7	106.97	72.6	111.69	93.16
Tyumen Region	59.34	135.43	−468.34	−5.11	118.63
Udmurtian	151.25	72.86	75.28	157.31	101.4
Ulyanovsk Region	93.45	61.7	125.38	114.62	126.21
Vladimir Region	101.42	56.83	86.73	184.38	87.08
Volgograd Region	91.37	120.4	88.6	120.87	88.07
Vologda Region	125.08	24.43	162.34	112.39	100.19
Voronezh Region	94.14	109.72	75.46	111.92	87.15
Yaroslavl Region	101.47	156.01	57.19	147.53	94.84

	TRTB					HCS				
	2002	2003	2004	2005	2006	2002	2003	2004	2005	2006
Adygeya	83.1	156.23	82.51	68.63	103.87	61.05	109.32	104.82	225.02	51.17
Altai	99.12	81.45	87.46	69.14	101.93	113.29	94.16	96.36	162.25	81.47
Altai Territory	90.58	105.64	112.53	70.98	111.66	82.19	83.83	95.22	99.75	47.7
Amur Region	112.39	109.45	108.31	78.65	110.8	66.08	99.21	86.08	110.1	74.77

(continued)

(continued)

	TRTB					HCS				
	2002	2003	2004	2005	2006	2002	2003	2004	2005	2006
Arkhangelsk Region	96.69	112.17	105.11	73.77	109.33	78.18	101.54	94.58	79.59	133.2
Astrakhan Region	85.4	107.63	101.56	73.81	98.81	88.51	108.18	85.39	128.81	87.17
Bashkortostan	105.74	104.9	97.33	78.6	88.01	81.92	105.09	109.81	108.84	89.14
Belgorod Region	99.7	109.12	107.83	81.47	98.64	62.75	90.25	232.03	81.66	124
Bryansk Region	112.82	97.79	110.1	75.39	99.14	111.22	94.9	97.29	106.66	65.93
Buryatia	84.32	98.61	109.4	85.06	106.2	82.55	91.41	99.72	54.01	71.68
Chechen						73.36	74.37	53.67	269.45	215.9
Chelyabinsk Region	92.73	112.42	102.35	94.07	99	66.98	125.13	85.16	114.56	103.4
Chita Region	96.35	112.39	98.05	84.29	96.49	96.63	89.49	98.41	113.7	78.63
Chuvash	91.76	108.35	111.73	72.3	107.58	59.33	104.18	98.17	146.91	60.47
Dagestan	117.95	104.85	106.45	73.64	110.64	94.05	97.71	92.68	172.23	66.16
Ingushetia	56.59	103.91	118.78	64.31	117.94	96.82	93.51	84.96	209.5	82.64
Irkutsk Region	101.63	107.91	101.9	82.82	101.07	84.37	90.78	115.1	92.2	101
Ivanovo Region	95.92	112.66	115.59	70.31	95.79	86.21	108.55	86.86	104.97	73.24
Jewish Autonomous Region	137.16	108.18	112.91	89.14	98.4	81.82	90.19	76.15	95.89	93.7
Kabardino-Balkaria	65.46	95.37	113.91	63.63	100.33	70.27	112.44	103.5	107.79	32.22
Kaliningrad Region	96.62	105.59	108.84	76.79	98.6	86.26	89.08	92.85	135.34	95.55
Kalmykia	103.2	143.45	116.65	47.24	129.3	95.96	89.87	96.31	130.42	104.2
Kaluga Region	98.05	103.47	117.25	74.1	106.59	91.12	82.25	90.42	123.08	81.2
Kamchatka Region	90.17	95.58	116.51	74.51	109.01	107.04	104.23	79.74	152.99	90.45
Karachayevo–Circassia	69.35	126.29	126.66	79.32	93.54	86.19	88.76	88.93	141.58	81.39
Karelia	91.98	98.67	115.67	91.71	95.69	74.07	95.25	88.18	119.42	83.9
Kemerovo Region	87.87	122.09	104.15	78.42	102.66	67.17	115.09	106.36	111.23	82.24
Khabarovsk Territory	101.47	98.05	109.01	78.84	106.69	86.17	83.87	83.43	131.45	83.38
Khakassia	100.77	93.59	102.34	79.72	104.21	60.26	108.57	101.95	85.99	110.3
Kirov Region	95.98	90.4	109.67	65.47	105.88	75.6	101.34	115.71	61.87	89.68
Komi	96.07	99.87	100.23	82.27	99.16	79.37	98.06	89.83	85.77	80.72
Kostroma Region	78.26	112.8	114.82	75.09	102.99	58.73	136.62	83.23	103.99	85.17
Krasnodar Territory	89.1	102.76	102.4	80.61	103.05	59.56	106.5	91.71	123.38	108
Krasnoyarsk Territory	92.08	109.31	103.41	82.25	104.44	123.74	93.97	93.14	99.43	86.29
Kurgan Region	88.01	112.4	106.33	76.14	113.71	78.75	101.88	89.97	100.88	83.18

(continued)

	TRTB					HCS				
	2002	2003	2004	2005	2006	2002	2003	2004	2005	2006
Kursk Region	99.1	108.71	107.3	83.95	103.09	65.71	99.49	91.08	100.11	72.84
Leningrad Region	102.58	102.94	102.5	82.14	101.77	88.82	101.67	83.58	108.01	83.52
Lipetsk Region	90.33	110.19	103.58	86.42	101.51	100.25	88.87	88.62	163.94	65.87
Magadan Region	106.63	93.99	113.57	69.44	107.77	71.14	68.64	105.43	201.44	104.6
Mari El	87.41	116.16	105.49	73.72	109.95	77.34	131.16	106.19	128.73	81.1
Mordovia	91.95	227.52	125.46	71.31	79.44	76.01	100.53	82.34	206.27	70.28
Moscow	118.47	92.82	103.48	96.31	102.88	102.49	92.58	104.81	154.66	115.8
Moscow Region	98.18	109.02	94.69	88.45	97.4	78.42	85.78	99.51	123.33	72.8
Murmansk Region	82.81	107.42	102.62	91.18	93.49	70.4	99.74	112.84	64.21	155.4
Nizhny Novgorod Region	96.9	116.63	103.08	79.16	105.74	93.09	98.2	75.54	116.7	97.68
North Ossetia–Alania	80.39	80.21	119.95	63.61	111.69	87.98	131.07	95.98	134.68	97.9
Novgorod Region	98.3	102.66	108.46	86.77	98.75	81.38	105.57	87.02	112.74	97.58
Novosibirsk Region	104.62	107.99	109.46	79.88	108.12	95.52	106.55	97.75	90.55	83
Omsk Region	95.75	123.16	115.83	82.3	75.95	73.39	114.94	76.86	111.57	85.18
Orel Region	90.78	101.3	111.66	69.81	104.81	113.1	114.36	97.71	53.43	97.48
Orenburg Region	102.21	102.92	99.13	75.82	101.02	83.82	101.32	104.11	86.15	102.7
Penza Region	85.67	102.57	147.01	62.57	101.71	82.41	90.27	80.69	105.68	85.74
Perm Region	96.37	113.89	99.28	88.67	99.35	89.77	95.42	88.31	81.99	109.5
Primorye Territory	95.71	105.11	114.22	78.05	101.39	66.52	110.63	92.35	98.93	102.5
Pskov Region	83.53	115.78	120.22	76.27	103.88	109.3	91.86	101.52	89.58	61.63
Rostov Region	98.44	98.97	104.75	74.55	105.68	91.23	95.69	98.89	113.31	104.3
Ryazan Region	98.16	97.67	110.99	78.49	102.43	68.15	78.67	124.34	138.45	69.54
Sakha (Yakutia)	107.74	107.1	118.41	73.17	100.6	102.6	101.15	106.72	135.13	79.63
Sakhalin Region	95.78	100.71	111.77	86.64	97.87	77.75	112.96	86.95	92.14	89.75
Samara Region	105.74	98.16	99.42	82.05	108.78	106.32	113.34	86.3	98.65	79.61
Saratov Region	82.85	117.02	97.15	82.64	102.31	79.2	112.27	107.58	63.65	85.71
Smolensk Region	94.62	111.59	103.14	79.97	104.28	88.24	91.75	88.72	91.52	84.5
St. Petersburg	100.42	114.57	97.09	77.27	109.57	108.19	97.12	73.31	197.05	109.4
Stavropol Territory	76.41	126.96	97.45	73.35	111.5	94.3	91.63	101.75	129.11	62.99
Sverdlovsk Region	97.69	105.46	112.66	85.56	103.16	72.26	105.59	97.92	102.87	96.16
Tambov Region	87.49	104.02	119.18	66.84	102.07	70.96	99.56	118.66	146.8	60.39
Tatarstan	100.81	104.18	104.62	82.07	111.06	147.82	88.88	77.3	187.57	109.2

(continued)

(continued)

	TRTB					HCS				
	2002	2003	2004	2005	2006	2002	2003	2004	2005	2006
Tomsk Region	102.07	99.24	99.27	82.68	105.01	73.77	109.08	98.52	98.64	95.85
Tula Region	85.26	99.04	118.14	76.93	97.98	86.42	88.71	79.88	107.84	78.02
Tuva	119.83	110.56	115	75.08	118.81	86.79	79.93	88.56	53.97	126.7
Tver Region	103.59	96.61	108.26	69.21	105.26	66.06	110.08	100.99	96.66	97.41
Tyumen Region	105.18	95.86	121.73	69.01	91.22	75.62	90.29	116.59	94.59	93.06
Udmurtian	90.21	109.62	103.44	77.67	99.04	69.63	94.16	103.75	93.3	91.22
Ulyanovsk Region	100.64	103.35	93.01	78.04	88.51	83.69	100.41	89.48	92.76	66.44
Vladimir Region	90.66	112.18	104.86	75.42	103.62	78.53	81.18	95.52	97.12	106.8
Volgograd Region	102.24	93.92	99.96	76.38	107.79	82.87	90.05	93.8	121.04	83.91
Vologda Region	96.5	119.64	105.45	84.15	94.62	83.17	110.62	93.75	82.93	115.9
Voronezh Region	89.35	106.41	107.77	74.52	106.01	84.13	108.53	69.47	106.35	112.8
Yaroslavl Region	101.61	104.38	102.81	77.15	99.28	73.31	103.1	101.14	94.9	92.29

	HCS 2001 %				
	2002	2003	2004	2005	2006
Adygeya	61.05	66.74	69.96	157.41	80.55
Altai	113.29	106.68	102.8	166.78	135.88
Altai Territory	82.19	68.9	65.6	65.44	31.21
Amur Region	66.08	65.56	56.43	62.13	46.45
Arkhangelsk Region	78.18	79.38	75.08	59.76	79.62
Astrakhan Region	88.51	95.75	81.76	105.32	91.81
Bashkortostan	81.92	86.08	94.52	102.88	91.72
Belgorod Region	62.75	56.63	131.4	107.3	133
Bryansk Region	111.22	105.55	102.69	109.53	72.22
Buryatia	82.55	75.46	75.25	40.64	29.13
Chechen	73.36	54.55	29.28	78.9	170.37
Chelyabinsk Region	66.98	83.81	71.37	81.77	84.51
Chita Region	96.63	86.48	85.1	96.76	76.08
Chuvash	59.33	61.81	60.68	89.14	53.9
Dagestan	94.05	91.89	85.16	146.67	97.04
Ingushetia	96.82	90.54	76.92	161.14	133.18
Irkutsk Region	84.37	76.58	88.15	81.27	82.09
Ivanovo Region	86.21	93.57	81.28	85.32	62.49
Jewish Autonomous Region	81.82	73.8	56.2	53.89	50.5
Kabardino-Balkaria	70.27	79.01	81.78	88.15	28.4
Kaliningrad Region	86.26	76.84	71.34	96.56	92.26
Kalmykia	95.96	86.24	83.06	108.32	112.86
Kaluga Region	91.12	74.95	67.77	83.41	67.73

(continued)

	HCS 2001 %				
	2002	2003	2004	2005	2006
Kamchatka Region	107.04	111.57	88.96	136.11	123.11
Karachayevo–Circassia	86.19	76.5	68.03	96.32	78.4
Karelia	74.07	70.55	62.21	74.29	62.33
Kemerovo Region	67.17	77.31	82.23	91.46	75.22
Khabarovsk Territory	86.17	72.27	60.3	79.26	66.09
Khakassia	60.26	65.42	66.7	57.35	63.24
Kirov Region	75.6	76.62	88.66	54.85	49.19
Komi	79.37	77.82	69.91	59.96	48.41
Kostroma Region	58.73	80.24	66.78	69.45	59.15
Krasnodar Territory	59.56	63.43	58.17	71.77	77.51
Krasnoyarsk Territory	123.74	116.28	108.3	107.68	92.92
Kurgan Region	78.75	80.24	72.19	72.83	60.58
Kursk Region	65.71	65.38	59.55	59.61	43.42
Leningrad Region	88.82	90.3	75.47	81.51	68.08
Lipetsk Region	100.25	89.09	78.95	129.43	85.26
Magadan Region	71.14	48.83	51.48	103.71	108.52
Mari El	77.34	101.44	107.72	138.67	112.46
Mordovia	76.01	76.42	62.92	129.79	91.22
Moscow	102.49	94.89	99.45	153.82	178.08
Moscow Region	78.42	67.27	66.94	82.55	60.1
Murmansk Region	70.4	70.22	79.23	50.87	79.08
Nizhny Novgorod Region	93.09	91.42	69.06	80.59	78.72
North Ossetia–Alania	87.98	115.32	110.69	149.08	145.94
Novgorod Region	81.38	85.91	74.76	84.29	82.25
Novosibirsk Region	95.52	101.77	99.48	90.08	74.76
Omsk Region	73.39	84.36	64.84	72.34	61.63
Orel Region	113.1	129.35	126.39	67.53	65.83
Orenburg Region	83.82	84.93	88.42	76.17	78.23
Penza Region	82.41	74.39	60.02	63.43	54.39
Perm Region	89.77	85.65	75.64	62.02	67.9
Primorye Territory	66.52	73.6	67.97	67.24	68.94
Pskov Region	109.3	100.41	101.93	91.31	56.28
Rostov Region	91.23	87.29	86.32	97.81	102.04

	HCS 2001 %				
	2002	2003	2004	2005	2006
Ryazan Region	68.15	53.61	66.66	92.29	64.18
Sakha (Yakutia)	102.6	103.78	110.75	149.66	119.17

(continued)

(continued)

	HCS 2001 %				
	2002	2003	2004	2005	2006
Sakhalin Region	77.75	87.82	76.36	70.36	63.15
Samara Region	106.32	120.51	104	102.59	81.67
Saratov Region	79.2	88.92	95.66	60.88	52.19
Smolensk Region	88.24	80.96	71.83	65.73	55.54
St. Petersburg	108.19	105.07	77.03	151.8	165.99
Stavropol Territory	94.3	86.42	87.93	113.52	71.5
Sverdlovsk Region	72.26	76.3	74.71	76.85	73.9
Tambov Region	70.96	70.64	83.82	123.05	74.31
Tatarstan	147.82	131.38	101.55	190.48	207.97
Tomsk Region	73.77	80.47	79.27	78.2	74.95
Tula Region	86.42	76.66	61.24	66.04	51.52
Tuva	86.79	69.37	61.43	33.16	42.02
Tver Region	66.06	72.72	73.44	70.99	69.15
Tyumen Region	75.62	68.28	79.6	75.3	70.07
Udmurtian	69.63	65.56	68.02	63.46	57.88
Ulyanovsk Region	83.69	84.04	75.19	69.75	46.34
Vladimir Region	78.53	63.75	60.89	59.14	63.13
Volgograd Region	82.87	74.63	70	84.73	71.09
Vologda Region	83.17	92	86.25	71.53	82.89
Voronezh Region	84.13	91.31	63.43	67.46	76.12
Yaroslavl Region	73.31	75.58	76.44	72.54	66.95

A3.4b Values of the dependent (criterion) variables

No	Region	Summary of II on "quality of the population"				
		2002	2003	2004	2005	2006
1	2	3	4	5	6	7
1	Adygeya	4.952	4.894	5.093	5.280	5.158
2	Altai	4.756	4.417	4.681	4.383	4.393
3	Altai Territory	4.529	4.503	4.418	4.520	4.280
4	Amur Region	4.986	4.611	4.306	4.161	3.811
5	Arkhangelsk Region	4.749	4.506	4.593	4.684	4.507
6	Astrakhan Region	5.230	5.162	5.348	5.319	5.218
7	Bashkortostan	5.724	5.623	5.620	5.804	5.696
8	Belgorod Region	4.375	4.610	4.465	4.791	4.922
9	Bryansk Region	3.947	4.004	3.980	4.299	4.091
10	Buryatia	5.525	5.036	4.809	5.243	4.850
11	Chelyabinsk Region	4.756	4.663	4.752	4.950	4.698
12	Chita Region	4.630	4.149	3.874	4.254	3.967
13	Chukotka Autonomous Area	6.961	6.311	5.136	4.857	4.828

(continued)

No	Region	Summary of II on “quality of the population”				
		2002	2003	2004	2005	2006
1	2	3	4	5	6	7
14	Chuvash	5.111	4.810	4.787	5.231	4.906
15	Dagestan	6.642	6.488	6.990	7.075	7.073
16	Irkutsk Region	4.740	4.452	4.544	4.548	4.162
17	Ivanovo Region	3.132	3.176	3.251	3.494	3.434
18	Jewish Autonomous Region	4.577	4.371	3.860	3.759	3.419
19	Kabardino-Balkaria	6.444	6.545	6.442	6.526	6.413
20	Kaliningrad Region	4.425	4.164	4.163	4.211	4.008
21	Kalmykia	5.832	5.833	6.430	6.353	6.353
22	Kaluga Region	3.635	3.900	3.933	4.116	4.063
23	Kamchatka Region	6.157	6.097	5.957	6.009	5.934
24	Karachayevo–Circassia	5.765	5.682	6.014	6.402	6.207
25	Karelia	4.186	3.795	4.069	4.442	4.135
26	Kemerovo Region	4.220	4.098	4.182	4.105	3.930
27	Khabarovsk Territory	5.232	4.859	4.730	4.909	4.445
28	Khakassia	4.478	3.987	4.576	4.033	4.108
29	Kirov Region	3.979	3.884	3.942	4.381	4.127
30	Komi	5.370	5.040	5.094	5.204	4.951
31	Kostroma Region	3.375	3.324	3.587	3.692	3.728
32	Krasnodar Territory	4.848	4.969	5.189	5.266	5.199
33	Krasnoyarsk Territory	4.809	4.866	4.972	4.891	4.721
34	Kurgan Region	4.437	4.222	4.099	4.487	4.194
35	Kursk Region	3.961	3.862	4.006	4.245	4.102
36	Leningrad Region	2.893	3.160	3.541	3.688	3.672
37	Lipetsk Region	4.290	4.163	4.511	4.841	4.734
38	Magadan Region	6.375	6.332	5.454	5.859	5.410
39	Mari El	4.645	4.393	4.194	4.787	4.291
40	Mordovia	4.652	4.484	4.719	5.178	4.897
41	Moscow	4.719	5.749	6.210	6.527	6.843
42	Moscow Region	4.093	4.415	4.536	4.892	4.781
43	Murmansk Region	6.204	5.963	5.905	6.032	5.755
44	Nizhny Novgorod Region	3.519	3.565	3.664	3.992	3.790
45	North Ossetia–Alania	6.049	5.848	6.196	6.577	6.248
46	Novgorod Region	2.842	2.608	3.059	3.264	3.162
47	Novosibirsk Region	5.141	5.125	5.202	5.338	5.034
48	Omsk Region	5.187	4.949	5.188	5.304	5.047
49	Orel Region	3.936	4.089	4.055	4.478	4.193
50	Orenburg Region	4.467	4.636	4.667	4.566	4.597
51	Penza Region	4.432	4.476	4.327	4.601	4.442
52	Perm Territory	4.238	4.078	4.183	4.150	3.874
53	Primorye Territory	5.152	4.635	4.663	4.640	4.197

(continued)

(continued)

No	Region	Summary of II on "quality of the population"				
		2002	2003	2004	2005	2006
1	2	3	4	5	6	7
54	Pskov Region	2.328	2.656	2.634	2.895	2.934
55	Rostov Region	4.563	4.669	4.972	5.262	5.150
56	Ryazan Region	3.159	3.256	3.683	4.094	3.947
57	Sakha (Yakutia)	6.698	6.398	6.641	6.904	6.593
58	Sakhalin Region	5.378	4.830	4.663	4.223	3.986
59	Samara Region	5.147	5.294	5.360	5.614	5.455
60	Saratov Region	5.313	5.252	4.990	5.380	5.170
61	Smolensk Region	3.188	3.173	3.304	3.477	3.273
62	St. Petersburg	3.916	4.244	4.666	5.135	4.956
63	Stavropol Territory	5.201	5.235	5.426	5.762	5.590
64	Sverdlovsk Region	4.390	4.433	4.730	4.951	4.793
65	Tambov Region	3.253	3.363	3.611	3.779	3.722
66	Tatarstan	5.472	5.600	5.751	5.981	5.854
67	Tomsk Region	5.787	5.636	5.924	5.871	5.763
68	Tula Region	2.770	2.777	2.947	3.252	3.101
69	Tuva	3.647	3.251	4.123	3.912	3.946
70	Tver Region	2.819	2.797	2.971	3.268	3.028
71	Tyumen Region	7.049	7.150	7.380	7.137	7.188
72	Udmurtian	5.039	4.770	4.797	5.029	4.711
73	Ulyanovsk Region	4.396	4.200	4.549	4.759	4.418
74	Vladimir Region	3.396	3.551	3.689	3.941	3.908
75	Volgograd Region	4.360	4.547	4.709	4.982	4.975
76	Vologda Region	3.888	3.583	3.967	4.311	4.186
77	Voronezh Region	4.658	4.509	4.722	5.136	4.804
78	Yaroslavl Region	3.551	3.604	3.862	4.236	4.153

No	Region	Summary of II on "level of material well-being"				
		2002	2003	2004	2005	2006
1	2	3	4	5	6	7
1	Adygeya	4.545	3.770	4.004	4.374	4.071
2	Altai	2.793	2.462	2.903	2.967	2.908
3	Altai Territory	3.361	3.380	3.656	3.923	3.900
4	Amur Region	2.741	2.762	2.964	3.185	3.055
5	Arkhangelsk Region	3.259	3.231	3.631	3.419	3.512
6	Astrakhan Region	5.136	4.980	4.550	4.564	4.393
7	Bashkortostan	5.450	5.187	5.188	5.245	5.117
8	Belgorod Region	5.857	5.490	5.479	5.677	5.503
9	Bryansk Region	4.165	4.030	4.200	4.361	4.350

(continued)

No	Region	Summary of II on "level of material well-being"				
		2002	2003	2004	2005	2006
1	2	3	4	5	6	7
10	Buryatia	3.120	2.720	2.693	2.964	2.967
11	Chelyabinsk Region	4.231	4.165	4.411	4.716	4.804
12	Chita Region	2.466	2.784	3.271	3.323	3.297
13	Chukotka Autonomous Area	3.971	4.480	4.308	4.212	3.452
14	Chuvash	4.315	4.465	4.430	5.072	5.016
15	Dagestan	1.699	2.054	2.742	2.838	3.061
16	Irkutsk Region	3.248	2.980	3.322	3.387	3.289
17	Ivanovo Region	2.062	2.232	2.924	3.661	3.810
18	Jewish Autonomous Region	2.926	2.839	3.296	3.261	3.320
19	Kabardino-Balkaria	4.130	4.004	4.259	4.416	4.460
20	Kaliningrad Region	4.638	4.977	5.175	5.393	5.644
21	Kalmykia	2.340	1.945	2.312	2.803	2.918
22	Kaluga Region	4.010	4.112	4.485	4.417	4.306
23	Kamchatka Region	3.053	3.178	3.670	3.337	3.133
24	Karachayevo–Circassia	3.368	3.344	3.838	3.984	4.161
25	Karelia	4.071	3.949	4.205	4.061	4.017
26	Kemerovo Region	4.110	4.019	4.260	4.200	4.280
27	Khabarovsk Territory	3.603	3.368	3.687	3.632	3.622
28	Khakassia	3.764	3.529	3.687	3.667	3.597
29	Kirov Region	3.368	3.155	3.626	3.628	3.628
30	Komi	4.194	3.977	4.067	3.835	3.686
31	Kostroma Region	3.559	3.351	3.713	3.845	3.752
32	Krasnodar Territory	4.895	4.587	4.661	4.780	4.850
33	Krasnoyarsk Territory	3.941	3.676	3.871	3.941	3.835
34	Kurgan Region	2.779	2.508	3.260	3.485	3.553
35	Kursk Region	4.187	4.169	4.628	4.911	5.015
36	Leningrad Region	3.862	3.792	4.581	4.812	4.902
37	Lipetsk Region	5.265	5.203	5.448	5.623	5.746
38	Magadan Region	3.530	3.611	3.885	3.556	3.475
39	Mari El	2.820	2.588	3.022	3.822	3.938
40	Mordovia	3.312	3.450	3.790	4.150	4.157
41	Moscow	8.521	8.172	8.052	7.914	7.765
42	Moscow Region	6.278	6.132	6.444	6.805	6.789
43	Murmansk Region	3.430	3.361	3.828	3.578	3.587
44	Nizhny Novgorod Region	4.649	4.423	4.688	4.744	4.671
45	North Ossetia–Alania	4.392	4.609	5.243	5.230	5.223
46	Novgorod Region	4.130	4.081	4.329	4.439	4.450
47	Novosibirsk Region	3.253	3.595	4.106	4.231	4.316
48	Omsk Region	3.926	3.994	4.485	4.524	4.542

(continued)

(continued)

No	Region	Summary of II on "level of material well-being"				
		2002	2003	2004	2005	2006
1	2	3	4	5	6	7
49	Orel Region	4.741	4.476	4.562	4.679	4.491
50	Orenburg Region	4.033	3.958	4.176	4.221	4.164
51	Penza Region	3.732	3.605	4.074	4.294	4.283
52	Perm Territory	4.052	3.878	4.042	4.125	4.112
53	Primorye Territory	2.627	2.905	3.526	3.647	3.589
54	Pskov Region	4.426	4.557	4.894	4.692	4.702
55	Rostov Region	4.538	4.412	4.669	4.288	4.237
56	Ryazan Region	4.306	4.309	4.391	4.710	4.689
57	Sakha (Yakutia)	4.242	3.957	3.594	3.329	3.178
58	Sakhalin Region	3.184	3.357	3.679	3.427	3.431
59	Samara Region	5.162	5.039	5.132	5.059	4.778
60	Saratov Region	3.921	3.765	4.008	4.211	4.210
61	Smolensk Region	4.714	4.679	4.778	5.046	4.768
62	St. Petersburg	6.453	6.817	6.980	7.100	6.883
63	Stavropol Territory	3.850	3.813	4.155	4.420	4.370
64	Sverdlovsk Region	4.139	4.248	4.552	4.558	4.541
65	Tambov Region	4.653	4.575	4.692	4.819	4.934
66	Tatarstan	5.751	5.558	5.516	5.508	5.617
67	Tomsk Region	4.179	3.933	4.058	4.010	3.979
68	Tula Region	4.553	4.489	4.588	4.567	4.534
69	Tuva	1.866	1.406	1.463	1.320	1.369
70	Tver Region	3.839	3.924	4.547	4.834	4.923
71	Tyumen Region	6.065	5.416	5.239	4.971	5.022
72	Udmurtian	4.322	3.862	4.138	4.280	4.190
73	Ulyanovsk Region	3.413	3.502	3.752	4.216	4.262
74	Vladimir Region	3.999	3.981	4.176	3.987	3.833
75	Volgograd Region	4.149	4.030	4.404	4.243	4.372
76	Vologda Region	4.414	4.320	4.510	4.246	4.135
77	Voronezh Region	4.681	4.726	4.877	5.060	4.960
78	Yaroslavl Region	4.517	4.490	4.606	4.444	4.495

Region	Summary of II on "quality of social services"									
	II QSS					ln II QSS				
	2002	2003	2004	2005	2006	2002	2003	2004	2005	2006
1	2	3	4	5	6	7	8	9	10	11
Adygeya	6.88	6.67	6.62	6.78	6.61	1.93	1.90	1.89	1.91	1.89
Altai	4.65	5.14	5.05	4.37	4.46	1.54	1.64	1.62	1.47	1.50

(continued)

Region	Summary of II on "quality of social services"									
	II QSS					ln II QSS				
	2002	2003	2004	2005	2006	2002	2003	2004	2005	2006
1	2	3	4	5	6	7	8	9	10	11
Altai Territory	4.72	5.53	5.68	5.41	5.03	1.55	1.71	1.74	1.69	1.62
Amur Region	5.93	5.87	5.69	5.39	4.82	1.78	1.77	1.74	1.68	1.57
Arkhangelsk Region	6.63	6.66	6.4	6.43	6.61	1.89	1.90	1.86	1.86	1.89
Astrakhan Region	6.42	6.69	6.2	6.16	6.23	1.86	1.90	1.82	1.82	1.83
Bashkortostan	6.92	7.08	6.96	6.94	6.69	1.93	1.96	1.94	1.94	1.90
Belgorod Region	7.97	7.92	7.64	7.84	8.07	2.08	2.07	2.03	2.06	2.09
Bryansk Region	7.23	7.21	7.01	7.05	6.79	1.98	1.98	1.95	1.95	1.92
Buryatia	4.87	5.24	4.92	4.7	4.64	1.58	1.66	1.59	1.55	1.53
Vladimir Region	7.18	7.28	6.86	7.03	6.74	1.97	1.99	1.93	1.95	1.91
Volgograd Region	7.11	7.1	6.83	6.91	6.93	1.96	1.96	1.92	1.93	1.94
Vologda Region	6.95	6.66	6.44	6.52	6.88	1.94	1.90	1.86	1.87	1.93
Voronezh Region	7.73	7.59	7.47	7.67	7.54	2.05	2.03	2.01	2.04	2.02
Dagestan	4.87	5.3	5.68	6.01	5.67	1.58	1.67	1.74	1.79	1.74
Jewish Autonomous Region	4.34	4.96	5.29	5.22	5.11	1.47	1.60	1.67	1.65	1.63
Ivanovo Region	6.28	6.3	6.3	6.77	6.4	1.84	1.84	1.84	1.91	1.86
Ingushetia										
Irkutsk Region	3.57	3.76	3.88	3.79	3.73	1.27	1.32	1.36	1.33	1.32
Kabardino-Balkaria	5.97	6.48	6.17	6.17	5.8	1.79	1.87	1.82	1.82	1.76
Kaliningrad Region	6.47	6.28	6.16	6.53	6.81	1.87	1.84	1.82	1.88	1.92
Kalmykia	5.79	5.62	5.19	5.29	5.13	1.76	1.73	1.65	1.67	1.64
Kaluga Region	7.63	7.43	7.08	6.9	6.89	2.03	2.01	1.96	1.93	1.93
Kamchatka Region	5.9	6.4	6.65	6.16	6.37	1.77	1.86	1.89	1.82	1.85
Karachayevo–Circassia	6.38	6.27	6.05	6.13	5.49	1.85	1.84	1.80	1.81	1.70
Karelia	6.45	6.52	6.47	6.69	6.4	1.86	1.87	1.87	1.90	1.86
Kemerovo Region	5.26	5.26	5.5	5.53	5.6	1.66	1.66	1.70	1.71	1.72
Kirov Region	6.75	6.96	6.59	6.86	6.76	1.91	1.94	1.89	1.93	1.91
Komi	5.52	5.4	5.47	5.31	5.48	1.71	1.69	1.70	1.67	1.70
Kostroma Region	6.77	6.9	6.91	7.38	7.23	1.91	1.93	1.93	2.00	1.98
Krasnodar Territory	7.42	7.25	7.14	7.38	7.14	2.00	1.98	1.97	2.00	1.97
Krasnoyarsk Territory	5.66	5.99	6.11	5.95	5.87	1.73	1.79	1.81	1.78	1.77
Kurgan Region	5.87	6.04	5.62	5.71	5.66	1.77	1.80	1.73	1.74	1.73
Kursk Region	6.81	6.91	6.83	6.86	6.58	1.92	1.93	1.92	1.93	1.88
Leningrad Region	6.5	6.35	6.23	6.66	6.42	1.87	1.85	1.83	1.90	1.86
Lipetsk Region	7.3	7.55	7.38	7.57	7.71	1.99	2.02	2.00	2.02	2.04
Magadan Region	6.28	6.68	6	6.17	5.97	1.84	1.90	1.79	1.82	1.79
Mari El	6.43	6.28	6.1	6.11	5.95	1.86	1.84	1.81	1.81	1.78
Mordovia	7.48	7.45	7.13	7.01	6.84	2.01	2.01	1.96	1.95	1.92
Moscow	8.05	7.89	7.94	7.98	8.11	2.09	2.07	2.07	2.08	2.09
Moscow Region	7.41	7.36	7.21	7.4	7.39	2.00	2.00	1.98	2.00	2.00

(continued)

(continued)

Region	Summary of II on "quality of social services"									
	II QSS					ln II QSS				
	2002	2003	2004	2005	2006	2002	2003	2004	2005	2006
1	2	3	4	5	6	7	8	9	10	11
Murmansk Region	6.39	6.91	6.94	6.65	6.89	1.85	1.93	1.94	1.89	1.93
Nizhny Novgorod Region	7.56	7.58	7.23	7.41	7.22	2.02	2.03	1.98	2.00	1.98
Novgorod Region	6.91	6.99	6.75	7.01	7	1.93	1.94	1.91	1.95	1.95
Novosibirsk Region	6.4	6.12	6.22	5.99	6.15	1.86	1.81	1.83	1.79	1.82
Omsk Region	7.1	6.88	6.9	6.92	6.83	1.96	1.93	1.93	1.93	1.92
Orenburg Region	6.25	6.07	5.87	5.8	5.93	1.83	1.80	1.77	1.76	1.78
Orel Region	7.04	7.21	7.12	7.3	7.23	1.95	1.98	1.96	1.99	1.98
Penza Region	6.95	7.43	7.43	7.47	7.42	1.94	2.01	2.01	2.01	2.00
Perm region	4.51	5.44	5	4.95	4.74	1.51	1.69	1.61	1.60	1.56
Primorye Territory	5.16	5.03	5.17	5.24	5.01	1.64	1.62	1.64	1.66	1.61
Pskov Region	7.35	7.35	7.04	6.95	7.05	1.99	1.99	1.95	1.94	1.95
Rostov Region	7.38	7.22	7.26	7.38	7.29	2.00	1.98	1.98	2.00	1.99
Ryazan Region	7.32	7.42	7.32	7.38	7.42	1.99	2.00	1.99	2.00	2.00
Samara Region	5.35	5.07	5.09	5.29	5.14	1.68	1.62	1.63	1.67	1.64
St. Petersburg	7.3	7.08	7.02	7.18	7.3	1.99	1.96	1.95	1.97	1.99
Saratov Region	7	7.07	7.15	7.39	7.26	1.95	1.96	1.97	2.00	1.98
Sakha (Yakutia)	5.99	6.27	6.09	5.97	5.14	1.79	1.84	1.81	1.79	1.64
Sakhalin Region	5.92	6.06	5.85	6.16	5.8	1.78	1.80	1.77	1.82	1.76
Sverdlovsk Region	5.49	5.44	5.68	5.75	5.8	1.70	1.69	1.74	1.75	1.76
North Ossetia–Alania	7.73	7.17	6.81	6.93	7.01	2.05	1.97	1.92	1.94	1.95
Smolensk Region	6.95	7.07	6.46	6.68	6.57	1.94	1.96	1.87	1.90	1.88
Stavropol Territory	7.12	7.69	7.44	7.43	7.08	1.96	2.04	2.01	2.01	1.96
Tambov Region	7.48	7.22	7.07	7.24	7.09	2.01	1.98	1.96	1.98	1.96
Tatarstan	7.04	7.44	7.35	7.22	7.34	1.95	2.01	1.99	1.98	1.99
Tver Region	6.84	6.77	6.6	6.58	6.39	1.92	1.91	1.89	1.88	1.85
Tomsk Region	6.21	5.93	5.88	5.98	6.25	1.83	1.78	1.77	1.79	1.83
Tula Region	7.36	7.43	7.18	7.54	7.45	2.00	2.01	1.97	2.02	2.01
Tuva	3.78	3.59	3.5	3.41	3.29	1.33	1.28	1.25	1.23	1.19
Tyumen Region	5.28	6.22	6.14	6.29	6.33	1.66	1.83	1.81	1.84	1.85
Udmurtian	6.32	6.61	6.31	6	5.83	1.84	1.89	1.84	1.79	1.76
Ulyanovsk Region	6.25	6.32	6.1	6.31	6.13	1.83	1.84	1.81	1.84	1.81
Khabarovsk Territory	5.76	5.45	5.73	5.32	4.97	1.75	1.70	1.75	1.67	1.60
Khakassia	5.29	5.84	5.79	5.77	5.67	1.67	1.76	1.76	1.75	1.74
Chelyabinsk Region	5.8	6.12	5.98	6.34	6.33	1.76	1.81	1.79	1.85	1.85
Chechen										
Chita Region	4.13	4.31	4.27	3.74	4	1.42	1.46	1.45	1.32	1.39
Chuvash	6.91	7.31	6.94	6.66	6.5	1.93	1.99	1.94	1.90	1.87
Yaroslavl Region	6.81	7.23	6.88	7.03	7.52	1.92	1.98	1.93	1.95	2.02

№	Region	Summary of II on "quality of the ecological niche"				
		2002	2003	2004	2005	2006
	2	3	4	5	6	7
1	Adygeya	5.515	7.142	4.990	7.133	5.498
2	Altai	5.692	7.261	4.926	6.798	5.335
3	Altai Territory	3.719	4.578	2.679	3.455	2.433
4	Amur Region	3.702	4.218	2.790	3.665	2.333
5	Arkhangelsk Region	3.819	4.641	2.712	3.717	2.772
6	Astrakhan Region	4.186	5.007	3.119	4.114	3.069
7	Bashkortostan	3.890	4.773	2.902	3.820	2.821
8	Belgorod Region	3.511	4.373	2.460	3.397	2.487
9	Bryansk Region	3.990	4.909	3.028	3.927	3.036
10	Buryatia	5.036	5.553	4.182	5.611	4.299
11	Chelyabinsk Region	3.652	4.613	2.840	3.887	2.803
12	Chita Region	3.743	2.648	2.746	3.668	2.749
13	Chukotka Autonomous Area	4.039	4.381	2.980	3.438	4.191
14	Chuvash	4.246	5.166	3.172	4.150	3.134
15	Dagestan	3.760	4.744	2.692	3.575	2.604
16	Irkutsk Region	4.018	4.633	2.990	3.904	2.841
17	Ivanovo Region	3.864	4.684	2.882	3.723	2.787
18	Jewish Autonomous Region	4.084	4.650	3.078	3.752	2.815
19	Kabardino-Balkaria	5.657	7.465	4.919	6.909	5.349
20	Kaliningrad Region	3.700	4.543	2.646	3.488	2.502
21	Kalmykia	5.504	5.035	3.973	3.492	3.696
22	Kaluga Region	4.678	5.924	3.940	5.330	4.231
23	Kamchatka Region	5.467	6.869	4.753	6.885	5.070
24	Karachayevo–Circassia	5.366	7.001	4.643	6.656	5.090
25	Karelia	4.255	5.205	3.372	4.407	3.405
26	Kemerovo Region	2.686	4.307	3.031	4.286	3.241
27	Khabarovsk Territory	4.081	4.883	3.031	3.638	3.082
28	Khakassia	4.421	5.561	3.567	4.824	3.849
29	Kirov Region	3.650	4.401	2.605	3.449	2.516
30	Komi	4.153	5.079	3.068	3.912	2.849
31	Kostroma Region	3.937	4.776	3.000	3.664	3.096
32	Krasnodar Territory	4.714	5.690	3.573	4.647	3.381
33	Krasnoyarsk Territory	2.763	3.885	2.473	3.601	2.541
34	Kurgan Region	3.532	4.274	2.293	3.384	2.424
35	Kursk Region	3.558	4.445	2.510	3.444	2.533
36	Leningrad Region	3.119	3.866	2.112	3.163	2.143
37	Lipetsk Region	3.532	4.448	2.619	3.518	2.601
38	Magadan Region	2.381	3.161	2.778	3.569	2.526
39	Mari El	4.285	5.294	3.304	4.228	3.285
40	Mordovia	4.369	5.325	3.372	4.366	3.344
41	Moscow Region	3.494	4.148	2.481	3.366	2.432

(continued)

(continued)

№	Region	Summary of II on "quality of the ecological niche"				
		2002	2003	2004	2005	2006
	2	3	4	5	6	7
42	Murmansk Region	3.632	4.318	2.506	3.109	2.157
43	Nizhny Novgorod Region	3.705	4.582	2.687	3.547	2.578
44	North Ossetia–Alania	5.323	7.190	4.691	6.555	5.042
45	Novgorod Region	4.186	5.260	3.255	4.428	3.379
46	Novosibirsk Region	3.552	4.318	2.517	3.419	2.223
47	Omsk Region	3.501	4.204	2.402	3.305	2.343
48	Orel Region	4.475	5.586	3.403	4.524	3.427
49	Orenburg Region	3.338	4.161	2.359	3.171	2.255
50	Penza Region	3.696	4.437	2.683	3.455	2.535
51	Perm Territory	4.052	4.896	3.039	4.015	3.012
52	Primorye Territory	4.293	4.798	3.339	4.452	3.374
53	Pskov Region	4.530	5.473	3.213	4.010	2.661
54	Rostov Region	3.669	4.499	2.629	3.258	2.531
55	Ryazan Region	4.765	6.042	3.947	5.210	4.072
56	Sakha (Yakutia)	3.637	4.415	2.521	3.472	2.690
57	Sakhalin Region	3.926	4.554	2.848	4.153	2.942
58	Samara Region	4.332	5.327	3.294	4.234	3.186
59	Saratov Region	3.765	4.527	2.605	3.322	2.469
60	Smolensk Region	4.261	5.255	3.479	4.413	3.408
61	Stavropol Territory	3.756	4.619	2.615	3.390	2.515
62	Sverdlovsk Region	3.818	4.531	2.749	3.604	2.574
63	Tambov Region	3.666	4.470	2.644	3.502	2.597
64	Tatarstan	3.877	4.705	2.697	3.474	2.573
65	Tomsk Region	3.489	4.109	2.391	3.007	2.429
66	Tula Region	3.411	4.184	2.340	3.171	2.240
67	Tuva	4.058	5.191	3.343	4.367	3.647
68	Tver Region	3.816	4.276	2.779	3.563	2.562
69	Tyumen Region	3.619	4.305	3.145	3.455	2.672
70	Udmurtian	3.891	4.704	2.831	3.532	2.711
71	Ulyanovsk Region	3.656	4.492	2.573	3.373	2.453
72	Vladimir Region	5.015	6.407	4.217	5.512	4.340
73	Volgograd Region	3.174	4.050	2.223	2.919	1.924
74	Vologda Region	3.956	4.956	3.113	4.172	3.146
75	Voronezh Region	3.856	4.852	2.866	3.712	2.836
76	Yaroslavl Region	4.288	5.430	3.466	4.631	3.591

A4 Appendices to Chapter 4

A4.1 Some auxiliary information on methods and algorithms of statistical processing of non-quantitative characteristics

In the study of determinant characteristics, we are faced with the need to statistically analyse multidimensional random variables X of so-called mixed nature. We are talking about situations characterized by the fact that among the vector components X can be both quantitative and qualitative (rank) and classification (nominal) characteristics. What are qualiy and classification random variables and how they are set? On which characteristics of their relationships (simila to pair correlations coefficients that characterize the relationship of quantitative characteristics) do we base the appropriate methods of multivariate statistical analysis? W are the typical problems of multivariate statistical analysis of non-quantitative characeristics an what methods and algorithms do we use to slve them? Which approach d we use to unify the alorithms of statistical analsis of characteistics of mixed nature? All these questions, unfortunately, are not studied enough. This appendix is dedicated to th description of some options of formalization of these issues and approaches to solving thm.

A. What are quality and classification random variables and how are they set?

So let the finite set of objects under analysis be $\mathbf{O} = \{O_i\}_{i=\overline{1,n}}$ and let $X_i = (x_i^{(1)}, x_i^{(2)}, \ldots, x_i^{(p)})$ be the values of set of characteristics of object O_i. At the same time among the components $x^{(1)}, x^{(2)}, \ldots, x^{(p)}$ there may be quantitative, qualitative and classification characteristics. Their definition and nature, as well as the basic forms of recording "observed meanings", are shown in Table A4.1.

Table A4.1. Definition and meaning of quantitative and qualitative sets of characteristics and forms of recording them

Name (nature) of characteristic	Definition (essence) of characteristic	Basic form of recording "observations"	General (unified) form of recording "observations"
Quantitative	Allows quantitative measurement of the degree of a certain characteristic (quality) of the object on a determinate numeric scale	$x_i^{(l)}$ – generally speaking, it can be any real number	$x_i^{(l)} = \begin{pmatrix} x_i^{(l.1)} \\ x_i^{(l.2)} \\ \vdots \\ x_i^{(l.m_l)} \end{pmatrix}$, where m_l is the total number of gradations (intervals) of grouping or not distinguishable by the analysed property of the group of characteristics
Qualitative ordinal (rank or categorized)	Allows organization of the analysed objects by the		

(continued)

Table A4.1. (continued)

Name (nature) of characteristic	**Definition (essence) of characteristic**	**Basic form of recording "observations"**	**General (unified) form of recording "observations"**
	degree of manifestation of certain properties. The scale, which could be used to quantitatively measure the degree of manifestation of this property is missing or is not known to researchers. If we set the number of possible uniform classes that could be sorted by $x^{(l)}$ (the number of gradations m_l), then the sign is called *ordinal categorized*, otherwise – *ranking*	$x_i^{(l)} = \begin{cases} R_i^{(l)}, & \text{if variable } x^{(l)} \text{ is ranked;} \\ v_i^{(l)}, & \text{if variable } x^{(l)} \text{ is categorized,} \end{cases}$ where $R_i^{(l)}$ is a *rank of i*-th observation, i.e. a number of the objects position O_i in the rank of all *n* statistically surveyed objects, sorted (from the best to the worst) by the characteristic $x^{(l)}$, and $v_i^{(l)}$ is the number of the class gradation), part of which is object O_i by the characteristic $x^{(l)}$ $(v_i^{(l)} = 1, 2, \ldots, m_l)$	$x^{(l)}$, and $x_i^{(l,v)}$ equals 1 or 0 depending whether the "value" of the characteristic $x^{(l)}$ in object O_i belongs to the v-th gradation
Qualitative and classification	Allows the break of the analysed population of objects into homogeneous (by analysed property) classes that cannot be ordered. The scale is set by the possible "values" (gradations, categories, numbered classes) of analysed variable	$x_i^{(l)} = v_i^{(l)}$, where $v_i^{(l)}$ is number of the class to which object O_i belongs; $v_i^{(l)}$ can assume "values" $1, 2, \ldots, m_l$, where m_l a number of possible gradations (categories) of the characteristic $x^{(l)}$	

To set a random variable means to describe a probabilistic mechanism, whereby an object randomly extracted from the general analysed population under study becomes a part of a gradation (or "category") of this variable.[1]

This description of the probabilistic mechanism is implemented, as is known, by means of some form of the law of probability distribution. At the same time, in relation to the quantitative characteristics as a space of elementary events is the area of the possible values of the analysed characteristic (as a result of a coarse version of grouping we can consider as a space of elementary events a finite set of intervals of grouping). For each non-quantitative characteristic, we would have to build a specific space of elementary events. Ultimately, however, we are interested not in the probability distribution law but in the probability distribution induced by it on all possible outcomes of allocating n of objects $O_1, O_2, \ldots, O_n$ taken from the analysed general population for possible gradations (categories, intervals of grouping, quality levels or homogeneous classes). It is easy to calculate that in the case of a grouped qualitative and quantitative categorized variable the total number of such outcomes $L(n, m_l)$ will be equal to $\sum_{k=1}^{m_l} k! \cdot \varphi(n, k)$ and in the case of the classification variable the number $L(n, m_l)$ of all possible outcomes of the placement n of objects according to m_l classes will result in $\sum_{k=1}^{m_l} \varphi(n, k)$. Here, $\varphi(n, k)$ are Stirling numbers of the second kind and the statement of the problem involves the possibility of empty intervals of grouping, gradations or classes.

Below are several options for the description of the probabilistic mechanism of formation of the "values" of a random variable of an arbitrary nature. We chose them due to their ability to facilitate the construction of mathematical and statistical constructions.

Option 1. We set the probabilities $p_k = P\{x^{(l,k)} = 1\}$, $k = 1, 2, \ldots, m_l$, i.e. the probability that an object randomly taken from the general analysed population seems to belong (by characteristic $x^{(l)}$) to the k-th gradation.

Option 2. We set probabilities p_ν^* $(\nu = 1, 2, \ldots, L(n, m_l))$ that objects $O_1, O_2, \ldots, O_n$ randomly taken from the general analysed population are placed according to m_l gradation using ν process (the total number of methods $L(n, m_l)$ is described above).

Option 3. Let there be a "true" (but not known to us) placement of the analysed objects $O_1, O_2, \ldots, O_n$ on m_l gradations. We are trying to recreate this true placement,

1 Let's not confuse the notions of the analysed general population, an abstract concept, interpreted as a totality of all **theoretically** possible objects of the type we are interested in, with the very specific notion of the analysed population $\mathbf{O} = \{O_i\}_{i=\overline{1,n}}{}^{\mathrm{T}}$, interpreted as a sample of the analysed general population. Specific applied goals of our study prompt us to forgo a common description that, in particular, is expressed in (a) the transition to a grouped data by quantitative characteristics and (b) postulating a finite number of possible gradations (grouping intervals) of analysed characteristics.

but doing it "fuzzily" with random errors. The probabilistic mechanism of this fuzziness of reproduction (i.e. random variable $x^{(l)}$) is set by the matrix $\alpha = \alpha(x^{(l)})$ with elements $\alpha_{ij}(x^{(l)})(i, j = 1, 2, \ldots, m_l)$, interpreted as the probability of an object referred to the graduation j, while in fact it belongs to graduation i.

Obviously, identity matrix

$$\alpha = \begin{pmatrix} 1 & & & 0 \\ & 1 & & \\ & & \ddots & \\ 0 & & & 1 \end{pmatrix}$$

sets "perfectly accurate" reproduction of an unknown placement of objects over the gradations (or the "most informative" variable $x^{(l)}$), while the matrix with all elements equal to $1/m_l$ sets a virtually meaningless reproduction of the desired placement (or "totally uninformative" variable $x^{(l)}$).

Option 4. Let's consider the situation where the classes (gradations), used to break up the given population $\mathbf{O} = \{O_i\}_{i=\overline{1,n}}$, can be described in advance in terms of fuzzy variables. We will find ourselves in this situation, for example, during the analysis of the classification variable determining the partition of the migrant population into classes according to a predetermined set of motives for migration, "policy", "social", "economic", "natural and ecological", "cultural", "personal", etc. With regard to the quantitative variable, such a situation arises, for example, in cases where the measurement of an analysed characteristic is performed with a significant random error.

Then, for the mathematical description of random variables, it is natural to use the mechanism of multivariate fuzzy subsets. In particular, the random variable $x^{(l)}$ is assumed to be given, if m_l-dimensional system of fuzzy subsets of the analysed set O, i.e. the $(m_l \times n)$-matrix is set

$$\mu(x^{(l)}) = (\mu_{ij})_{\substack{i = \overline{1, m_l} \\ j = \overline{1, n}}}.$$

Here, m_l is the number of gradations or classes (set in advance) into which the elements of the population **O**are distributed, and $\mu_{i1}, \mu_{i2}, \ldots, \mu_{in}$are values of the so-called function of belonging to the i-th class $(i = 1, 2, \ldots, m_l)$, normalized for convenience in a way that μ_{ij}could be interpreted as the probability of attributing the j-th element of the i-th gradation (and thus $\sum_{i=1}^{m_l} \mu_{ij} = 1$ for all $j = 1, 2, \ldots, n$).

This description of a random variable of common nature is productive not only in terms of building elements of statistical theory on classifications but also in terms of building specific classification algorithms.

Option 5. In situations where the real mechanism of classification uses as the initial data the results of *pairwise comparisons* based on the concepts of proximity or similarity of the analysed objects, the most convenient method of formalized description of the random variable of the general nature should be the use of the concept of fuzzy relations of tolerance or just fuzzy tolerances.

Fuzzy tolerance defined on the population of objects $\mathbf{O} = \{O_i\}_{i=\overline{1,n}}$ is set by the $(n \times n)$-matrix $Q^{(x(l))} = (q_{ij}(x^{(l)}))_{i,j=\overline{1,n}}$. The elements of this matrix $q_{ij}(x^{(l)})$ are the metrics of the "degree of similarity" of the i-th and j-th objects and at the same time can be interpreted as probabilities of "indistinguishability" of the i-th and j-th objects during their experimental comparison.

Thus, the experimental realization of a certain tolerance Q is a regular (not fuzzy) tolerance,[2] which is naturally set by $(n \times n)$-matrix $Q^{(x(l))} = (q_{ij}(x^{(l)}))_{i,j=\overline{1,n}}$, where $s_{ij} = 1$ if i-th and j-th objects were indistinguishable from each other and $s_{ij} = 0$ if otherwise. Since the information contained in the matrix $\hat{S}(Q)$ identifies a particular partition of the set **O** into classes [Айвазян, Бежаева, Староверов, 1974], and the probabilistic mechanism of reproduction of this partition, in turn, is defined by the fuzzy tolerance $Q^{(x^{(l)})}$, and thus we describe another way of formal definition of random variable $x^{(l)}$of general nature.

B. On the characteristics of proximity (comparison) of non-quantitative characteristics. As you know, the starting point for the statistical analysis of multivariate schemes is the pairwise comparison matrix. In the case of multivariate random variables of quantitative nature, this is represented by the covariance matrix. In the statistical analysis of non-quantitative characteristics, we use pairwise relations of the following coefficients as elements of the pairwise comparison matrix:

- while processing categorized qualitative and classification variables, i.e. such variables whose gradations (classes) are pre-conditioned by the meaning of the problem: ranking correlation coefficients and coefficients calculated by the so-called conjugate matrices (see, e.g. [Айвазян, Мхитарян, 2001];
- while processing classification variables of a general nature (the number of classes or their substance are not known to the researcher in advance): coefficients calculated based on the distance of Kemeny – Snell and Tanimoto types.

C. Verification of some statistical hypotheses of the nature of classification characteristics. Among the problems considered in this book we use as classification characteristics separate components of the vector of descriptive characteristics X, as well as parametric population of partitions (classifications) $S_Y(\theta)$ and $S_X(\Omega)$ taking place in the spaces of behavioural (**Y**) and descriptive (**X**) characteristics. In the course of their statistical analysis at various stages of research, there is a need to test the statistical hypotheses of the following form.

The hypothesis H_1 of statistically insignificant difference between the two (or more) analysed classification characteristics, such as $x^{(l_1)}$ and $x^{(l_2)}$ (or $S_Y(\theta_1)$ and $S_Y(\theta_2)$,

2 Equation on the set O is a tolerance if it is reflexive and symmetrical.

$S_X(W_1)$ and $S_X(W_2)$ or $S_Y(\theta)$ and $S_X(W)$). For a rigorous formulation of the problem in this case, it will be convenient to use "option 5" of the formalized setting of the classification problem. Then, if Q_i $(i = 1, 2, \ldots, T)$ is the matrix of fuzzy tolerances, setting the i-th analysed characteristic, then hypothesis H_1 is formulated as a statement of the fact that

$$Q_1 = Q_2 = \cdots = Q_T = Q\cdot$$

Substantially, hypothesis H_1 means the uniformity or consistency of the observed partitions, and its verification is based on the statistics $\hat{d}_{lt} = \sum_{1 \le i < j \le n} |\gamma_{ij}^{(l)} - \gamma_{ij}^{(t)}|$, where γ_{ij}^{ν} are the elements of Boolean matrix Γ, which is an experimental implementation of ν-th analysed classification variable (i.e. Γ_ν sets the ν-th analysed partition of the considered family of objects **O** into classes). A detailed description of the criterion for statistical testing of hypothesis H_1 is contained in [Aivazian, 1976].

Hypothesis H_2 is about the random nature of the existing set of partitions $S_1, S_2, \ldots, S_T$, or which is the same, about the absence of any relationship between the analysed classifications $x^{(l_1)}, x^{(l_2)}, \ldots, x^{(l_T)}$.

In terms of "option 1" of the formalized setting of the classification variable, hypothesis H_2 is formulated as a statement of the fact that

$$p_1 = p_2 = \cdots = p_{m_l} = \frac{1}{m_l}$$

(here, in accordance with the previously selected symbols, p_i is the probability to include a randomly taken object by the analysed characteristic $x^{(l)}$ to the i-th gradation, and m_l is the total number of gradations on this characteristic).

In terms of "option 2" of the formalized setting of the classification variable, the same hypothesis H_2 can be formulated as an equation

$$p_1^*(x^{(l)}) = p_2^*(x^{(l)}) = \cdots = p_{L(n,m_l)}^*(x^{(l)}) = \frac{1}{L(n,m_l)} \quad \text{for any } l = 1, 2, \ldots, T.$$

In terms of "option 3",

$$H_2 : \alpha_{ij}(x^{(l)}) = \frac{1}{m_l} \quad \text{for any } l = 1, 2, \ldots, T \text{ and at all } j = 1, 2, \ldots, m_l.$$

In terms of "option 4",

$H_2 : \mu_{ij}(x^{(l)}) = \frac{1}{m_l}$ for any $l = 1, 2, \ldots, T$ and at all $j = 1, 2, \ldots, m_l$ and $j = 1, 2, \ldots, n$.

In terms of "option 5",

$$H_2 : q_{ij}(x^{(l)}) = \frac{1}{2} \quad \text{for any } l = 1, 2, \ldots, T \text{ and at all } i, j = 1, 2, \ldots, n.$$

With regard to the problem of finding a sustainable partition by analysing a parametric family of classification algorithms $S_Y(\theta)$ or $S_X(W)$, a confirmation of hypothesis H_2 means either a complete uninformative character of the used classification variables (algorithms of partition) in terms of revealing the true distinct stratification structure of the analysed population of objects **O** or an objective absence of such structure.

The criteria for statistical testing of such hypotheses are based on statistics such as Hamming distance between partitions. Their description is given in particular in [Aivazian, 1976, 1977].

Hypothesis H_3 of "the homogeneity of two samples" is taken from the general population of classification characteristics. We are talking about the next problem. Let in the result of the ν_1-time realization of the classification algorithm $S_Y(\theta)$on the analysed population of objects $O_1, O_2, \ldots, O_n$ we obtained partitions $S_{Y.1}, S_{Y.2}, \ldots, S_{Y.\nu_1}$, and in the result of ν_2-time realization of the classification algorithm $S_X(W)$ on the same population of objects we obtained partitions $S_{X.1}, S_{X.2}, \ldots, S_{X.\nu_2}$. And let the obtained partitions be set using respective tolerance matrices. In terms of "option 5" of the formalized setting of random classification variable, the probabilistic mechanism of application of algorithms $S_Y(\theta)$ and $S_Y(\theta)$ will be conditioned by setting of matrices of fuzzy tolerance Q_1 and Q_2 generally not known to us.

Then hypothesis $H_3$of homogeneity (consistency) of the two analysed groups of specific partitions of the same population of objects **O** into classes can be formulated in the form

$$H_3 : \; Q_1 \equiv Q_2.$$

A statistical procedure of testing of this hypothesis provides a preliminary estimation of the matrix elements Q_1 and Q_2(using the elements of corresponding experimental tolerance matrices), and then the use of the statistical of type χ^2 (for more on this, see [Aivazian, 1976]).

We have discussed here only mathematical/statistical (still relatively poorly studied) aspects of the analysis of classification characteristics.

You will find a fairly complete description of algorithmic aspects of such analysis, relating, in particular, to implementation of the ideology of factor analysis on non-quantitative characteristics in, for example [Миркин, 1976].

Here, we will focus only on some moments of algorithmic aspects of the analysis of non-quantitative characteristics. These moments are prompted by the objectives of the research presented in Section 4.1.

D. On certain algorithms for the analysis of multivariate variables of general (mixed) nature. Consider the problem of partitioning a population of objects **O** in the space $\tilde{X}$ characteristics of mixed nature. Let's assume that "observations" are presented in the unified form, i.e. as a vector of corresponding measurements with binary components, and that we resolved the issue of the selection of metric $\rho_{\tilde{X}}$ in the analysed space. We will describe here the algorithm borrowed from

[Hartigan, 1975], which is a kind of generalization of the well-known algorithm of classification of "N-means" in the case of non-quantitative variables.

The algorithm consists in the consecutive recalculation of the group centres and building of partitions of population **O** according to these centres. For a given number of classes N on the first step of the algorithm, arbitrary N of various objects are declared centres. Then, using these centres, we construct the partition $S_{\tilde{X}.0} = (S^{(1)}_{\tilde{X}.0}, \ldots, S^{(N)}_{\tilde{X}.0})$ of the population **O** in such way that the k-th group includes objects for which among all centres the k-th centre is the nearest. According to this partition, we recalculate new centres. Using the new set of centres, we build a new partition, and so on. The rule of recalculation of the centres is as follows. For the k-th centre, we take the vector with the values of each coordinate equalling the modes of the corresponding component of vectors $S^{(k)}_{\tilde{X}.0}$. The algorithm stops when the centres of the groups no longer change.

This algorithm is, obviously, a standard algorithm (see [Aivazian, Bezhaeva, Staroverov, 1974], 104). Indeed, on the t-th iteration of the centres $E^t, \ldots, E^t$ (standards), we build a partition $S_{\tilde{X}.t} = (S^{(1)}_{\tilde{X}.t}, \ldots, S^{(N)}_{\tilde{X}.t})$ as follows:

$$S^{(k)}_{\tilde{X}.t} = \{\mathbf{X}_i : \rho_{\tilde{X}}(\mathbf{X}_i, \mathbf{E}^t_k) < \rho_{\tilde{X}}(\tilde{X}_i, \mathbf{E}^t_l), \ \ l = 1, 2, \ldots, N; \ \ l \neq k\}.$$

If $\rho_{\tilde{X}}(\tilde{X}_i, E^t_k) = \rho_{\tilde{X}}(\tilde{X}_i, E^t_l)$, then vector $\tilde{X}_i$ belongs to the class with the lowest ordinal number. Then, using partition $S_{\tilde{X}.t}$, we find a new system of standards $(E^{t+1}_1, \ldots, E^{t+1}_N)$. The k-th standard is vector E^{t+1}_k such that

$$\sum_{\tilde{X}_i \in S^{(k)}_{\tilde{X},t}} \rho_{\tilde{X}}(E^{t+1}_k, \tilde{X}_i) = \min_{\tilde{X} \in S^{(k)}_{\tilde{X},t}} \sum_{\tilde{X}_i \in S^{(k)}_{\tilde{X},t}} \rho_{\tilde{X}}(\tilde{X}, \tilde{X}_i) = \min_{\tilde{X} \in S^{(k)}_{\tilde{X},t}} \sum_{\tilde{X}_i \in S^{(k)}_{\tilde{X},t}} \sum_{l=1}^{p'} \varepsilon_l(\tilde{X}, \tilde{X}_i),$$

where

$$\varepsilon_l(\tilde{X}, \tilde{X}_i) = \begin{cases} w_l, & \text{if } \ l\text{–th coordinates of vectors } \tilde{X}, \ \ \tilde{X}_i \text{ do not coincide;} \\ 0, & \text{if } \ l\text{–th coordinates of vectors } \tilde{X}, \ \ \tilde{X}_i \text{ do not coincide.} \end{cases}$$

We denote $m_k(\tilde{X}_i, l)$ the number of vectors from $S^{(k)}_{\tilde{X},t}$, the l-th coordinate of which coincides with the l-th coordinate of vector $\tilde{X}_i$. Then, for $w_l \equiv 1$,

$$\sum_{\tilde{X}_i \uparrow S^{(k)}_{\tilde{X},t}} \rho_{\tilde{X}}(\tilde{X}, \tilde{X}_i) = \sum_{l=1}^{p'} \sum_{\tilde{X}_i \in S^{(k)}_{\tilde{X},t}} \varepsilon_l(\tilde{X}, \tilde{X}_i) = \sum_{l=1}^{p'} (n_k - m_k(\tilde{X}, l)),$$

and the minimum $\sum_{\tilde{X}_i \in S^{(k)}_{\tilde{X},t}} \rho_{\tilde{X}}(\tilde{X}, \tilde{X}_i)$ is attained at the vector E^{t+1}_i, the value of each coordinate of which is the mode of this coordinate of vectors from $S^{(k)}_{\tilde{X},t}$.

As follows from the results of [Методы анализа данных, 1985] for the described algorithm, the value $\sum_{k=1}^{N} \sum_{\tilde{X}_i \in S_{\tilde{X},t}^{(k)}} \rho_{\tilde{X}}(E_k^{t+1}, \tilde{X}_i)$ (total intra-class variation relative to the centres of classes) monotonously doesn't increase with the increase of t and the algorithm reaches a steady state after a finite number of steps.

Assuming further processing of large amounts of data, we tried in our work consecutive algorithm, which is an analogue of the McKean algorithm for non-numeric characters. In the algorithm, a number of classes of N partition is set where the first N points of the population are declared centres. Then, the points of the population are exposed consecutively. When showing the next point, we need to find the closest centre to it. If the point has a few closest centres, then it belongs to a centre with a minimum ordinal number. After referring the point of population to the k-th centre, we recalculate this centre. We take vector E_k as the new centre, which minimizes the value $\sum_{\tilde{X}_i \in S\tilde{X}^{(k)}} \rho_{\tilde{X}}(E, \tilde{X}_i)$. Here, we take the sum over all the points of the population **O** previously viewed together, referred to the centre with the number k. For quantitative characteristics with Euclidean metric between vectors, vector $E_k$satisfying this rule equals the average of all vectors previously assigned to the k-th centre (as it is calculated in the McKean algorithm). In the case of non-numeric characteristics, $E_k$is a vector with its each coordinate equalling to the mode of this coordinate of all the vectors referred earlier to the k-th centre. The case, where a coordinate has a few modes when recalculating centres, is resolved by a special agreement as to the choice of one of the modes (for example, all possible values of the coordinate are ordered and a mode with the minimum ordinal number is taken).

Once all the population points are viewed, the entire population is divided into classes with respect to the centres calculated during the last step. Each point refers to the class whose centre it is closest to. As before, if there are several centres equally close to the point, it refers to the centre of the minimum ordinal number.

To identify the most informative descriptive characteristics in terms of the typology of consumption, we can suggest the following procedure (see [Aivazian, Bezhaeva, Staroveov, 1974], 193). Let's consider a set of coordinates in the space descriptive characteristics. Let $I_{p'} = (i_1, i_2, \ldots, i_{p'})$be the numbers of the selected coordinates. We will denote the space of descriptive characteristics with $X(I_{p'})$based on this set. Let $S_{X(I_{p'})}$ be the partition of population of families in the space $X(I_{p'})$obtained using the procedure described above. We assume that the most informative in terms of the typology of consumption is the set of descriptive characteristics I^*, which maximizes the minimum distance between the classes, i.e. expression

$$\min_{1 \le k_1, k_2 \le N} \Delta(k_1; k_2; X), \tag{A4.1}$$

where $\Delta(k_1, k_2, X)$ is defined by equation (4.14).

It is possible to find that the most informative in this sense can be a set of descriptive characteristics, sorting all the possible sets I of coordinates of space, computing the value of functional (A4.1) and selecting a set of coordinates in which this value is maximal. However, other than that, we can use various methods to reduce sorting, such as the Monte Carlo method, the method of random search with adaptation.

The Monte Carlo method, or a method of random search, is different from the exhaustive sorting by the fact that the values of (A4.1) are not available for all subspaces of the space X, but only for some. These subspaces are selected randomly from all the subspaces on the assumption that all subspaces are equal.

The method, called random search with adaptation, is a modification of the Monte Carlo method. The ideas of the method of random search with adaptation can be used as follows. First, we conduct a random search of the subspace $X(I_k^*)$ of dimension k, maximizing (A4.1) for all the subspaces $X(I_k)$ the dimensions k. According to the method of random search with adaptation, the search $X(I_k^*)$ is performed with the "encouragement" and "punishment" of the individual characteristics of the space X. For this purpose, in the beginning of the search, we set the probabilities $q_1^{(1)}, q_2^{(1)}, \ldots, q_p^{(1)}$ of selecting each of the characteristics of the space X. Independently of each other, we select r sets for k characteristics $\tilde{I}(1) = \{I_k^{(1)}(1), I_k^{(2)}(1), \ldots, I_k^{(r)}(1)\}$. Each set of characteristics is selected in accordance with probabilities $q_1^{(1)}, \ldots, q_p^{(1)}$. If before the search there is no information about the preferred choice of any characteristic, then we assume $q_1^{(1)} = q_2^{(1)} = \cdots = q_p^{(1)} = 1/p$.

For each set of characteristics r, we calculate the value (A4.1) and among the obtained values r, we select minimum and maximum ones. $I_k^{(\min)}(1) = (i_1^{\min}(1), i_2^{\min}(1), \ldots, i_k^{\min}(1)$ and $I_k^{(\max)}(1) = (i_1^{\max}(1), i_2^{\max}(1), \ldots, i_k^{\max}(1))$ are the set of numbers of the coordinates of the space X, which delivered minimum and maximum values, respectively (A4.1). After this comes the "punishment" of the coordinate of the space X with the numbers $I_k^{(\min)}(1) = (i_1^{\min}(1), \ldots\ldots, i_k^{\min}(1))$ and "encouraging" of the coordinates with the numbers $I_k^{(\max)}(1) = (i_1^{\max}(1), \ldots.$ $\ldots, i_k^{\max}(1))$. "Encouragement" boils down to $q_i^{(2)} = q_i^{(1)} + h$, if $i \in I_k^{\max}(1)$ and when "punishment" $q_i^{(2)} = q_i^{(1)} - h$ for all $i \in I_k^{\min}(1)$. For all other coordinates, $q_i^{(2)} = q_i^{(1)}$.

After that, we select r subspaces $X(I_k^{(1)}(2)), \ldots, X(I_k^{(r)}(2))$ of the space X again. And every time to get another subspace, we randomly select k of p coordinates in accordance with probabilities $q_1^{(2)}, q_2^{(2)}, \ldots, q_p^{(2)}$. Then, the sets of numbers $I_k^{(\min)}(2)$ and $I_k^{(\max)}(2)$ are found for which, respectively:

$$\pi_{\min}^{(2)}(k) = \hat{\Delta}_{\min}\left(x^{(i_1^{\min})}(2), \ldots, x^{(i_k^{\min})}(2)\right) = \min_{i_1, i_2, \ldots, i_k} \Delta_{\min}(x^{(i_1)}, x^{(i_2)}, \ldots, x^{(i_k)}),$$

$$\pi_{\max}^{(2)}(k) = \hat{\Delta}_{\min}\left(x^{(i_1^{\max})}(2), \ldots, x^{(i_k^{\max})}(2)\right) = \max_{i_1, i_2, \ldots, i_k} \Delta_{\min}(x^{(i_1)}, x^{(i_2)}, \ldots, x^{(i_k)}),$$

where the value $\hat{\Delta}_{\min}(X)$ is determined by equations (A4.1), (4.14) and (4.14').

If $\pi_{\max}^{(2)}(k) > \pi_{\max}^{(1)}(k)$, then the coordinates with numbers from $I_k^{(\max)}(2)$ are "encouraged", but if $\pi_{\min}^{(2)}(k) < \pi_{\min}^{(1)}(k)$, then the coordinates with numbers from $I_k^{(\min)}(2)$ are "punished" by the method mentioned above. After that vector $q_1^{(3)}, q_2^{(3)}, \ldots, q_p^{(3)}$ is formed, subspaces $X(I_k^{(1)}(3)), \ldots, X(I_k^{(r)}(3))$ are selected and the whole procedure is repeated again.

The search ceases on the value of t_k for which $\pi_{\max}^{(l_k)}(k) \geq \pi_{\max}^{(t_k+l)}(k), l = 1, 2, \ldots, T$. Assuming $I_k^* = I_k^{(\max)}(t)$, we have a set of subspaces $X(I_k^*)$, $k = 1, 2, \ldots, p$. The choice of the most informative set of coordinates $I_{p'}^*$ focuses on the search for the set $I_{p'}^*$, for which

$$\pi_{\max}^{(t_{p'})}(p') = \max_k \pi_{\max}^{(t_k)}(k).$$

The procedure of isolating the most informative characteristics described here is a modification of the similar procedure of G. S. Lbov (see [Aivazian, Bezheva, Staroverov, 1974]). We have introduced modifications justified by the content side of the problem and a greater adaptability to handle large data sets.

Note that the same logistics of random search with adaptation can be used in solving the problem of determining the optimal "weights" C involved in setting metrics (4.18).

After identifying the set of descriptive characteristics, the most informative in terms of consumer typology (identifying key typological characteristics), we need to describe each type of consumption in the space of this set of typological characteristics. This description consists, firstly, in the calculation of average characteristics for space X for each consumption type. The average value of each characteristic in the space X is a mode of the coordinate of this characteristic. In addition, analogues of correlation matrix for numeric characteristics for each class of consumers in the space X can be explored. To this end, for each class of consumers, we can build the matrix coefficients of Pearson and Chuprov and matrices of distances between pairs of attributes defined by Tanimoto and Kemeny – Snell distances.

Histograms for each coordinate of space X are also a convenient tool for content analysis. For each k-th class, we can also naturally estimate $p_k(\tilde{X})$ – the probability that a family that belongs to this k-th type of consumption will have a set of descriptive characteristics equalling $\tilde{X}$. An estimate $\hat{p}_k(\tilde{X})$ of the value $p_k(\tilde{X})$ equals a share of the families from the analysed k-th class in which the vector of descriptive characteristics equals $\tilde{X}$.

These characteristics give the expert an opportunity of meaningful description of the types of consumption. In addition, using these characteristics, we can build procedures that assigned each specific family with its well-known objective conditions to one of the selected and described types of consumption.

A4.2 Probability of the household survey evasion as a function of some of its characteristics (analysis results)

The model used to investigate the probability (p) of the household survey evasion as a function of the logarithm of its total per capita expenditures ($2^{(1)}$), characteristics of HH residence $(z^{(2)})$ η and level of education of the family head is described in 4.2.2 of the book.

Before we present the results of the statistical analysis of this model, we will *characterize the nature and structure of the initial statistical data used as an information base* of that analysis.

In each round of RLMS of second wave (i.e. since 1994), interviewers visit all the households of the initial sample (4718 households) and record whether a survey was conducted in the household or not, and if not, then why. The recorded response codes are listed in Table A4.2.

Table A4.3 presents the frequency of evasion in relation to the total number of interviewees.

The ultimate goal of the analysis is to answer the question: "Does the probability of evasion from a sociological survey depend on the welfare of the family?" Based on the above evasion data in conjunction with data on income and expenditures of the households given in the main files of RLMS, we formed econometric models with a binary dependent variable (evasion/participation), and household expenditures and several other characteristics as explanatory variables.

Obviously, if the household refused to participate in the survey in a given round, the data on its expenditures cannot be obtained. However, since the RLMS data are *panel*, i.e. the same household participates in different rounds (at least potentially involved, and other household cannot participate in this sample), the information on the level of family wealth can be obtained from other rounds, considering that the welfare of the family in different rounds stays approximately constant. This assumption can be criticized on the basis that income mobility in Russia is quite high (see, e.g. [Богомолова, Тапилина, Ростовцев, 2000]). According to our information, mobility is not critically high: inter-annual variance of log consumption expenditure RLMS sample is between 0.018 (i.e. deviation from the average is less than 2%) and up to 1.32 (i.e. consumer expenditures change 3.7 times), averaging 0.25 with a median of 0.21 (i.e. consumer expenditures fluctuate around a "permanent" level of about 25%). It should be noted, incidentally, that the HH income is relatively more variable than its expenditures (consumption smoothing).

Table A4.2. Codifier of the results of the visits

01	Surveyed	27	Actions against the interviewer
	Objective reasons	28	Other
02	Uninhabited premises		**Motives for refusal**
03	No one lives in the apartment (house)	41	Unmotivated refusal
		42	Claim of business
04	Apartment not accessible	43	"Really busy"
05	Apartment rented to foreigners	44	"I don't open to anyone"
06	No one is home (three visits)	45	"These questions do not add or change anything"
07	Do not open the door (do not engage into conversation)	46	"I don't want to tell anyone about my life"
08	Could not be surveyed due to illness		
09	Could not be surveyed due to handicap	47	"I have a right not to answer"
		48	"I want to rest"
10	No adults at home	49	"I don't want my data in the computer"
11	Person who opened the door is intoxicated	50	"I have just taken part in another survey"
14	Family is absent during the survey period	51	"We are just here temporarily"
15	Family is home only late at night	52	Family circumstances
		53	Not interested in the survey profile
16	Family is residing in another place	54	Sick of politics
18	Other	55	Refusal out of protest
	Refusals	56	Fear of interview consequences: unwillingness to provide information about own political views
30	Refusal to participate		
	Conversation situation		
21	Refusal through the closed door		
22	Refusal after the door is opened	57	*Fear of interview consequences: unwillingness to provide information about own well-being*
23	Refusal of the respondent		
24	Refusal of another family member		
25	Refusal during the interview	58	Do not believe the interviewer
26	Refusal-lie	59	Other

Table A4.3. Frequency of the refusals from survey participation

	Round 5	Round 6	Round 7	Round 8
Survey is not conducted	743	963	1118	1254
Number of refusals	410	539	489	701
Refusal due to unwillingness to provide information on well-being			17	19
Survey conducted	3973	3781	3750	3831

In our calculations, we have used the average value of deflated expenditures for all periods of observation of the household. This value is interpreted as the "fixed expenditures" ("fixed consumption"), viewed in the context of the hypothesis of Friedman for earnings over the life cycle [Deaton, 1992]. Using other measures of well-being (median costs for available periods, imputed expenditures,[3] the first principal component of expenditures) leads to meaningful similar results, with the elasticity of the probability of refusal on income (in appropriate units) varying within a few percent. Moreover, as the continued application of the weights calculated by the multifactorial logit models, accounting probability of evading survey amends all the relevant values (mean, standard deviation, poverty rates) by a few percent, so the excessive concentration on the methodology for calculating the weights seems inefficient waste of time. In the future, we will use the average value of the expenditures as the most understandable value.

The household consumption expenditures deflated to a single initial period (variable totexpr* for different years, in the RLMS database they used a deflator from the "Review of the Russian Economy", constructed by the Russian-European Centre for Economic Policy), as well as the level of urbanization of the area, which is home to the household *($2^{(2)}$)*, and level of education of the head of household (or rather, a member of the household with the highest income *($2^{(3)}$)*) – all of the above serve as

3 The software STATA that we currently use helps to restore (impute) missing values of expenditure characteristics of the household based on the linear regression model. For each structure pattern of gaps, we build its own regression model, which constitutes the base for restoring missing values [[Little, Rubin, 1987; STATA, 2009]. In other words, for each observation with missing values of a variable being restored, we form a set of covariates whose values are not missing in this observation, we estimate linear regression and we construct a forecast value of the target variable. With further use of the data recovered in the above manner as covariates, the evaluations of the corresponding coefficients are likely to shift due to the prediction error. As it is seen in practice, they often shift in the direction of zero.

It is worth noting that the use of imputed expenditures can only partially solve the problem of the refusal of the household to answer individual questions (item "non-response"). In contrast to the refusal to answer all the questions (unit "non-response"), which is compensated by weighing, refusing to answer specific questions is offset by imputation.

base variables in the analysis of the probability of evasion from survey depending on income. The metric used for household wealth is the mean value of logarithms of deflated expenditures in available periods (the maximum – four periods, rounds 5–8). Using the average of several periods allows us to get closer to the value of the "constant consumption" for the periods indicated. As the dependent variable, we used an observed fact of whether the household refused to participate in at least one of the four rounds of RLMS. We also carried out an analysis using the category of evasion "I do not want to report information on income", but this category is not numerous (about 0.5%), while the logit model (like the vast majority of the methods of the analysis of binary dependent variables) works well with the share of successes (in this case, the evasions for the above reason) within 10–90%.

Estimates of several specifications of the logit model, based on the data described above, are given in Table A4.4. The dependent variable of the model, the indicator of the model, shows whether the HH evaded any survey for a random reason (795 out of 4239 observations, the latter figure is the highest frequency of participation in the survey Table A4.3 due to the fact that this panel includes all households, at least once participated in the survey, that is a superset of households for each period). The variables that characterize the level of household well-being is the median or the average of the four rounds of the logarithms of real (deflated to 1992) expenditures. The base category with the introduction of corresponding dummy variables is the gradation "the city" (U; the letters in parenthesis refer to the chart below). Educational categories are built on the accumulation and not the indicator system as follows: the basic category is education below average (L), followed by secondary education (the coefficient of this category indicates the difference from the category "education is below average"), and further on is vocational, technical, higher education in addition to the secondary (i.e. we measure the difference of these categories from persons with secondary education; secondary education does not exclude technical or higher education, so the indicator S does not exclude the indicator P, T or H). The weight that we used below is calculated by the model with the highest likelihood ratio for one degree of freedom, i.e. the latest.

In Figure A4.1, we are showing the charts of predicted values of probability of evasion from the survey for several categories of households (the scale of total per capita expenditures is given on a logarithmic scale). Since the model has four geographical and five education categories, the total number of individual logistic curves in the graph should be 20. Figure A4.1 shows some of the **most "populated" and representative of these curves.**

Table A4.4. The results of the analysis of a multi-factor model for the probability of evasion from the survey*

	(1)	(2)	(3)	(4)
Median expenditures	0.396(0.084)**	0.355(0.075)**		
Average expenditures			0.429(0.089)**	0.399(0.079)**
Capital regions (M)		1.052(0.206)**		1.043(0.203)**
Rural areas (R)		−1.583(0.292)**		−1.576(0.291)**
Townships (P)		−0.876(0.310)**		−0.878(0.308)**
Secondary education (S)		−0.862(0.156)**		−0.868(0.156)**
Vocational High School (P)		−1.826(0.184)**		−1.825(0.182)**
Technical Associate Degree (T)		−1.268(0.212)**		−1.277(0.213)**
College education (H)		−0.857(0.142)**		−0.880(0.142)**
Constant	−4.532(0.653)**	−3.140(0.588)**	−4.788(0.691)**	−3.464(0.632)**
Number of observations	4239	4239	4239	4239
Wald testEmpirical level of relevance	Wald(1) = 22.050.00	Wald(8) = 317.860.00	Wald(1) = 23.390.00	Wald(8) = 334.780.00

*In parentheses, you see the standard deviations of the coefficients calculated with an adjustment for clustering of observations (stratified sample).
**Significant at 1%.

The obtained results, although interesting in themselves, are intended only to clarify the sampling weights of households, i.e. the probability of participation of the specific household in sample survey. In the interim report on the project, we used a single-factor model with per capita expenditures as the sole explanatory variable. As you can see, the use of the multi-factor model allows us to refine the results, which certainly should have a positive impact on future results. Since the elasticity of the probability of evasion from the survey in terms of total per capita expenditures is not statistically distinguishable for the cases of specification of the model only by expenditures or by expenditures and additional factors, we can consider the overall outcome for the dependence of the probability of evasion on the household expenditures reliably verified.

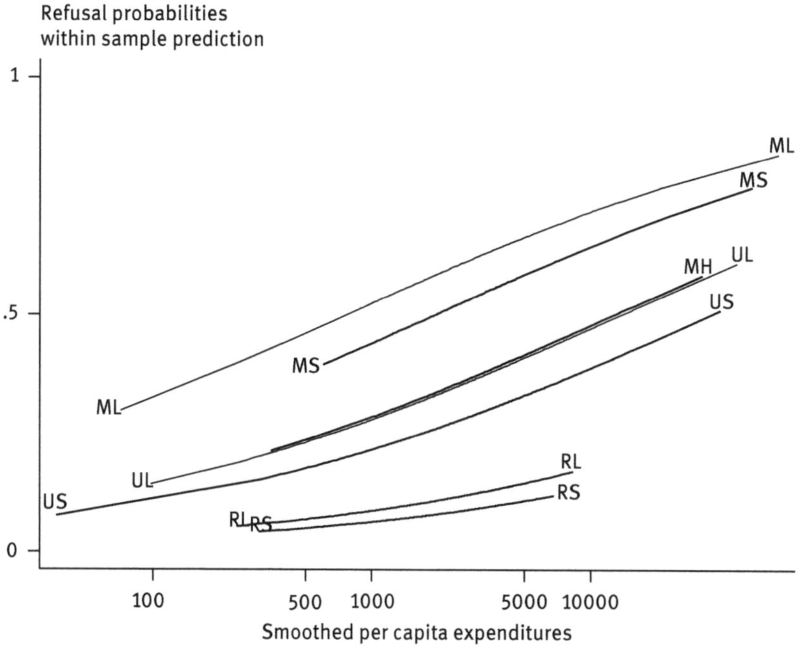

Figure A4.1. Family of curves describing the dependence of the probability *p* of total per capita expenditures *x*

A4.3. The results of the statistical analysis of the models of mixtures of distributions within the statistical observed range of values of total per capita expenditures of households

1. On the statistical methodology of estimating the parameters of the models of mixtures of distributions (EM algorithms and their modifications)

We will briefly describe the procedure for the statistical estimation of parameters

$$\Theta(k) = (\tilde{q}_1, \ldots, \tilde{q}_k; a_1, \ldots, a_k; \sigma_1^2, \ldots, \sigma_k^2)$$

of probability density function

$$\tilde{\varphi}_k(z|\Theta) = \sum_{j=1}^{k} \tilde{q}_j \, \varphi \, (z|a_j; \sigma_j^2) \tag{A4.2}$$

on a random sample (4.43″) using the method of maximum likelihood for a *set* value of the number of components of the mixture k (in equation (A4.2), $\varphi(z|a_j; \sigma_j^2)$ means the density function of a normal distribution with the mean a_j and variance σ_j^2).

The task is to find such values

$$\hat{\Theta}(k) = (\hat{\tilde{q}}_1, \ldots, \hat{\tilde{q}}_k; \hat{a}_1, \ldots, \hat{a}_k; \hat{\sigma}_1^2, \ldots, \hat{\sigma}_k^2)$$

in which the log-likelihood function

$$l_k(\Theta(k)) = \sum_{j=1}^{k} \omega_i \left[\ln \sum_{j=1}^{k} \tilde{q}_j \varphi\,(z_i|a_j;\, \sigma_j^2) \right] \tag{A4.3}$$

reaches its maximum, i.e.

$$\hat{\Theta}(k) = \arg\max_{\Theta(k)} l_k(\Theta(k)) \tag{A4.4}$$

(in equation (A4.3), z_i are elements of a sample (4.43''), ω_i is the weight of observation defined by formula (4.43') and n is the volume of the available sample).

The iterative EM *algorithm* (algorithm "Expectation-Maximization"), with which we solve problem (A4.4), is based on the following logistics (see [Day, 1969; Dempster, Laird, Rubin, 1977]):

(i) log-likelihood function (A4.3) is represented as

$$l_k(\Theta(k)) = \sum_{i=1}^{n} \omega_i \sum_{j=1}^{k} g_{ij} \ln \tilde{q}_j + \sum_{i=1}^{n} \omega_i \sum_{j=1}^{k} g_{ij} \ln \varphi(z_i|a_j;\, \sigma_j^2) - \sum_{i=1}^{n} \omega_i \sum_{j=1}^{k} g_{ij}\,, \tag{A4.5}$$

where the values

$$g_{ij} = \frac{\tilde{q}_j\, \varphi\,(z_i|a_j;\, \sigma_j^2)}{\tilde{\varphi}_k(z_i|\Theta(k))} \tag{A4.6}$$

in accordance with rule of calculating conditional probabilities, we determine the probability of observing class j, provided that as the i-th element of the sample we have an observation z_i (called a posteriori probabilities of observation of class j);

(ii) *phase "Expectation":* let the t-th step of the iterative procedure obtain the value

$$\hat{\Theta}^{(t)}(k) = \left(\hat{\tilde{q}}_1^{(t)}, \ldots, \hat{\tilde{q}}_k^{(t)}; \hat{a}_1^{(t)}, \ldots, \hat{a}_k^{(t)}; (\hat{\sigma}_1^2)^{(t)}, \ldots, (\hat{\sigma}_k^2)^{(t)} \right)$$

of the estimation of parameter $\Theta(k)$; substituting into (A4.6) these values, we obtain values $g_{ij}^{(t)}$, which we then insert into the right-hand side of (A4.5) instead of values g_{ij};

(iii) *phase "Maximization":* for the next iteration, we will maximize by $\hat{\Theta}^{(t)}(k)$ the expression

$$l_k\left(\hat{\Theta}^{(t)}(k)\right) = \sum_{i=1}^{n}\omega_i\sum_{j=1}^{k}g_{ij}^{(t)}\ln\hat{\bar{q}}_j^{(t)} + \sum_{i=1}^{n}\omega_i\sum_{j=1}^{k}g_{ij}^{(t)}\ln\varphi\left(z_i|\hat{a}_j^{(t)};\ (\hat{\sigma}_j^2)^{(t)}\right) - \sum_{i=1}^{n}\omega_i\sum_{j=1}^{k}g_{ij}^{(t)} \quad \text{(A4.7)}$$

at *fixed* $g_{ij}^{(t)}$ values. As a result, we get the following solutions:

$$\hat{\bar{q}}_j^{(t+1)} = \sum_{i=1}^{n}\omega_i g_{ij}^{(t)},$$

$$\hat{a}_j^{(t+1)} = \frac{1}{\hat{\bar{q}}_j^{(t+1)}}\sum_{i=1}^{n}\omega_i g_{ij}^{(t)} z_i,$$

$$(\hat{\sigma}_j^2)^{(t+1)} = \frac{1}{\hat{\bar{q}}_j^{(t+1)}}\sum_{i=1}^{n}\omega_i g_{ij}^{(t)}\left(z_i - \hat{a}_j^{(t+1)}\right)^2,$$

$$j = 1,\ 2, \dots, k.$$

Then, we return to phase "Expectation", i.e. using the values $\hat{\bar{q}}_j^{(t+1)}$, $\hat{a}_j^{(t+1)}$ and $\left(\hat{\sigma}_j^2\right)^{(t+1)}$ $(j = 1,\ 2, \dots, k)$; we calculate values $g_{ij}^{(t+1)}$using formula (A4.6); the value is inserted into the right-hand side of (A4.7), to go to phase "Maximization", etc. In [Dempster, Laird, Rubin, 1977], and other later works,[4] they proved (under fairly general assumptions, the most stringent of which is *the requirement of limited log-likelihood function*) the beneficial properties of the EM algorithms and, in particular, their convergence (in probability) to the desired solution (A4.4).

Modifications of the EM algorithm used in our calculations are of a technical nature. They consist in assigning weights z_i to observations, as well as in the use at the initial stage of the algorithm of auxiliary ("zero") so-called background component, which is in contrast to other (normal) components considered *evenly* distributed, roughly speaking, on the entire range of changes in sample data. A detailed description of the version of the EM algorithm, implemented in the package "Klass-master", can be found in [Jakimauskas, Sushinkas, 1996].

Above we have described a procedure for determining the estimates $\hat{\Theta}(k)$ of maximum likelihood of parameters $\Theta(k)$ for a *known* value of the number of mixture k components. Now, we give a brief description of **the procedure for estimating the number of components in the mixture** k, *that is, the number of components that can be statistically identified within the range of the observed values of average per capita total household expenditures.*

4 In fact, the general scheme of algorithms, later called the EM algorithms, was apparently first proposed in M. *Schlesinger,* On the distinction between spontaneous images//Reading machines, 1965. P. 38–45. The basic properties of these algorithms are also researched there. However, this work is difficult to access and little known among foreign experts.

This procedure was a consistent verification of a simple hypotheses of the type $H_0 : k = j$, when the alternative type $H_1 : k = j + 1, -j = 1, 2, \ldots$, with the critical statistics $\gamma(j) = -2 \ln \frac{l_j(\hat{\Theta}(j))}{l_{j+1}(\hat{\Theta}(j+1))}$.

The first value $j = \hat{k}$, at which hypothesis H_0 was not rejected, was taken as the estimate of the number of components in mixture (A4.2). This procedure was complemented by the procedure described in [Aivazian, 1996] of an approximate determination of the number of clusters, based on the method of focused projection (Projection Pursuit Method), as well as the content analysis of the resulting classes.

To solve the same problem, we used together with the above-described modified EM algorithm (implemented in the software package "Klassmaster") a program created by S. O. Kolenikov using the devices from the STATA package (in particular, its internal "maximizer"; see *W. Gould, W. Sribney*. Maximum Likelihood Estimation with STATA. Stata Press, 1999). This maximization algorithm implemented in STATA can be briefly summarized as follows:

1. the initial values are found: if the user did not offer some initial values, then randomly;
2. in the vicinity of the initial values, we randomly search the best values;
3. a one-dimensional optimization is performed for each of the model parameters;
4. a multidimensional optimization algorithm is triggered:
   - it is determined whether the likelihood function is convex on this point (first and second derivatives are calculated numerically);
   - if the function is convex, the iteration is carried out with the Newton – Raphson method;
   - if not, then we use the gradient method of steepest descent;
5. the completion of the work combines:
   - stabilization of the likelihood function (by default, 10^{-6}; if necessary, it can be changed);
   - stabilization of the coefficient values (by default, the relative change is less than 10^{-7});
   - the gradient of the likelihood function is small in value (set as a parameter; the value used is 10^{-3});
   - there are too many iterations (by default, 16,000);
   - inability to calculate the derivatives (plate of the likelihood function) causes an emergency stop.

If maximization is successful, the STATA displays a table of coefficients' estimators of the model with standard deviations and confidence intervals (see below). In addition to this, the output results are complemented by values of different criteria for evaluation quality: information statistics (AIC, ICOMP) and test results χ^2 on the agreement with the model distribution. The number of intervals of grouping N (degree of freedom chi-square), into which the entire range of the analysed characteristic is divided, is selected taking into account the recommendations $N \approx \log_2 n$, where n is the total number of observations. In accordance with hypothesis H_3 (in the form H_3', see 4.1), the described procedures are implemented under the simplifying condition $\sigma_1^2 = \sigma_2^2 = \ldots = \sigma_n^2 = \sigma^2$, where $\sigma_j^2 = D(\ln \xi)$.

The results of applying these procedures to the RLMS, round 8 (in Russia), as well as to each of the regional data sets (Komi Republic, Volgograd and Omsk regions) are presented in Tables A4.5–A4.8.

Table A4.5. The results of estimation of the parameters in the model of mixtures of normal laws describing distribution of Russia's largest population logarithm of total per capita expenditures ($n = 9716$, $m = 14$)

k	Statistics $\chi^2(\nu(k))$ for the goodness of fit	The significance level of goodness of fit	$\hat{\sigma}^2$	$\hat{a}_1$	$\hat{\hat{q}}_1$, %	$\hat{a}_2$	$\hat{\hat{q}}_2$, %	$\hat{a}_3$	$\hat{\hat{q}}_3$, %	$\hat{a}_4$	$\hat{\hat{q}}_4$, %	$\hat{\mu}_{\text{mod.}}$, thousands of rubles
1	152.5	$<10^{-7}$	0.865	6.343	100	–	–	–	–	–	–	0.876
2	96.4	$<10^{-6}$	0.826	6.370	98.90	3.914	1.10	–	–	–	–	0.874
3	**58.3**	**$<10^{-5}$**	**0.756**	**6.340**	**95.80**	**8.282**	**2.30**	**4.159**	**1.80**	–	–	**0.927**
4	58.4	$<10^{-5}$	0.716	6.297	90.96	7.618	6.54	4.235	2.29	9.790	0.21	0.953

Chosen model with five components.
k – number of mixture components in distribution function;
$\nu(k)$ – number of degrees of freedom for the critical statistics $\chi^2(\nu(k))$;
$p(k)$ – number of estimated parameters of the model;
m – number of grouping intervals for the input data;
$\nu(k) = m - p(k) - 1$.

Table A4.6. The results of estimation of the parameters in the model of mixtures of normal laws describing distribution of Republic of Komi's largest population logarithm of total per capita expenditures ($n = 1089$, $m = 12$)

k	**Statistics $\chi^2(v(k))$ for the goodness of fit**	**The significance level of goodness of fit**	$\hat{\sigma}^2$	$\hat{a}_1$	$\hat{\hat{q}}_1$, %	$\hat{a}_2$	$\hat{\hat{q}}_2$, %	$\hat{a}_3$	$\hat{\hat{q}}_3$, %	$\hat{a}_4$	$\hat{\hat{q}}_4$, %	$\hat{a}_5$	$\hat{\hat{q}}_5$, %	$\hat{\mu}_{mod.}$, **thousands of rubles**
1	32.08	0.0002	0.654	−0.842	–	–	–	–	–	–	–	–	–	0.598
2	16.54	0.0208	0.610	2.114	0.50	−0.857	99.50	–	–	–	–	–	–	0.630
3	23.67	0.0003	0.475	−0.162	22.88	−1.058	76.79	2.527	0.33	–	–	–	–	0.636
4	**9.77**	**0.4615**	**0.285**	**−1.976**	**8.99**	**0.010**	**25.41**	**−1.038**	**65.17**	**2.338**	**0.43**	**–**	**–**	**0.628**
5	5.90	0.0150	0.168	0.890	3.07	−1.082	55.02	2.756	0.27	−0.115	29.47	−2.029	12.17	0.633

Chosen model with five components.

k – number of mixture components in distribution function;

$v(k)$ – number of degrees of freedom for the critical statistics $\chi^2(v(k))$;

$p(k)$ – number of estimated parameters of the model;

m – number of grouping intervals for the input data;

$v(k) = m - p(k) - 1$.

Table A4.7. The results of estimation of the parameters in the model of mixtures of normal laws describing distribution of Volgograd region's population on the total value of the logarithm of per capita expenditures ($n = 1263$, $m = 12$)

k	**Statistics $\chi^2(v(k))$ for the goodness of fit**	**The significance level of goodness of fit**	$\hat{\sigma}^2$	$\hat{a}_1$	$\hat{\hat{q}}_1$, %	$\hat{a}_2$	$\hat{\hat{q}}_2$, %	$\hat{a}_3$	$\hat{\hat{q}}_3$, %	$\hat{a}_4$	$\hat{\hat{q}}_4$, %	$\hat{a}_5$	$\hat{\hat{q}}_5$, %	$\hat{\mu}_{mod.}$, **thousands of rubles**
1	43.63	$<10^{-5}$	0.723	−1.259	100	–	–	–	–	–	–	–	–	0.408
2	42.78	$<10^{-5}$	0.673	0.116	2.58	−1.295	97.42	–	–	–	–	–	–	0.414
2	42.72	$<10^{-5}$	0.716	−1.254	99.77	−3.044	0.23	–	–	–	–	–	–	0.407
3	37.53	$<10^{-5}$	0.577	−2.456	4.10	−1.280	90.57	0.021	5.34	–	–	–	–	0.414
4	32.694	$<10^{-5}$	0.180	0.613	4.10	−2.833	7.00	−0.647	38.56	−1.661	50.34	–	–	0.413
5	**12.03**	**0.0005**	**0.099**	**−2.943**	**6.16**	**−1.927**	**28.58**	**−0.481**	**27.89**	**0.650**	**4.21**	**−1.266**	**33.16**	**0.411**

Chosen model with five components.
k – number of mixture components in distribution function;
$v(k)$ – number of degrees of freedom for the critical statistics $\chi^2(v(k))$;
$p(k)$ – number of estimated parameters of the model;
m – number of grouping intervals for the input data; $v(k) = m - p(k) - 1$.

Table A4.8. The results of estimation of the parameters in the model of mixtures of normal laws describing distribution of Omsk region's population on the aggregate value of the logarithm of per capita expenditures ($n = 1244$, $m = 12$)

k	**Statistics $\chi^2(v(k))$ for the goodness of fit**	**The significance level of goodness of fit**	$\hat{\sigma}^2$	$\hat{a}_1$	$\hat{\hat{q}}_1$, %	$\hat{a}_2$	$\hat{\hat{q}}_2$, %	$\hat{a}_3$	$\hat{\hat{q}}_3$, %	$\hat{a}_4$	$\hat{\hat{q}}_4$, %	$\hat{a}_5$	$\hat{\hat{q}}_5$, %	$\hat{\mu}_{mod.}$, **thousands of rubles**
1	23.83	0.0050	0.656	−0.838	–	–	–	–	–	–	–	–	–	0.600
2	13.75	0.0550	0.602	0.671	2.34	−0.875	97.66	–	–	–	–	–	–	0.612
2	25.63	0.0006	0.591	−2.915	1.48	−0.807	98.52	–	–	–	–	–	–	0.592
3	**13.47**	**0.0190**	**0.382**	**−0.960**	**84.72**	**−2.911**	**2.20**	**0.294**	**13.09**	–	–	–	–	**0.607**
4	23.30	<0.0001	0.351	2.192	0.18	0.165	16.95	−2.908	2.28	−0.998	80.59	–	–	0.613
5	14.77	0.0020	0.278	−3.003	2.06	−1.260	46.09	0.548	8.67	−0.563	43.18	–	–	0.606
	18.42	<0.0001	0.211	−3.047	2.01	0.436	11.87	−1.433	34.27	−0.662	51.68	2.302	0.18	0.612

Chosen model with five components.

k – number of mixture components in distribution function;

$v(k)$ – number of degrees of freedom for the critical statistics $\chi^2(v(k))$;

$p(k)$ – number of estimated parameters of the model;m– number of grouping intervals for the input data;

$v(k) = m - p(k) - 1$.

A4.4. Results of nonparametric analysis of the distribution of population of Russia's regions and the country as a whole on the value of total per capita expenditures

1. **Russian Federation (RLMS data, eighth round, November 1998, total number of observations *n* = 11,397)**

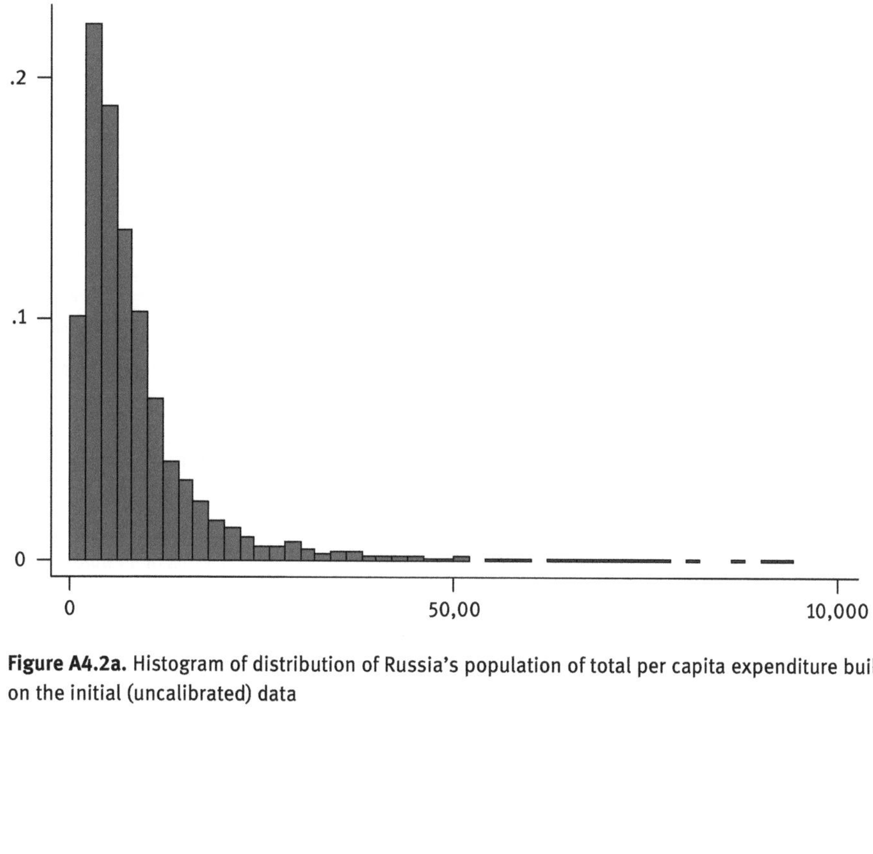

Figure A4.2a. Histogram of distribution of Russia's population of total per capita expenditure built on the initial (uncalibrated) data

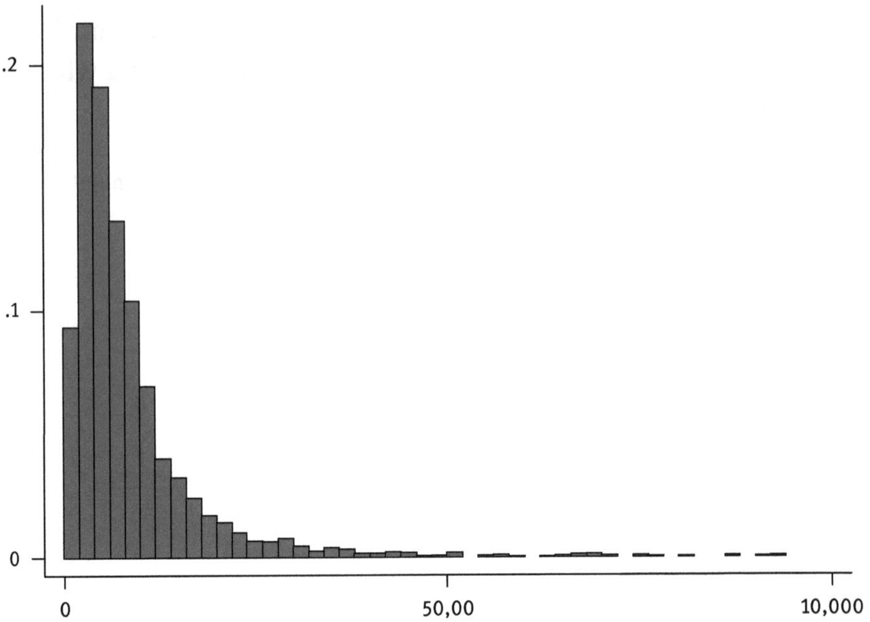

Figure A4.2b. Histogram of distribution of Russia's population of total per capita expenditure built on the calibrated data

Table A4.9. Sampled values of the principal quantitative characteristics of distribution of Russia's population of total per capita expenditure

Number of variable	Name of characteristic, thousand rub.	Values of sample characteristics	
		On initial data	On calibrated data
1	Average value of $(\hat{\mu})$ of total per capita expenditures	0.914	0.932
2	Standard deviations (S)	1.765	1.682
3	Minimum value in sample $(x_{\min})$	0.008	0.008
4	Maximum value in sample $(x_{\max})$	49.344	49.344
5	Bottom decile $(\hat{x}_{0,1})$	0.200	0.209
6	Top decile $(\hat{x}_{0,1})$	1.699	1.763

2. Komi Republic (data of selective sampling of households, second quarter of 1998, total number of observations $n = 1089$)

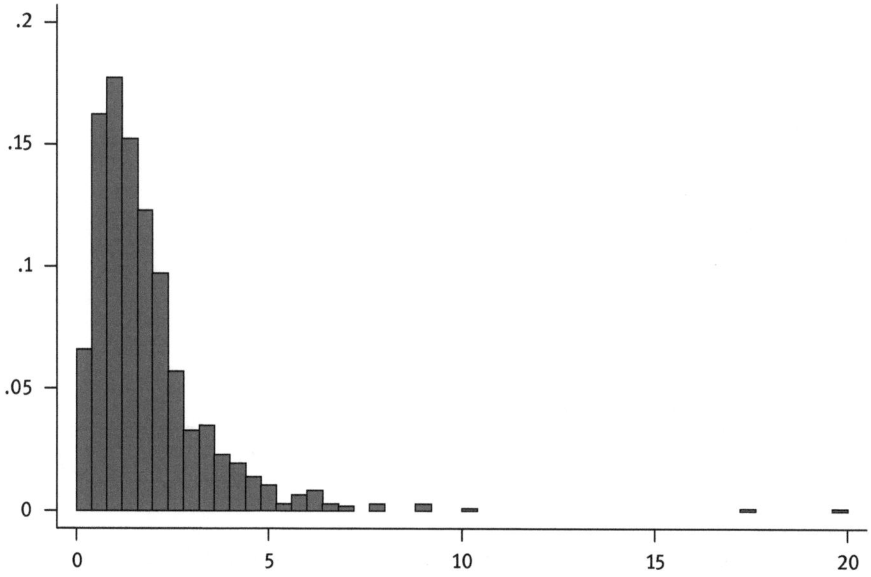

Figure A4.3a. Histogram of distribution of Komi Republic's population of total per capita expenditures built on the initial (uncalibrated) data

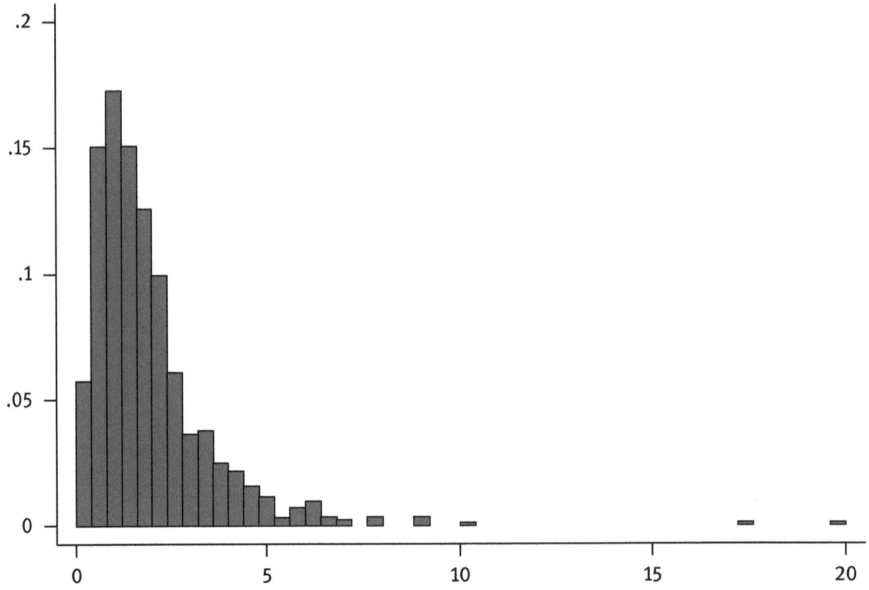

Figure A4.3b. Histogram of distribution of Komi Republic's population of total per capita expenditures built on the calibrated data

Table A4.10. Sampled values of the principal quantitative characteristics of distribution of Komi Republic's population of total per capita expenditures

Number of variable	Name of characteristic, thousand rub.	Values of sample characteristics	
		On initial data	On calibrated data
1	Average value of $(\hat{\mu})$ of total per capita expenditures	0.633	0.686
2	Standard deviations (S)	1.087	1.249
3	Minimum value in sample $(x_{\min})$	0.054	0.054
4	Maximum value in sample $(x_{\max})$	24.797	24.797
5	Bottom decile $(\hat{x}_{0,1})$	0.154	0.163
6	Top decile $(\hat{x}_{0,1})$	1.208	1.302

3. Volgograd region (data of selective sampling of households, second quarter of 1998, total number of observations n = 1263)

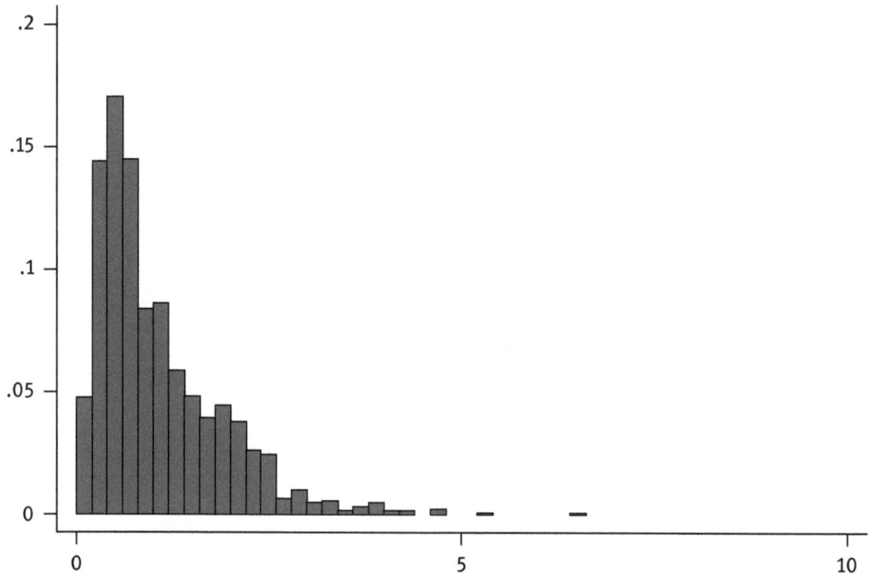

Figure A4.4a. Histogram of distribution of Volgograd region's population of total per capita expenditure built on the initial (uncalibrated) data

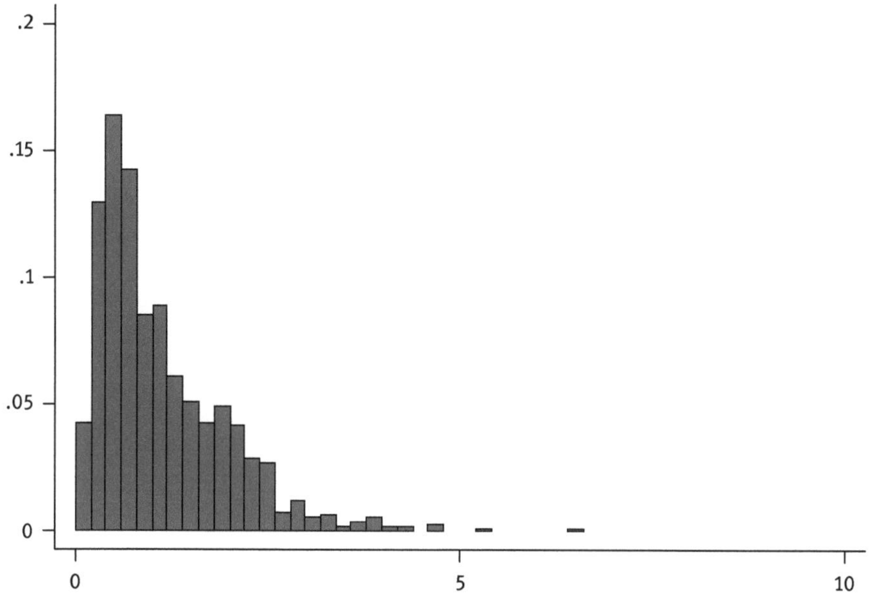

Figure A4.4b. Histogram of distribution of Volgograd region's population of total per capita expenditure built on the calibrated data

Table A4.11. Sampled values of the principal quantitative characteristics of distribution of Volgograd region's population of total per capita expenditures

Number of variable	Name of characteristic, thousand rub.	Values of sample characteristics	
		On initial data	On initial data
1	Average value of $(\hat{\mu})$ of total per capita expenditures	0.412	0.433
2	Standard deviations (S)	0.458	0.479
3	Minimum value in sample $(x_{\min})$	0.017	0.017
4	Maximum value in sample $(x_{\max})$	6.101	6.101
5	Bottom decile $(\hat{x}_{0,1})$	0.101	0.110
6	Top decile $(\hat{x}_{0,1})$	0.766	0.794

4. Omsk region (data of selective sampling of households, second quarter of 1998, total number of observations n = 1244)

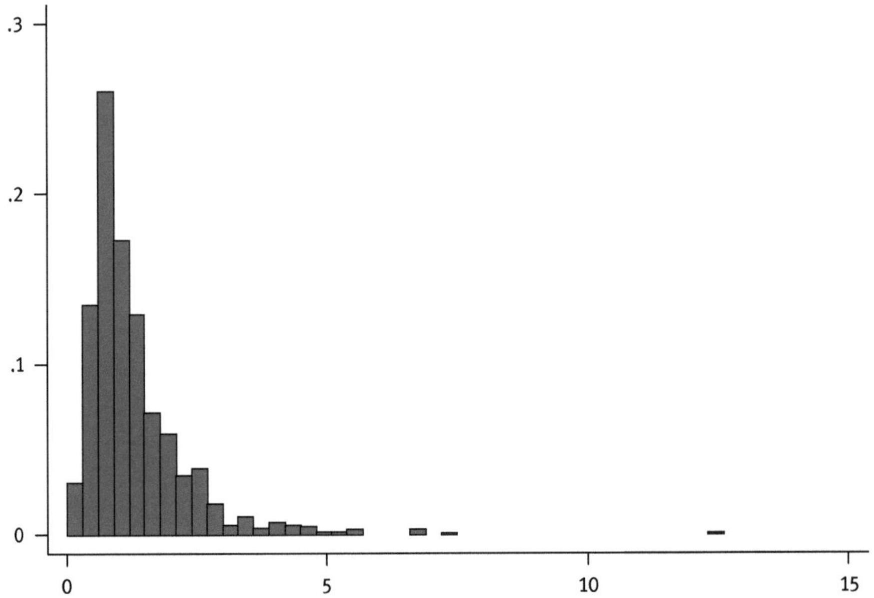

Figure A4.5a. Histogram of distribution of Omsk region's population of total per capita expenditure built on the initial (uncalibrated) data

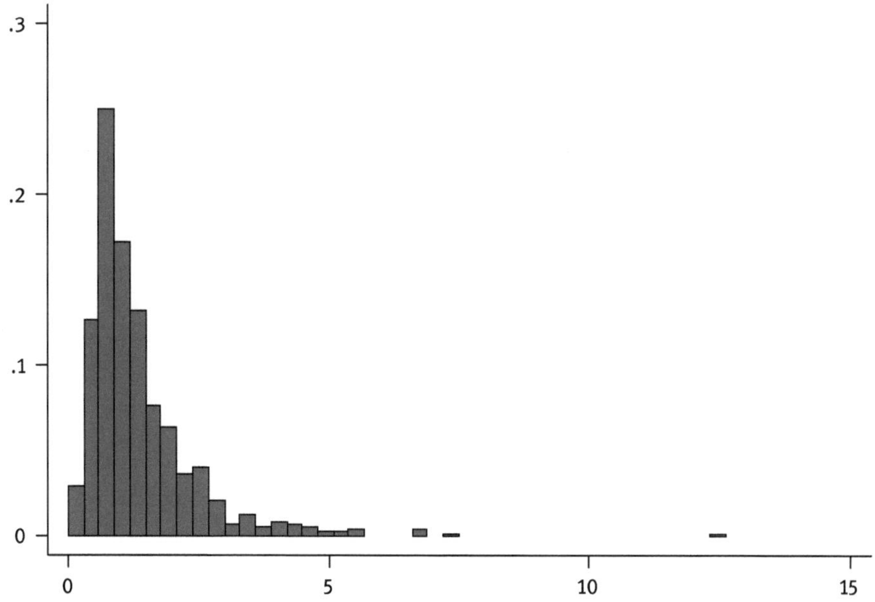

Figure A4.5b. Histogram of distribution of Omsk region's population of total per capita expenditure built on the calibrated data

Table A4.12. Sampled values of the principal quantitative characteristics of distribution of Omsk region's population of total per capita expenditure

Number of variable	**Name of characteristic, thousand rub.**	**Values of sample characteristics**	
		On initial data	**On initial data**
1	Average value of $(\hat{\mu})$ of total per capita expenditures	0.611	0.641
2	Standard deviations (S)	0.708	0.761
3	Minimum value in sample $(x_{\min})$	0.034	0.034
4	Maximum value in sample $(x_{\max})$	11.809	11.809
5	Bottom decile $(\hat{x}_{0,1})$	0.160	0.163
6	Top decile $(\hat{x}_{0,1})$	1.211	1.238

References

Айвазян С.А. Моделирование семейных доходов//Экономика и математические методы. 1970. Т. 6. № 2.

Айвазян С.А. Об опыте применения экспертно-статистического метода построения неизвестной целевой функции//Многомерный статистический анализ в социально-экономических исследованиях. М.: Наука, 1974.

Айвазян С.А. Многомерный статистический анализ в социально-экономических исследованиях// Экономика и математические методы. 1977. Т. XIII. № 5.

Айвазян С.А. Модель формирования распределения населения России по величине среднедушевого дохода//Экономика и математические методы. Т. 33. № 4. С. 74–86.

Айвазян С.А. Интегральные индикаторы качества жизни населения: их построение и использование в социально-экономическом управлении и межрегиональных сопоставлениях. М.: ЦЭМИ РАН, 2000.

Айвазян С.А. Россия в межстрановом анализе синтетических категорий качества жизни населения. Часть I: Методология анализа и пример ее применения//Мир России. 2001. Т. X. № 4.

Айвазян С.А. Роль социально-экономической политики и институционального развития в повышении качества жизни: результаты межстранового эконометрического анализа// Россия в глобализирующемся мире. М.: Наука, 2008.

Айвазян С.А. Методы эконометрики. М.: Магистр, 2010.

Айвазян С.А., Бежаева З.И., Староверов О.В. Классификация многомерных наблюдений. М.: Статистика, 1974.

Айвазян С.А., Герасимова И.А. Социальная структура и социальное расслоение населения Российской Федерации (по материалам выборочного обследования населения трех регионов РФ). М.: ЦЭМИ РАН, 1998.

Айвазян С.А., Колеников С.О. Уровень бедности и дифференциация по расходам населения России//EERC. 2001. WP № 1.

Айвазян С.А., Мхитарян В.С. Прикладная статистика и основы эконометрики. Т. 1: Теория вероятностей и прикладная статистика. М.: Юнити, 2001.

Айвазян С.А., Рабкина Н.Е., Римашевская Н.М. Методика расчета ожидаемого распределения рабочих и служащих по размерам заработной платы. М.: НИИ труда ГКСМ СССР по вопросам труда и заработной платы, 1967.

Андерсон Т. Введение в многомерный статистический анализ. М.: Физматгиз, 1963.

Бирман И. Уровень русской жизни (а также американской). М.: Экономика, 2007.

Бобков В.Н., Масловский-Мстиславский П.С., Маликов Н.С. Качество жизни: вопросы теории и практики. М.: ВЦУЖ, 2000.

Богомолова Т.Ю., Тапилина В.С., Ростовцев П.С. Рост мобильности по доходам в изменении неравенства в распределении доходов//EERC/РПЭИ, 2000.

Бородкин Ф.М., Айвазян С.А. Социальные индикаторы. М.: Юнити, 2006. 608 с.

Брейтуэйт Дж. Адресность и относительно длительная бедность в России. Доклад на семинаре Всемирного банка 19 апреля 1999г.

Великанова Т., Колмаков И., Фролова Е. Совершенствование методики и моделей распределения населения по среднедушевому доходу//Вопросы статистики. 1996. № 5.

Великанова Т.Б., Фролова Е.Б. Методология прямой оценки величины уровня бедности, основанная на распространении результатов выборочного обследования на генеральную совокупность//Материалы Управления статистики уровня жизни населения Госкомстата РФ, 1999.

Вербик М. Путеводитель по современной эконометрике. М.: Научная книга, 2008.

Волкова Г., Мигранова Л., Римашевская Н. Вопросы методики оценки дифференциации доходов населения//Вопросы статистики. 1997. № 2.

Волконский В.А. Об объективной математической характеристике народного потребления//Народнохозяйственные модели. Теоретические вопросы потребления. М.: АН СССР, 1963.

Волконский В.А. Экономико-математические модели согласованного планирования платежеспособного спроса и розничных цен//Экономика и математические методы. 1973. Т. 9. № 4.

Волконский В.А. Стохастическая модель поведения потребителя и изучение зависимости спроса от цен//Доходы и потребительский спрос. М.: Статистика, 1974.

Демография, экономика, экология: некоторые аспекты сравнительного анализа регионов Российской Федерации/И.А. Герасимова, М.Б. Денисенко, Н.М. Калмыкова, В.Э. Цитовцев, В.В. Шакин. М.: ЦЭМИ РАН, 1997.

Доклад Комиссии по оценке экономических результатов и социального прогресса//Вопросы статистики. 2010. № 11–12; 2011. № 2–3. (Полные англоязычная и франкоязычная версии доклада размещены на сайте: www.stiglitz-sen-fitoussi.fr.)

Доклад о развитии человеческого потенциала в российской Федерации. М.: Права человека. 1996; 1997; 1998; 2000; 2001.

Доманский В.К., Крепс В.Л., Калягина Л.В. Построение целевой функции для оценки экономических событий на основе экспертного ранжирования//Вестник КрасГАУ. 2006. № 11.

Ершов Э.Б., Майер В.Ф. Методологические и методические проблемы определения уровня, объема и дифференциации доходов населения//Материалы к заседанию ВЦУЖ 28 декабря 1998.

Известия. 1995. 27 декабря. Сколько в России богатых (по материалам Центра комплексных исследований и маркетинга «Круглого стола бизнеса России»).

Инструментарий обследования бюджетов домашних хозяйств и структура файлов первичных данных для руководства к использованию итогов обследования/Госкомстат. М., 2002.

Качество населения. Демография и социология. Вып. 6. М., 1993.

Квалиметрия жизни/ А.Г. Азгальдов, В.Н. Бобков, В.Я. Ельмеев, Ю.С. Перевощиков, В.А. Беляков. М.: ВЦУЖ, 2006.

Кейнс Д. Общая теория занятости, процента, денег. М.: Прогресс, 1978.

Киселев Н.И. Экспертно-статистический метод определения функции предпочтения по результатам парных сравнений объектов//Алгоритмическое и программное обеспечение прикладного статистического анализа. М.: Наука, 1980.

К методике оценки распределения генеральной совокупности населения по величине среднедушевого дохода на основании эмпирических данных/Материалы Госкомстата РФ (тезисы к докладу на Ученом совете в ЦУЖе). М., 1999.

Колеников С.О. Методы анализа качества жизни. Серия РЭШ «Лучшие студенческие работы». М.: ЦЭМИ, 1999.

Колосницын И.В., Макарчук Н.И. О построении целевой функции потребления//Экономика и математические методы. 1982. Т. XVIII. № 4.

Корчагина И., Овчарова Л., Турунцев Е. Система индикаторов уровня бедности в переходный период в России. Научный доклад № 98/04 по секции «Микроэкономика-2 (поведение домохозяйств)» Российской программы Консорциума экономических исследований. РПЭИ/Фонд Евразия, 1999.

Коссов В.В. Динамика качества жизни в субъектах Российской Федерации и направленность социальной политики//Мир России. 2001. Т. 10. № 2.

Макаров В.Л. Социальный кластеризм. Российский вызов. М.: Издательский дом «Бюджет», 2010.

Маликов Н.С. К вопросу о содержании понятия «качество жизни» и его измерению//Уровень жизни регионов России. 2002. № 2.

Маслоу А.Г. Мотивация и личность/Пер. с англ., С-Пб.: Евразия, 1999.

Медков В.М. Качество населения: сущность, содержание, критерии//Народонаселение. Современное состояние научного знания.//М., 1991.

Методологические положения по статистике. Вып. 1. М.: Госкомстат РФ, 1996.

Методы анализа данных: подход, основанный на методе динамических сгущений. М.: Финансы и статистика, 1985.

Минц А.А. Экологическая оценка естественных ресурсов. М.: Мысль, 1972.

Миркин Б.Г. Анализ качественных признаков. М.: Статистика, 1976.

Митоян А.А. Потребительское поведение семей: дифференциация, динамика, классификация. М.: Экономика, 1990.

Национальная оценка и распространение информации. Предварительный отчет в рамках проекта Всемирного банка по теме «Структурная перестройка системы социальной защиты населения» № SPIL-2.2.2/11. М.: ГУ-ВШЭ, 1999.

Об организации работ по созданию Единой государственной системы экологического мониторинга. Приказ Минприроды России от 27.04.94

№ 121, 1994.

Определение основных показателей агрегирования данных обследования бюджетов домашних хозяйств. Материалы Госкомстата РФ. М., 1999.

Перкинс Дж. Исповедь экономического убийцы. М.: Pretext, 2005.

Пилотные программы по введению адресной социальной поддержки малоимущих семей в Республике Коми, Воронежской и Волгоградской областях. 1999. Предварительные итоги. Министерство труда и социального развития МФ. М., 1999.

Прикладная статистика: классификация и снижение размерности/С.А. Айвазян, В.М. Бухштабер, И.С. Енюков, Л.Д. Мешалкин. М.: Финансы и статистика, 1989.

Римашевская Н.М. Человеческий потенциал России и проблемы «сбережения населения»//Российский экономический журнал. 2004. № 9–10.

Римашевская Н.М., Овсянников А.А. Потребительское поведение населения: теория и результаты моделирования//Экономика и математические методы. 1981. Т. 17. № 6.

Сафонова Е.В. Фактор экономики знаний в социально-экономическом развитии и качестве жизни населения//Экономика и математические методы. 2004. Т. 40.

Сен А. Развитие как свобода. М.: Новое издательство, 2004.

Слуцкий Е.Е. К теории сбалансированного бюджета потребителя//Народнохозяйственные модели. Теоретические вопросы потребления. М.: Изд-во АН СССР, 1963.

Соградов А.А. Теория и методы изучения качества населения. М.: Изд-во «Гуманитарный фонд», 1995.

Сопцов В.В., Николаева Н.И. Взаимосвязь потребительского и трудового поведения населения//Моделирование социальных аспектов народнохозяйственного планирования. М.: ЦЭМИ АН СССР, 1987.

Суворов А.В., Ульянова Е.А. Денежные доходы населения России: 1992–1996 гг.//Проблемы прогнозирования. М., 1997.

Типология потребления/ под ред. С.А. Айвазяна, Н.М. Римашевской. М.: Наука, 1978.

Фролова Е. О ходе работ по совершенствованию обследования бюджетов домашних хозяйств//Вопросы статистики. 1995. № 7.

Цели развития тысячелетия. Генеральная ассамблея ООН. М., 2000.

Шакин В.В. Биосистемы в экстремальных условиях. М.: ВЦ РАН, 1997.

Шевяков А.Ю., Кирута А.Я. Экономическое неравенство, уровень жизни и бедность населения России и ее регионов в процессе реформ: методы измерения и анализ причинных зависимостей. Финальный отчет по проекту EERC. М., 1999.

Aivazian S.A. Mixture-model cluster analysis using the projection pursuit method // *Computational Learning and Probabilistic Reasoning* / Ed. by A. Gammerman. N.Y.: Wiley, 1996. P. 277–286.

Aivazian S.A. On key factors of socio-economic policy and institutional development that determine the improvement of the quality of life // *Austrian Journal of Statistics*. 2008. Vol. 37. N. 1. P. 5–20.

Aivazian S.A. *Probabilistic-Statistical Modelling of the Distributary Relations in Society* // Private and Enlarged Consumption. Amsterdam-New York-Oxford: North-Holland, 1976. P. 235–247.

Andersen E.M. Population measurement of Health-Related Quality-of-Life (HRQoL) // *Quality of Life Newsletter*. 1999. Jan. 21.

Aron J. Growth and institutions: A review of the evidence // *World Bank Research Observer*. 2000. Vol. 15. N. 1. P. 99–153.

Asheim G. *Green National Accounting: Why and How?* / Department of Economics; University of Oslo, 1999.

Atkinson A.B. On the measurement of poverty // *Econometrica*. 1987. Vol. 55. N. 4. P. 749–764.

Baltagi B.H. *Econometric Analysis of Panel Data*. John Wiley and Sons, 1995.

Becker G.S. Investment to human capital: Theoretical analysis // *Journal of Political Economy*. 1962. Vol. 70.

Becker G., Philipson T., and Soares R. The quantity and quality of life and the evolution of world inequality // NBER Working Paper 9765. 2003.

Bourguignon F., and Fields G. Discontinuous loss from poverty, generalized measures, and optimal transfers to the poor. XI-th World Congress of the International Economic Association. Tunis, 1995.

Bourguignon F., and Fields G.S. Poverty measures and anti-poverty policy // *Recherches Economiques de Louvain*. 1990. Vol. 56. N(3–4). P. 409–427.

Campbell A., Converse Ph., and Rodgers W. *The Quality of American Life*. N.J., 1976.

Cobb C., Halstead T., and Rowe J. *The Genuine Progress Indicator: Summary of Data and Methodology*. San Francisco, CA: Redefining Progress, 1995.

Cummins R.A. *Assessing Quality of Life* // Quality of Life for People with Disabilities. Cheltenham: Stanley Thornes, 1997. P. 116–150.

Cummins R.A. *Comprehensive Quality of Life Scale*. Adult Manual, Fifth Edition. Melbourne: School of Psychology, Deakin University. 1997. P. 1–51.

Cummins R.A. The domains of life satisfaction: An attempt to order chaos // *Social Indicators Research*. 1996. Vol. 38. P. 303–328.

Cummins R.A. Objective and Subjective Quality of Life: An Interactive Model // *Social Indicators Research*. 2000. Vol. 52. P. 55–72.

Cummins R.A. On the trail of the gold standard for life satisfaction // *Social Indicators Research*. 1995. Vol. 35. P. 179–200.

Cummins R.A. *A Psychometric Evaluation of the Comprehensive Quality of Life Scale*, Fifth Edition // Urban Quality of Life: Critical Issues and Options. Singapore: University Press, 1999. P. 20–33.

Deaton Angus. *Understanding Consumption: Clarendon Lectures in Economics*. Oxford University Press; Clarendon Press, 1992.

Deming W.E. *Statistical Adjustment of Data*. N.Y.: John Wiley, 1943.

Dempster A., Laird G., and Rubin J. Maximum likelihood from incomplete data via the EM-algorithm // *Journal of Royal. Statistical Society*. 1977. Vol. B.39. P. 1–38.

Deville J.-C., and Särndal C.-E. Calibration estimators in survey sampling // *Journal of the American Statistical Association*. 1992. Vol. 87. N. 418. P. 376–382.

Deville J.-C., Särndal C.-E., and Sautory O. Generalized raking procedures in survey sampling // *Journal of the American Statistical Association*. 1993. Vol. 88. N. 423. P. 1013–1020.

Diener E. Subjective well-being // *Psychological Bulletin*. 1984. Vol. 93. P. 542–575.

Diener E. A value based index for measuring national quality of life // *Social Indicators Research*. 1995. Vol. 36. P. 107–127.

The East Asian Miracle. Economic Growth and Public Policy. World Bank, Oxford University Press, 1993.

Easterlin R.A. *Happiness and Economics*. Edward Elgar, 2002.

Esteban J.-M., and Ray D. On the measurement of polarization // *Econometrica*. 1994. Vol. 62. N. 4. P. 819–851.

Estes R.J. Social development trends in Africa // *Social Development Issues*. 1995. Vol. 17. N. 1. P. 18–47.

Estes R.J. Social development trends in Asia // *Social indicators Research*. 1996. Vol. 37. N. 2. P. 119–148.

Estes R.J. Social development trends in Europe. 1970–1994: Development prospects for the new Europe // *Social Indicators Research*. 1997. Vol. 42. N. 1. P. 1–19.

Estes R.J. Social development trends in transitional economies. 1970–1995 // *Challenges of Transformation and Transition from Centrally Planned to Market Economies*. United Nations: United Nations Centre for Regional Development, 1998. P. 13–30.

Estes R.J. *The Social Progress of Nations*. N.Y.: Praeger, 1984.

Estes R.J. *Trends in World Social Development*. N.Y.: Praeger, 1988.

EuroBarometer. Brussels: Commission on the European Communities, 1998. Online access: http://europa.eu.int/en/comm/dg10/infcom/epo/eb.html.

Fajnzulber P., Lederman D., and Loayza N. Inequality and violent crime. *The research project "Crime in Latin America" of the World Bank*, 1999.

Ferrans C., and Powers M. Quality of life index: Development and psychometric properties // *Advances in Nursing Science*. 1985. Vol. 8. P. 15–24.

Foster J., Greer J., and Thorbeck E. A class of decomposable poverty measures // *Econometrica*. 1984. Vol. 52. N. 3. P. 761–766.

Foster J.E., and Shorroks A.F. Poverty orderings // *Econometrica*. 1988. Vol. 56. N. 1. P. 173–177.

Fridman M.A. *Theory of the Consumption Function*. Princeton, NJ: Princeton University Press, 1957.

Gerdtham U.-G., and Johannesson M. The relationship between happiness, health and socio-economic factors: Results based on Swedish micro data // *Working Paper Series in Economics and Finance 207*. Stockholm School of Economics, 1997.

Gould W., and Sribney W. *Maximum Likelihood Estimation with Stata*. Stata Corp., 1999.

Granger C.W.J. Investigating causal relations by econometric methods and gross-spectral methods // *Econometrica*. 1969. Vol. 37. P. 424–438.

Guterbock T.M. Why money magazine's "best places" keep changing // *Public Opinion Quarterly*. 1997. Vol. 61. P. 339–355.

Hagenaars A. A class of poverty indices // *International Economic Review*. 1987. Vol. 28. P. 583–607.

Hagerty M.R., Cummins R.A., Ferriss A.L., Land K., Michalos A.C., Peterson M., Sharpe A., Sirgy J., and Vogel J. Quality of life indexes for national policy: Review and agenda for research // *Social Indicators Research*. 2001. Vol. 55. P. 1–96.

Hamilton J. *Time Series Analysis*. Princeton University Press, 1994.

Hartigan J.A. *Clustering Algorithms*. London, 1975.

HDR – Human Development Report. United Nations Development Programme. N.Y., 2000; 2001; 2002; 2003. (www.undp.org).

Hennessy C.H., Moriarty D.G., Zak M.M., Scherr P.A., and Brackbill R. Measuring health-related quality of life for public health surveillance // *Public Health Reports*. 1994. Vol. 109. P. 665–672.

Jakimauskas G., and Sushinkas J. Computational aspects of statistical analysis of gaussian mixture combining EM algorithm with non-parametric estimation (one-dimensional case). *Preprintas N96-6*, Matematicos ir Informaticos Institutas. Vilnius; Lietuva, 1996.

Johnston D.F. Toward a comprehensive "quality life" index // *Social Indicators Research*. 1988. Vol. 20. P. 473–496.

Kalton G., and Flores-Cervantes I. Weighting methods // *Journal of Official Statistics*. 2003. Vol. 19. N. 2. P. 81–97.

Kanbur S.M.R. Measurment and alleviation of poverty // *IMF Staff Papers*. 1987. 34. P. 60–85.

Kotler P., Haider D., and Rein I. *Marketing Places*. N.Y.: The Free Press, 1993.

Land K. *Social Indicators* // Encyclopedia of Sociology. Revised Edition. N.Y., 2000.

Lane R.E. Quality of life and quality of person: A new role for government // *Political Theory*. 1994. Vol. 22. N. 2.

Lewis A.W. *The Theory of Economic Growth*. London: Allen and Unwin, 1955.

Little R.J.A., and Rubin D.B. *Statistical Analysis with Missing Data*. Wiley, 1987.

Mangahas M. *Dimensions of Poverty* // *SWS Occasional Paper*. Quezon City: Social Weather Stations, Inc., 1986.

Mangahas M. *Measuring Philippine Development* (ed.) Mamila: Development Academy of the Philippines, 1976.

Mangahas M. Review of surveys on crime // *Social Weather Bulletin*. 1996. Vol. 11. N. 12 (June). Quezon City: Social Weather Stations, Inc.

Mangahas M. *Rural Poverty, Social Surveillance and Democratic Development* // SWS Occasional Paper. Quezon City: Social Weather Stations, Inc., 1987.

Mangahas M. Self-rated poverty in the Philippines, 1984–1992 // *International Journal of Public Opinion Research*. 1995. Vol. 7. N. 1. P. 40–55.

Mangahas M., and Guerrero L.L. *Self-sustained quality of life monitoring: The Philippine social weather reports* // *SWS Occasional Paper*. Quezon City: Social Weather Stations, Inc., 1998.

Michalos A.C. Combining social, economic and environmental indicators to measure sustainable human well-being // *Social Indicators Research*. 1997. Vol. 40. P. 221–258.

Michalos A. *What Makes People Happy: Social Indicators and Quality of Life*. Oslo, 1987.

Micklewright J. What can child anthropometry reveal about living standards and public policy? An illustration from Central Asia // *Review of Income and Wealth*. 2001. Vol. 47.

Miringoff M.L., and Miringoff M.L. *The Social Health of the Nation: How America is Really Doing*. Oxford University Press, 1999.

Morris M. *Measuring the Condition of the World's Poor: The Physical Quality of Life Index*. N.Y., 1979.

Mroz T., Popkin B., Mancini D., Glinskaya T., and Lokshin V. *Monitoring Economic Conditions in the Russian Federation: The Russia Longitudinal Monitoring Survey 1992–1996* // Report submitted to the U.S. Agency for International Development. Carolina Population Center. University of North Caroline at Chapel Hill. 1997. February.

Myers D. Community-relevant measurement of quality of life. A focus on local trends. *Urban Affairs Quarterly*. 1987.

Noll H.H., and Wiegand E. System Sozialer Indikatoren für dieBundesrepublik Deutschland-Zeitreihen 1950–1991. Tabellenband. ZUMA (Mannheim), 1993.

Noll H.H., and Zapf W. *Social Indicators Research: Societal Monitoring and Social Reporting // Trends and Perspectives in Empirical Social Research*. Berlin; N.Y., 1994. P. 1–16.

Nubler I. The Human Development Index Revisited // *Intereconomics*. 1995. (July/August). P. 171–176.

Osberg L., and Sharpe A. An Index of Economic Well-Being for Canada // Research Paper R-99-3 E, Applied Research Branch. Human Resources Development Canada, 1998.

Peterson M., and Malhotra N.K. Comparative marketing measures of societal quality of life: Substantive dimensions in 186 countries // *Journal of Macromarketing*. 1997. Vol. 17. N. 1. P. 25–38.

Peterson M., Malhotra N.K., and Wagner J. Country quality of life and foreign direct investment decisions // *Global Outlook*. 1999. Vol. 11. P. 1, 51–62.

Pigou A.C. *The Economy of Welfare*. Fourth Edition with eight new appendices. London: MacMillan, 1952.

Polterovich V., and Popov V. *Democratization, Quality of Institutions and Economic Growth* // Working Paper N.56. M.: New Economic School, 2006.

The Quality of Life / ed. by M. Nussbaum and A. Sen. Oxford: Clarendon Press, 1995.

Ravallion M. *Poverty Comparisons*. Chur: Switzerland. Harwood Academic Publishers, 1994.

Riecen G., and Yavas U. Improving quality of life in a region: A survey of area residents and public sector implications // *The International Journal of Public Sector Management*. 2001. Vol. 14. N. 6/7.

RLMS The Russia Longitudinal Monitoring Survey: "Family questionnaire", and "Sample of Russian Federation" (Rounds V, and VI). Technical Report, 1996, 1997.

Rodric D., Subramanian A., and Trebbi F. *Institutions Rule: The Primacy of Institutions over Geography and Integration in Economic Development*. http://www.ksg.harvard.edu/rodrik/papers.html, 2005.

Rudzkis R., and Radavicius M. Statistical estimation of a mixture of Gaussian distributions // *Acta applicandae Mathematical*. 1995. Vol. 38. N. 1.

Schultz T. Capital formation by education // *Journal of Political Economy*. 1960. Vol. 68.

Schwartz S.H. Universals in the content and structure of values: Theoretical advances and empirical tests in 20 countries // *Advances in Experimental Social Psychology* (ed. M. Zanna), 20; Orlando, Florida: Academic Press, 1992.

Scott S., Strode M., and Chinganya O. Estimating annual expenditure from monthly survey data // *Inter-Stat*. 2000. P. 107–126.

Sen A. *Development Thinking at the Beginning on the Beginning on the 21st Century* // Economic and Social Development into the XXI Century. Washington, DC, 1997.

Sen A. Development: Which way now? // *The Economic Journal*. 1983. Vol. 93. N. 372.

Sen A. *The Living Standard* // Oxford Economic Paper. 1984. Vol. 36. Issue Supplement: Economic Theory and Hicksian Themes.

Smith A. *An Inquiry into the Nature and Causes of the Wealth of Nations*. Republished, London: Home University Library, 1776.

Stata Reference Manual. Release 12. Stata Press, 2009.

Tielbout Ch. A pure theory of public expenditures // *Journal of Political Economy*. 1956. Vol. 64. P. 416–424.

UNHDP. United Nations Human Development Programme: 1990, Human Development Report, Oxford University Press, 1990.

University of Virginia Center for Public Service. *Virginia's Local Economies: Statistical Update*. Charlottesville, VA, 1996.

University of Virginia Center for Public Service. *Virginia Statistical Abstract*, Charlottesville, VA, 1996.

Veenhoven R. *Conditions of Happiness*. Dordrecht: Reidel, 1984.

Veenhoven R. The cross national pattern of happiness. Tests of predictions implied in three theories of happiness // *Social Indicators Research*. 1995. Vol. 34. P. 33–68.

Veenhoven R. Happy life expectancy: A comprehensive measure of quality of life in nations // *Social Indicators Research*. 1996. Vol. 39. P. 1–58.

Veenhoven R. Is happiness a trait? Tests of the theory that a better society does not make people any happier // *Social Indicators Research*. 1994. Vol. 32. P. 101–160.

VeenhovenR. Quality-of-life and happiness: Not quite the same // *Health and Quality-of-Life*. Il Pensierro Scientifico. Rome, 1998.

Veenhoven R., and Ouweneel P. Livability of the welfare-state. Appreciation-of-life and length-of-life in nations varying in state-welfare-effort // *Social Indicators Research*. 1995. Vol. 36. P. 1–49.

Vogel J. Coping with the European welfare mix. Welfare delivery systems, family formation and material inequality between types of families in the European Union // Paper presented at the 1998 Siena Group Meeting on Families at the end of the Twentieth Century. Sidney, Australia. 1998 b. 12. P. 7–9.

Vogel J. Welfare surveys and social reporting in the Nordic countries // *Statistik im Dienste der Öffentlichkeit*. Bundesamt fuer Statistik. Bern, Schweiz, 1998a.

Vogel J., and Häl L. Välfärd och ojämlikhet I 20-arsperspektiv 1975–95 ("Living conditions and inequality 1975–95") // Report 91. 1997. Series "Living Conditions". Statistics, Sweden.

WCY. *The World Competitiveness Yearbook*. Edition IMD International, Lausanne, Switzerland. 1997–2011.

The WHOQOL Group. *WHOQOL Study Protocol*. Geneva: WHO. Division of Mental Health, 1993.

The WHOQOL Group. *WHOQOL-100: Domains and Facets Questionnaire*. Geneva: WHO, Division of Mental Health, 1995a.

The WHOQOL Group. *WHOQOL-100: The 100 Questions with Response Scales*. Geneva: WHO, Division of Mental Health, 1995 b.

The WHOQOL Group. *WHOQOL-100: Facet Definitions and Questions*. Geneva: WHO, Division of Mental Health, 1995c.

The WHOQOL Group. *WHOQOL-100: Australian Version*. Geneva: WHO, Division of Mental Health, 1995d.

The WHOQOL Group. The World Health Organization quality of life assessment (WHOQOL): Position paper from the World Health Organization // *Social Science and Medicine*. 1995e. Vol. 41. P. 1403–1409.

The WHOQOL Group. *WHOQOL-BREF: Field Trial Version. Questionnaire*. Geneva: WHO, Division of Mental Health, 1996a.

The WHOQOL Group. *WHOQOL-BREF: Field Trial Version: Introduction, Administration, Scoring and General Version of the Assessment*. Geneva: WHO, Division of Mental Health, 1996 b.

The WHOQOL Group. The World Health Organization quality of life assessment (WHOQOL): Development and general psychometric properties // *Social Science and Medicine*. 1998a. 46. P. 1569–1585.

The WHOQOL Group. Development of the World Health Organization WHOQOL-BREF quality of life assessment // *Psychological Medicine*. 1998 b. Vol. 28. P. 551–558.

Zagorski K. and MCDonell J.S. "Consumer confidence" Indexes as social indicators // *Social Indicators Research*. 1995. Vol. 36. P. 227–246.

Zapf W. Soziale Indikatoren– eine Zwischenbilanz // *Sozialpolitik und Sozialberichterstattung*. Frankfurt a.M., N.Y. Gampus Verlag. 1977. P. 231–246.

ZieschangK.D. Sample weighting methods and estimation of totals in the consumer expenditure survey // *Journal of the American Statistical Association*. 1990. Vol. 85. P. 986–1001.

Information sources

I1 Регионы России: стат. сборник. М.: Росстат, 2002–2009.

I2 Российский статистический ежегодник. М.: Росстат, 2004–2009.

I3 Социальное положение и уровень жизни населения России: стат. сборник. М.: Росстат, 1998–2009.

I4 Демографический ежегодник России. М.: Росстат, 2002–2009.

I5 Труд и занятость в России. М.: Росстат, 2002–2009.

I6 Социально-экономическое положение России. М.: Росстат, 1998–2009.

I7 Охрана окружающей среды в России: стат. сборник. М.: Изд-во Росстата, 2001–2004.

I8 Официальные материалы Росстата (в авторизированном доступе), 2005: (www.gks.ru).

I9 Статистика по особо охраняемым территориям (http://reserves.biodiversity.ru).

I10 Статистика по ВИЧ-инфицированным (http://www.hivrussia.org).

I11 Россия – университетская информационная система. М.: НИВЦ МГУ, 2004. (www.cir.ru).

I12 Социальный атлас российских регионов (http://atlas.socpol.ru).

I13 Web-Атлас России и ее регионов (www.sci.aha.ru/ATL).

I14 Опора России. Условия и факторы развития малого предпринимательства в регионах России: результаты общероссийского исследования с ВЦОМ, 2006. (http://www.opora.ru/ana lytics).

I15 МВД. Показатели социальной патологии по регионам РФ за 2002–2006.

I16 Пенсионный фонд, 2005. – Web-сайт Пенсионного фонда России: (www.pfrf.ru).

I17 Официальный сайт Министерства финансов РФ: (http://www1.minfin.ru/ru).

I18 Официальный сайт Федерального казначейства Министерства финансов Российской Федерации: (www.roskazna.ru).

I19 Сайт Министерства экономического развития, инвестиций и торговли СО (http://www.econo my.cam-reg.ru).

I20 ИНДЕМ. Региональные индексы коррупции (проект TI–Russia), 2008.

I21 ИНДЕМ. (http://www.anti-corr.ru/rating_regions/index.htm).

I22 МЕРКАТОР ИГРАН. Электоральный справочник «Российские выборы в цифрах и картах, 1995–2007.» (http://www.radioportal.ru/elections/mercator.shtml).

I23 Mercer – 2008. «Москва попала в 20-ку самых грязных городов планеты» (*рейтинг Mercer Human Resource Consulting*). (http://www.inline.ru/gonews.asp?NewsID=131714).

I24 Социально-экономическое положение России. Январь–сентябрь 1996, Госкомстат, 1996.

Bei Fragen zur Produktsicherheit wenden Sie sich bitte an:
If you have any questions regarding product safety,
please contact:

Walter de Gruyter GmbH
Genthiner Straße 13
10785 Berlin
productsafety@degruyterbrill.com